A History of Farming Systems Research

Edited by

M. Collinson

Published by
Food and Agriculture Organization of the United Nations
and
CABI *Publishing*

A catalogue record for this book is available from the British Library, London, UK.

Library of Congress Cataloging-in-Publication Data
A history of farming systems research / edited by M.P. Collinson.
p. cm.
Includes bibliographical references (p.).
ISBN 0-85199-405-9 (alk. paper)
1. Agricultural systems--Research History. I. Collinson, M.P. (Michael P.)
S494.5.S95H57 1999
630'.72--dc21 99-42827
CIP

Transferred to print on demand 2002

Published jointly by:
CABI Publishing, a division of CAB International
CABI Publishing, CAB International, Wallingford, Oxon OX10 8DE, UK
Tel: +44(0)1491 832111, Fax: +44(0)1491 833508, Email: cabi@cabi.org
CABI Publishing, 10 E 40th Street, Suite 3203, New York, NY 10016, USA
Tel: +1 212 481 7018, Fax: +1 212 686 7993, Email: cabi-nao@cabi.org

Food and Agriculture Organization of the United Nations (FAO),
Viale delle Terme di Caracalla, 00100 Rome, Italy
Tel: +39 06 57051, Fax: +39 06 57053152, Email: webmaster@fao.org
Website: http://www.fao.org

ISBN 0 85199 405 9 (CABI) – hard cover edition
ISBN 92 5 104311 6 (FAO) – paperback edition

Typeset by Columns Design Ltd, Reading.
Printed and bound by Antony Rowe Ltd, Eastbourne

Contents

Contributors' Biographies

EDITOR AND COMPILER

Mike Collinson. The first farm economist appointed into the British colonial agricultural research services (1960), Mike worked with resource-poor African farmers for 25 years. From 1975, as regional economist with CIMMYT, he helped build capacity for on-farm research with systems perspective in the agricultural research institutions of eastern and southern Africa. He retired to the UK in 1996 after 9 years as social science adviser in the secretariat of the Consultative Group for International Agricultural Research (CGIAR), in Washington, DC.
Email: mikecollinson1@compuserve.com

CONTRIBUTORS

Ponniah Anandajayasekeram. Ananda, as he is universally known in farming systems research circles, has directed FSR training programmes and advised on methods and institutional capacity building in Africa since 1982. He was a Past President of the Southern African Association for FSR-E and the immediate Past President of the global Association for Farming Systems Research/Extension (AFSR/E).
Email: pananda@farmesa.co.zw

Michel Benoit-Cattin. Michel is in charge of the international scientific exchanges in CIRAD, in Montpellier, France. An agronomist with a State Doctorate in economics he has worked for 30 years on agricultural development problems at the farm, regional and national levels in France, Algeria, Ivory Coast and Senegal. He now supervises theses in these topics by young researchers from all parts of the world.
Email: benoitca@cirad.fr

Julio A. Berdegué. Based in Chile, Julio has been the Coordinator of RIMISP, the International Farming Systems Network, since its foundation in 1986. Between 1992 and 1995 he was the Technical Director of Chile's Agricultural Development Institute (INDAP). Julio is the co-editor of four books on different aspect of FSR-E, and is the owner and manager of a seed and vegetable producing farm.
Email: berdegue@reuna.cl

Joseph Bonnemaire, Jacques Brossier and B. Hubert. All three are scientists with Systemes Agraire et Development (SAD), the unique agrarian systems division of INRA in France. Jacques Brossier was a co-organizer of the 1994 Montpellier AFSRE Symposium and is Chair of the INRA Regional Research Centre of Dijon. B. Hubert is a veterinarian and a department chief in INRA, Joseph Bonnemaire is an animal husbandry specialist, a professor in ENESAD and a 'charge de mission' with INRA.
Email: brossier@dijon.inra.fr

Cornelia Butler-Flora and Charles Francis. Cornelia is Director of the North Central Regional Center for Rural Development, which conducts research and extension programmes in rural development in the 12 midwestern American states, and is Professor of Sociology at Iowa State University. Charles is Agronomist and Director of the Center for Sustainable Agricultural Systems, University of Nebraska, Lincoln. Cornelia and Charles have pioneered the application of a systems approach in extension in the American Midwest.

Emails: cflora@iastate.edu charles.francis@ipf.nlh.no

J.P. Deffontaines, Jacques Brossier, M. Barbier, M. Benoit, E. Chia, J.L. Fiorelli, M. Gafsi, F. Gras, H. Lemery and M. Roux. The authors are members of the SAD research team in INRA responsible for the Vittel study. Brossier and Deffontaines had scientific responsibility for the study. J.P. Deffontaines is the father of 'geo-agronomy' concerned with spatial connections of farming systems and played a leading role in the establishment of the SAD department in INRA in 1979. Lemery is a sociologist specializing in farmers' behaviour and social networking with a focus on the transformation of professional practices relating to agriculture. He is a professor in ENESAD, Dijon. Roux is an animal specialist, and a professor in ENESAD, Dijon. Barbier, Chia and Gafsi are economists, Benoit and Fiorelli are agronomists. Gras was researcher in biological pedology at CNRS, the University of Nancy and ORSTOM. He is now retired.
Email: brossier@dijon.inra.fr

John Dixon. John is Programme Coordinator, Farm-Level Agricultural Research Methods for East and Southern Africa, Harare, Zimbabwe. He has worked in farming systems research and development for more than 25 years, half of which has been spent in Ethiopia, Iran, Nepal, Thailand and Zimbabwe. He was also based in Rome for several years implementing FAO's farming systems programme.
Email: john.dixon@farmesa.co.zw

German Escobar. German is the Research Director of the International Network on Farming Systems Research Methodology (RIMISP), based in Chile. A Colombian researcher, German trained as an Agricultural Economist and FSR practitioner and has worked in several countries as researcher, project designer and advisor to national research and rural development institutions.
Email: ge@rimisp2.cl

John Farrington. John heads the Rural Policy and Environment Group at the Overseas Development Institute, London, UK, and is Visiting Professor at the Agricultural Extension and Rural Development Department of Reading University.
Email: j.farrington@odi.org.uk

Elon Gilbert. A Visiting Research Fellow with the Overseas Development Institute, Elon is an agricultural economist with more than 20 years experience in Sub-Saharan Africa and, more recently, in South Asia. He wrote, with others, the original American Journal of Agricultural Economics article on FSR in 1980, and was a member of the ISNAR team examining national programme experience with FSR in the mid 1980s.
Email: elonsm@aol.com

Robert Hart. While working at CATIE, Bob, an ecologist, proposed the multilevel systems hierarchy now accepted as a conceptual framework for FSR. He is currently the Director of the USAID-funded global Sustainable Agriculture and Natural Resource Management (SANREM) Collaborative Research Support Program coordinated by the University of Georgia.
Email: rdhart@arches.uga.edu

Richard Harwood. The founder and Head of IRRI's Asian Cropping Systems Program from 1967 to 1972, Richard developed the Asian Cropping Systems Network. At Rodale, he directed the Research Center and its integrated systems studies until 1985. As Director of Asian Programes for Winrock International from 1985 to 1990, he supported farming systems projects in Indonesia, Nepal and the Philippines. Currently the C.S. Mott Chair of Sustainable Agriculture at Michigan State University, he is also a member of the CGIAR Technical Advisory Committee.
Email: rharwood@pilot.msu.edu

Richard Hawkins. Following his PhD studies on maize growth and yield in the Kenya highlands, Richard spent 10 years on FSR projects in Central America (CATIE), Nepal and Java. Richard has spent the last 10 years putting this experience to good use, first as Coordinator of the ICRA Anglophone training programme in research for development, and more recently with the Postgraduate College in Mexico.
Email: rhawkins@infosel.net.mx

Peter Hildebrand and Dennis Keeney. Peter is Professor, Food and Resource Economics Department, Institute of Food and Agricultural Sciences, University of Florida, Gainesville. He has worked in farming systems research for more than 25 years, including 13 years in Colombia, El Salvador and Guatemala. He has coordinated the University of Florida Farming Systems Research-Extension Program for 15 years and was the founding President of the global Association for Farming Systems Research-Extension. Dennis Keeny, a past President of the American Agronomy Society, is currently Director of the Leopold Center for Sustainable Agriculture, Iowa State University.
Email: hildebrand@fred.ifas.ufl.edu

Janice Jiggens. A past President of the International Association for Farming Systems

Research and Extension, Janice is currently Professor of Human Ecology at the Swedish Agricultural University. She has spent most of her interdisciplinary career working on agriculture and rural development, mainly in Africa and South Asia. She has a particular interest in combining participatory and 'high tech' approaches and tools, and in ensuring that gender issues, and women professionals, are fully present in the agricultural sciences.
Email: janice.jiggens@lbutv.slu.se

Stuart Kean and M. Creasy Ndiyoi. Stuart is a social scientist with interests in agriculture, rural development and community-based natural resource management. He worked for 11 years in Zambia and was the first National Coordinator of the Adaptive Research Planning Team (ARPT). He is currently coordinating the Northern Namibia Environmental Project featuring multi-agency community-based natural resource management initiatives.
Email: skean@iafrica.com.na

Ndiyoi is the Chief Agricultural Research Officer, Farming Systems and Social Sciences, Ministry of Agriculture Food and Fisheries, Zambia. He has worked in the field of farming systems and small farmer development for the last 17 years. He is the current President of the Southern African Association for Farming Systems Research and Extension (SAAFSRE).
Email: mndiyoi@pop3.zamnet.zm

Clive Lightfoot. Clive, an agronomist, worked in on-farm research for many years in Africa and in Asia. He has put much of the flesh on the bones of agroecosystems research and is Chairman of the International Support Group (ISG) for stakeholder reconciliation.
Email: clive.lightfoot@agropolis.fr

Hal MacArthur. A past President of the International Association for Farming Systems Research, Hal has worked extensively in FSR programmes in both Africa and Asia.
Email: hmcarthur@hawaii.edu

Constance McCorkle. A sociologist with wide experience in Africa, Asia and Latin America, Constance is a world authority on the study and application of ethnoveterinary medicine.
Email: constancemckorkle@msn.com

Evaristo Eduardo de Miranda. Born in Brazil, Evaristo has had over 100 works published at home and abroad. He is a Professor at São Paulo State University (USP), researcher at the EMBRAPA Remote Sensing Monitoring Center (www.nma.embrapa.br) and also heads the NGO ECOFORCE Research and Development (www.ecof.org.br). He has applied remote sensing and geographic information systems to the assessment and monitoring of development and environmental protection programmes.
Email: mir@nma.embrapa.br

David Norman. David is Professor, Department of Agricultural Economics, Kansas State University. Since the mid 1960s he has spent a total of 20 years working in national agricultural research systems in Africa (Nigeria and Botswana) and continues to take on short-term assignments in Africa, Asia and the South Pacific. He was among those who initiated the annual farming systems research and extension symposia at Kansas State University, which later evolved into the global Association for Farming Systems Research-Extension (now the International Farming Systems Association). He is also a past President of the Association.
Email: dnorman@agecon.ksu.edu

James Olukosi. James is a professor of Agricultural Economics with the Institute of Agricultural Research at Ahmadu Bello University in Nigeria, and has worked widely in FSR in West Africa. He is the Coordinator of the West African Farming System Research Network (WAFSRN).
Email: ICRISAT-W-NIGERIA@cgiar.org

Roberto A. Quiroz, Carlos Leon-Velarde and Walter Bowen. Roberto is the Head of the Production Systems and Natural Resource Management Department at the International Potato Center (CIP). He received his MS and PhD degrees from North Carolina State University. His interests are in FSR and in the integration of GIS and remote sensing with simulation models in land use studies. Carlos is a systems scientist at the CIP and the International Livestock Research Institute (ILRI). He holds a PhD from the University of Guelph in animal breeding and genetics. His research has focused on the development and application of simulation models in mixed crop–livestock systems, and natural resources. Walter is a systems scientist, specializing in soil fertility, with the International Fertiliser Development Center (IFDC), outposted to the CIP. He earned his MS and PhD degrees from Cornell University. His research activities have centred on the development and application of crop growth models.
Email: R.Quiroz@cgiar.org

Nimal Ranweera. Nimal is a past president of both the Asian Association for Farming Systems Research and the International Association for Farming Systems Research-Extension. He is a

senior administrator in the Ministry of Agriculture, Sri Lanka. He hosted and organized the 1996 international FSRE symposium.
Email: minagr@slt.lk

Bill W. Shaner. Bill is Professor Emeritus, Colorado State University. He was team leader and senior author of the book, *Farming Systems Research and Development: Guidelines for Developing Countries*. He occasionally consults overseas on projects related to economic development.
Email: interdev@lamar.colostate.edu

Hilary Sims Feldstein. Hilary is the Training Specialist at the International Center for Research on Women, Washington, DC. For 9 years, she was the Program Leader, Gender Analysis, for the Gender Program at the CGIAR. In recent years she has worked with the Kenya Agricultural Research Institute and with Uganda's National Environmental Management Authority to integrate gender analysis into their organizations' technology development and natural resource management activities.
Email: Hilary_Feldstein@dai.com

Louise Sperling and Jacqueline Ashby. Louise Sperling, an anthropologist, is a senior scientist at CIAT and facilitates the Plant Breeding Group of the CGIAR Systemwide Program on Participatory Research for Gender Analysis and Technical Innovation. She has worked for 20 years in Africa and Asia on innovative breeding and seed system strategies to benefit small farmers. Jacqueline Ashby is a PhD in development sociology who has published extensively on participatory research and social ecology. She is currently Director of Research for Natural Resource Management at the International Center for Tropical Agriculture (CIAT) where she previously developed approaches to applied participatory research which are now taught and practised in several countries across the world.
Email: L.Sperling@cgiar.org J.Ashby@cgiar.org

Willem Stoop, Omar Niangado, Demba Kébé and Toon Defoer. Willem is an agronomist and soil scientist. Since 1992 he has been external advisor to L'Institut d'Economie Rurale in Bamako, Mali, and its respective ESPGRN teams, including Sikasso. He was formerly with the Royal Tropical Institute in Amsterdam and is now an independent consultant. Omar Niangado is a plant breeder with particular expertise in millet. He was Director General of L'Institut d'Economie Rurale in Bamako, Mali, from 1993 to 1998. Demba Kébé is agricultural economist at L'Institut d'Economie Rural in Bamako and was formerly Head of the ESPGRN team in Sikasso. Toon Defoer, Senior Scientist at the Royal Tropical Institute (KIT) Amsterdam, specializes in famer participatory methodology development for natural resources management. He was the technical advisor to the ESPGRN team in Sikasso from 1993 to 1996. *Email:* nadamo@wxs.nl

Ann Stroud and Roger Kirkby. Ann, a Californian, has resided in East Africa since 1982. She has worked for a number of organizations specializing in weed management, pesticide issues, farming systems research and more recently natural resource management. Her experience in systems agronomy is derived from field research and extensive advising and capacity building with NARS in the East and southern African region. She is currently the coordinator of the African Highlands Initiative, an ecoregional programme at ICRAF. Roger is an agronomist who has carried out on-farm experiments in eastern Africa and Latin America since 1969. His PhD from Cornell University included looking at ways to integrate experimentation by farmers and the formal sector; other degrees are from the Universities of Wales and Cambridge. He is currently CIAT's Coordinator for Africa.
Email: A.Stroud@cgiar.org R.Kirkby@cgiar.org

Alistair Sutherland and John N. Kang'ara. John Kang'ara is a researcher in livestock production and nutrition with the Kenya Agricultural Research Institute, based at Embu Research Centre. He was the livestock researcher on the DAREP project from 1993 to 1997 and the project coordinator from 1995. Alistair Sutherland is a social anthropologist with the Natural Resources Institute of the University of Greenwich, UK. He was the anthropologist and technical advisor on the DAREP project from 1993 to 1997.
Email: alistair.sutherland@nri.org

Robert Tripp. Robert is an anthropologist and currently a research fellow at the Overseas Development Institute (ODI) in London, UK. He was previously with the CIMMYT Economics Program, where he worked on training and methods for on-farm research.
Email: r.tripp@odi.org.uk

Toon van Eijk. Toon is a farming systems agronomist who has worked in Mozambique, Kenya, Tanzania and Zambia over the last 20 years. Having recently completed his thesis at Wageningen University on FSR and Spirituality he is now a freelance advisor on rural development based in his home in Dar-es-Salaam, Tanzania.
Email: tvaneijk@ud.co.tz

Abbreviations and Acronyms

AAFSRET	African Association of Farming Systems Research Extension and Training
AEZ	Agroeconomic zone
AFSA	Asian Farming Systems Association
AFSRE	Association for Farming Systems Research and Extension
ARFSN	Asian Rice Farming Systems Network
ARPT	Adaptive Research Planning Team
AVRDC	Asian Vegetable Research and Development Center
CARDI	Caribbean Agricultural Research and Development Institute
CATIE	Tropical Agricultural Center for Research and Training
CGIAR	Consultative Group on International Agricultural Research
CIAT	International Center for Tropical Agriculture
CID	Consortium for International Development
CIDA	Canadian International Development Assistance
CIMMYT	International Maize and Wheat Improvement Center
CIP	International Potato Center
CIRAD	Centre International de Recherches Agronomiques pour le Developpement
CONDESAN	Consortium for the Sustainable Development of the Andean Ecoregion
DAREP	Dryland Applied Research and Extension Project
ECOGEN	Ecology, Community Organization and Gender programme (Clark University)
EMBRAPA	Institute for Agricultural and Livestock Research (Brazil)
FAO	Food and Agriculture Organization of the United Nations
FM/FI	Farmer managed/farmer implemented
FPR	Farmer participatory research
FPR-E	Farmer participatory research and extension
FSR	Farming systems research
FSR-E	Farming systems research and extension
FSSP	Farming systems support project
GIS	Geographical information systems
IADS	International Agricultural Development Service
IARC	International Agricultural Research Centre
ICA	Colombian Agricultural Institute
ICARDA	International Center for Agricultural Research in the Dry Areas
ICLARM	International Center for Living Aquatic Resource Management
ICRA	International Centre for Development Oriented Research in Agriculture
ICRAF	International Centre for Research on Agroforestry
ICRISAT	International Crops Research Institute for the Semi-Arid Tropics

ICTA	Institute of Agricultural Science and Technology (Guatemala)
IDRC	International Development Research Center
IFAD	International Fund for Agricultural Development
IFSA	International Farming Systems Association (formerly AFSRE as above)
IHH/FSR-E	Intra-Household and FSR-E Case Studies Project
IICA	Interamerican Institute for Cooperation on Agriculture
IITA	International Institute of Tropical Agriculture
IK	Indigenous knowledge
ILCA	International Livestock Center for Africa
ILRAD	International Laboratory for Research on Animal Diseases
INRA	National Institute for Agricultural Research (France)
IPM	Integrated pest management
IRAT	Institute for Tropical Agronomic Research
IRD	Integrated rural development
IRRI	International Rice Research Institute
ISNAR	International Service for National Agricultural Research
NARE	National Agricultural Research and Extension
NARI	National Agricultural Research Institute
NARS	National Agricultural Research Service
NRI	Natural Resources Institute
NRM	Natural Resource Management
ODA	Overseas Development Administration (now DFID)
OF	On-farm
OFE	On-farm experimentation
OFR	On-farm research
ORSTOM	Institut Français de Recherche Scientifique pour le Developpement en Cooperation
OS	On-station
OSR	On-station research
PRA	Participatory rural appraisal
PSNRM	Production Systems and Natural Resource Management
PSP	Production Systems Programme
RD	Recommendation domain
RIMISP	International Farming Systems Research Methodology Network
RISPAL	Latin American Animal Production Systems Research Network
RMD	Resource management domain
RM/FI	Researcher managed/farmer implemented
RM/RI	Researcher managed/researcher implemented
RRA	Rapid rural appraisal
RS	Remote sensing
SAAFSR-E	Southern African Association of Farming Systems Research and Extension
SACCAR	Southern African Centre for Co-operation in Agricultural and Natural Resources Research
SAD	Department for Research on Agrarian Systems and Development (within INRA, France)
SANREM CRSP	Sustainable Agriculture and Natural Resource and Environment Management Collaborative Research Project
SARE	Sustainable Agriculture Research and Extension
SIDA	Swedish International Development Assistance
USAID	United States Agency for International Development
WAFSRN	West African Farming Systems Research Network
WARDA	West Africa Rice Development Association
WIAD	Women in Agricultural Development program (University of Florida)
WIRFS	Women in Rice Farming Systems

Foreword

Janice Jiggens, Past President of the International Association for Farming Systems Research and Extension

As President of the Association for Farming Systems Research and Extension (AFSRE – now the International Farming Systems Association (IFSA)) when the book was commissioned, I am delighted to contribute a foreword to this history of farming systems research (FSR) and its applications, seeing it as an opportunity to offer a personal account of my own love affair with FSR. It mirrors, in many respects, the sequence of the text itself – a retrospective on my own baptism; a focus on what are, for me, key aspects of FSR; the progress made and challenges remaining; and my own perceptions of some key lessons learned.

A RETROSPECTIVE

I strayed into FSR at the end of the 1970s when I was working as a social scientist in the northern and central provinces of Zambia. The challenge was to find ways to develop technologies for, and supply services to, impoverished small-scale farmers in areas of high male outmigration. I became fascinated by the experiments of some women famers to improve crops of a traditional green leaf vegetable, grown between the main cereal crop both for home consumption and sale in the local market. The leaves are rich in minerals, dry well and form an important seasonal additive to the relish which accompanies the starchy main meal, as well as providing cash for household necessities. But when I persuaded an agronomist from the provincial research station to visit the farmers to learn more about their experiments and perhaps give some advice, we ended up in a blazing argument about wasting his time just to show me some women growing weeds! I was forced to think deeply about the specific value of the vegetables to crop production and food systems, about the gender-specific roles of men and women, and about the nature of a science-based training in agriculture which could so easily set aside farmers' knowledge and a crop that was essential to the livelihood of the women and to the nutrition of their families.

A second formative experience brought into question agricultural survey research methods. My team had developed a questionnaire in the local language to prepare a statistical sample of households in an area of shifting cultivation. However, after a few days in the field I realized that my male Zambian colleagues were establishing less formal relations with the women in the village than that of interviewer and respondent. It clearly did not make any communicative sense to turn up a few hours later with a questionnaire in hand. Yet the long, drawn-out methods of the anthropologist were not practical: what tools and techniques could we use in the 3 weeks we had to ensure some reliable degree of rigour and representivity yet were based on a more natural process of enquiry? With hindsight, I wish I had paid more attention to the refinement of the concept of 'recommendation domains' and methods of informal survey that Mike Collinson and his colleagues were applying at the time in the central province.

Much of my field work in Zambia turned into an exploration of alternative methods, culled from whatever source book or experienced person then available to me. But it did not feel like 'good' research. I was learning more than I had ever done before, but how could I present this knowledge in a way that would convince my own peer group?

Towards the end of my stay in Zambia Robert Chambers and I worked together on a Basic Needs mission sponsored by the International Labour Organization. The long trek up to the shores of Lake Bangweulu gave us ample time for discussion of these questions, which Robert himself was also pondering and exploring, along with many others, as I later realized. A hazardous canoe trip across the lake brought us to the old 'goat woman'. She remains in my memory as our tutor in what later became known as participatory appraisal methodology. We worked with her for a day using techniques still regarded as innovative, analysing the management of her goats which were renowned for their twins and good health, and which she sustained through the careful recycling of waste and the use of traditional herbs she grew herself.

These formative experiences added in a small way to the river of accomplishment documented in this book by bringing together FSR perspectives, gender analysis and participatory methods.

KEY ASPECTS: FSR-E, GENDER ANALYSIS AND PARTICIPATORY METHODS

As the experiences of researchers around the world during the 1980s demonstrated, there is much to be gained by marrying these three ways of learning and cooperating. On gender analysis Feldstein and Jiggins[1] concluded that using gender as a focus resulted in a better description of the system as a whole and opened the door to a greater understanding of the opportunities to technical innovation. Gender adds a little complexity for a lot of insight, while participatory process and techniques enable farming systems researchers to engage more effectively with members of farming communities.

The marriage of FSR-E, gender analysis and participatory methods has, to a considerable extent, become common practice. Four strengths stand out. First, the quality of the information is better because it is richer, more deeply contextualized and yet amenable to aggregation. It is focused yet cost-effective across scale, where 'scale' is understood as a recommendation domain. Second, in combination they can lead to the rapid discovery of contradictions such as the points where experience diverges, where information is inconsistent and where interpretations vary. Where there is convergence, consistency and agreement, one can proceed with confidence along well-established pathways; where there are contradictions, assumptions are challenged and further investigation is required. This is the opportunity for genuinely new theoretical and practical knowledge to emerge. Review of experience suggests that the combination of FSR-E plus gender analysis plus participatory methods, prompts discovery by offering three different 'windows' into complex situations[2]. Third, the combination of perspectives and methods focuses attention on constraints and opportunities, rather than problems. In my view, the emphasis on problems in agricultural research has been a hindrance to development, if only because it provides such poor inspiration for effort and for specification of the potential for change in agricultural reality. Finally, the application of these methods has drawn attention to the important and necessary technology-led gains that can be achieved with poor people living in variable, diverse and uncertain environments.

However, the combination does have a number of weaknesses. At the theoretical level, thinking about systems does not have to be systemic to be useful. But at the practical level, if the research and technology development objective is in some way to change the system, then the methodological toolbox must include the tools of researching farming as an human activity. Best practice research is generating a rich and constructive case book of the participatory methodologies essential to systemic change. More commonly however, these methodologies seem to be applied mechanisticially or in an extractive manner, giving rise to failures in the change process[3]. While lip-service might be paid in research proposals to the role of women in farming systems, the sad fact is that this remains a male-dominated area and FSR is still failing in the proper handling of this essential ingredient.

There is a third area in which FSR-E practice falls short of its potential, perhaps because of its strong historical roots in farm management economics. Research has highlighted the extent to which an accomplished end-of-season system 'design' is the desired outcome of responses to events unfolding through the season. Wherever the degree of uncertainty is high, the tendency to assess farming in terms of performance is particularly marked[4], but this tendency is also to be found in more highly controlled production environments[5]. Given the importance to resource-poor farmers of managing uncertainty, greater attention should be paid to the overall implications of dryland farming.

Best practice points the way, for example through examination of strategies for coping with varying seasonal conditions and the rules which guide farming choices. Cox *et al.*[6] conducted elegant research among dryland wheat farmers in northern Queensland which reveals much about the nature of contingent decision making in conditions of uncertainty. They found decisions to be based on a rather small number of simple rule sets which were: nested; triggered by events; interconnected; linked to additional sets, stable in response to stress (such as prolonged drought); adaptive to long-term trends in system states; interpretative; and supportive of simultaneous management of multiple indicators of system performance.

A focus on the management of uncertainty also suggests a need for greater emphasis in FSR-E practice on collaboration between farmers and scientists[7]. Best practice has, in fact, already moved in this direction, a movement reinforced by emerging concerns about the relation between on-farm developments and landscape scale resource management. FSR-E is now being challenged to investigate the relationships among on-farm systems development, ecological systems management and agricultural policy effects[8]; and apply participatory applied research at farm and community levels to natural resource management.

A final problem lies in the field of FSR-E education. For many years I shared the frustrations of field personnel in trying to turn the human products of specialist university degrees into systems thinkers with at least some competence in working with farmers on system development. My early efforts at the University of Guelph in Ontario to take the lessons of the field back into academia to produce a generation of professionals competent in FSR were positive at the human level. The students reacted enthusiastically to participatory methods, interdisciplinary learning and systems thinking. But, despite the goodwill and support of key individuals, undoubted barriers remained in the rigidities of departmental structures, the defence of intellectual territory and the problems of reconciling systems-oriented courses and the needs of students within the existing study programme. At Guelph, many of the difficulties of rigidity between departments have been eased by the recent creation of an interdisciplinary PhD offered through a new Faculty of Environmental Design and Rural Development.

As one who is directly involved as a newcomer to university life at the Swedish University of Agricultural Sciences, what most strikes me is the irrelevance of much of what is on offer at universities. Many students respond by finding their own pathways of learning through *ad hoc* self-study reading groups and by making off-campus links to community- and farmer-based action. The regular programme is what they have to do to qualify, not what they want to do to learn. Meanwhile, collaborative initiatives among coalitions of those with a personal commitment to change processes are creating new institutional structures and networks which bypass existing structures[9].

It is encouraging to find that even in the financially hard-pressed educational environment of eastern and southern Africa, such innovations are occurring. For example, a consortia of non-government organizations whose activities focus on various forms of ecological farming in partnership with farmers and in collaboration with the University of Zimbabwe, have now developed a degree course which supplements classroom study informed by systems thinking with periods of field work with the participating NGOs.

SOME EXAMPLES AND LESSONS FROM BEST PRACTICE

Learning Together, by Hagmann, Murwira and Chuma in 1996[10], documents the development and extension of soil and water conservation technologies in Masvingo and Chivi, Zimbabwe. This example of a new approach was called *kuturaya* (to try) by the farmers – a translation of 'research' into Shona. It was based on dialogue, on farmers' own real time, on whole-system experiments and on a strengthening of self-organizational capacity at community level. After two seasons each participating farmer, besides tied ridging, had at least two other trials ongoing, selected from among experiments suggested by project staff, local research stations and farmer innovators, or arising out of discussion of farmers' indigenous knowledge. More than 10 options have emerged from this joint process, including mechanical, agronomic, biological and water saving/irrigation methods and technologies. Within three seasons from 1992–93, at least 80% of the total of 1136 households within one administrative unit in Chivi District were practising soil and water conservation. The important lessons include the need to focus on integrated land husbandry since individual techniques cannot overcome the diversity of conditions nor alone generate sufficient economic benefit, the value of farmer involvement right from the start in extending, enriching and validating the portfolio of experimentation and emerging options, and the necessity of supporting institutional and organizational development (within communities but also within research and extension agencies) in order to support participatory process.

Learning to Learn with Farmers, by Hamilton in 1995[11], focuses on a project in southern Queensland. This provided invaluable input into research on the development, use and effects of providing farmers with better tools for monitoring and interpreting system states and trends, as the basis for informed decision making with regard to fallow management. The project was based in a region where 1.8 million ha of the total cultivated area of 2 million ha was designated as a 'needs protection' area in the face of widespread soil erosion. In the space of 4 years, the interventions raised the percentage of dryland wheat farmers in the vulnerable areas who had adopted one or more fallow management practice from 30% to 75% – some 1600 farmers. This success was the more remarkable for being achieved through a period of deepening drought and economic hardship. An interdisciplinary team of scientists and extension advisers worked with farmers on joint systems analysis, and through periods of sometimes painful and conflictual reflection on what was being learned and how the learning process was occurring. A series of tools were devised, again largely in collaboration with farmers, to enhance individual and shared learning about systems states and performance. These included: a rainfall simulator, a soil corer, How Wet (a computer-aided decision support tool), the Fallow Management Game (which allows players to expand on and interrogate scenarios generated by the use of the other three tools) and With and Without (a user friendly comparative economic analysis tool). Three lessons stand out: the importance of paying explicit attention to FSR processes, the power of stimulating shared knowledge creation and the need for science leaders and policy makers to accept that the process will not lead to adoption of uniform or standardized resolutions across an ecosystem. Rather, a mosaic emerges adapted to the systemic requirements at unit levels (the farm, field and crop).

FINALE

Despite the growing number of examples of good practice with demonstrably cost-effective results, there is much still to be learned at the cutting edge of FSR-E and a continuing need for vigilant quality control in everyday practice. However, to end on a pessimistic note would give a false picture of the contribution that systems research in agriculture and resource management is making to the resolution of urgent human problems. In my experience, it is a field of endeavour that attracts dedicated scientists, researchers and development workers of exceptionally high calibre broadly united in a commitment to the betterment of human existence and the life systems which support it. In the inclusive direction in which it is evolving, FSR-E provides a framework for understanding, and the processes and tools for pursuing the agenda for human survival captured by Goethe, who might be regarded as an early member of the FSR-E family, in the following stanza:

Es ist nicht genug zu wissen
Man muss es auch anwenden;
Es ist nicht genug zu wollen
Man muss es auch tun.

COMMISSIONING THE BOOK

In 1992 I was honoured to be elected as President of the IAFSRE. One of my main tasks during my term as President, apart from a permanent struggle with financing, was the organization of the 14th International Symposium in Montpellier, France, alongside our French hosts. One issue had been taxing the Board of the Association and its members since 1989 – the writing of a history of the Association, and perhaps a history of FSR in general. FAO, in the person of Karl Friedrich, then Head of the Farm Management and Production Economics Branch, had offered support for the history within the context of FAO promotion of an FSR-based approach to development, but possible authors and editors were all were too busy 'FSR-ing' to take on the job. Then, in December 1994, at Montpellier, it all came together. Karl Friedrich and I met with Mike Collinson. Although an FSR veteran and enthusiast, Mike's committments over the last 10 years had inhibited his involvement in AFSRE and he was attending only his third or fourth (he can't remember!) symposium of the 14 that had been held. Now, however, he was due to retire and he committed himself to the compilation and editing of an history of FSR on his retirement. He finally retired in early 1996 and has devoted much of his time since to finding contributors and to coaxing their contributions from them. This is the result – 40 contributions from 50 of the world's leading professionals, from some 20 countries – an inclusive sweep of the spectrum of professions and continents involved in FSR-E.

REFERENCES

1. **Feldstein, H. & J. Jiggins (Eds), 1994**. Tools for the Field. Gender Issues in Farming Systems Research and Extension. West Hartford, Kumarian Press.
2. **Jiggins, J. & K. Raman, 1994**. Decennial Review, 1984–1994. Eastern India Farming Systems Research Programme. New Delhi, The Ford Foundation.
3. **Bawden, R., 1995**. On the Systems Dimension in FSR. Journal of Farming Systems Research and Extension, 5(2), 1–18.
4. **Richards, P., 1985**. Indigenous Agricultural Revolution. Hutchinson, London.
de Steenhuijsen Piters, B., 1995. Diversity of Fields and Farmers. Explaining Yield Variations in Northern Cameroon. Published PhD thesis, Wageningen, Agricultural University.
5. **Leeuwis, C., 1993**. Of Computers, Myths and Modelling. The Social Construction of Diversity. Knowledge, Information and Communication Technologies in Dutch Horticulture and Agricultural Extension. Published PhD thesis, Wageningen, Agricultural University.
6. **Cox, P.G., A.D. Shulman, P.E. Ridge, M. Foale & A.L. Garside, 1995**. An interrogative approach to systems diagnosis: an invitation to the dance. Journal of Farming Systems Research and Extension, 5(2), 67–83.
7. **Sperling, L. & P. Berkowitz, 1994**. Partners in Selection. Bean Breeders and Women Bean Experts in Rwanda. CGIAR, Washington, DC.
Sperling, L. & U. Schiedegger, 1995. Participatory Selection of Beans in Rwanda: Results, Methods and Institutional Issues. Gatekeeper series no. 51, IIED, London.
8. **Caldwell, J. & E. Akobundu, 1997**. Agricultural Systems and Policy. Blacksburg, Virginia Polytechnic Institute and State University for AFSRE.
9. **Jiggins, J. & D. Gibbons, 1997**. What does interdisciplinary mean? Experiences from SLU. Paper presented to Session 5: Agricultural Knowledge and Information Systems. 13th European Seminar on Extension Education, 1–6 September. University of Dublin, Dublin.
10. **Hagmann, J., K. Murwira & E. Chuma, 1996**. Learning together: development and extension of soil and water conservation in Zimbabwe. Quarterly Journal of International Agriculture, 35(2), 1–14.
11. **Hamilton, N.A., 1995**. Learning to Learn with Farmers. Published PhD thesis, Wageningen, Agricultural University.

Chapter 1
Introduction

Mike Collinson

1.1 FSR – TERMINOLOGY AND DEFINITION

Even within the choir of advocates there has long been controversy on terminology in farming systems research (FSR)[1]. It raised its head again during the preparation of this book. I hope I have outflanked the controversy by referring to *FSR and its applications*. FSR itself is defined as a diagnostic process; a basket of methods for researchers to elicit a better understanding of farm households, family decisions and decision-making processes. Its applications use this understanding to increase the efficiency in the use of human and budgetary resources for agricultural development, including research, extension and policy formulation. These are important applications, both for those countries which rely on the traditional agricultural sector to drive their economic development, and for other countries where that sector is small in terms of population, but where a social conscience demands measures to combat rural poverty.

I have tried to give the book diversity through the number and origins of its contributors, and coherence through its structure. While the application of FSR in developed country agriculture is occasionally illustrated, the book is primarily focused on FSR in its original role, with small, resource-poor farmers in developing countries. The origins of contributors are sometimes deceptive. Europeans and North Americans write about experiences in Africa, Asia and Latin America, for expatriates indeed dominate the early history of FSR, itself perhaps a factor in the resistance to change in institutions in many developing countries. An expanding professional capacity there began to make itself felt in FSR's application and evolution in the 1980s, yet institutional change is still perhaps the single biggest constraint to wider application. Similarly, the early days of FSR are male dominated but the number of contributions in the book from women demonstrates how they have increasingly asserted themselves in agricultural development.

The book is divided into five parts (each with an editorial introduction) and 12 chapters, each with several contributors. Part I of the book tries to capture the origins and the essence of FSR; its conceptual framework and some of the methods central to the understanding of the farming of resource-poor communities. It begins with contributions from a group of pioneers – fondly labelled 'the old dogs'. Part II examines the application of FSR understanding to the choice and development of technology, to the planning and evaluation of extension, and to policy formulation. Part III focuses on efforts made to incorporate FSR into agricultural research and extension systems in Africa, Asia and Latin America. It also covers the essential companion to institutionalization; the training of professionals in FSR. Part IV looks at the organization of FSR professionals, with contributions on the growth of associations and networks in Africa, Asia and Latin America, as well as on the Association for Farming Systems Research and Extension (AFSRE), subsequently renamed the International Farming Systems Association (IFSA). These accounts are complemented by commentaries from professionals in agronomy, farm management and rural sociology on the interaction of these disciplines with FSR. The fifth and final part of the book turns to the future. Current practitioners discuss cutting edge methods and applications in FSR and the final chapter looks at the lessons of the past and the possibilities for the future. It sets out how FSR has moved toward its original goal – a better understanding of small farmers – and, as systems applications in agriculture proliferate, asks whether it still has a distinct role. The editorial introductions to each of the five parts outline the contributions and offer a personal commentary on the theme covered. Where appropriate, this summarizes the evolution of that theme, highlighting both progress and unresolved issues. Three unresolved issues pervade the editorial introductions and take centre stage in Chapter 12; the scope of FSR, its place in the R & D process, and strategy for institutional change.

1.2 THE ISSUE OF SCOPE

FSR was one of a number of threads from systems thinking that reached into agricultural R & D in the late 1960s and early 1970s. Crop modelling, dominated by the disciplines of physiology and agronomy, was another innovative thread, as was cropping systems research, recalled by Dick Harwood in Chapter 2 as underpinning the origins of FSR in Asia. Eagerly grasped by a variety of constituencies, the early, tight focus of FSR rapidly widened. Texts on systems and agricultural development, including those by Penning de Vries, Teng and Metselaar in 1993, Dent and Macgregor in 1994 and CIRAD in 1996[2], demonstrate the growing range of systems applications in agriculture. It has become unclear, perhaps even confusing, to practitioners, how FSR is best viewed within that spectrum. Proliferating constituencies for systems applications in agriculture, and confusion over the scope of FSR have arguably distracted from its practice and institutionalization.

FSR was an innovation in the research process, emerging from field practitioners, an early effort to bridge the gap between the needs and capacities of small, resource-poor farmers and publicly funded agricultural research establishments. Early in the book, founder members of the FSR family talk about its origins. The common threads through the different accounts leave no doubt that in the 1960s and early 1970s the same problem was widely identified across the developing world; technologies recommended as a result of agricultural research investments were, in general, inappropriate to the priorities and circumstances of small farmers. Field practitioners recognized the importance of the problem and targeted a better understanding of small farmers and the way they make decisions, as a path to its solution. Their concern for appropriate improvements for small-scale, illiterate and resource-poor farmers was the origin of FSR and remains its foundation.

But FSR has also been elaborated, and for some confounded, by the scrutiny of academics. Development theorists, often economists, have criticized the narrowness of conceptual frameworks pinned together by practitioners preoccupied by technology adoption. These originally ignored such issues as intra-household equity, population dynamics, intergenerational equity and sustainability, and the wider macro and policy linkages that these imply. 'Imported' methods, driven mainly by academics doing research to add to theory, or to test out methods in new circumstances, have sometimes diverted professional attention from the operational circumstances of developing countries, the modest institutional capacities and thin budgets with which FSR professionals were wrestling. A notorious example in farm management was the quest to apply linear programming to the small-farm sector in the 1960s. Promoted by the 'have tool will travel' brigade, usually from academia in the USA, it has not yet made a significant operational impact in developing country agriculture. Its failure has been due to the intensive data collection efforts required, and the very high costs of bringing the results of programming to bear on farm units with such low levels of income that even major improvement would offer little return for the costs of the research and advisory process.

1.3 FSR AS AN INNOVATION IN THE R & D PROCESS

Still today, a generation on, in many of the countries where the small-farm sector remains crucial to both the national economy and to the environment, the research/farmer interface remains a critically weak link in the development process. Thus, despite a 25-year history, FSR remains an innovative component in the process for agricultural R & D. The prolonged gestation for FSR reflects the forces governing innovation – particularly innovation in public institutions – in developing countries, and is itself a lesson for both governments and aid agencies. There has been great difficulty in fitting FSR into agricultural institutions. Is this a failing in FSR as an innovation, or are the power dynamics and the entrenched institutional and professional interests in national agricultural R & D too formidable for change? Has the timing of its introduction been inappropriate? The book examines these important ongoing issues. Indeed, the history of FSR is a case study of the dynamics of institutional innovation in developing countries.

The introduction of FSR has been complicated by:

- The need for changes in professional attitudes and institutional orientation and organization.
- The biases of the inherited, often colonial, establishments, in both agricultural education, research and development; expatriate-driven, Western mind-sets, isolated from the small-farm sector, with inappropriate criteria for success.
- Differences between commercial farmers, often driving public programmes in many developing countries, and resource-poor farmers.

Small farmers do not behave like commercial farmers. They are not organized to interact with the wider market economy, nor are they politically articulate like commercial farmers. These had attracted a set of service institutions, for example in credit and insurance, for protection against the vagaries of weather and the market. These older institutional processes, oriented to and organized for large farmers, cannot operate cost-effectively with small farmers who, in the absence of an appropriate enabling infrastructure, must manage their environment directly by their own decisions and by their activities both on and off the farm. Small farmers often cannot use the technologies appropriate for commercial farmers and always need explicit consideration in agricultural R & D. These insights have given rise to the development of new investigative methods to manage the different circumstances of resource-poor farmers under conditions of scarce professional and financial resources. A start has been made in reorganizing agricultural R & D institutions to implement the new methods and to adjust higher agricultural education to achieve congruity between the mind-sets of peasant farmers and professionals to encourage mutual respect and partnership in agricultural improvement.

A parallel feature of the last 15 years, and one which holds great hope for the future, has been the growth of FSR professional associations. FSR associations attract people from a range of disciplines, from agronomy, ecology and plant breeding to economics, anthropology and rural sociology. The growth of these pioneering associations has received much of its impetus from the leadership of university professionals, who established an annual symposium for FSR-E in the USA in the early 1980s. This evolved into the AFSRE and associations and institutional networks now exist at the continental level in the USA and Asia, and at the regional level in Africa, Latin America and Europe. Several contributions to this book document the evolution of these associations which promote interdisciplinary interaction around key problems, encourage independence for professionals in developing countries and complement allegiance to discipline with allegiance to people in a refocusing of the R & D process in agriculture. In Africa, Asia and Latin America FSR associations are moving professionals out from under the spell of developed country fora, finding their feet in their own context, and helping to bring both education and development processes into line with the needs of local people. It is good to be able to record progress towards these goals. But it is important to record that these gains remain fragile and there is a danger that governments, courted by the dynamics of growth at any price, may despair of their smallholder constituencies as an engine to achieve it.

Appropriate intervention for farm improvement remains the heart of FSR. Experience has widened the portfolio of interventions beyond the early emphasis on technology development. Accumulating insights into the nature of the traditional agricultural sectors of developing countries have shaped the evolution of an FSR process for their successful development and deployment. The early insights included:

- Recognition that vast numbers of small farms dominate agricultural sectors in many developing countries under widely diverse circumstances.
- Recognition that on one small farm, a major improvement of productivity, even 100%, is a small absolute benefit, and costs of achieving it must be low.
- Recognition that appropriately qualified agricultural professionals are an extremely scarce resource.

The scope of FSR and the strategy for promotion and institutionalization, perhaps the fundamental issues of FSR, are revisited in the final chapter. I hope this book will provide a foundation on which a second, or now perhaps a third, generation of farm systems practitioners can build.

REFERENCES

1. **Merrill-Sands, D., 1986**. Farming systems research: clarification of terms and concepts. Experimental Agriculture, 22, 87–104.

   Simmonds, N.W., 1985. Farming Systems Research – a Review. World Bank Technical Paper no. 43. World Bank, Washington D.C.

2. **Penning de Vries, F., P. Teng, & K. Metselaar (Eds), 1993**. Systems Approaches for Agricultural Development. Kluwer, Dordrecht, and IRRI, Los Baños, Philippines.

   Dent, J.B & M.J. Macgregor (Eds), 1994. Rural and Farming Systems Analysis – European Perspectives. CAB International, UK.

   CIRAD, 1996. Systems Oriented Research in Agriculture and Rural Development. Lectures and Debates from an International Symposium, Montpellier, France. CIRAD-SAR, Montpellier.

Part I

FSR – Understanding Farmers and Their Farming

EDITORIAL INTRODUCTION

Mike Collinson

In my general introduction in Chapter 1, I skirted the historical controversy on terminology by distinguishing the process of farming systems research (FSR) from its applications in technology development, in extension, and in policy formulation. Part I of the book, in two chapters, deals with the development of the FSR process in its role of understanding farming systems.

THE CONTRIBUTIONS

Chapter 2 features personal accounts by five 'old dogs' of the experiences which drew them into the development and promotion of FSR in the late 1960s and early 1970s. They recapture the insights which convinced them that FSR could improve the relevance of conventional agricultural research to the situation of countless small farmers in the developing world. The contributions vary from personal, even anecdotal, to semi-formal. Each offers lessons and many of the issues raised from the 1960s and 1970s remain issues today.

The contributions are from German Escobar, working in Latin America, Pete Hildebrand in central America, Dick Harwood in south-east Asia, David Norman in west Africa, and myself in east Africa. German, David and myself are farm economists, with Dick the only thoroughbred agronomist. I have to mark Pete down as 'hybrid'; an agricultural economist by training, much of his best known work has been in the analysis of stability in biophysical parameters important to farmers. Some of the most telling points made in Chapter 2 are listed in Box I.1, all arise in more than one account, and some arise in all five.

Even in the early days linkages were important and some of these commonalities can be attributed to the interactions that occurred across continents. German Escobar mentions the influence of Hans Ruthenburg[1], another 'old dog', and both he and Dick Harwood acknowledge the value of interacting with David Norman in formulating an approach to the problem of non-adoption by small farmers. Yet much was clearly spontaneous. All five 'old dogs' overlapped in the timing of their 'conversions' to FSR, all were focused firmly on the small resource-poor farmer and agricultural technology. It is as though the 'bones' of the process were buried around the world but were dug up by the 'old dogs' from one continent to the other, sometimes in a different sequence.

Chapter 3 also has five contributions. These address the conceptual framework and four aspects of methodology I judged as central to an in-depth understanding of farm systems: typology or characterization, diagnostic methods, gender analysis as a neglected dimension of diagnostics, and the crucial step from problem identification to an understanding of the causes. Robert Hart analyses the evolution of the conceptual framework for FSR. My contribution on typologies highlights how new methods for reconciling physical and human attributes are contributing to a revolution in typologies for agricultural development. John Farrington reviews the evolution of diagnostic methods and asks how far farmer participatory research (FPR), as a less extractive, more collegial approach, is a development of FSR or an alternative to it. Hilary Feldstein describes the gradual acceptance of gender analysis, a diagnostic method to capture the understanding of gender roles in the household and on the farm. Finally, Robert Tripp, a colleague with the economics staff of the International Maize and Wheat Improvement Center (CIMMYT) in the 1970s and 1980s, delves into causal analysis, a step in the diagnostic process to help ensure that on-farm experimentation, the most expensive step in the on-farm research (OFR) sequence, is attacking the most relevant issues in the most appropriate way.

None of the contributions here is a manual on 'how to do it', there are plenty of these. Each is a commentary on the evolution of ideas in these defined areas of the FSR process, aimed towards the understanding and improvement of small-farmers' systems.

A COMMENTARY

Perhaps the most telling change over time is the shift from the farm system *per se* as a framework, to a hierarchy of systems within which the farming system is one of a number of levels, a change pioneered by Robert Hart[2]. In his contribution here he uses communities and watersheds as examples of wider systems in a hierarchy.

One important consequence of the wider framework is the diversity of perspectives embraced. Decision makers at all levels in the hierarchy have different views on a given issue, shaped by their varied roles: communities may take a different view from individual farmers on those on-farm activities with impacts beyond the farm boundaries – externalities in the economist's jargon. Policy makers need to reconcile the perspectives of many such communities while pursuing aggregated interests at the local or national government levels. The diversity of perspectives is synonymous with the concept of multiple stakeholders. It illustrates why FSR has moved from formal to informal farmer surveys, to embrace farmer participatory methods, stakeholder analysis and conflict resolution. The hierarchy of systems also responds to earlier criticisms of FSR, including its failure to embrace long-term sustainability and to address the important interaction with policy. FSR needs to reconcile the wider perspectives at higher levels of the hierarchy with the ways in which it helps farmers improve their welfare. Again this highlights its scope, an issue taken up in the concluding chapter of this book.

For those operationalizing such frameworks over the years the phrase 'the farming systems approach' has been a valuable 'shorthand'. On reflection, in some circumstances, the phrase has perhaps been counterproductive. Certainly it has been an anathema to the ears of the research establishment, carrying as it does the implication that FSR is an alternative to the traditional process. In serious circles FSR has always been perceived as a supplement, or complement, to the existing process. It is clearly no substitute for reductionist experimentation. Yet, partly because its promotion was sometimes perceived as confrontational, its role in focusing and interpreting more formal experimentation, both on and off the research station, has remained controversial. Insensitive promotion may have been one of the most serious hurdles to institutionalization. Use of the word 'approach' perhaps originated in the early competition between agencies in promoting their

Box 1.1. Common insights from the early days of FSR.

- The limited adoption of research recommendations by small farmers, yet their obvious skills in managing their environment with limited resources
- Priorities in research set from a researcher's perspective, based on the need to get the most out of the commodity under local biophysical conditions
- The contrasts between farming methods on research stations and on the surrounding small farms, and between the evaluation criteria used by researchers and by small farmers
- Small farmers' willingness to learn and to change, and the way they make these changes; by trying out new ideas on a small scale and adapting these, often quite radically, to their own circumstances
- Low professional/farmer ratios and the need to deal with numbers of farmers together
- The farming system as a basic unit for agricultural R & D
- The trade-off between coverage and intensity of effort, and the search for cheaper methods; qualitative versus quantitative investigation to provide understanding
- The difficulties of interdisciplinary research and capacity building in new methods
- The problem of social scientists, often new, junior, recruits, criticizing the products of senior members of the established research disciplines with which they need to collaborate
- The value of learning through experience

own paradigms. Although FSR's application to technology development has four agreed stages – diagnosis, design, implementation and evaluation – there have been contrasting approaches to its operationalization and practitioners have attempted to categorize these, as in Norman and Collinson, 1985[3]. One illustrative contrast is between the Tropical Agricultural Center for Research and Training (CATIE) and CIMMYT.

The CATIE 'approach' was labelled 'formal, quantitative and rigorous' in the early 1980s. It was characterized by intensive data collection, parameter measurement and computer modelling. Analysis held all options open across the system and all model parameters were potentially variable – embracing the dynamics of system evolution and the opportunities for changes in policy, and avoiding some of the criticisms made of less sophisticated approaches. In contrast the CIMMYT 'approach' was labelled 'informal and qualitative'. This purposefully put to one side large areas of the system as it focused in to detail on maize or wheat. It moved from an understanding of the climate, markets and policies facing local farmers, through an understanding of farmers' strategies in managing their system, to analyse how the specific practices used to manage the maize or wheat enterprises were dictated by this wider context. The CIMMYT approach was pre-focused, closing down options as understanding grew of how maize or wheat management was shaped by its system context. The CATIE/CIMMYT comparison also illustrates extremes in the search for acceptable compromise between theoretical desirability and operational possibility.

Characterization is an important first step in the FSR process of description and analysis. In the current literature the term 'characterization' often seems to replace 'diagnosis', but properly alludes only to the description, or profiling, of discrete units, such as the agroecology and the farm enterprise pattern. It does not, on the whole, assume an understanding of the farming system. Many development programmes are still implemented by administrative units, even though it has been clear since the early 1960s that surveying of any sort using administrative boundaries, cannot, except serendipitously, differentiate types of farming. Since surveying across farming systems confuses description and thoroughly confounds understanding, the grouping of farmers into types by profiles of the systems they operate takes on particular importance. Essentially a means of stratification, it seeks to maximize differences between types and minimize sources of variation within them. No stratification device is perfect and the farming system is no exception. Beyond this, important parameters such as yields will often express variation more strongly at the farm rather than the farming system level, particularly across systems within the same agroecology. From year to year these yield variations have a major impact on the relative performance of farmers operating the same system.

Historically, biophysically derived zones have dominated typing in agricultural R & D, mainly in the identification of uniform zones of land and crop potential. While it is true that a relatively narrow set of activities offers the best physical potential for any area, a much wider set of economic production opportunities typically exists for the area. From a purely land use perspective, the physical performance of this wider set may be relatively poor, some may even threaten the integrity of the resource base. Nevertheless small farmers choose activities that are economically superior from the wider set. CIMMYT's 'recommendation domains' (RDs) of the late 1970s were a pioneering step in defining groups of farmers for whom the same changes would be relevant. Since then widening acceptance of the link between poverty and environmental degradation is forcing further reconciliation between traditional physically based definitions of zones, in terms of climate and soil, and people-based definitions.

In diagnosis proper, the unravelling of the complexity of the household has been an important step towards better understanding. What had historically been seen as 'the farmer' was overtaken, as understanding improved, by the interaction, and indeed negotiation, between household members for access to resources and control over output flows. The credit for unravelling this tapestry and for earning gender analysis a place in the FSR, goes to a relatively small group of intrepid women who kept their eyes, and their actions, firmly on the unacknowledged role of their gender in traditional agriculture. Prominent among the group, both for their articulation of the issue, and persistence in its pursuit in the field, are Hilary Sims Feldstein, Janice Jiggens, Joyce Moock and Susan Poats.

Despite the increasing pervasiveness of market forces, non-market objectives – putting food on the table day in day out, with family food preferences more or less satisfied depending on the vagaries of the season – continue to drive many of the actions of the majority of small farmers. The direct management of risk through the enterprise pattern and farm management practices, rather than through market institutions offering overdrafts, credit and insurance, as they do to commercial farmers, dominates management strategy. Thus, even in farming systems where market access offers cash earning opportunities, subsistence and survival goals often take priority in the allocation of family land and labour. I believe there are still relatively few situations in the traditional farm sector where market opportunities are so valuable that 'basic needs' goals and strategies are subordinated to their exploitation. Understanding the specifics of these goals and strategies for each important farming system is the key to identifying interventions which are valuable to farmers, to designing relevant extension and credit programmes, and to the sensitive formulation of policy.

The search for cost-effective methods of gaining understanding led to the development of rapid rural appraisal, informal surveys in which representative farm families and farmer groups are engaged in conversations with researchers that are allowed to flow, often guided by the farmers. Initial conversations in the process are descriptive, later ones, usually with other families, are analytical, seeking verification of hypotheses set up from the earlier descriptions in order to unlock family priorities and farmer management strategies[4]. The evolution of diagnostic methods for FSR was accompanied by strong professional debate at each stage. Two social science schools emerged, the hard data modelling school, exemplified by the CATIE approach within FSR, and more broadly by researchers using formal economic models or seeking to contribute to development theory. The place of modelling within FSR remains an issue and is discussed further below. The scientific credibility of the informal approach which emerged in FSR was questioned by natural scientists in agricultural research institutions. Battle was engaged between natural scientists critical of the qualitative research methods, and FSR practitioners horrified by the preoccupation with precision and apparent unconcern for relevance in agricultural research. Qualitative understanding has been legitimized by systems writers[5]. Further, occasional rigorous comparisons of results from informal and formal survey work showed no significant distortions of reality from an informal approach[6].

Bridging the gap between diagnosis and action has long been acknowledged as a weakness both in FSR and in FPR. Clive Lightfoot (Chapter 11.5) has termed it the 'so what' syndrome – we have an insightful description of local farming, but where do we go from here? I regard causal analysis as another of the key contributions from the International Agricultural Research Centres (IARCs). This came from a collaboration between the CIMMYT economics team of the 1980s and CIAT agronomists[7]. It is a technique to carry the process from the diagnosis of problems to the identification of solutions, in a sense, the fulcrum of OFR. It may involve experimentation by natural scientists to determine physical causes, or research by social scientists to identify economic and cultural ones. In his contribution Robert Tripp also distinguishes proximate and ultimate causes, and, while warning against pursuing the causal chain *ad infinitum*, brings out the value of understanding a number of links to multiply the strategies and options for solution. It is a technique which I believe has taken its place among the 'best practices' of contemporary diagnosis in FSR.

CONCLUSIONS

The better understanding of farming systems has both improved old paths for farm improvement and identified new ones:

- Shaping new technologies to the cultural circumstances and resource constraints of the existing farming system.
- Creating a better understanding of development dynamics, particularly the watershed between increasing the area cultivated and yield intensification as land scarcity grows, and the benefits of wider market access allow greater relevance in interventions.
- Reinforcing farmers' own strategies for managing uncertainty and raising productivity.

- Seeking greater flexibility of action for farmers in the face of climatic and market uncertainty.
- Looking beyond the system itself to those factors influencing farmers decisions about their system. Breaking resource constraints by action off the farm, at a higher level in the hierarchy.
- Revising community or policy decisions to change farmers' incentives and encourage practices that reduce resource degradation. New institutional orientations can be particularly effective where degrading practices are a response to market and price uncertainty.
- Developing strategies for the introduction of new market production opportunities.

Three issues persist with the process for understanding small-farm systems. The first, much the most threatening, is superficial diagnosis. Continuing weakness in current diagnostic practice jeopardizes the credibility of the process and inhibits its mobilization in developing country institutions. Superficial diagnosis is also closely related to the second issue; whether sound systems understanding adds value in the diagnostic process over and above collegial farmer participation. Both this and the value of quantitative models in FSR are questions clouding a conclusion on best practice.

Superficial diagnosis

In the early days of FSR some agronomists moving off-station into on-farm experimentation simply brought their old programme of station-based experiments with them, ignoring diagnosis. Diagnosis was also often decried by extension and development professionals who insisted that they knew all there was to know about their farmers. It is a tendency which has persisted and one which is encouraged by superficial problem identification from FSR teams; 'low yields', 'poor soil fertility' or 'crop disease' are scarcely insightful. Poor training, partly a result of the rush to climb aboard the FSR bandwagon – or perhaps gravy train – of the early 1980s, has been responsible and the fact that there are still relatively few professionals aware of what qualitative diagnosis can provide is a reflection of the slow pace of change in university curricula. Good diagnosis depends on a sound grasp of the principles of agronomy, production economics and farm management, and on disciplinary interaction.

It is worth looking briefly at the sort of insights which emerge from good diagnosis. Examination of background information on climate and soils, markets and institutions is an essential basis. It builds up an understanding of what local farmers must manage, and helps identify the facets of the farmers' environment that create problems. Risk management plays a large role in resource-poor farmers' decisions and is a key area for diagnosis. A good example is within season rainfall variability in dryland agriculture. In southern Zimbabwe, CIMMYT collated rainfall data for a 30-year period by pentads[8]. Analysis identified three patterns of within season drought: some 30% of years showed a significant delay in the onset of rains, some 30% showed an early finish, and some 40% a 3-week mid-season drought occurring anytime within the 3-month period from early December to the end of February. Some years showed more than one type of drought, others showed none.

The main starch staple for the system was maize. The analysis of rainfall data provided a sound basis for discussion with farmers' on how they managed the uncertainties these patterns created. Losses from drought were a well understood risk in local farming and farmers responses on the occurrence of the three types of drought were related back to the rainfall data for verification. Farmers reported making a series of two, three, four or more plantings over the period November to February, increasing the chances that one or more plantings would not be caught by drought in a critical period of growth. Later plantings were made with early maturing varieties. Those early plantings critically damaged by delayed onset, or by early mid-season drought, were replanted with earlier maturing material. These demonstrate both pre-emptive and reactive risk management strategies. Farmers routinely make several plantings with maizes of differing maturity period in anticipation of drought. Replanting, on the other hand, is a reaction to drought occurrence. These strategies benefited from the policy of government purchase of maize at a fixed price, common to many African countries until the late 1980s. Farmers could not over-produce maize, it was a win–win situation for them. Strengthening their strategies for drought management in maize, both pre-emptive and reactive, offered an important focus for an improvement programme, an importance enhanced by the opening of markets under structural adjustment initiatives.

A vital thread running through the diagnostic process is the interaction between biological and social scientists. Interdisciplinarity is well established between breeders, pathologists, physiologists and agronomists in classic agricultural research – the new element in FSR is the social scientist. One early rendering of the interaction ran like this:

> The biologist brings to the diagnostic process a perception of the ideal technical management for crops in the conditions of climate and soil under which farmers are operating. The social scientist brings an understanding of farmers' priorities and the constraints operating on them, limiting the ways in which they can adjust their management. The biologist evaluates the background information on natural conditions to assess crop potential and management practices likely to be important. The social scientist evaluates background information on economic, cultural and institutional conditions. The biologist learns about farmer management practices and identifies changes which would better exploit biological potential. The economist learns why farmers are doing things the way they are and identifies when resources are available and when limiting. In interpreting the survey work the biologist puts forward ideal changes from a technical point of view and estimates the likely improvement in yield. The social scientist assesses their possible profitability and their compatibility with the ways farmers currently allocate their resources to realise family priorities[9].

Although they remain 'offstage' here, interaction with farmers is central to the process.

Participation versus systems understanding

Public attention and much funding has shifted from FSR to participatory research in which farmers are seen as full, indeed dominant partners. Most of us follow Robert Chambers in advocating empowerment for small-scale farmers[10]. Empowerment, however, has an affinity with community development. Both concepts are abstract and require operational goals and processes beyond themselves for their implementation. Technology generation includes just such a process and was adopted as one vehicle for implementing the participatory concept. FSR best practice has absorbed many methods from the participation portfolio. However, at the extreme it can be argued that much of the process adopted as a vehicle for empowerment by participators has been usurped by them. There is an insistence that farmers and communities make the decisions and that agencies service these. Like early FSR this seems to neglect the need to reconcile local, national and indeed global interests. It also seems, rather like the extreme advocates of indigenous knowledge, to deny that outside knowledge will have a key part to play in providing sustainable livelihoods. Recently cracks have appeared between rhetoric and reality in participation. Some were highlighted by Rhoades in 1998: 'the social scientist who attempts to raise analytical points about stratification, differential access to power and resources, and other social shaping dynamics are accused of being top-down and then are marginalised by turf guarding NGOs'[11]. It is an issue to which we return in further editorials and in the final chapter.

Quantitative modelling as a diagnostic tool

Although the value of qualitative understanding is now widely accepted, modellers continue to press their case for the use of quantitative socioeconomic models within an FSR context. Against my own history of moving away from quantitative to informal surveys to reduce costs and gain coverage in terms of the systems researched for a given tranche of manpower and budget, this almost seems sacrilegious. Modelling apart, experience shows there are a number of valuable roles that hard data can play; verification of informal survey conclusions, including quantification of the incidence of farmer priorities and strategies across the population. Evidence of impact is valuable, even the simple recording of adopters over time. However, as Roberto Quiroz and his colleagues demonstrate in Chapter 11.3, and as Dent and others have long argued, models can make a valuable contribution to understanding, particularly in *ex ante* evaluation of the impact of interventions on the existing system. Questions inevitably arise: can modelling be reconciled with a low cost approach to small-farm improvement still necessary in many developing countries where professional manpower remains limited? With what specifications of objective function does formal modelling improve on a sound qualitative understanding of the system? Is the degree of improvement worth the extra cost and effort of intensive data collection?

My own conclusion is that the application of quantitative modelling has advanced a great deal since the 1970s, particularly in the ease of analysis. The data requirements, however, remain heavy and expensive for local application in an FSR context where professional manpower is scarce and budgets low. Also, for those small-farm systems still dominated by subsistence production, representation of the objective function remains weak. For me it is too 'hands off' and has too many pitfalls to advocate its incorporation in routine FSR diagnosis. That said, it is important to recall that many early practitioners had a farming systems perspective imbued through an involvement in socioeconomic modelling. The important problem of superficial diagnosis in current FSR practice suggests modelling might have a valuable role in training. As establishments mature, the idea of using the staff deployed in the field for experimentation to collect the detailed data for modelling is attractive. The goal might be a portfolio of models for major farming systems for training in both farm improvement and policy analysis, and for policy analysis itself. There is a good case for universities developing and using locally relevant models to ingrain a farming systems perspective into their agricultural students.

REFERENCES

1. **Ruthenburg, H., 1980**. Farming Systems in the Tropics, 3rd edn. OUP, Oxford.
2. **Hart, R.D., 1982**. An ecological conceptual framework for agricultural research and development. In: Shaner, W.W., Phillipp, P.F. & Schmehl, W.R. (Eds) Readings in Farming Systems Research and Development. Westview Press, Boulder, Colorado, pp. 44–58.
3. **Norman, D. & M. Collinson, 1985**. Farming systems approach to research in theory and practice. In: Remenyi, J.V. (Ed.) Agricultural Systems Research for Developing Countries. ACIAR, Canberra.
4. **Collinson, M., 1974**. Transferring technology to developing economies: the example of applying farm management economics in traditional African agriculture. World Development, 2(2), 93–7.

   Hildebrand, P., 1976. Generating technology for traditional farmers: a multidisciplinary methodology. ICTA, Guatemala City.

   Rhoades, R., 1981. The Art of the Informal Agricultural Survey. Training Doc. 1982–2, Social Science Department, CIP, Lima.
5. **Checkland, P.B., 1981**. Systems Thinking, Systems Practice. John Wiley, Chichester.
6. **Franzel, S. & E. Crawford, 1987**. Comparing formal and informal survey techniques for farming systems research: a case-study from Kenya. Agricultural Administration and Extension, 27, 13–33.
7. **Tripp, R. & J. Woolley, 1989**. The Planning Stage of On-Farm Research. Identifying Factors for Experimentation. CIMMYT and CIAT, Mexico, D.F. and Cali.
8. **CIMMYT & Government of Zimbabwe, 1981**. Report no. 5. Demonstrations of an Interdisciplinary Approach to Planning Adaptive Agricultural Research Programmes. CIMMYT, Nairobi.
9. **Collinson, M., 1981**. A low cost approach to understanding small farmers. Agricultural Administration, 8, 433–50.
10. **Chambers, R., 1994**. The origin and practice of Participatory Rural Appraisal. World Development, 22 (7), 953–69.
11. **Rhoades, R., 1998**. Participatory Watershed Research and Management; Where the Shadow Falls. Gatekeeper series no. 81. IIED, London.

Chapter 2
FSR: Origins and Perspectives

2.1 MY INITIATION INTO FSR IN LATIN AMERICA

German Escobar

Neglect of the peasant sector brought an emphasis on the use of FSR methods in the field by organizations with a social conscience. As a young professional 22 years old, as you can imagine, I found the opportunity to share my theoretical training and political ideas with very small farmers living in poor conditions very stimulating.

2.1.1 Introduction

The introduction of the farming system research (FSR) approach in Latin America and the Caribbean (LAC) was not an organized, monolithic event, but a rather haphazard, serendipitous process. It was introduced as a part of a rural development effort to reach a target population of small farmers who were traditionally neglected by the agricultural research system. The initial elements used were on-farm experiments rather than formal systems approaches, and most projects were implemented by agricultural researchers trained in traditional disciplinary methods with no knowledge of systems theory. FSR began in a range of research and rural development projects within existing institutions, and these, like the manpower, were organized to operate the more traditional approach. These institutions pursued the conventional process; field research activities were identified and designed by senior researchers working on an experiment station. Extension agents provided small farmers with technical recommendations based on nationwide research results which, in the best scenario, included a trial from the local region.

It can rightly be claimed that FSR in LAC is characterized by heterogeneity in its application, its methodology and its institutional context. The initial introductions of FSR were concentrated in the Puebla project in Mexico, the Caqueza project in Colombia and at the Tropical Agricultural Center for Research and Training (CATIE) in Guatemala. I was fortunate enough to have personal experience in each of these three: I was a young professional in the formation of the Caqueza team and went to the Puebla project for in-service training. Later, following the Caqueza project and graduate work, I moved to CATIE which was assembling a strong team of systems researchers. This paper is a personal interpretation of the initial development of an FSR approach in Latin America and it is a difficult task for a practitioner used to writing up experiences in the most formal way possible; a defensive strategy adopted to preempt the permanent criticism from biological researchers of the lack of 'science' in FSR applications!

The political context of development in Latin American countries played a strong role in how FSR was introduced. Neglect of the peasant sector brought an emphasis on the use of FSR methods in the field by organizations with a social conscience. As a young professional 22 years old, I found the opportunity to share my theoretical training and political ideas with very small farmers living in poor conditions very stimulating. In those early days of the 1970s,

FSR relied on readily available methods and there was limited interest in conceptualization or the evolution of theory. Some description of the political context is important to understand the fragmented and *ad hoc* emergence of FSR, followed by an account of some of the experiences in the three projects which made an impression on me. Finally, some comment on the evolution of the FSR concept and its methodology is offered. This is perhaps the most difficult part, since the rapid evolution of the concept often contradicts the interpretation a young technician made some 25 years ago.

2.1.2 The context of agricultural research in LAC

Although FSR became a phenomenon in the early 1970s, changes in the LAC agricultural sector during the 1960s influenced its application. The national agricultural research institutes (NARI) established in the 1960s internalized a strong influence from the USA 'land grant system' through technicians trained in the USA and in research cooperative programmes supported by the USA in Mexico and Colombia in the 1950s[1]. Agricultural research was disciplinary and crop-oriented, and its institutions were designed to operate through research stations. From the beginning, priorities were set from the supply side, mainly in response to the import substitution policies of governments and the scientific interest of researchers. The publicized success of the Green Revolution supported by the International Agricultural Research Centres (IARCs) reinforced this traditional approach.

In the 1960s the contribution of agricultural products to the gross national product (GNP) and the increasing importance of the multilateral agencies made agricultural research an important area for national policy. These policies were driven by the need to produce cheap food for the urban and industrial sectors and support services such as agricultural credit, marketing and some physical infrastructure, were strengthened to complement agricultural research and extension. During the 1960s land reform and agricultural development projects, implemented in most LAC countries, highlighted the gap between USA agriculture and LAC, on the one hand, and the traditional small-farmer subsector and the commercial subsector on the other. Strengthening the extension services and the NARIs with the collaboration of the international community was not sufficient to close these gaps. NARIs did reinforce their research on rural issues that were known to influence production and technical change among farmers. Programmes such as agricultural economics, communications, adult education, social organization and home economics were incorporated into their structure. This required the skills of social scientists and other growing professions that could contribute new information about farm organization, farm production strategies, land tenure and related issues, farmers' organizations, their priorities and values, their communication systems and their off-farm activities. This new information improved the analytical capacity of NARIs, while better definition of the target population improved their responses.

Both land reform and agricultural development projects put extensionists and some researchers in the field with farmers. The organization of these projects brought together, perhaps for the first time in LAC, most of the available agricultural support services as well as technical teams from different institutions and disciplines. Though it could not be hailed as a coordinated effort, technicians in this working environment vigorously identified with the small producers, land reform beneficiaries and settlers. With hindsight it was probably a key step in transforming research to reach small farmers. At the same time, new information on farmers' production strategies, technical practices, the social organization of production and local markets brought new evidence of the productivity gap and the missing link between agricultural research results and the capacity of this traditional subsector to assimilate and utilize technical information.

The influence of multilateral agencies was more evident on the design and implementation of the Integrated Rural Development (IRD) projects initiated at the end of the 1960s. These were specifically aimed at improving production and living conditions of small traditional farmers through agencies and policies designed to help the consolidation of the industrial sector. In a number of countries these IRDs were important components of national development policy, and in some cases, as in Brazil, Colombia, Ecuador,

Honduras and Mexico, this lasted until the mid 1980s. Most IRD schemes posted interdisciplinary technical teams in the field as well as support institutions. Although rural development components other than technical change were considered important, increasing agricultural production and productivity usually became the central focus, supported by credit programmes which were redesigned for this purpose. Implementation increased the evidence of the limited use of the existing technology in the production conditions of small farmers.

These donor-supported projects provided a context which allowed a number of technicians to question the capacity of research institutions to generate technical recommendations for this target population. Conceptually, the rationality and efficiency of small farmers were already demonstrated[2]. Pragmatically, the rejection of technical recommendations by small producers was a daily reality for the field teams. Some began to test different arrangements of known technological packages at the farm level, sometimes maintaining practices and components already used by producers[3].

The pioneer work in the late 1960s and the early 1970s in Mexico, Colombia and Central America was pragmatic; standard research designs, farm trials and biophysical analyses were utilized. In response to national policies, staple foods (grains) were preferred to high revenue crops. Farmers' crops were taken as the basis for designing technological alternatives and the emphasis was on providing support services, particularly agricultural credit. Field teams were formed in most cases by young technicians with no experience. Expatriate technical advisors were often involved and the rigid institutional organization was not changed.

2.1.3 Initial FSR applications

Puebla, Mexico

Most LAC practitioners recognize the Puebla Project in Mexico created in 1967 as the first agricultural research activity closely related to farmers' production patterns. Although the examples of Borgo-a-Mozano, the Comilla Project and the Intensive Agricultural Districts Program in India were known, Puebla introduced the on-farm adaptation of production technology developed at experiment stations which considered farmers' cultural and capital limitations to adoption. The project was designed and technically supported by CIMMYT, the International Maize and Wheat Improvement Center, in partnership with the Chapingo Graduate School.

Caqueza, Colombia

In 1970, the Colombian Agricultural Research Institute (ICA) initiated the Caqueza project with the collaboration of the International Development Research Center of Canada (IDRC)[4]. The field team of the project was formed by young technicians: agronomists, a veterinarian, a sociologist, a home economist, two advisors from IDRC and a number of field assistants trained at the intermediate level. Activities related to on-farm research, extension, planning and evaluation were established from the beginning. Some researchers from experiment stations, graduate and undergraduate students, foreign volunteers and some consultants carried out research activities in the project, producing over 100 publications in the first 6 years.

Recruited as a team member, I was instructed to go to the field and 'get the region developed'. This missionary approach reflected two key problems. First, senior managers had no clear idea of the type of project they wanted. Second, the integration of traditional researchers in the project's field activities was not systematically organized and became a major difficulty. Several efforts were made by senior project managers to plan joint research in the field, but these were weakly implemented throughout the first 10 years of the project. The plans for senior researchers to assist the technical team were never realized, and indeed these researchers became critical of the validity of on-farm research. I was sent for in-service training to the Puebla Project in Mexico which was very effective in providing training for a number of young technicians from different countries. The Puebla approach and its components were extrapolated to many other projects, including on-farm technology trials and the analytical research orientation through students and university faculty members. The training instilled a much stronger conceptual framework for subsequent activities in on-farm research in Caqueza.

Institutionally, in Caqueza, the major preoccupation was to provide the capacity to adapt technology tested in experiment stations to farmer's circumstances and to obtain high adoption rates among direct beneficiaries. Despite the great efforts made by some members of the team to develop a framework for the entire project, the farming systems concept was never explicit. However, terms such as multiple cropping, on-farm research, multidisciplinary teams, farmers' constraints, farm types, farm production components and the understanding of the farm as a decision unit were all used in the implementation of the project. Most Latin American practitioners at that time wrote for field colleagues, but their writings were not reviewed by their peers, formally published or translated into other languages. Conceptual contributions to FSR began to be made only after some years of field experience[5]. In the mid 1970s Richard Harwood's writing on his experience in Asia, and David Norman's on his time in Africa, offered useful frameworks which were, in turn, complemented by learning from other applications in Latin America[6].

The concepts of team and institution, the institutional challenges and the strong identification with small farmers were, in my view, the key motivations for the small group of young technicians that began their professional activities in the Caqueza Project. The level of involvement of every one was remarkable, substituting for the conceptual elaboration and development theory unavailable to them. The possibility of constructing a different institutional model, the open-ended opportunity to learn about small farmers, the search for solutions to real problems, alternative institutional instruments and, above all, applied research methods, were powerful reasons to maintain the professional interest and the motivation required to initiate a long exploration of the application of FSR to agricultural development.

It has to be said that the socioeconomic dimension still lacked methodological integrity. More than a year was needed to complete a regional diagnostic that made little contribution to the understanding of farmers' production strategies. Indeed, a number of on-farm experiments were concluded and an extension strategy was in place before the results of diagnosis were available to the technical team. Technical integration between biology and social technicians was a difficult process. A common view of the project was shared by all team members but different working methods in the field caused important discrepancies. These differences frequently discriminated against social research and, in some cases, resulted in personal conflicts.

One important result of the Caqueza Project was the recognition of the need to adapt institutions to the circumstances of small farmers. The project generated the so-called 'buffer institutions' as pilot institutional programmes to facilitate adoption of recommended technology, and some changes to institutional services were introduced when the IRD projects and, some years later, the national fund to support rural development activities, were put in place.

While I attended graduate school in the mid 1970s I became better acquainted with the literature. There were still very few publications classified under FSR but it was a great opportunity to make sense of theory, working approaches and farmers' reality. It confirmed the need to develop methods and institutional arrangements to respond to farmers' real conditions.

CATIE, Costa Rica

In the early 1970s, CATIE initiated on-farm research programmes that evolved into major applications of the systems approach with the support of the United States Agency for International Development (USAID) and largely through the influence of Richard Bradfield. These efforts lasted more than 10 years and influenced most research programmes at CATIE and a number of institutions, researchers and graduate students in Central America.

My work with the team at CATIE was the third major influence in drawing me into advocacy for FSR. The CATIE technical team was well qualified and equipped. It was formed by well-trained agronomists, animal scientists, agricultural economists and ecologists with a support team for biostatistical data processing. A number of papers, training materials and MSc theses were published both internally and externally. In comparative terms, CATIE publications combined empirical experiences with conceptual development and constituted the state of the art on the application of FSR in Latin America in the early 1980s. A peculiarity of the early CATIE FSR team was the decision to work independently from the ideas developed at Puebla and in Guatemala. Strong personal attitudes dictated the approach developed. It did not incorporate the Guatemalan 'sondeo' approach, or the ideas of farmer organization and the provision of support services. Since CATIE is mandated for research and training only, extension, technology transfer and institutional adaptation were virtually absent from the CATIE work schedule. The first exchange of concepts and some experience from El Salvador led to the design of a big central experiment, testing numerous interactions and different topological arrangements, at Turrialba station. This generated a number of novel ideas among the research team but was so complex that analysis of the experiment was never completed!

After a team member visited the International Rice Research Institute (IRRI), the conceptual framework being used there by Harwood and Zandstra was introduced into a regional project that CATIE developed in Central America. Basic system concepts were applied to agricultural development, allowing the construction of a systematic body for teaching and training purposes[7]. This conceptual base brought about the development of different methodological phases for applying the systems approach on the field: area selection, diagnosis or characterization, technological alternatives, design, on-farm research, validation and dissemination, including a feedback mechanism to generate results. In every case field work was initiated as an empirical test of a conceptual elaboration[8]. The improvement of production systems based on farmers' practices captured most design efforts, the programme included better understanding of the biophysical relations among plant, soil, water and crop management. Later on, agroforestry practices for resource conservation and rationalization were developed and added as technological alternatives for improving farmers' production systems.

Economic and social analysis became an integral part of the CATIE approach. The economic analysis of agronomic trials introduced by CIMMYT was complemented with a range of analytical tools, putting emphases on two aspects:

- Farmers' constraints due to both farm and off-farm circumstances.
- The impact on farm activities as a whole and the adoption possibilities.

These analyses were useful to understand the production system as a decision unit. The application of discrete and linear programming models to agroecosystem analyses helped technicians and students understand the basic economic relationships involved in farming systems.

2.1.4 Early evolution of the FSR approach

Since CATIE has a regional mandate, a number of institutions in almost every country were involved in the application of FSR as well as in the development of methodology. A considerable number of technicians were trained on those concepts and their application, together with students from South America. Projects following the systems approach were also implemented in a number of countries, influenced by either these pioneer examples or by donors that adopted the approach for the development of small farming areas in Latin America. Such projects were found for example in Ecuador, Honduras, El Salvador, Bolivia, Peru and Brazil. Thus the initiation of the FSR application in LAC was extremely dynamic and volatile. Changes in the development process, multiple applications to different conditions, the involvement of the scientific community and a vigorous exchange of information contributed to rapid evolution in the first 10 to 15 years. These early changes can be summarized in four points:

- Wider options to be considered in the design of technological alternatives for a farming system.
- The adaptation of some research and extension programmes to the FSR approach and the movement of a number of researchers off the experiment station.
- The introduction of socioeconomic analyses to consider the complete farm as the focus within a hierarchy of systems.
- The evolution of the idea that recommendations might include non-agronomic interventions based on comparative advantage, commercial strategies or farm management techniques.

In the early 1980s, several attempts to institutionalize the FSR approach took place all over the continent. NARIs in El Salvador, Panama, Guatemala, Colombia, Ecuador, Peru and Brazil, among others, formally announced their decision to adopt this research approach. Nonetheless, research institutions have maintained, in general terms, a centralized organization that favours the traditional research approach. On the other hand, it has to be recognized that the concept of FSR has been accepted for most agricultural researchers and rural development agents, who have used some of its principles to design their research work. Personally, the 5 years at CATIE consolidated my opinion: by the time I joined IDRC as a programme officer I was a firm advocate for FSR, convinced that a systems approach is a powerful tool to analyse agricultural development at any level.

REFERENCES

1. **Moreno, R.A., 1984**. Estado actual de la investigacion en sistemas de finca en America Latina y El Caribe. In: Compton, P. & DeWalt, B. (Eds) El sorgo en sistemas de produccion en America Latina. CIMMYT-INSTORMIL, Mexico.
2. **Schultz, T.W., 1964**. Transforming Traditional Agriculture. Yale Press.
3. **Moreno, R.A., 1984**. (Op. cit.).
4. **Zandstra, H., K. Swanberg, C. Zulbert & B. Nestel, 1979**. Caqueza: Experiencias en desarrollo rural. Centro Internacional de Investigaciones para el Desarrollo, Bogota.
5. **Escobar, G. & K. Swanberg, 1972**. Metodologia para investigacion del nivel de vida como componente de una estrategia de desarollo. ICA. Reg. 1. Proyecto de Desarrollo Rural del Oriente de Cundinamarca. Mimeo, Bogota.
   Myren, D.T., 1971. The Puebla Project: a Development Strategy for Low Income Farmers. Seminario sobre Estrategias de Desarrollo del Pequeo Agricultor. Ohio State University, Ohio.
   Instituto Colombiano Agropecuario, 1972. Justificacion de la estrategia y un modelo para el desarrollo rural. ICA, Reg. 1. Proyecto de Desarrollo Rural del Oriente de Cundinamarca. Mimeo, Bogota.
6. **Hildebrand, P.E. & E.C. French, 1974**. Un sistema salvadore de multicultivos. Su potencial y sus problemas. In: Conferencia Sobre Sistemas de Producci Agricola Para el Tropico. CATIE, Turrialba, Costa Rica.
   Zulberti, C.A., 1974. Rural Development Information Requirements. IDRC: Simposio del Personal de Campo de Ciencias de Alimentos y Nutricion. Ottawa, Canada.
7. **Hart, R., 1981**. Agroecosystems. CATIE, Turrialba, Costa Rica.
8. **Escobar, G. & R. Moreno, 1984**. Desarrollo de Tecnologa de Producci Agricola: Enfoque Metodologico y Aplicaci Empirica. Trabajo presentado en el Taller Internacional de Sistemas Agricolas. FAO. CATIE, Santiago de Chile.

2.2 A PERSONAL HISTORY IN FSR

Peter E. Hildebrand

During the 'Sondeo', or reconnaissance for the survey, we found that most of the farmers were on the steep, rocky hillsides, not in the valleys where the experiment station was and where the first 'on-farm' trials were being conducted.

2.2.1 Early days

For me, FSR-E started with multidisciplinary research as soon as I started my professional career as an agricultural economist. At the Range Management Department of Texas A & M University back in 1959, I worked very closely with plant and other biological scientists. Moving to Colorado State University in 1961, I was part of a joint project with the Agronomy Department, alongside William Schmehl of the Shaner *et al.* 1982 publication on Farming Systems[1]. This groundwork was reinforced by my first overseas assignment from

1964 to 1967 with an engineering firm, working for the West Pakistan Water and Power Development Authority in Lahore. As Chief Economist and then Chief of Planning, it became evident to me that a coordinated, multidisciplinary effort, with a common objective, in this case irrigation reclamation project design, was a powerful approach to R & D.

Towards the end of my third overseas post, assigned by the University of Nebraska to the ICA in Colombia from 1968 to 1972, I began to work with small, very limited-resource farmers in an on-farm research and demonstration project[2]. It convinced me that small farmers are neither backward nor against change, but are very willing to adopt new technologies when these are appropriate to their conditions.

These experiences were good preparation for my assignment by the University of Florida to work with Centro National de Tecnologia Agropecuaria (CENTA) in El Salvador in 1972 on the design and testing of technology expressly for the many small, limited resource farmers of that country. Here, the influence of Richard Bradfield's multiple cropping system[3] was strong and was partly responsible for the establishment of the multidisciplinary Multiple Cropping (Multicultivos) Program in the Agricultural Economics Department of CENTA which I headed. One objective of the programme was to design technology for scarce land and abundant labour conditions. National policy was to reduce the level of vegetable imports, mostly from Guatemala. So the programme emphasized increased vegetable production without reducing the output of maize and beans, the main staple crops in the country. Multicultivos moved into on-farm trials in its second year, 1973, becoming a national programme in 1976.

The influence of Agricultural Economics on Multicultivos was due to two factors. First, the system was developed by the Department of Agricultural Economics. Second, CENTA recognized that the combination of crops into systems to improve resource use was an area of farm management. It was logical, therefore, to base Multicultivos in the Department[4].

2.2.2 ICTA, Guatemala

At the same time, the Institute of Agricultural Science and Technology (ICTA) in Guatemala was being created as a semi-autonomous agricultural research institute. It was charged with developing technologies for basic grains grown by the small and medium-sized farmers who produced two-thirds of the country's requirements, but who had been by-passed by previous technology development efforts[5]. In May 1974, I met with Directors of ICTA and Joe Black, Director of Social Sciences of the Rockefeller Foundation, to discuss the responsibilities of a position of Coordinator of Rural Socio-economics (SER).

They explained the situation. The Guatemalans felt that if the small farmers of the country were to be included in economic development, it was essential to have a better understanding of their needs and limitations – something that the social sciences should be able to provide. The aim was an institute in which social sciences were integrated with biological sciences to ensure that research was, in fact, oriented towards the needs of the small farmers. There was general agreement at that meeting on the possible scope of SER's work, but there was virtually no discussion of the methodology to be used[6].

In 1974, the Rockefeller Foundation appointed me the Coordinator of Rural Socio-economics of ICTA and in January 1975, SER presented a seminar to the Institute outlining our role as we perceived it, and our plans for the year[7]. We planned to help the commodity programmes understand the needs of small and medium farmers through surveys and by helping these programmes design and analyse their trials. SER had no resources to conduct trials on its own. Our contribution was to influence the substance of the trials being undertaken and was important for ICTA. Early results, however, were not without conflict.

One of our first projects was a survey in the eastern part of Guatemala. During the 'Sondeo' or reconnaissance for the survey, we found that most of the farmers were on the steep, rocky hillsides, not in the valleys where the experiment station was and where the first 'on-farm trials' were being conducted. In collaboration with the bean programme, we proposed trials on the poorest (and only sloping) part of the new station, using bullocks to plough rather than tractors to keep the conditions as close as possible to those of our small-farmer clients.

The Regional Director, however, wanted to homogenize the station. Our planned use of bullocks, and our plans to do without fertilizer, did not fit with this concept, so in March 1975, we decided, in consultation with the Technical Director and the Regional Director to move off the station and conduct our farm trials on rented land. It was the best thing that could have happened[8].

The result was the first on-farm trial to be conducted under the real, very difficult rocky hillside conditions of the farmers in that part of the country. Although the land was rented, local farmers were very much involved in the design of the trial and resulting technology. The site was visited by an impressive list of individuals, who, along with the Institute, learned a great deal about the nature of 'on-farm trials' from the experience. This showed that it was necessary to conduct trials under these conditions if we were really to understand how and why the farmers did what they did. Eventually it was recognized that SER should have a budget to conduct its own on-farm trials. The difficulties of working under the trying conditions of the hillsides also forced the Institute to get to grips with the reality of working for the small-scale farmer.

In January 1976, SER conducted an agrosocioeconomic survey around Tecpan in the central highlands, with the help of one person from the bean commodity programme and one from the local 'Technology Testing Team' as a foundation for on-farm trials that year. The trials were again conducted on rented land, but with ample input from local farmers. Three classes of farmers had been identified in the survey and treatments were designed for each. In the second year SER initiated 'Farmers' Tests' with the technology generated in the trial.

Our experience in Tecpan demonstrated the benefits of a multidisciplinary team gathering information based on local survey results before the trials began. And it helped us understand the value of using local farmers both as advisers and as sources of labour in the field trials. Tecpan showed the value of simple technology, based on that already found in an area, with only minimal changes[9].

2.2.3 The *Sondeo*

By January 1977 we were beginning to doubt the need for the full-scale surveys being conducted by SER in conjunction with the commodity and production teams. Because they were participating in the field surveys, the field teams were analysing and interpreting during the survey. The purpose of the surveys, to provide information for the regional teams to use in orienting and planning their work, was accomplished by the time the field work had been completed. Little was gained by a written report delivered several months later. We were finding that the first impressions gained during the 'Sondeo' or preliminary survey, to obtain an initial understanding of the area and design an appropriate questionnaire, were correct. Although we continued to do the full surveys during 1977, we conducted 'Sondeos' only in two areas. We found the *Sondeo* provided a great deal of useful information for planning on farm experiments in both instances.

We conducted just one survey and three *Sondeos* in 1978 and, for the first time, a *Sondeo* in Moyuta in south-eastern Guatemala was conducted without any plan to follow it with a survey. We found that much more information was gained than if a *Sondeo* was carried out as a prelude to a survey like that done in Jalapa in the Eastern Highlands that same year. In September 1978 in Zacapa, we began to firm up the *Sondeo* methodology.

While ICTA was primarily a crop research institute, it did have a small livestock component. Our first livestock survey in early 1978 was problematic. For the first time we were unable to understand the farming system being studied. By concentrating on the livestock and not considering either the crops, or the interaction of the two, the livestock practices did not make sense. A follow-up survey including both was successful. After that we began looking at both crops and livestock in every *Sondeo*, regardless of whether one or the other was our primary focus. By 1979, the 5-week long Sondeos, conducted early in the year, had become the accepted method for obtaining preliminary information for an area.

In addition to increasing confidence in the results of *Sondeos*, another reason to eliminate questionnaire-based surveys was the farm record project, begun in 1975 as an additional method of obtaining crop production information in areas where 'Technology Testing Teams' were assigned[10]. The project was conceived as a crop, not a farm record programme. Data were kept on individual crops or crop associations, not on the farm as a whole. The crop record project grew to a full national project that has continued over the years. Between 1975 and 1978 it expanded from one area to 11, from three crops or crop associations to 34, from 40 individual records to 583 and from 390 ha to over 1400[11].

2.2.4 Index of acceptability

Perhaps one of the most important features of the organization and management of research at ICTA was charging the socioeconomic unit with the responsibility for the technical evaluation of its research.

As ICTA's Five Year Plan of 1975 said:

> Technical evaluation of ICTA will be in the charge of SER. The reason for putting this group in charge of evaluation is to assure an orientation not only of the agronomic factors, but also of the socio-economic factors of the farmers. By doing this, the institute hopes to have an orientation directed towards resolving the problems of the small and medium farmers of the country and avoid investing in projects that would have little potential for increasing the income of the clients or increasing national production. ... The evaluation process will begin with the development of new projects, continue during the execution to assure that it is being done under conditions relevant to the farmers, include the evaluation of recommendations and of the results of the technology when it is placed in the farmers' hands by determining the acceptability of the technology and finally will close the circle with recommendations based on an analysis of the previously described process.[12]

To help meet this responsibility, SER developed an Index of Acceptability which is still in use today[13]. This index, which measures acceptability – not acceptance – of a technology, indicates the extent to which farmers were convinced to continue and widen the testing of a technology tried in on-farm tests the previous season. As finally developed, the index calculates the percentage of farmers who tested a technology the previous year, multiplied by the percentage of area of that crop on which they are using the technology, divided by 100. In 1976/77 the index for maize in one project area clearly showed that:

- Technologies with a number of components were essentially never acceptable in total.
- Some components such as soil insect control and planting date were not acceptable.
- Improved seed and planting distance (mechanized) were acceptable.

As the number of components included in trials dropped from eight to four, the average Acceptability Index increased from 19.8 to 47.6. It is interesting that fertilizer use, included in the trials and recommended for farmers, was clearly not acceptable, a finding confirmed from farm records[14].

2.2.5 University of Florida, Gainesville

Ken McDermott, the project manager for the USAID-funded Shaner *et al.* report, made occasional visits to ICTA in Guatemala. On one of these I was showing him around the country and explaining the differences in farming systems as we drove. He said, 'Pete, I can see that you understand the differences in these systems that to me look very much alike. Can you train others to do the same thing?' This question and challenge stayed with me and while on sabbatical leave from the Rockefeller Foundation at the University of Florida (UF) in 1980, I decided to accept the challenge. Rather than return to the Foundation, I chose to stay on at UF and became the Coordinator (a non-title) of the Farming Systems Program (a non-entity) in IFAS (the Institute of Food and Agricultural Sciences) at UF.

The programme had a number of facets. We created a short course training programme which became part of the USAID-funded Farming Systems Support Project (FSSP) headquartered at UF. When the FSSP terminated, the short course moved into the International Training Division (ITD) which I directed for 3 years. We taught the first Farming Systems Research–Extension Methods course at the graduate level in spring semester 1980, and it has been taught every year since. With funding from the Vice President for Agricultural Affairs, we initiated Farming Systems Assistantships which brought many fine students to the graduate programmes in agronomy, agricultural economics, soil science, forestry, and agricultural education and communication. Funding from the US Department of Agriculture and IFAS helped us create the North Florida Farming Systems Project in a six county area. Farming systems is now the largest minor for graduate students in the College of Agriculture, while a concentration (major) in farming systems at the MS degree level in the Agricultural Education and Communication Department has a number of graduates and current students.

Even though the FSSP ceased to exist in 1987 and the North Florida Farming Systems Program shortly after that, the graduate programme in farming systems is still going strong and the short course programme is still running. After nearly 20 years, the Farming Systems Program is flourishing at UF and farming systems methodology remains an important tool for many of the graduate students currently involved in conservation and sustainable development curricula in many departments and colleges on campus.

REFERENCES

1. **Shaner, W.W., P.F. Philipp & W.R. Schmehl, 1982**. Farming systems research and development: guidelines for developing countries. Westview Press, Boulder, CO.
2. **Andrew, C.O. & P.E. Hildebrand, 1993**. Applied agricultural research: foundations and methodology. Westview Press, Boulder, CO.
3. **Bradfield, R., 1966**. Toward more and better food for the Filipino people and more income for her farmers. ADC Paper, the Agricultural Development Council, New York.
4. **Hildebrand, P.E. & E.C. French, 1974**. Un sistema salvadore de multicultivos. Su potencial y sus problemas. In: Conferencia Sobre Sistemas de Producci Agricola Para el Tropico. CATIE, Turrialba, Costa Rica.
5. **Waugh, R.K., 1975**. Four years of history. Instituto de Ciencia y Tecnologia Agricolas (ICTA), Guatemala.
6. **Hildebrand, P.E., 1979**. Incorporating the social sciences into agricultural research: the formation of a national farm systems research institute. Instituto de Ciencia y Tecnologia Agricolas, Guatemala and The Rockefeller Foundation, New York.
7. **Hildebrand, P.E., 1975**. El papel de socioeconomia rural in el Instituto de Ciencia y Tecnologia Agricolas (ICTA). ICTA, Guatemala.
8. **Hildebrand, P.E., 1979**. (Op. cit.).
9. (Ibid.).
10. **Shaner, W.W., P.F. Philipp & W.R. Schmehl, 1982**. (Op. cit.).
11. **Hildebrand, P.E., 1979**. (Op. cit.).
12. (Ibid.).
13. **Hildebrand, P.E. & F. Poey, 1985**. On-farm agronomic trials in farming systems research and extension. Lynne Rienner Publishers, Boulder, CO.
14. **Hildebrand, P.E., 1979**. (Op. cit.).

2.3 THE EVOLUTION OF FSR-E IN ASIA THROUGH THE MID 1970S: A VIEW FROM IRRI

Richard Harwood

The major resource allocation problem which faced systems research in the 1970s, as it does now, is that systems understanding is one step removed from actual impact, and is not as obvious in its contributions to eventual improvement.

Introduction

The framework for Asian FSR and extension in the 1970s can be traced directly from the work of Dr Richard Bradfield. His research took place within the context of a centuries-old evolution of intensive cropping systems by Asian farmers and was influenced along the way by those researching farming systems, particularly in Africa. Such authors as Nye in 1961[1] on organic matter and nutrient cycles under moist tropical forest, Papadakis in 1965[2] on crop ecology and Ruthenberg in 1971[3] on farming systems integration, not only influenced Bradfield, but had a major influence on my own thinking. FSR in Asia became increasingly collaborative with Asian scientists in several countries under Bradfield in the late 1960s and then through the IRRI-based Asian Network programme through the mid 1970s. Underlying assumptions, methodologies and key areas of focus evolved as participation broadened. This summary captures some of that evolution.

Bradfield wrote in 1964 that his first contact with rice growing in Asia dated to the arrival of Felix Ponnamperuma on the Cornell campus from Ceylon in the early 1950s. Shortly after, in 1955–56, Robert Chandler and Bradfield travelled throughout Asia as representatives of the Rockefeller Foundation to make recommendations on how to improve agricultural production across the region. In his 1956 final report to the Rockefeller Foundation, Bradfield said that 'rice yields in all countries of the area visited, with the possible exception of Japan and Taiwan, could be at least doubled by further research and education and that in many of the tropical areas with adequate water supply, food

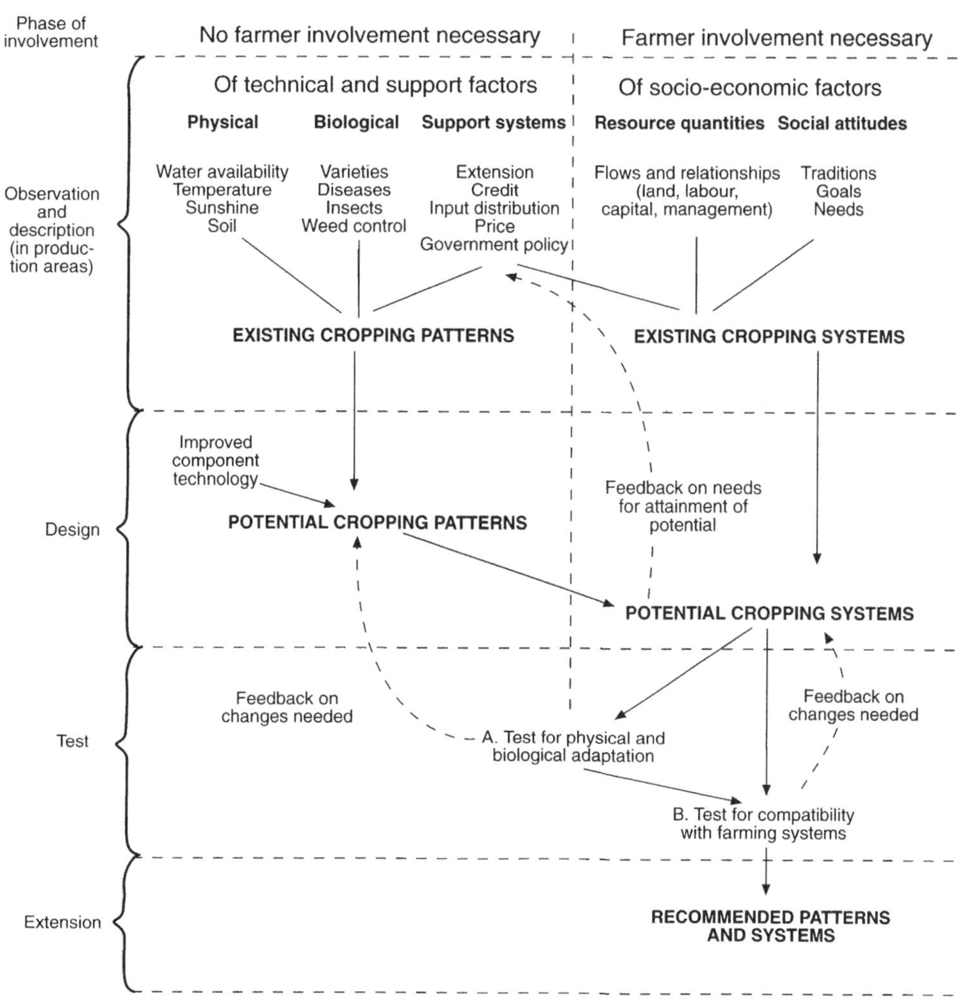

Fig. 2.3.1. A framework for cropping systems research and extension (from IRRI, 1975, p. 11).

production could be greatly diversified and again doubled in quantity by more intensive use (of other crops) during the dry season'[4]. This report recommended the establishment of IRRI.

2.3.2 Bradfield at IRRI

In support of his rationale for increased yield potential, Bradfield began to compare water availability, solar energy and temperatures between Ithaca, NY, his research base for several decades, and Los Baños in the Philippines. His research showed that resources in tropical Asia were grossly under-utilized. By 1964 the Bradfield programme at IRRI was based on the following assumptions:

- Family farms will continue to predominate in Asia, but will increase in size to 4 or 5 ha.
- Rice will continue as the basic staple.
- Water availability will limit non-rainy season rice production.
- Water can be developed for year-round use in most of Asia.
- Average rice yields can be increased 2–5 times over the next two human generations.
- Farmers' need for income will require a doubling of farm size, the yield of rice increased 2–4 times, and non-rice crops grown in dry season.
- Multiple cropping intensity must be doubled.
- Field operations must be minimized and mechanized.

From these assumptions, Bradfield developed his system which, by 1971, he termed the 'maximum cropping project'[5], making maximum use of geophysical resources, crop rotations and the integration of mechanization[6]. Bradfield had modified the Taiwanese ridge–furrow system to fit a very versatile hand tractor, the Landmaster[7]. The Taiwanese system captured in the colour slides that he gave me on my arrival at IRRI in 1972, bore a remarkable resemblance to his mechanized 'IRRI' system, but used hand labour and animal traction. The Landmaster could successfully do all of the soil shaping and tillage needed for the intensive production system shown in Table 2.3.1. In his most intensive cereal system, shown in Fig. 2.3.1, the rice–sorghum system with ratoon crops produced at its maximum 30 t of grain ha^{-1} in a 12-month period in 1968.

Table 2.3.1. Typical maximum cropping pattern (from Bradfield, 1970, p. 239).

Crop	Date of: Planting	Date of: Harvest	No. of days	Tillage operations	Mean yield Mt ha^{-1}: Crop	Mean yield Mt ha^{-1}: By-products*	Price (US$)	Gross income ha^{-1} (US$)
Rice	1 June	10 Sept.	102	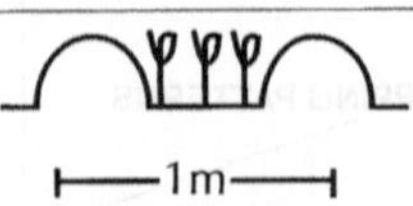	5	5	100	500
Sweet potato	15 Sept.	4 Dec.	100	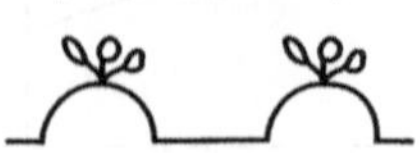	25	20	10	1000
Soybeans (dry)	27 Dec.	17 Mar.	85		2.5	–	100	250
Maize	1 Mar.	5 May	66	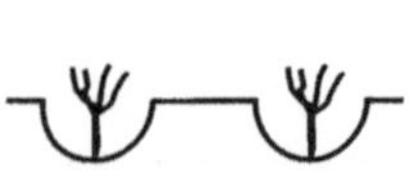	40,000 (ears)	15	0.02	800
Soybeans (vegetable)	1 May	1 July	60		6 (green pods)	6	100	600
								Total 3150

* By-products: straw, vines and stalks.

Apparently this yield was never repeated. Twenty-five tons was more commonly achieved both by Bradfield and by those of us who followed. Bradfield also organized training classes for field researchers from Indonesia and Ceylon in the late 1960s. Prompted by his visits and training, the Maha Illiphalama site in Ceylon became the first national training centre of the 1970s era in Asia.

Bradfield should be credited for three major accomplishments:

- He proved that the yield potential of cereal and feedgrain, on an annual basis in the tropics, could equal or surpass that of temperate areas, dispelling the myth that tropical areas had low production potential.
- He designed an integrated system of soil, water and crop management around a very particular kind of power source, all intended to *maximize* use of soil, water, temperature and light. The system was designed for a farm of 3 to 4 ha in size, having full access to inputs and markets. The trials were all done on field-scale plots of 0.25 ha. To my knowledge this was the first instance of such complete system research design and test to utilize a prescribed resource base.
- He recognized that specialized systems training was necessary for workers in integrated systems which were designed around resource-use concepts.

Bradfield was wrong, however, in several of his assumptions. He did not foresee the rapid industrialization and labour shortages, even with mechanization, which have actually decreased cropping intensity in several Asian countries, most notably Taiwan. His assumption of increasing farm size was correct in a way, again as in Taiwan – with structural transformation the numbers engaged in agriculture fell and average farm size increased.

2.3.3 Making progress

When I followed on from Bradfield at IRRI in 1972, I became convinced that his system design capability, learned over a life time, could be conceptualized and taught. The need for that capability and for the excellence in technologies required by integrated, high productivity systems, guided the IRRI programme of the 1970s. Working with Gordon Banta, an agricultural economist, IRRI's work for the next 4 years consisted of:

- Mastering the Bradfield maximum production methods, quantifying results, and reviewing underlying assumptions (during 1972).
- Increasing linkages to collaborators in Asian national programmes – the Asian Cropping Systems Working Group was formed.
- Developing a process for cropping systems research.
- Developing component technologies for intensive systems.
- Assembling an expanded research team to deal with the multidisciplinary demands of Asian systems.
- Selecting multiple sites and methodology development for systems research.
- Training for outreach site coordinators.

The first four activities were underway in 1972. By 1973 a conceptual framework was taking shape and training had started. By 1974 multilocation testing occupied most of our time, with methodologies for system research in outreach sites becoming the central focus.

The Bradfield maximum cropping method required complete control of water, both for irrigation and drainage. Uniformity of slope across a field was essential. The Landmaster power source was key to working the soil in alternative ridges and furrows to achieve adequate tilth, weed control and, most importantly, water control. Timing of field operations was critical.

During 1972 we learned much from trials with 'maximum' cropping. First, the ability to conduct any given operation on time – when it needed to be done – was an essential design consideration. Systems designed for clay soils in high rainfall areas showed increasing levels of 'management instability' if their soil and water management requirements were complex[8]. Our attempts at maximum cropping in farmers' fields in two nearby barrios was a near disaster because of the inability to manage the crops in a timely way. Component technologies were an essential building block for productive integrated systems. Locally adapted cultivars and the technologies to manage them were essential. A knowledge of specialized techniques for relay planting and intercropping seemed necessary. Weed and insect control was important. Economic analysis showed that good crop management and high yields of a limited number of crops gave greater net return than wide diversity in inter and relay crops[9].

Second, Bradfield's assumptions for the long-term need for maximum productivity, whether or not true for the very long term, were not a useful guide for short-term economic gain for farmers. Stepped increases in productivity were more appropriate and economical[10]. Bradfield did not anticipate the competing demands for labour that would actually reduce cropping intensity in every country but China well before maximum intensities were achieved. He did foresee a requirement for mechanization.

Third, the availability and type of power source was crucial for timeliness of operations, but costs varied by source. In the early 1970s, economic conditions in the Philippines were such that hand and animal-powered operations were competitive with mechanization for tillage, despite compromising the timeliness of operation. It was obvious, however, that power sources and their cost were critical factors in systems design[11].

2.3.4 National links

It had become increasingly clear that collaborative links with national programmes were essential for cropping systems development. The problems of conceptualizing system design, the complexity of the research process and the specifics of local environments, all dictated a need for learning together. There had to be buy-in from each country into the process. Collaboration was achieved through training at IRRI, through many weeks of my own travel during 1972 and through national multiple cropping events such as that in Bogor, Indonesia, in 1973 and in Sri Lanka.

In 1973 the IDRC, the programme donor, agreed to my argument that we should be reviewed by our clients, the heads of national programmes in Asia, on the condition that two other external reviewers could be named. This arrangement became pivotal to future programme success. Dr David Norman from Nigeria and Dr Bert Krantz from India joined what was called the Asian Cropping Systems Working Group. That group not only reviewed programme direction, but came to develop many of the methodologies of the Asian network. The group was to meet twice a year, once at IRRI and once hosted by a national programme. Dr Virgellio (Pexy) Carangal joined the IRRI team to coordinate the network activities. In retrospect, the Asian Network, later to become the Farming Systems Network, was one of the major contributions of the work of the early 1970s. At its first meeting at IRRI in March of 1974 a process for cropping systems research and extension was hammered out in a truly democratic fashion.

It cannot be overemphasized how important this process model was, and the collaborative way in which it was developed, to the longevity and impact of Asian farming systems work. The sequence of observation and description, of design, test and extension, seems trite today, but it was critical to methodology development[12] and to an orderly sequence of site establishment and research design. The process model was supported by recommendations on methods and organization and on the relationships of researchers and farmers. Farmer responsibility for technology access and test became a central theme.

2.3.5 A changing conceptual framework

A conceptual framework for the development of cropping systems was still not clear. By early 1973 we had discarded the notion of maximum use of land, water and sunlight. In travelling across Asia it became clear to me that there were strong physical determinants of temperature and water and of soil type to cropping intensity. We commissioned a group of Asian scientists to develop an agroclimatic classification system for Asian rice-based farming in 1973. Their system was based on monthly duration of rainfall above and below 50, 100 and 200 mm month^{-1}.[13] Two hundred millimetres per month for 3 months was deemed adequate for one good rice crop, and for 5 months for two crops if the first could be direct-seeded. The Ilo Ilo site in the Philippines was chosen because of its 5 months of 200 mm month^{-1} rainfall, and its traditional single crop. Within 2 years the entire area had gone to double-cropping. Maps of climatic zones were subsequently created for Bangladesh, Indonesia and other countries by L.R. Oldeman. This classification, and its application to cropping systems work, is described in detail in Harwood, 1979 (pp. 45–62)[14]. It eventually evolved into a classification of cropping systems research and recommendation domains[15].

A second dimension to systems conceptualization was according to 'development stage', a dimension classified primarily by a farm family's degree of participation in a market economy[16]. System design changes as cash flow and market orientation change. Production systems go through rather predictable changes in structure as they shift from subsistence-oriented to a market economy and then towards production in support of an industrial economy[17].

2.3.6 Modelling controversy

In early 1973 a major controversy engulfed the programme. A review by the IRRI Board of Directors made a strong recommendation that systems modelling be used as a core for the programme, and that a systems design expert be hired. It was recommended that this work be patterned after the systems modelling work supported by the Ford Foundation in Thailand and being used at the International Center for Tropical Agriculture (CIAT) in Colombia and the International Institute of Tropical Agriculture (IITA) in Nigeria. Gordon Banta and I were adamant in our refusal to take this direction, for several reasons:

- The three existing programmes were very expensive, and did not seem to be producing useful results.
- They used the mainframe computer hardware and software of the day, which had restricted capability and were unavailable to non-computer systems scientists.
- Farmers allow no set of parameters to interact over a wide enough range of conditions to isolate response from 'background' noise. Farmers, for example, diverted fertilizer to other crops or switched from planting maize to other crops at both high and low productive potential. Models could not be verified with the computing capabilities of the 1970s.

The controversy, and the pressures for modelling, had subsided somewhat by the mid 1970s as each of the other modelling efforts was eventually terminated. The commitment required to oppose allocation of major resources to premature modelling may have distracted us too far from developing a conceptual framework for cropping systems to underpin modelling research in the future. We focused on farmer-based testing, perhaps, with hindsight, to the detriment of the upstream–downstream balance.

2.3.7 Beyond modelling

The need for effective component technologies quickly became a high priority in the Asian programme. The availability of high yielding crop varieties and of adequate seed for them was of major importance to IRRI's research in the Philippines and to the network. The region-wide collection and uniform trials of cultivars of the major field crops had become a major activity, and by November 1975 was given high priority as an IRRI function by the Working Group[18].

A second area of component technology work was that of relay and intercropping. A series of studies carried out at IRRI between 1972 and 1975 quantified the results from alternative management practices. Relay cropping was the easiest to document, producing rather straightforward results for the timing of overlap for various crops and was the most useful information for Asian network collaborators. Intercrop studies were more complex. We concentrated on combinations of maize, mung bean and groundnut because of their yield advantage, disease and insect control, and on maize, upland rice and cassava for its nutrient use characteristics. Three key points emerged:

- Relay intercrops of 2–3 weeks, under ideal conditions for stand establishment, do not effect yields of either crop, but management of the overseeded crop will often be compromised[19].
- The land equivalent ratio (LER) – the amount of land needed to produce in monocultures the same produce obtained from one unit of intercrop – began to be used in IRRI literature starting with Bantilan and Harwood, and Herrera and Harwood, both in 1973[20].

- Though an intercrop mix will overproduce with a land equivalent ratio of up to 1 : 6, the increased production is almost entirely a result of the longer duration of cropping, with the period of maximum competition coming at a time when the lower-canopied rice crop is least sensitive. Intercrops are thus not physiologically more efficient than a well-grown monoculture during the period when the crops are together[21].

Many relay and intercrop studies provided a wealth of information on the possible role and weaknesses of traditional practices in modern systems, as well as an exciting context for graduate research. However, they did not make the major contribution that cultivar selection, distribution and testing was to make to Asian agriculture.

By 1974 the IRRI programme had shifted in a major way to focus on a few carefully selected outreach sites in the Philippines, selected according to agroclimatic zone classification and farming system type. All had systems oriented towards a market economy and all covered at least three townships within the same zone. Research staff took up residence and protocols were developed for rapid site characterization, research problem identification and conduct of trials. As the Asian network was formed the procedures became more and more a product of Asia-wide discussion[22].

Training programmes began in the late 1960s to train field workers on agronomic techniques for maximum cropping. By 1975 these had evolved into the IRRI 6-month course for 'site coordinators', covering five general areas; cropping systems concepts, site selection and operations, site characterization, economic analysis and statistics, component production technologies and extension methods. Degree training was carried out at the University of the Philippines, Los Baños.

2.3.8 Conclusion

Bradfield's contribution to cropping systems work has been enormous in its demonstration of the design of integrated systems. The greatest subsequent impact of the programme of the early 1970s was its network operation with buy-in by national programme leaders. Generations of site coordinators trained in the programme were to become national research directors in many countries. Having an IRRI core set of sites, and intensive crop rotation experiments both at IRRI and other sites, has provided the backdrop for training in production methodologies as well as site and programme management.

With 20 years of hindsight, there was one major weakness in the programme. We did not make progress on a conceptual framework for use in the *ex ante* determination of systems change. This would have required more extended periods of interaction among senior national research scientists. For such a framework to have been successful, it, like the site research methodologies which were produced, would have been collaborative in development. Improved component technologies, effectively targeted through systems understanding, will be the building blocks for future change. The major resource allocation problem which faced systems research in the 1970s, as it does now, is that systems understanding is one step removed from actual impact, and is not as obvious in its contribution to eventual improvement. Systems understanding in turn, requires knowledge of process-level functions of social, economic, biogeochemical and crop and animal physiology dimensions. Science is moving towards definitions of many of those functions, but our grasp of many of the key pieces is, even today, quite qualitative[23].

Credit for changing rice production in Ilo Ilo, in the Philippines, from 2 to 10 t ha^{-1} $year^{-1}$ went to the newly introduced rice varieties, not to the systems understanding which brought them there. This dilemma remains with us 20 years later. But farmers and researchers alike need such understanding to help guide change in a world where the pace of change is engulfing us all, sometimes moving us away from the best interests of farming families, environmental health and economic stability.

REFERENCES

1. **Nye, P.H., 1961**. Organic matter and nutrient cycles under moist tropical forest. Plant and Soil, 13, 333–346.
2. **Papadakis, J., 1965**. Crop ecological survey in West Africa. FAO, Rome.
3. **Ruthenberg, H., 1971**. Systems with perennial crops. In: Farming systems in the tropics. Clarendon Press, Oxford, UK.
4. **Bradfield, R., 1964**. Some unconventional views about rice culture in southeast Asia. IRRI Seminar, 16 April, 1964.
5. **Bradfield, R., 1971a**. 1970 Research Program Review, maximum cropping project. 9 Feb., 1971.
6. **Bradfield, R., 1970**. Increasing food production in the tropics by multiple cropping. In: Research for the World Food Crisis, American Association for the Advancement of Science, pp. 229–42.
7. **Bradfield, R., 1971b**. Mechanized maximum cropping systems for the small farms of the rice belt of tropical Asia. Shin-Norinsha Company, Tokyo, Agricultural Machinery News Weekly and Farming Mechanization Monthly.
8. **Harwood, R., 1979**. Small Farm Development: Understanding and Improving Farming Systems in the Humid Tropics. Westview, Boulder, CO.
9. **Banta, G.R., 1973a**. Mechanization, labor and time in multiple cropping. Agricultural Mechanization in Asia, 4(1), 27–30.
10. **Banta, G.R., 1972**. The outlook for multiple cropping in the Philippines. Presented at the symposium, 'Toward More Progressive Barrios'. April, 1972. University of the Philippines, Los Baños.
11. **Banta, G.R., 1973b**. Comparison of power sources in multiple cropping. IRRI Saturday Seminar, August 11, 1973.
12. **IRRI, 1975a**. Report of the Cropping Systems Working Group. Los Baños, Philippines. March, 1975.
    Harwood, R.R. & E.C. Price, 1976. Multiple cropping in tropical Asia. In: Multiple Cropping. American Society of Agronomy, pp. 11–40.
13. **Coulter, J.K., J.F. Derting, L.R. Oldeman, M.M. Obradovich & T.B. Strattery, 1974**. An agro-climatic classification for evaluating cropping systems potentials in Southeast Asian rice growing regions. International Rice Research Institute. Los Baños, Philippines.
14. **Harwood, R., 1979**. (Op. cit.).
15. **Garrity, D.P., H.G. Zandstra & R.R. Harwood, 1978**. A classification of Philippine upland rice growing environments for use in cropping systems research. Philippines Journal of Crop Science, 3(1), 25–37.
16. **Harwood, R.R. & E.C. Price, 1976**. (Op. cit.).
17. **Harwood, R.R., 1992**. Biological principles and interactions in sustaining long-term agricultural productivity. In: Proceedings of the Regional Workshop on Sustainable Agricultural Development. Asian Development Bank and Winrock International, Manila, Philippines, pp. 27–49.
18. **IRRI, 1975b**. Report of the Cropping Systems Working Group, second meeting. Indonesia.
19. **IRRI, 1973**. Annual report for 1972. International Rice Research Institute. Los Baños, Philippines.
20. **Bantilan, R. & R. Harwood, 1973**. The influence of intercropping field corn (*Zea mays*) with mungbean (*Phaseolus aureus*) or cowpea (*Vigna sinensis*) on the control of weeds. Crop Science Society of the Philippines.
    Herrera, W. & R. Harwood, 1973. Effect of intercropping soybean (*Glycine max*) under different populations of field corn (*Zea mays*). Crop Science Society of the Philippines.
    IRRI, 1974. Annual report for 1973. International Rice Research Institute. Los Baños, Philippines.
21. **Sooksathan, I., 1976**. A comparative growth analysis of intercrop and monoculture plantings of rice and corn. Dissertation. University of the Philippines, Los Baños.
    IRRI, 1977. Annual report for 1976. International Rice Research Institute. Los Baños, Philippines.
22. **Garrity, D.P., H.G. Zandstra & R.R. Harwood, 1978**. (Op. cit.).
    Harwood, R., 1979. (Op. cit.).
    Garrity, D., R. Harwood, H. Zandstra & E. Price, 1981. Determining superior cropping patterns for small farms in a dryland rice environment: test of a methodology. Applied Science Publishers, pp. 269–83.
23. **Harwood, R.R. & M. Cavigelli, 1997**. Michigan Field Crop Ecology: Managing Biological Processes for Productivity and Environmental Quality. Extension Bulletin E2646, Michigan State University, East Lansing, MI.

2.4 FSR: A PERSONAL EVOLUTION

David Norman

Since the experiment station had been there for almost 50 years, the question that immediately came to mind was why none of the results or lessons from the research station had 'rubbed off' on neighbouring farms.

2.4.1 Introduction

My first professional work experience was as a full-time researcher linked to the Institute for Agricultural Research (IAR) at Ahmadu Bello University (ABU) in northern Nigeria. In 1965, a rural sociologist, a human geographer and myself were appointed to staff the Rural Economy Research Unit (RERU). The Unit had been set up with a substantial Ford Foundation grant to catalyse and support interdisciplinary social science research in the rural areas covered by IAR's mandate. IAR already had a well-deserved reputation in technical research and was well staffed with a highly motivated and qualified cadre of scientists. The rural sociologist, B.J. Buntjer, and myself were, however, the first social scientists appointed by the Institute and we enjoyed a great deal of freedom in designing and carrying out a social science research programme.

2.4.2 Early impressions

In retrospect two important initial impressions influenced my future career, and hopefully blunted my potential arrogance as a young, freshly trained PhD student with all the answers to the problems of agricultural development. The first was surprise at comparing what was happening on the research station with what farmers were doing on their own fields. The experiment station research plots were largely well tended and weed free; the crops were invariably vigorous and grown in sole stands. In contrast, just outside the experiment station fence, farmers' crops were usually grown in mixtures, were very variable as to vigour and sometimes suffered from competition with weeds. Since the experiment station had been there for almost 50 years, the question that immediately came to mind was why none of the results or lessons from the research station had 'rubbed off' on neighbouring farms. The second impression was the high quality of the scientific capacity represented in IAR, and a strong desire not to alienate these colleagues. Developing credibility as social scientists within what was historically a technical research institute, was crucial. We managed to develop a research programme that was not confrontational. In retrospect, this was perhaps lucky. During the 11 years that I was associated with IAR on a full-time basis there were never any real antagonisms from the technical scientists. They were also searching for ways to make research relevant to the needs of the farmers in the region. Although I cannot speak for my colleagues in RERU, I suspect they had similar impressions, since I do not recall any disagreements with the approach we laid out for our work.

2.4.3 A work plan for RERU

The basic work plan which began under RERU was continued under the Department of Agricultural Economics and Rural Sociology – which I headed for about 8 years. It concentrated almost exclusively on micro-level studies, and was divided into four phases as follows[1]:

- Positive phase – determining what farmers are doing.
- Hypothesis testing phase – determining why farmers do things in the way they do.
- Normative phase – determining what farmers might do.
- Policy phase – determining how the changes suggested under the normative phase could be brought about. This also could involve consideration of the hypotheses tested to determine whether the suggested policy changes would conflict with farmers' reasons for doing things in the traditional way.

Much of the research work during the period 1965–71 concentrated on the first two phases. With the foundation derived from these 'basic studies' emphasis shifted towards 'change studies' concentrating on the second two phases. The experiences in the 'basic study' phases led to my questioning a 'top-down' approach to developing technologies for illiterate limited-resource farmers. An appreciation of the complexities of developing a process for improved technologies for such farmers evolved in the course of the 'change studies'.

2.4.4 Cost-route survey methods

The 'basic studies' were a series of studies in 10 villages in four different areas of northern Nigeria. Sample farming households were interviewed twice weekly to obtain a detailed picture of the stock and the flow of inputs and outputs relating to the farm family and their business. Spending considerable amounts of time in the field with farmers to administer these very complex and detailed surveys helped boost the credibility with the technical scientists in IAR. After some time, however, I developed three main concerns about the cost-route methodological approach being used in the early surveys:

- They were very costly in time and other resources, not only in terms of the data collection phase but also in terms of analysis. In fact it took 8 years for the results from one set of village studies to be finally distributed.
- Much of the data collected were never completely analysed or exploited. Legitimate questions could also be asked about combining in the same analytical framework some variables measured accurately and others for which only very inaccurate estimates were available. Unfortunately, in the field situation, the criterion for how accurate the data were was often dependent on how easy it was to collect, rather than the degree of accuracy required from the point of view of analysis.
- The commitment required for such studies and to ensure complete farm records, often meant sacrificing spontaneous interaction with farmers. Thus the farmer tended to become an object from which data were extracted, rather than a colleague from whom one can learn.

The major limitation of dependence on extracting data from farmers via enumerators was brought home to me when, after years of painstaking data collection and analysis, I concluded farmers were rational in growing crops in mixtures[2]. I then thought, 'Why don't I ask farmers why they grow crops in mixtures?' After 1 week I obtained answers similar to those from the detailed surveys. Unlike Mike Collinson who learnt much earlier, it took me some years to develop a healthy scepticism for detailed farm management cost-route surveys, and to learn to place greater reliance on meaningful direct interaction with farmers. One of my most vivid memories of this development was a particularly interesting discussion with a farmer who explained in great detail why he grew certain crops in mixtures. He had, for example, recognized the beneficial effect of mixing cereals and legumes, although of course he did not know about the nitrogen fixing qualities of the latter. However, his keen visual and mental perceptions greatly impressed me, convincing me that the skills and experiences of farmers should somehow be put to constructive use in the search for improvements.

Important lessons were learned during the data collection and analysis stages which have had a major influence on my evolution towards advocating a farming systems approach. Four of these lessons were:

- It is vital to try to understand the farming system with which one is working. This can only be done in the field through interacting with those responsible for operating that system. If this is not done then there is a real danger of misleading interpretations with 'outside' eyes.
- There was great heterogeneity in the complex farming systems being implemented because of differences, not only in the biophysical factors, but also in the socioeconomic factors, faced by different farming households. The strategy of recommending blanket technology packages did not, therefore, seem relevant.
- I also developed a healthy respect for the farmers' skills in operating in such complex production environments and for the rationality of their behaviour. Examination of issues relating to mixed cropping in particular, influenced my rapidly developing respect.
- Because of the seasonal nature of rainfed agriculture in northern Nigeria, labour shortages at certain times of year were the

norm. At the same time, few improved technologies were used. Access to, and availability of, purchased inputs was very limited. Given these constraints and realities were the technologies being recommended at the time likely to be relevant?

2.4.5 Evaluating technologies with farmers

The knowledge and impressions built up during the 'basic studies' led naturally to the next step of evaluating, with farmers, some technological packages being recommended by extension from work done at IAR. This involved farmer volunteers testing technological packages for sorghum, maize, cowpeas and cotton, in a farmer managed and farmer implemented (FM/FI) mode. Detailed records were kept as farmers grew test plots and compared their results with their own adjacent control plots. Again this was a major learning experience and, in addition to reinforcing many earlier insights, produced new ones.

Very few farmers tested the packages in the way researchers had intended and many deviated significantly from the recommended methods. Sometimes they were unable to do exactly what was recommended because, for example, labour was not available for weeding when it was required. However, deliberate decisions were made to deviate, with some giving priority to getting food crops established before planting cotton, for example. Because of these deviations I was constantly consulting with technical scientists on the next best alternative for the farmers. For example, should a top dressing of fertilizer be put on the field if the weeding operation was inadequate? This later led me to argue for recommendations to specify the best stepwise adoption if a technology package is involved. I also argued that recommendations should include conditional and targeting information giving guidelines about what would be best to do if deviations occur, and for what types of situations the recommended technology would be most suitable. The objective was to make the recommended technologies at least partially relevant to as many farmers as possible, and to exploit to the maximum extent, the expertise and knowledge of the technical scientists themselves.

The evaluation criterion in developing the technologies on the experiment station centred on yields per hectare which was often incompatible, or at least not completely congruent, with the criteria used by farmers in their fields. Given the importance of labour as a resource in the seasonal agricultural cycle, the package only increased the value of returns per unit area *and* per unit of labour used when applied to maize – not a major traditional crop. The other packages increased the net return per unit area but not the return to labour[3]. In retrospect, it is not surprising that later, when the World Bank appeared with their integrated rural development projects – with efficient input distribution and product marketing systems – maize areas expanded and production increased dramatically[4]. These area increases occurred largely at the expense of sorghum and even legumes. The obvious lesson from this was that irrelevant evaluation criteria on experiment stations inevitably lead to the recommendation of irrelevant technologies.

Given the somewhat negative connotation of such findings there was the possibility of confrontation between the social and technical scientists. Given the years of effort the technical scientists had spent developing the technologies this would be understandable. To safeguard against possible confrontation, efforts were made to consult technical scientists before the testing programme was implemented. The most rewarding experience was working with the cotton scientists who showed a definite desire to collaborate actively to test their technological package with farmers. This, as it happened, was very significant because the recommendation called for early planting of cotton to maximize cotton yields. Within 1 month of collaborating in the field the cotton scientists realized this was an unreasonable expectation. This led to very constructive collaboration between the technical and social scientists. Consequently, the potential value of later planted cotton was examined, together with replacing the water demanding and labour intensive spraying system with an ultra-low volume (ULV) (i.e. oil-based) insecticide spraying regime[5]. These experiences taught us to consult before embarking on something that might be sensitive to others and, if possible, to collaborate actively with the other parties so that the eventual findings are more acceptable. If progress in meeting the needs of farmers was to occur, I became increasingly convinced that technical and social scientists had to collaborate. This also stemmed from the realization that evaluating anything effectively on the experiment station that was not technical in nature was difficult for scientists. Evaluating yields in terms of per unit of area is, therefore, much easier than in terms of per unit of labour.

Perception of the huge gap between what technical scientists do and can do on the experiment station and what farmers do, and can do, on their farms, led to a conviction that major changes were necessary in the way in which technologies were designed and evaluated. This conviction became focused on the notion that continued *ex post* evaluation of technologies by social scientists was not very efficient, and could, in fact, be considered counterproductive. Instead, attention needed to be focused on interdisciplinary collaboration between technical and social scientists to develop and evaluate technologies in more of an *ex ante* operational mode. It was also becoming obvious to me that farmers had to become actively involved in that process, because of the intimate knowledge they possessed about local production environments and the fact that they could actively and constructively contribute to the development and evaluation of relevant technologies. These convictions were supported and nurtured by two other influences. One was a growing awareness that workers in other areas were also going through something of the same transition. For example, I remember being particularly impressed with some of the work of the Unite Experimentales in Senegal with its philosophy that one learns through 'perturbing' the system. And in the early 1970s I became a member of the Asian Cropping Systems Network, coordinated by IRRI, which resulted in very useful trips to Asia where the same issues were under discussion.

2.4.6 Reflections

Looking back, the 11-year experience at IAR was truly rewarding and I now realize that, in many ways, what was happening there was ahead of its time. There was little confrontation between the social and technical scientists and the working climate was open to changes. There were, for example, formal annual meetings between scientists and extension staff to discuss research programmes and results. During the early 1970s the research programmes were reorganized into multidisciplinary commodity/subject matter teams (including a farming systems-based team), and experiment station research work started on mixed cropping. The working environment into which the social scientists were introduced in the mid 1960s was, therefore, supportive rather than antagonistic, enabling their contribution to be nurtured – a unique opportunity for those of us there at the time.

Perhaps the final step in embracing the farming systems approach came from a Ford Foundation farming systems workshop hosted by the Institut d'Economie Rurale in Bamako, Mali, in mid 1976. That workshop was significant for at least three reasons. The first was that it accepted that the description of farming systems should be viewed as a means to an end and not an end in itself. The second was that the workshop developed and approved an analytical framework[6] to improve the efficiency with which relevant improved technologies are developed and evaluated. That analytical framework, still in use today, was first drawn on a bedsheet, compliment of the Hotel de l'Amitie! The third was that this was called 'the farming systems research approach', the first time I personally recall this label being used.

Obviously since then, as a result of interacting with many farming system practitioners, nearly 9 years of further field experience in Botswana, and many short-term assignments in different parts of the world, my personal evolution in the application of the farming systems approach has continued. However, the basic principles that I embraced during the early years of my career are still as valid today. Perhaps the most important of these principles has been the belief that the farmer should be, and has a right to be, at the centre of the stage.

REFERENCES

1. **Department of Agricultural Economic and Rural Sociology, 1976**. Progress Report no. 10. Zaria, Institute for Agricultural Research, Ahmadu Bello University.
2. **Norman, D.W., 1974**. Rationalising mixed cropping under indigenous conditions: the example of northern Nigeria. Journal of Development Studies, 11(1), 3–21.
3. **Norman, D.W., E. Simons & H.M. Hays, 1982**. Farming Systems in the Nigerian Savanna: Research and Strategies for Development. Westview Press, Boulder, CO.
4. **Smith, J., A.D. Barau, A. Goldman & J.H. Mareck, 1994**. The role of technology in agricultural intensification: the evolution of maize production in the northern Guinea Savanna of Nigeria. Economic Development and Cultural Change, 42(3), 537–54.
5. **Norman, D.W., J.A. Hayward & H.R. Hallam, 1974**. An assessment of the cotton growing recommendations as applied by Nigerian farmers. Cotton Growing Review, 51, 266–80.
See also:
Norman, D.W., P. Beeden & J.A. Hayward, 1976. A comparative evaluation of ultra-low volume insecticide application on cotton farms in the north Central State of Nigeria. Nigerian Journal of Plant Protection, 2, 23–9.
6. **Gilbert, E.H., D.W. Norman & F.E. Winch, 1980**. Farming Systems Research: a Critical Appraisal. MSU Rural Development Paper no. 6. Michigan State University, East Lansing.

2.5 MY FSR ORIGINS

Mike Collinson

My involvement with the scientists on the research station impressed on me the overwhelming need to understand small farmers whose objectives clearly differed from those of crop researchers. They also differed from the objectives of commercial farmers, whose motivations underpinned the conventional western approaches to farm management.

2.5.1 Introduction

Two long periods of my professional life were devoted to FSR. The first, from 1960 to 1971, including nearly 10 years in Tanzania, focused on understanding the adoption of technologies by smallholders. The second, from 1975 to 1987, based in Kenya with CIMMYT, focused on institutionalizing FSR-based on-farm research in eastern and southern Africa. The first phase provided the conviction for the second, so I dwell on the 1960s in this description.

In 1961 I started my first career job at the Western Region Research Centre, Ukiriguru, in the north-west of Tanzania – then Tanganyika – joining a dozen 'hard' scientists; in plant breeding, agronomy, pathology, entomology and soil science, all trying to raise crop yields. Most were working on cotton, the main regional cash crop, others on important food crops, such as maize and rice, which were also sold in local markets. My task as the first social scientist on the research station (and, as it turned out, the first in the agricultural research institutions of East Africa) was to document the priorities of the farmers of the region and how they used their resources to meet them.

2.5.2 Professional baggage

I brought to the work a professional baggage acquired in UK universities. This baggage included the research and extension approach then used in the UK Farm Management Service in which constraints on the performance of the individual farm were identified by comparing data for that farm to standards for a group of farms of the same type[1]. In the late 1950s whole farm planning had just found a niche in the farm management arsenal. In 1958 in the USA Heady and Candler[2] detailed the application of linear programming (LP) to farm planning. Five years later, Clayton provided an early example of its use in peasant agriculture[3].

Both approaches involved a farm management professional in *one on one* contact with the individual farmer, to collect data for analysis and to give advice. Earlier work in farm management in developing countries was limited. In 1953 Jolly began to plan and develop holdings for small resource-poor farmers by helping selected families to build up 'Unit Farms' under controlled conditions at the college campus in Trinidad. Surveys of small farmers in the 1950s were being pioneered by Morgan-Rees in Zambia and Edwards in Jamaica; the only formal attempts I was aware of to bring farm management principles to the improvement of the small-farm sector in developing countries. Beyond conventional farm management, studies by anthropologists such as Richards in 1939[4], Conklin in 1957[5], and De Schlippe and Batwell in 1955[6] had convinced me of the value of understanding small farmers' systems. Jolly's work in Trinidad also proved particularly valuable to my research in Tanzania in an unexpected way and I return to this.

2.5.3 Conventional methods in crop improvement

My involvement with the scientists on the research station impressed on me the overwhelming need to understand small farmers whose objectives clearly differed from those of crop researchers. They also differed from the objectives of commercial farmers, whose motivations underpinned the conventional western approaches to farm management. Three aspects stood out:

- In the early 1960s, and in many communities today, cash income could not guarantee command over food supply. For most families food marketing was essentially local. In a bad year there was scarce food at home and nothing to buy in the local market. Family priorities were for a combination of foods produced on the farm itself that was reliable day in day out. These food security priorities dominated farm decisions on resource allocations.
- Small farmers managed the full range of external uncertainties; weather variation, degrading land, pests and diseases and unreliable markets and prices, not through the protective devices available to commercial farmers, such as insurance, but through their own enterprises and husbandry practices. Fully exposed to this diversity of uncertainties they used the production strategies evolved by their culture for survival.
- Small farmers using hand hoes, and even those using oxen for ploughing, have very limited power at their disposal. The returns to spending limited cash on increased power to extend the area cultivated were, and are, often better than the returns on purchasing inputs to intensify production on the existing area, especially given the vagaries of input supply, and the physical difficulties in procurement.

These three aspects in particular caused me to reflect on the commodity approach to crop improvement on the research station. The scientists there, in the classic reductionist tradition, were trying to identify the best germplasm and best husbandry to maximize yields of their commodity in the conditions particular to western Tanzania. Many characteristics of classical experimental methods and of research station operation isolated their results from the real world of the small farmer, and the current practices of those farmers were never the context in which the recommendations were identified. The criteria used by farmers in evaluating recommendations were rarely compatible with higher physical yield per unit area of land monopolizing the attention of the scientists. Heavy machinery on the station meticulously prepared the fields, creating a tilth impossible for the hoe to achieve. Machinery meant that acres could be prepared and sown after each planting rain, while farmers might be limited to the half acre their family could prepare and sow in the 2 or 3 days following that rain. Unlimited and unrecorded labour tended the experimental plots and high levels of non-treatment variables allowed the full expression of differences between treatments, yet created a multi-component technological package. Both non-treatment and treatment variables were usually new technology components as far as the farmers were concerned. Each one used their labour or cash in new ways and, through clashes with other enterprises competing for these resources, often jeopardized the family's priority – an assured, preferred food supply.

There was a stark contrast between the sophistication of design and statistical analysis for experiments, the ad lib resources used for their management, and the all too common irrelevance of the criterion for evaluating the results. Cotton research dominated the station and had first priority for equipment as the season started. Because of the threat of American bollworm building up on early-planted maize and migrating to the cotton crop, maize planting experiments on the station were delayed until some 30 days after cotton establishment. Farmers, in contrast, wanted a supply of new season food flowing from the farm as soon as possible. They enjoyed fresh maize, and, in years following bad harvests, their only option was to buy it locally at four times the normal price. I tried to highlight this kind of dilemma in an article I wrote in 1968[7] focusing on the then current recommendation to farmers to use their limited cash to purchase fertilizer for cotton. I argued that farmers in western Tanzania would gain more from fertilizing their maize and growing a smaller area to meet their fixed needs, thereby releasing labour to plant a larger area of earlier cotton. The systems perspective was nicely illustrated; the benefits from putting fertilizer on maize came from a larger acreage of cotton, and an increase in yields from planting some of it earlier.

The whole issue of evaluation criteria among small semi-subsistence farmers demonstrates a gross irrationality in the application of science which unfortunately still prevails. The social sciences remain a rarity in most agricultural research institutions. Where they have found a niche they are still fighting an uphill battle to influence the technical establishment to modify the traditional research process.

2.5.4 A trial farm

In 1962, following Jolly, I established a 'trial farm' on the research station operated by a local family and managed between the farmer and myself with advice from the station scientists[8]. It proved to be the greatest single influence in changing station scientists' attitudes towards small farmers and their systems. In the first year I discussed with the family their food preferences and the quantities they would need to feed them through the year. They would grow maize, rice, cassava, sweet potato, cowpea, groundnut and pumpkins for food, and cotton for cash. I turned to the scientists for the best way to grow these crops. The most striking aspect of their recommendations was that every commodity researcher expected his crop to be planted immediately after the onset of reliable rains – typically the first week in December. I pointed out that it required some 18 man days to prepare and plant an acre of seedbed with hoes. With less than two labour units, to prepare and plant the 3 acres needed for food crops would take 3–4 weeks and the 2.5 acres of cotton planned would also take 3 weeks. I asked the scientists to agree which crops the farmer could delay planting with least compromise on yield potential. Their responses were all the same: 'any other – but not mine!'

The trial farm adopted farmers' priorities as a starting point. During its 5 year life it required huge compromises on best practice for any single commodity in order to accommodate the farmers' cash and labour resources. Nevertheless, in its first 3 years it provided returns to available labour 250–300% higher than those of the average local farmer and 150–200% higher than those of the best 30% of local farmers. It was an education for me and for all the scientists on the Centre. The compromises, however, caused reflection and extensive controversy on future priorities:

- Which crops had the greatest tolerance for delayed planting?
- Was there diversity in varietal tolerance to delayed planting within species which the breeders could exploit?
- Identifying shorter-term materials for crops to be planted later in the 6-month rainy season.

The strong and well-established interactions between time of planting, variety, plant density and fertilizer levels also demanded a re-examination of husbandry practices for crops for which planting was delayed.

It was years before these research themes were taken up, for example, with maize, sunflower and cassava in eastern and southern Africa. Even today it remains difficult for many plant breeders to accept that it may be better for the farmer to grow a maize with a shorter maturity period and lower yield potential:

- If it allows new food supply earlier in the season from the farm.
- If it allows a second crop to be planted.
- If it allows an extra field to be planted to maize months after the start of the rains when earlier planted fields have suffered for one reason or another.

2.5.5 Scarce professionals and countless farmers

I started by applying the cost-route approaches to survey learned in university but was brought up short by early experiences in my own work and exposure to wider efforts to develop smallholder agriculture in the early 1960s. An early lesson came from the breakdown of the Village Settlement Scheme introduced by the Government of Tanzania and the World Bank in 1964. An International Bank for Reconstruction and Development (IBRD) report to the Tanganyika Government in 1960 listed the entire graduate strength of the Tanzanian extension services as 58 people, and advocated new approaches to improve their effectiveness. The scheme which emerged proposed 69 settlements with an average of 250 families in a village and machinery provided to work the 2500–3750 acres planned for growing village crops. After implementing seven of the settlements it was clear that the management needs of 69 such villages would drain *all* the graduate and diplomate staff from the ministries' extension services. This scheme would have focused the nation's entire professional manpower on some 15,000 families – a small proportion of even the annual increase in the rural population. Further, the special circumstances within these schemes, particularly the full tractor mechanization of cultivation, implied that the innovations would be irrelevant to the rural population at large.

My involvement with these schemes crystallized a key dilemma in evolving an approach to small-farmer improvement. The scarcity of professional manpower and the large numbers of small farms precluded one on one interaction between professional and farmer. It ruled out the conventional approaches used by farm management professionals for farm family income improvement in the USA and the UK. The large number of farmers, the diversity of farms and their small size with low levels of production, limited the returns from investment in professional time to improve individual farm units. This pushed me towards faster and cheaper methods of data collection. I was seeking *wider coverage* with *sufficient accuracy* to understand farmers' aims and the way they used their resources to achieve these. The search had four phases.

2.5.6 Lowering the cost of understanding small farmers

An initial concern was with stratification. Common practice in survey work in the early 1960s was to stratify on the basis of administrative areas; districts or ginnery zones; units for which official decision making was already institutionalized. In western Tanzania most farming systems were providing full family subsistence. However, the dominant starch staple differed widely: maize, plantain, cassava, rice or sorghum, from one area to another. The main starch staple usually occupied as much as 70% of the area cultivated and absorbed most of the labour. The crop calendar, cultural practices, and therefore resource demands, differ across these staples and each system had to be sampled separately to avoid confounding an understanding of farmers' priorities, strategies and production decisions. This began a search for cost-efficient ways to identify a typology of farms.

An early preoccupation in data collection was to get away from expensive cost-route studies in which the measurement of production, expenditure and labour use dominated survey design and analysis, requiring frequent visits to each farm in the sample to collect daily data. The issue was the trade-off between visit frequency and data accuracy. Sixty observations within a farming system consistently gave an acceptable sampling error which could be managed with a degree of information about the population being surveyed – the real issue was observational or enumerator error.

In the course of some 20 field surveys on small farmers between 1961 and 1966, I incorporated a series of data collection experiments for important parameters. Collection intervals ranged from daily to yearly visits by enumerators and different intervals were used with different samples, sometimes subsamples from the same population. My guidelines for the experiments came from Zarkovich[9] whose later characterization of data as 'events'[10] brought clarity to a complex issue. He drew a distinction between regular and irregular events, and between frequent and infrequent events. He concluded, particularly for regular events, that respondents often answered out of experience rather than recalling the specific event. Exploiting this idea, I paid more attention to labour data. All new technologies, through changes in timing or the intensity of labour required, have important implications for labour redistribution. Consequently they impact the priorities of farm families which dictate the existing patterns of labour use. Data collection experiments showed that drawing on farmers' knowledge gave rates of work per unit area on the main crop operations not significantly different from labour rates based on frequent visit data collection techniques[11].

In the single visit method each main enterprise was covered separately. The farm family, with the women responding for the operations for which they were responsible, took the survey enumerator to a field previously planted with a crop for which labour data was needed. The enumerator measured the field. Standing in the chosen field, families were first asked to calendar the operations done on the specified crop in this field in a typical year. Then a sequence of three questions was asked about each operation from the beginning to the end of the season:

- Which members of the family, and/or hired labour, would normally work on this operation?
- For each worker, what time would work start, what time would it finish, on a typical day while working on this operation?
- With this group of people working, how many days would it take to finish the operation on this field?

The method provided labour data in a single visit to the farm. Variation was relatively low, and 30 observations within a farming system gave sufficient accuracy, allowing coverage of the five or six main enterprises by enumerating two or three on each of the 60 units within the sample. The data were used to build a labour calendar for the farming system to evaluate the impact of new technologies on labour distribution and on family priorities, including food supply, reflected by this allocation of labour. Later work for the World Bank[12] confirmed that farmer estimates, properly enumerated, are as accurate as crop cutting for the measurement of production levels.

Despite relatively cheap collection techniques, returns to labour are still largely ignored in evaluating potential innovations. The farmer's bottom line is a reliable, year in year out, return to household labour, even where land is very limited.

As survey experience accumulated it became clear that, within a farming system, there were many attributes common to the population as a whole. Often culturally determined, these attributes could be investigated by discussion with individuals, or with groups of farmers. Describing these attributes gave an understanding of the important dimensions of the system, as well as insights for better planning of more formal surveys. The *'pre-survey'* evolved[13]. Informal but carefully organized interactions with relatively few farmers provided qualitative information on a wide range of attributes of the farming system. These included the crop calendar, husbandry methods, rotational practices, changes in enterprises, land acquisition and tenure rights, preferred dishes and their substitutes, seasonal eating patterns and obligations to the community with respect to land, labour and livestock. This pre-survey evolved into the informal survey which was developed in FSR to provide a qualitative understanding of the farm as a whole, often as the foundation for soft system modelling[14].

2.5.7 Technology adoption by small farmers

In 1970, after almost 10 years' work in Tanzania, I went back to the UK, to the University of Reading to do a PhD. Using a linear programming package I modelled a farming system for which I had 3 years of detailed data. The 32K Elliot computer used batch processing and occupied two rooms in the applied statistics department – very different from today's laptop! My initial aim was to let traditional farm enterprises and improved enterprises, based on research results, compete for farmers' resources to best satisfy family food and cash priorities:

- Preferred constituent foods to be mixed in dishes favoured in different seasons.
- Preferred, in-season fresh foods supplied direct from the field.
- Insurance foods (cassava, sweet potato) grown to manage the risks of preferred food failure.
- Cash.

Optimizing the model offered a 300% rise in cash income. Yet it was clear that the managerial changes in the farm system required for this improvement were hugely complex and, given the limited information on season-to-season replicability of the results of the improved techniques, the results were also uncertain. For me the revelation was the fact that although I now had a target, an optimal system which looked good, I also had two far more complex questions to work on:

- What sequence of innovations would allow farmers to improve on their current system, move towards the target system, and minimize the risks of family food scarcity?
- Would the target remain the same as farmers progressed towards it?

Acknowledging the dynamics of both food and cash-crop markets, the answer to the second question had to be no.

2.5.8 Conclusions

That work left me with the conclusion that the optimal system is an illusion. The key challenges were to identify the immediate steps forward that would be most acceptable to farmers[15], identify homogeneous groups operating the same system to use scarce R & D resources efficiently, and to find cost-effective methods in understanding farmers' systems.

I came to FSR as a result of the shortcomings of my original professional baggage when faced by the circumstances of African smallholders. By the early 1970s I was confident the approach would have an important future[16] and determined, under a CIMMYT umbrella, to promote an FSR-based approach to technology development in eastern and southern Africa.

Recalling this history brings me back to a question posed in introducing the book: over a generation later how do African, Asian and Latin agricultural professionals differ in the tools they now bring to the job? Furthermore, are they better equipped? If so, can some of the improvement be attributed to the FSR movement? Where today's professionals are *not* better equipped, does the problem lie in FSR itself or in prevailing institutional conditions? Many of the other contributions in this book shed light on this important and ongoing issue.

REFERENCES

1. **Blagburn, C.H., 1961**. Farm Planning and Management. Longmans, London.
2. **Heady, E.O. & W. Candler, 1958**. Linear Programming Methods. Iowa State University Press.
3. **Clayton, E., 1963**. Economic Planning in Peasant Agriculture. Department of Economics Monographs, University of London, Wye College, UK.
4. **Richards, A.T., 1939**. Land Labour and Diet in Northern Rhodesia. OUP, Oxford.
5. **Conklin, H.C., 1957**. Hanunoo Agriculture: a Report on an Integral System of Shifting Cultivation in the Phillipines. FAO, Rome.
6. **De Schlippe, P. & B.L. Batwell, 1955**. Preliminary study of the Nyangwara system of agriculture in the southern Sudan. Africa, XXV(4), 321–350.
7. **Collinson, M.P., 1968**. The evaluation of innovations in peasant agriculture. East African Journal of Rural Development, 1(2), 50–9.
8. **Collinson, M.P., 1969**. Experience with a trial management farm in Tanzania, 1962–65. East African Journal of Rural Development, 2(1), 28–43.
9. **Zarkovich, S.S., 1961**. Sampling for Censuses and Surveys. FAO, Rome.
10. **Zarkovich, S.S., 1964**. The Quality of Sample Statistics. FAO, Rome.
11. **Collinson, M.P., 1972**. Farm Management in Peasant Agriculture. Praegar, New York.
12. **Murphy, J., D.J. Casley & J.J. Curry, 1990**. Farmers' estimations as a source of production data. World Bank Technical Paper 132. World Bank, Washington, DC.
13. **Collinson, M.P., 1972**. (Op. cit.).
14. **Rhoades, R., 1981**. The Art of the Informal Agricultural Survey. Training Doc. 1982–2. Social Science Department, CIP, Lima.
15. **Collinson, M.P., 1972**. (Op. cit.).
16. **Collinson, M.P., 1974**. Transfering technology to developing economies: the example of applying farm management economics in traditional African agriculture. World Development, 2(2), 93–7.

Chapter 3
FSR – Understanding Farming Systems

3.1 FSR'S EXPANDING CONCEPTUAL FRAMEWORK

Robert Hart

If concepts are generalizations, a conceptual framework is a set of interconnected generalizations. Agreeing on a conceptual framework is one of the most difficult aspects of interdisciplinary research, and farming systems research is no exception.

3.1.1 Introduction

Biological scientists, social scientists, farmers and policy makers all view the same reality from different perspectives. When farming systems research (FSR) was just getting started, newly formed interdisciplinary teams spent a long time agreeing on a common conceptual framework. Researchers who followed used the existing frameworks as a starting point and adapted them. Each new initiative brings together a new team that is never completely satisfied with past efforts, and a new conceptual framework is developed.

This chapter is a subjective retrospective, analysing the evolution of the conceptual frameworks that guided the research and development processes I observed at first hand. It is biased by my own disciplinary background in ecology and agronomy and, of course, by the approaches taken by the interdisciplinary teams with which I have worked. Specifically I should note the influence of colleagues at the Centro Agronomico Tropical de Investigacion y Ensensana (CATIE), the Caribbean Agricultural Research and Development Institute (CARDI), Winrock International and the Rodale Research Center. The conceptual frameworks that my colleagues and I constructed were strongly biased towards an ecological perspective. The frameworks we constructed were never static and were constantly evolving, even over the life of a given project.

Three general tendencies are obvious:

- There has been an expansion in the scale of the target systems from an early emphasis on crop populations towards a later interest in farm system and watersheds.
- There has been an expansion in the criteria used to evaluate system performance from an early emphasis on productivity towards a later interest in stability and criteria related to sustainability.
- There has been an expansion in the target beneficiaries from an early emphasis on 'small farmers' towards a later emphasis on women and gender issues, urban as well as rural poor and the current interest in sustainability and the benefits that will be received by future generations.

These three tendencies are summarized in Table 3.1.1 and described in detail in the first three sections of this chapter. In the fourth section I have tried to analyse the operational implications of this expansion of the conceptual framework used to guide the different FSR initiatives.

Table 3.1.1. Changes in the FSR conceptual framework.

	1970s	1980s	1990s
System scale	Cropping systems	Cropping systems Livestock systems Farming systems	Cropping systems Livestock systems Farming systems Community systems Watersheds
Performance criteria	Productivity	Productivity Stability	Productivity Stability Sustainability
Target beneficiaries	Small farmers	Small farmers Women	Small farmers Women Next generation

3.1.2 Expanding target systems

FSR has many mothers. Social scientists, ecologists and agronomists all have a legitimate claim to having played a role in its conception. Some people are likely to cite the influence of anthropologists and agricultural economists who in the 1960s highlighted the fact that many traditional farmers manage complex multi-species cropping systems – for example Norman, 1968[1]. Others will note the role of ecologists and agroecologists who pointed out the virtues of higher diversity – Margalef, for example, in 1968[2].

Agronomists working with perennial crops and pasture species had been interested in multi-species cropping systems for many years. In the 1960s and early 1970s there was a dramatic increase in interest in annual crop-based multi-species cropping systems and hundreds of agronomic papers on intercropping were published. But financial support for this type of research was, in general, limited because the multi-species cropping systems used by resource-poor farmers were viewed by many mainstream development institutions as part of the problem, contributing to low farm-level productivity. This changed dramatically when Richard Bradfield began his research with multiple cropping systems at the International Rice Research Institute (IRRI)[3]. His prior international reputation as a soil scientist helped to legitimize research on multi-species cropping systems and donors began to provide the type of financial support needed for interdisciplinary research.

In the 1970s most cropping systems teams were looking at two-crop systems like rice/wheat rotations and intercropped maize and beans. Today the emphasis is on the analysis of complex watershed-level systems. This expansion in FSR's conceptual framework was no orderly chronological expansion in the limits of the systems under consideration. Naturally, while many people were doing plot-level research with 'simple' cropping systems, others were analysing watershed hydrology and regional land use patterns. But in most cases, even though this research with larger systems was going on concurrently, FSR teams doing the plot level on-farm research in the 1970s did not include these larger systems within their conceptual frameworks.

But the expansion in system scale was not completely unsystematic. A common characteristic of the conceptual frameworks of many cropping system research initiatives was an emphasis on the hierarchical relationship among systems. This seems to have occurred independently in different countries – unsurprising given the importance of the concept of hierarchy in systems theory. The authors who most stressed the importance of hierarchy as an organizing principle were, perhaps, Gordon Conway[4], Louise Fresco[5] and myself[6]. Many cropping systems teams conceptualized cropping systems as subsystems of farms, that could, in turn, be viewed as subsystems of communities or watersheds or subregional systems functioning within the larger national and global systems (see Fig. 3.1.1). It was almost inevitable that once researchers started to see the importance of understanding the suprasystem in which their target system functioned,

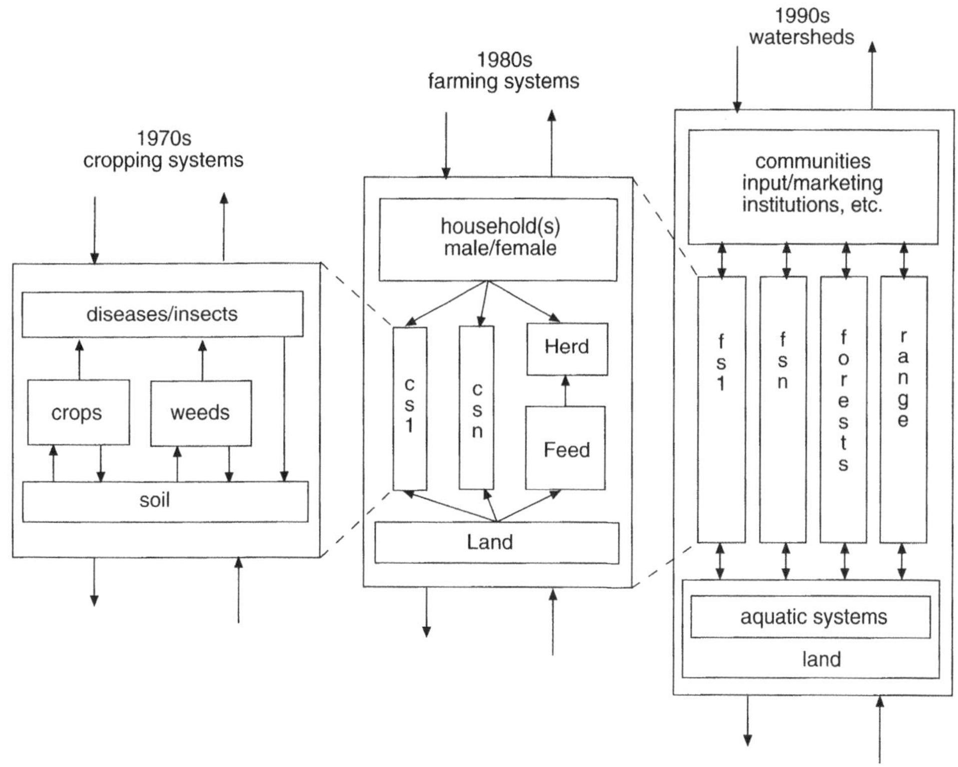

Fig. 3.1.1. FSR's expanding target systems: FSR's conceptual framework has expanded from an initial emphasis on cropping systems in the 1970s, to an emphasis on farming systems in the 1980s, to an emphasis on watershed-level systems in the 1990s.

they would begin to 'climb' the systems hierarchy ladder in their quest to understand the environment determining the structure and function of their target systems.

Cropping systems

Even after many years of cropping systems research, it is still not always clear how different FSR teams defined a 'cropping system'. The two most common approaches were:

- To limit the components of a cropping system to the crop plant populations that interact in space or time.
- To include within a cropping system, in addition to the crop populations, the soil and soil organisms, weeds, insects, pathogens, etc., that interact with the crops.

The first approach evolved mainly from crop physiology and autecology. The second evolved out of ecosystem ecology.

Prior to the evolution of cropping systems research, the conceptual framework for agricultural research and development was strongly influenced by plant breeders. Agronomists manipulated the environment so that new varieties could take advantage of their 'improved' genetic potential. Situations that undermined this, such as competition from other crop populations, low soil fertility and pressure from herbivores, were to be avoided. Cropping systems research turned these concepts upside-down. Its proponents suggested that cropping systems should be improved by changing the arrangements of crop populations in space and time and by using varieties that could fit into these patterns. They also suggested that genetic improvement should be directed towards developing germplasm that could be used to develop more productive cropping systems rather than towards the development of higher yielding varieties.

Farming systems

Once efforts were underway to develop better cropping systems and better agricultural ecosystems, teams soon began to expand research to include interactions between crops and livestock and between crops and trees. This tendency was strongly influenced by the increasing emphasis on on-farm research supported by complementary field station experiments. It became clear that farmers did not manage cropping systems in isolation; they manage farms in which the cropping systems of interest to the researchers are only one of many farm subsystems.

Animal science has a long tradition of taking a systems approach to the improvement of livestock production. While agronomists were moving towards on-farm research and farm-level analysis, animal scientists in many countries were developing production modules integrating feed production with herd management. However, the merger of agronomic and animal science approaches was hindered by the fact that agronomists were working with smaller farmers and animal scientists tended to work with medium- to large-sized farmers and ranchers. In the 1980s, agronomists and animal scientists had to agree on a common conceptual framework in order to work on issues such as the use of crop residue as livestock feed, and animal traction and manure for cropping systems.

The evolution from crop/livestock systems to an interest in improved farming systems changed the relationships between agronomists and animal scientists with economists. Economists had always played a key role on both cropping systems and animal production systems teams, but this was primarily to use data gathered by biological scientists for economic analyses. The economists rarely had the resources to do much more than make brave assumptions about labour use, opportunity costs of inputs, and so on. With growing recognition of farms as real systems with their own unique structure and function, it became increasingly clear that off-farm employment and the complex objectives of farm families are as important as agronomic considerations. Social sciences and biophysical sciences were forced to develop a common conceptual framework in order to work together.

Regional systems

While the evolution towards farming systems research brought together economists, agronomists and animal scientists, the interest in regional systems brought in other groups of biophysical scientists, such as silviculturists and hydrologists, and another group of social scientists, such as sociologists, geographers, anthropologists and those interested in community development. The farming systems and regional systems specialists had 'discovered' each other.

The development of a conceptual framework acceptable to these various disciplines at the regional level continues to be the challenge for the 1990s. It is complicated by the fact that teams are trying to integrate expanding system performance criteria and an expanding population of target beneficiaries. Further, the institutional arrangements and information management systems are not yet in place to support watershed-level systems research.

3.1.3 Expanding system performance criteria

Alongside the expanding scale of target systems, FSR teams have been expanding the criteria used to evaluate the performance of the systems they are developing, moving from a predominance of system productivity towards a balancing of multiple criteria that include criteria factors associated with stability and system/resource interactions. I have subdivided this evolution into productivity, stability, and sustainability (see Fig. 3.1.2).

Produ)ctivity

The primary challenge for agronomists in the 1960s and early 1970s was to find a way to measure system yield, rather than individual crop yield. Two important breakthroughs were the development of the land equivalent ratio (the amount of land to be planted in monoculture plots to equal the production obtained from a polycultural system) and the land-time equivalent ratio that looked at production over an equal time period as well as area. In the 1970s cropping systems teams saw that system productivity should be measured in relation to the system's primary limiting factors and that these were not always land area and time[7]. Yields began to be reported in kg mm^{-1} of rainfall or in tons $person\text{-}day^{-1}$ of labour. It began to be clear that farmers sometimes use different criteria to evaluate different crops within the same system. In many cases farmers evaluate one crop in terms of yield $area^{-1}$ while another is evaluated in terms of yield $volume^{-1}$ of seed planted. On the whole, these new perspectives on productivity are still seen as challenges by the research establishment.

Stability

Resource-poor farmers have no choice but to consider productivity and risk simultaneously. The risks that influence their decisions are both ecological and economic, including unpredictable market prices and the insecurity caused by unclear land tenure. Short-term family survival obviously must take precedence over potential long-term economic returns. Resource-poor farmers have no choice but to design crop and livestock systems that trade-off higher productivity for reduced risk, and those pushed into fragile environments with poor soils and unpredictable rainfall are particularly vulnerable.

Risk is easy to recognize, but not easily incorporated as a research criteria. For most research teams, the simplest proxy for risk was variability in yield or in efficiency. But on-farm research was only just starting and many years of data are needed to measure year-to-year variability. Many FSR teams resorted to using spatial variability as a proxy for temporal variability, measuring system production along a gradient of soil moisture, for example, as a proxy for the year-to-year variability in rainfall.

Gordon Conway, in 1985[8], made a distinction between what he called system stability (the degree to which production is constant in the face of small disturbances over time) and system sustainability (the ability of a system to maintain productivity in spite of major stress). Although his definition of sustainability was not commonly accepted, he made a very important contribution by suggesting that stability, sustainability, equity and productivity are all important system performance characteristics. In the 1980s FSR teams began routinely to apply system performance criteria in addition to productivity.

Sustainability

Sustainability has become one of those terms that will probably need to be abandoned because of its multiple meanings to different people. Most people who use the term are suggesting that long-term environmental costs need to be taken into consideration when evaluating short-term economic benefits. FSR is currently struggling with ways to operationalize this concept.

At least three different schools of thought on how to use the concept are evolving: some view sustainability as synonymous with a group of technologies such as the use of animal manure and integrated pest management viewed as being more 'ecological'. Others define it as a characteristic of the relationship between a

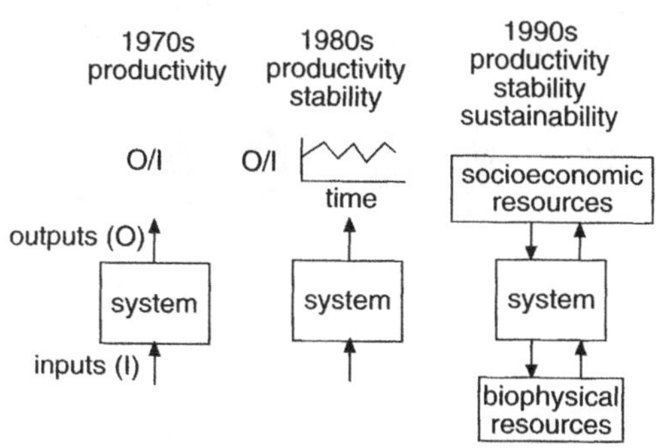

Fig. 3.1.2. FSR's expanding system performance criteria. FSR's conceptual framework has expanded from an initial emphasis on system productivity in the 1970s, to an emphasis on both productivity and stability in the 1980s, to an emphasis on a complex set of criteria related to sustainability in the 1990s.

system and the resources upon which the system depends, emphasizing the need to avoid environmental degradation. A third group view sustainability as a measure of intergenerational equity. All have an equal claim to the legitimacy of their definitions, but the three approaches have different implications for FSR's conceptual framework. The first will have a minimum impact on the FSR framework as the same technology can be appropriate in one system and completely unsustainable in another. The second use means that natural resources should be included explicitly within the FSR conceptual framework. The third use requires that future generations be included within the FSR target beneficiaries.

Clearly we cannot wait 50 years to decide if a system is sustainable, as the second or third approach would require. Possible indicators of 'sustainability' are currently being explored by many different institutions and these include:

- Measuring all inputs and outputs and calculating changes in system efficiency (the total factor productivity approach).
- Monitoring indicators of natural resource productivity and modelling the probable future impact of these changes, such as soil erosion.
- Setting up benchmark sites where many factors can be measured, in order to identify minimum data sets (indicators) for use by researchers working in similar environments.

3.1.4 Expanding target beneficiaries

While FSR was expanding the systems' scales and system performance criteria, it was also expanding and differentiating its target beneficiaries. Taking a systems approach to agricultural research makes sense regardless of farm size or farmer access to resources, but many people view cropping and farming systems research as synonymous with working with resource-poor farmers. This confusion occurred because in many countries the decision to adopt a systems approach in agricultural research and the decision to work with poor farmers occurred at roughly the same time. And, of course, it is easier to argue the merits of taking a systems approach when the target systems are more complex and less well understood.

Equity-related policy issues have always played a significant role in the evolution of FSR. In the 1970s when on-farm research began, most teams assumed that the primary beneficiaries of their efforts would be the farmers in the communities where the research was conducted. In the 1980s equity related policy again affected the evolution of FSR when gender issues were introduced and women became explicit target beneficiaries. In the 1990s the issues of sustainability and intergeneration equity added future generations to the list of target beneficiaries of FSR.

I have subdivided the phases of expanding beneficiaries into: small farmers, women and urban poor, and future generations, as depicted in Fig. 3.1.3.

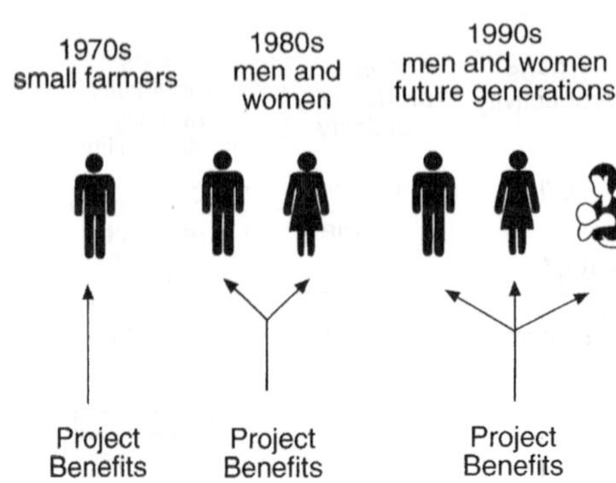

Fig. 3.1.3. FSR's expanding target beneficiaries. FSR's conceptual framework has expanded from an initial emphasis on small farmers as the beneficiaries in the 1970s, to an emphasis on both small farmers and women in the 1980s, to an emphasis on small farmers, women and the next generation in the 1990s.

Small farmers

In the 1970s one of the easiest ways to start an argument when FSR specialists got together was to ask someone to define what they meant by a 'small farmer'. What was clear to most FSR teams was that they were trying to develop better technologies for farmers with fewer resources. The selection of what type of resource-poor farmer to work with differed across different FSR teams. Institutions with a commodity focus, such as the Consultative Group on International Agricultural Research (CGIAR) centres, worked with national agricultural institutions in geographical regions where their commodity was important or where the environment meant that their commodity had high potential. In many countries, national policy directed this effort towards regions with resource-poor farmers. Cropping systems and farming systems were selected for experimentation and farmers in the region operating similar systems – or recommendation domains – were assumed to be the primary future beneficiaries.

Many of the FSR projects financed by external donors were often designed with a bias towards resource-poor farmers in general, rather than towards a particular crop or livestock enterprise, or, indeed, a particular farming system. In theory these teams had the option of deciding which crop or animal system, or non-agricultural household enterprise to work with, based on an analysis of the potential benefits to their target population. Two main problems made this difficult in practice: on-farm research often began after a very superficial appraisal of the local situation, and the biases of team members greatly affected the selection of the experimental focus. Too often, the intended beneficiaries gained little from the project.

Women

During the 'small farmer' phase described above, FSR teams assumed that local farm families were their beneficiaries. Most of the scientists were men and, particularly in Latin America, Africa and Asia, they tended to interact primarily with male farmers. In the late 1970s and early 1980s, as target systems expanded and teams began to look at farm and community systems, the involvement of anthropologists and sociologists grew. The issues of gender, equity and specific benefits to women began to be explicitly addressed.

Gender was incorporated into the FSR conceptual framework in different ways by different institutions. Many simply changed the way they wrote up their research proposals in order to increase their chances of getting money from donors that made gender an 'issue' and continued with their old approach. Other institutions began to include gender as a variable as they measured labour inputs and the flows of benefits. And some adopted an equity approach and set up special programmes targeted directly to women.

Regardless of motivation, the incorporation of gender issues led most FSR teams in the 1980s to subdivide 'the black box' usually labelled 'family' or 'household' in their farm systems diagrams into two categories: males and females. Many FSR teams began to recognize that gender differentiation meant more than just simply measuring which gender provides the most labour into a particular system. It also involves an analysis of who makes key decisions on resource allocations and marketing, and who receives and controls the money when commodities are sold. While it has not been easy to channel the benefits of FSR towards women (this is not surprising given that equity involves shifts in power) there is little doubt that changes in the FSR conceptual framework to incorporate gender have led to a better understanding of how farm systems work.

The next generation

In the 1980s the beneficiaries of FSR were subdivided by gender. In the 1990s they were further subdivided into present and future generations. The design of alternative systems and the development of appropriate technologies now becomes a question of how to predict benefit streams to different potential beneficiaries.

One interesting example is the analysis of the Plan Sierra Project in the Dominican Republic carried out by Alain de Janvry[9]. In a report written for the International Fund for Agricultural Development (IFAD), de Janvry and his colleagues analysed data from the project, suggesting that intergenerational equity could be a good way to measure sustainability. He conducted an appraisal of the development project from two perspectives: one from the standpoint of the present generation at one point in time and one from the standpoint of the next generation at another point in time. This differed from the typical approach by including externalities or charging a project for the depreciation of natural resource stocks. Basically, de Janvry's approach requires the actions of the present generation to maintain resources for use by the next generation.

This approach requires at least 40 years of data, assuming 20 years for each generation. While this is seldom available, approaches can be developed that build on the concept. Since the bridge between generations is the maintenance of the productive potential of natural resources, changes in resource productivity can, perhaps, be used as a proxy for intergenerational equity. One thing is very clear: the concept of time must be more explicitly taken into consideration along with measurements of the changing productive potential of natural resources.

3.1.5 Operational implications

The expansion of FSR's conceptual framework has made it almost impossible for a single institution to implement FSR. The expansion in system scale, performance criteria and target beneficiaries demanding the involvement of so many disciplines over such a long time period, makes it highly unlikely that one institution could organize and manage an FSR initiative using the expanded conceptual framework. The involvement of multiple disciplines from multiple institutions using multiple performance criteria with benefits flowing to multiple beneficiaries, makes FSR implementation even more complex.

However, the common sense sequence in FSR from analysis to design, evaluation, and dissemination stages, has not changed with the expansion in the conceptual framework.

To implement FSR within the expanded conceptual framework it is obvious that different types of institutional arrangements will be necessary. It is equally obvious that the involvement of multiple institutions and larger teams with representatives from more disciplines will require more efficient information management processes.

Institutional arrangements

All the following types of institutions have a role to play:

- Farmers and farmer organizations.
- Community level institutions such as municipalities and marketing associations.
- Watershed-level institutions such as irrigation management organizations and forest management associations.
- Research organizations involved in agronomic, soil, livestock and forestry research.
- Extension research organizations involved in agronomic, soil, livestock and forestry extension.
- National, state or provincial institutions involved in policy decisions affecting the watershed.
- International donor agencies.

What type of institutional arrangements need to be developed? Assuming that various institutions have indicated that they share a common goal and that a minimum of financial resources are available, the institutional arrangement must provide:

- Representative governance.
- Efficient administration.
- Effective programmes (Fig. 3.1.4).

All participating institutions must have confidence in the governing body. The administrative structure must ensure that programmes are following the policy guidelines of the governing body, making efficient use of financial resources. Programmes must integrate the disciplinary expertise of all participating team members to ensure that resource use meets the criteria defined by the governing body.

These institutional arrangements can be called consortia, coalitions, alliances, associations or networks. The name does not matter. What is important is that they develop representative governance, efficient administration and effective programmes. All three of these areas are a significant challenge, but, for some reason, the question of how to develop representative governance has been a particular problem. This seems to stem from the fact that donors providing financial support find it easier to ask larger international or regional institutions to manage the funds. They find it difficult to release the funds to institutions over which they have no control.

Information management processes

One lesson learned and re-learned by FSR teams is that the product or output of FSR cannot be summarized in concise recipes or static technological packages. FSR is a process that has no neat beginning and end.

The agricultural systems found in any given watershed or community are managed by farmers, community leaders, directors of watershed-level organizations, policy makers and so on. Rather than defining the operational goal of FSR as the development of new systems, perhaps the objective of FSR should be operationally defined as to provide decision makers with good information and the knowledge necessary to use this information to make better decisions (Fig. 3.1.5). This information and knowledge is what local people demand from FSR teams. The challenge is to respond to this demand.

Cooperative research, and in particular interdisciplinary research, is impossible without an information management system that captures, archives, processes and disseminates information. FSR's analysis, design and evaluation cycles capture information from multiple sources. This information is used by individuals from different disciplinary backgrounds to conduct *ex ante* design and *ex post* evaluation of alternative resource use systems. But these internal processes are of no practical value unless they respond to a real demand for information and if they do not lead to a dissemination of information to farm, community, region and national decision makers.

As in the case of new institutional arrangements, the development of efficient information management processes to support FSR has only recently been recognized as of critical importance. Most FSR initiatives use their computers to analyse their field-level data, to write reports and technical papers and to develop their project budgets. Very few are developing databases and using them to capture information from multiple sources, systematically evaluating alternatives using multiple criteria, or systematically transferring information to multiple potential beneficiaries.

3.1.6 Concluding remarks

Over the years I have had many opportunities to discuss these issues with colleagues from past projects as well as with other FSR practitioners, and many of the ideas that I have outlined probably originated with them.

An interesting question, of course, is, what will happen next? I have mentioned the problems facing institutions trying to operationalize the

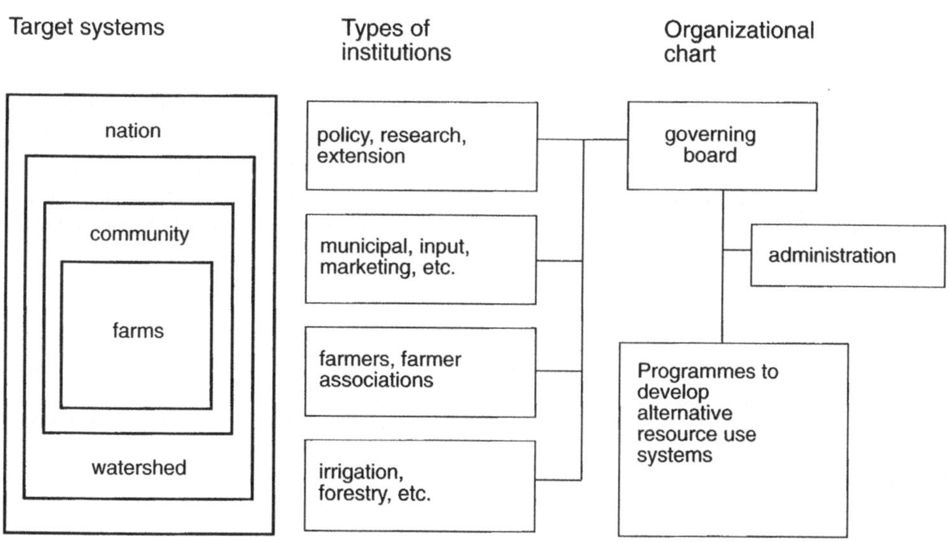

Fig. 3.1.4. FSR's institutional arrangements. In order to implement FSR using an expanded conceptual framework, institutional arrangements that include a broad range of stakeholder institutions must be developed.

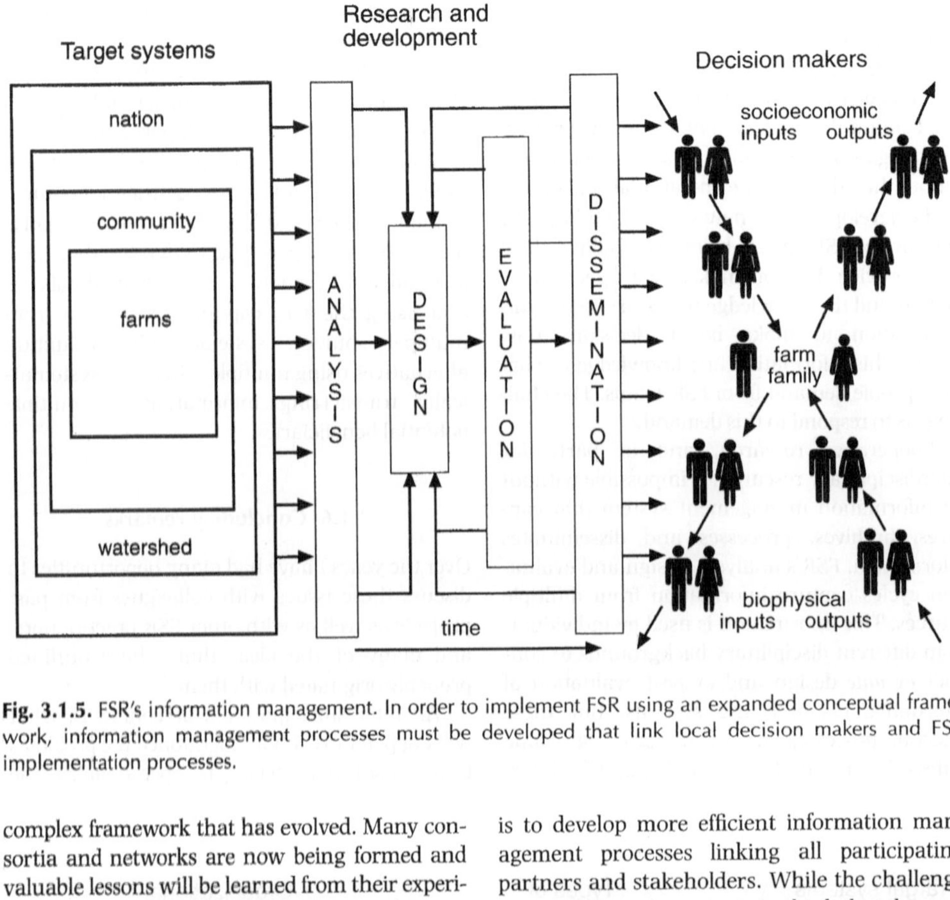

Fig. 3.1.5. FSR's information management. In order to implement FSR using an expanded conceptual framework, information management processes must be developed that link local decision makers and FSR implementation processes.

complex framework that has evolved. Many consortia and networks are now being formed and valuable lessons will be learned from their experiences. A major problem (in particular in this post cold-war era where money for agricultural development is scarce) is how to set up and manage the complex institutional arrangements that are a prerequisite for FSR implemented within its expanded conceptual framework.

I feel strongly that the key to making these complex institutional arrangements operational is to develop more efficient information management processes linking all participating partners and stakeholders. While the challenge is great, there is no reason to think that the next generation of FSR practitioners will not be creative enough to find a way to make things work. Fortunately the information revolution has arrived at exactly the right time to help create new information management tools. The next 10 years in the evolution of FSR should be very interesting.

REFERENCES

1. **Norman, D.W., 1968.** Why practice intercropping? Samaru Agricultural Newsletter, 10 (6), 107–16.
2. **Margalef, R., 1968.** Perspectives in Ecological Theory. University of Chicago Press, Chicago.
3. **Bradfield, R., 1972.** Maximizing food production through multiple cropping systems centred on rice. In: Rice, Science, and Man. International Rice Research Institute, Los Baños, pp. 143–63.
4. **Conway, G.R., 1985.** Agroecosystem analysis. Agricultural Administration, 20, 31–5.
5. **Fresco, L.O. & E. Westphal, 1988.** Hierarchical classification of farming systems. Experimental Agriculture, 24, 399–419.
6. **Hart, R.D., 1982.** An ecological systems conceptual framework for agricultural research and development. In: Shaner, W.W., P.F. Philipp & W.R. Schmehl (Eds) Readings in Farming Systems Research and Development. Westview Press, Boulder, Colorado, pp. 44–58.

7. **Zandstra, H.G., E.C. Price, J.A. Litsinger & R.A. Morris, 1981.** A Methodology for On-Farm Cropping Systems Research. The International Rice Research Institute, Los Baños.
8. **Conway, G.R., 1985.** (Op. cit.).
9. **de Janvry, A., E. Sadoulet & B. Santos, 1994.** Project Appraisal for Sustainable Rural Development: Notes for IFAD's Operational Guidelines. International Fund for Agricultural Development, Rome.

3.2 EVOLVING TYPOLOGIES FOR AGRICULTURAL R & D

Mike Collinson

The growing recognition of the link between poverty and environmental degradation is forcing a much-needed reconciliation between the traditional physically based definitions of zones and people based definitions. The farming system can be an effective interface between these two traditions and form a basic unit for agricultural research and development.

3.2.1 Introduction

Historically, useful typologies have been particular to purpose. As a means of stratification they reduce sources of variation and increase cost/effectiveness in sampling and extrapolation within a broad universe – in the case of most FSR the resource-poor farmers of the world. Two streams of activity have dominated typologies used in research and development for small-farmer agriculture. At the macro level, biophysically based agroclimatic and agroecological zoning and, at the micro level, farm classification. Agroecological zoning has been developed by land management professionals, primarily to classify land according to potential[1], whereas farm classification is a tool of the farm management profession, its primary purpose to compare the performance of managers operating similar farms[2].

The growing recognition of the link between poverty and environmental degradation is forcing a much-needed reconciliation between the traditional physically based definitions of zones and people-based definitions. The farming system can be an effective interface between these two traditions and form a basic unit for agricultural research and development.

3.2.2 Agroecological zoning

FAO's pioneering work

In the late 1970s and early 1980s, the Food and Agriculture Organization of the United Nations (FAO) completed agroecological zoning for the major regions of the world based on climate, soil and terrain criteria. The main purpose was to measure agricultural potential and define optimal land use. In one innovative application, zoning was used to show which countries of the world should be able to feed themselves into the future[3]. The productive potential of the zones was related to population and its growth, estimates of carrying capacity and, beyond that, calculations of future self-sufficiency at the country level. For the first time countries with long-term problems of food self-sufficiency were identified through a logical, analytical process.

Issues in the use of biophysical based typologies

Biophysically based zoning has been widely used, and sometimes abused, as a policy planning tool for agricultural research and development[4]. Planners have gone so far as to designate the zones of highest physical potential as the *only* areas where growers are serviced by public marketing, extension and credit institutions. Examples in eastern Africa showed areas denominated as 'very suitable' for sorghum which are, on the ground, dominated by maize, the staple food chosen by local farmers.

Zoning has played, and continues to play, an important role in agricultural research for three key reasons:

- To identify, for plant breeders, sets of conditions with common climatic constraints within which they must work, and provide a framework for testing and comparing materials.
- To identify relatively homogeneous sets of climate and soil characteristics that can be dealt with by one applied research programme in natural resource management.

- To show where else applied research results will be relevant, and facilitate transfers between geographically separate but relatively homogeneous sets of climate and soil conditions.

These roles also create distortions because they don't allow for farmers' decisions. Agroecological zones offer a range of cropping opportunities to farmers and, at the same time, within each crop opportunity offer a set of possibilities; for example, maize is an opportunity, but the farmer may choose a maize which matures in 110–200 days. Farmers' choices from these 'within crop' sets are determined by their socioeconomic circumstances and articulated in their farming system.

Preoccupation with yield has kept applied research firmly focused on the upper boundary of the 'within crop' possibility set. For example, in an agroecological zone characterized by a 6-month rainfall season, a 180-day maize offering the greatest yield potential will be the logical choice for conventional breeding programmes. Clearly, maize with maturity periods *below* 180 days are cropping possibilities for the zone but applied research will pay little or no attention to them, even though they will often be the preferred choice of farmers, for at least three possible reasons:

- Low power resources may stop farmers opening the land until it is thoroughly wetted. The earliest feasible planting time may be 20 days after the start of the rains, suggesting a 160-day variety.
- Local maize prices may be 300–400% higher several weeks before the main harvest, suggesting a maturity period of around 145 days to exploit this profitable market.
- Farmers may give priority to planting another crop before or after the maize, suggesting a maturity period of 110–120 days.

Varieties and management practices identified as 'optimal' will often be totally irrelevant to the choices made by farmers as a result of local circumstances. Indeed, the ranking of crops and crop varieties on the basis of physical potential for an area can be stood on its head when farmers' own criteria are used, rather than those of researchers[5].

3.2.3 FSR and farm classification

Recommendation domains

As the Farming Systems Support Project of the University of Florida said in 1987[6]: 'At one extreme, we do not have sufficient resources to carry out a specific research program for every individual farmer. At the other extreme, it does not make sense to try to develop a single research agenda relevant to all farmers in a country. We must compromise between these two extremes and plan research relevant to groups of farmers'.

The tradition of grouping farms and farmers, already acknowledged in Anglophone countries and in France, was carried into developing-country agriculture in the 1970s. Upton, in 1973[7], focused on the need for a process dealing with groups rather than individual farmers. One early struggle was the move away from survey work based on administrative areas, which confounded understanding of farmers' decision-making by averaging data across more than one farming system[8]. The work took a major step forward in the mid 1970s when the International Maize and Wheat Improvement Center (CIMMYT) developed the concept of the 'recommendation domain' (RD); a group of farmers operating the same system and for whom the same new technologies would be appropriate. The RD concept recognized spatial and hierarchical dimensions, with hierarchical variation between systems arising from differences in farmer resource endowments.

Criticism arose from two perceptions. First that manpower-intensive local surveys are a prerequisite to system differentiation and are impractical over wide areas. Second, that rapid rural appraisal could provide only qualitative information and this was unattractive to agricultural economists trained in a quantitative tradition.

Collinson, writing in 1996[9], held that variation in six key parameter sets – climate, soils, topography, culture, market opportunity and population pressure – cause spatial diversity across farming systems. However, he saw the ideal minimum data set to identify RDs as having three groups of criteria to capture the effects of both spatial and resource endowment variation:

- The spatial pattern and scale of farm activities.
- The practices used to manage these activities.
- The calendar of the application of these practices over the year.

These three groups feature empirical information about the systems operated by farmers. The data required can be captured by low cost rapid rural appraisal methods where up-to-date secondary data is not available.

Issues on RDs

The strength of the RD concept is its firm focus on grouping farmers – the managers who will take the decisions on using new technologies. Aspects under continuing discussion include the acceptable level of compromise in reducing sources of variation between farms, and the issue of domain boundaries. Viewed more widely, there are clear shortcomings to the RD concept in the search for a conceptual framework for agricultural research and development.

Byerlee, in 1987[10], distinguished three types of variation within farming systems; between farms, within farms and between years. He noted: 'A point is soon reached where between-farm variability is less important than variability within farms and across years and the benefits of [further] sub-grouping rapidly diminish'. The acceptability of the compromise depends on the importance of sources of between-farm variation that are not captured by the criteria used in system definition. Fortunately much of this variability is caused by two sources which do not create 'between-system' variability: age and its effect on motivation is one, and differences in managerial competence or natural ability as a farmer is the other. Both cause between-farm variability but are sources common to all systems and both are particularly useful in identifying subgroups within farming systems on which scarce resources for research and development resources might be focused.

A second area of continuing discussion, which applies to all zoning, is that of system boundaries and the extrapolation of results. Perrot and Landais addressed the issue in 1993[11], saying: 'All segregative methods present the disadvantage of focusing too much attention on the limits separating each type, on the definition of type content and therefore of their specificity which is subordinated to the limits between them. In other words, the borders are all important and the centre is neglected, when the contrary would be more consistent with the constructivist approach'.

Just as with climate and market opportunities, one farming system usually merges into another. Where there is a stark change in a causal factor, a dramatic change in topography for example, a boundary may be clearly identifiable, more usually there is a graded transition and no clear boundary. It is a phenomenon which can be managed locally. Research and development activities need to target the clearly defined populations within domain boundaries, and programmes, such as farmer-led experimentation, need to be located unambiguously *within* farming systems. Social forces and the market will stimulate diffusion to the fringes and thus define where new technologies and policies lose their relevance[12].

The RD concept has, like zoning, sometimes threatened to overreach itself. Even early documentation from CIMMYT[13] tentatively suggests that a farmer may be a member of two domains. Similarly, the biophysical importance of differentiating landscape types provides the temptation to be all things to all people[14] but breaks the ground rule: one farmer, one domain. The original RD concept based on grouping farmers as decision makers has been further compromised by the application of the term 'RD' to a technology. Not only is it confusing to have two contrasting definitions, but the technology-based definition wholly ignores the human side of the adoption equation, which FSR has struggled to bring into full partnership with the biophysical.

Shortcomings in the RD concept

The evolution of conceptual frameworks shows how purpose in FSR has continued to widen. Three factors, in particular, have added to the complexity of this process:

- The dynamics of farming systems in responding to widening market opportunities and increasing population densities[15].
- The reconciliation of the goals of individual farmers, their community and society at large, particularly with respect to environmental sustainability.

- The strong interactions between technology adoption and policy manipulation[16].

These same factors have compounded the complexity required in typologies as a framework for agricultural research and development. The recommendation domain concept has not evolved to embrace them, but new concepts and techniques are carrying the challenge forward.

3.2.4 Recent developments and current status

1986 saw a landmark meeting on agroecological characterization[17]. Though many contributors at the meeting emphasized the importance of socioeconomic parameters in a framework for agricultural R & D, participants concluded: 'It was not agreed that the time was ripe to incorporate socioeconomic parameters into databases along with data on other environmental attributes.' Contributions included pioneering studies that combined the use of macro-level geographical information systems (GIS) with socioeconomic data from national census results[18], and from rapid rural appraisal methods[19].

The sustainability issue

During the late 1980s and early 1990s, the call for a balance in biophysical and socioeconomic considerations was strengthened by a growing awareness of the sustainability issue. The Brundtland Report in 1987 provided a strong, early stimulus and was followed by the UN Conference on Environment and Development in 1992. In 1989 the Technical Advisory Committee recommended a restructuring of the Consultative Group for International Agricultural Research (CGIAR) to bring more of its science to bear on natural resource management research. The basis of the restructuring was the acceptance that, beyond natural processes, human decisions at levels from the farm to the globe are primary causes of resource degradation. The shift in CGIAR programming brought new impetus to the search for balance between biophysical and socioeconomic dimensions of the resource management problem. Similarly, as the UN Task Manager for the implementation of Chapter 10 of Agenda 21, FAO moved away from its historical emphasis on climate and soils and towards a holistic people-based approach to the planning and management of land resources. The sustainability issue drew both geographers and ecologists into the widening disciplinary mix of the emerging paradigm for agricultural research and development.

Growth dynamics

Recent work has begun to erode the static nature of typologies. Research by Weber, Smith and Manyong[20] identifies four broad types of agricultural systems (named as 'research domains') – market intensive, market extensive, population intensive and population extensive – designating market access and population increase as two key drivers in farming systems. The authors perhaps go too far in claiming 'the most relevant information about a system and its sustainability is related to its evolutionary pathway rather than its current characteristics'. It is true, however, that much of the historical rejection of research-derived technology has been a failure to understand labour and cash as greater constraints than land on farmer livelihoods.

Systems hierarchies

Fresco, in 1995[21], made the important point: 'If we want to explain, rather than solely describe, a phenomenon, processes at both higher and lower scales must be studied also'. Systems hierarchies had already been recognized as valuable tools, for example in agroecosystems analysis in the 1980s[22].

In addition, farmers were increasingly recognized as *de facto* land managers. Both were accompanied by a strengthening perception of the need for linkages between farmers' actions, national environmental policies and global conventions – important factors highlighting the relevance of a hierarchical framework. In the example framework at Table 3.2.1 human decisions (farm, community, local, regional, national policy, international convention) influence activities at several levels in both economic and ecological hierarchies. The impact of their decisions not only has repercussions on the adjacent levels *within* the hierarchy but also, importantly, interacts *across* the two hierarchies. So economic decisions have ecological implications and vice versa.

Operationally the 'decision point' identifies institutional levels which offer leverage on both the economic and ecological hierarchies. The key points on the decision hierarchy are farmers

Table 3.2.1. Three hierarchies: a framework for agricultural R & D.

Economic	Ecology	Decision point
Global	Gaia	International conventions and organizations
	Agro climate Agro ecology	
Agroecological zone		
Continental		Country groupings
National		National policy National institutions
Regional	Major topographical features (river basin, forests, mountains)	Regional institutions
Local	Ecosystem	Local government NGOs, farmer associations
Resource management domain		
Village	Landscape unit Farm system	Village authority Community
Recommendation domain		
Farm enterprise	Farm Resource niches/fields	Farm family Family members

and policy makers. With effective institutions open to the market, these decisions help shape the production environment in which farmers make their decisions. With weak or 'captured' institutions, as with market distortions, policy decisions have less influence. Farmer and community decisions on change directly impact the landscape and farm levels. As more farmers and communities adopt the changes, their accumulating decisions have repercussions at higher levels[23]. Linking policy and farmers' decisions in the same framework embraces the interaction between policy formulation and technological innovation.

The surge of interest in sustainability has stimulated the notion of resource management domains (RMDs). These are variously defined and some retain the early emphasis on land potential. One more general and more flexible definition comes from Esawaran writing in 1996[24]: 'Resource management domains are landscape units delineated on the basis of similarity with respect to response to management'. Esawaran located these in a hierarchical framework at an intermediate level between agro-economic zones (AEZs) and RDs. He describes them as a set of biophysical parameters bounding a group of technical problems with causes peculiar to their characteristics and a parallel set of technical solutions. The farm system or RD is shown as common to both the economic and the ecological hierarchies, highlighting the fact that farmers' decisions and actions are a dominant influence on *both* hierarchies.

3.2.5 The current state of the arts

As Hart has pointed out, system hierarchies offer a valuable conceptual and operational framework for agricultural research and development. In addition, GIS and remote sensing (RS) techniques both make a strong contribution to useful application by allowing the overlaying of spatially referenced data sets at different hierarchical levels. Users may draw on those data sets needed for their particular application. This flexibility allows use of the data sets for a variety of purposes making the framework more general and less 'purpose specific' than older typologies. However, the historical priority given to database compilation in climate and soils limits the scope of current GIS applications and the challenge of accumulating new databases is a critical contemporary constraint.

One way to apply this type of hierarchical framework in agricultural research and development, particularly in the sustainability context, is to use Naisbett's 1994 line: 'act locally, think globally'[25]. It is widely accepted that resource degradation begins locally. It is unnecessary to research all levels of the hierarchies. The key is effective local diagnosis to identify the decision points driving local activities, particularly those causing degradation. The issue of 'full' or 'partial and focused' systems analysis, and the use of qualitative or quantitative data have been controversial since the early days of FSR. CATIE, for example, sought full quantification and considered all the parameters of the farming system as variable. CIMMYT[26], on the other hand, applied low-cost techniques providing largely qualitative information to gain an understanding of the wider system. Carrying a 'partial and focused' approach into this wider framework would identify the hierarchical levels and decision makers, many of which will be external to the farming system, yet crucial to changing the activities causing local degradation[27].

A hierarchical framework embracing ecology, economics and institutions has important implications for the organization of institutions and their coordination in its effective application. Tims in 1995[28] argued: 'Research should increasingly centre around studies of farm households, each with farming as one of its activities, but among a number of other options to use its scarce resources, and with trade-offs which affect the choices in the field of agriculture. Also account must be taken of the character of production decisions in a number of cases as derived decisions. ... Households interact with policies and markets and research cannot truly answer policy questions unless it traces those market relations, with price and income formation and the responses to those by households and by governments'. Tims' statement highlights the range of external influences on farmers' decisions. It points to an understanding of farmer decisions as a prerequisite to progress in agricultural development.

The farming systems movement is in the forefront of equipping professionals with that understanding and the wider hierarchical framework offers a more robust context for their activity. Permanent field teams are justified by the need to understand small farm households and apply that understanding at several levels. With their eyes on resource degradation Carter and others have said (rightly in my view): 'put simply how can we predict what we don't understand?'[29]

3.2.6 A scenario for the future

To mobilize current knowledge in an integrated way requires a combination of a wider conceptual framework, institutional change as well as improved information management techniques. Each institution, and indeed each individual, needs to know where they are, and how they contribute, in a strong operational framework. The ability to position themselves in this framework tells them who their logical partners and clients are, and how information needs to flow if they are to be successful in their role.

The improving resolution in remote satellite sensing and progress in digitizing images, means that the spatial pattern of agricultural activities may offer a global basis for typing farming systems with less need for field work. But this should not, however, induce the search for the 'holy grail'. Countries dependent on their small-farm sector, with large numbers of farmers and few professionals, will still need local expertise[30]. FSR teams remain important institutional innovations for many countries. NGOs, farmer associations and publicly funded agricultural services might be appropriate vehicles for their operation, with a core task of participatory technology identification.

Local FSR teams are the key to the effective exploitation of both human and biophysical information. They will understand how farmers exploit their RMDs in the operation of their systems. They will also understand which enterprises are important sources of degradation and where. Such teams need access to the widest range of technical findings on the management of the RMDs with similar profiles, particularly for those enterprises contributing most to degradation.

As a long-term goal, local FSR teams should be able to compare their local RD profiles with other profiles from a global database, drawing on those options that have succeeded under the same conditions elsewhere. We are a long way from this ideal. An ambitious first step will be to identify RMDs and the technical and policy opportunities associated with them. In the short term, RMDs would be defined on the classic climate and soil criteria for which databases already exist, supplemented by an accumulating database of technical research and policy findings organized by RMDs. Local teams draw on these for the RMDs in use by local farmers. FSR teams will first identify technical and policy opportunities for those RMDs in their local space, and pre-screen these for their apparent relevance to local RDs or farming systems. In dialogue with local communities, those seen as the most appropriate options are tested in a participatory mode in farmers' fields.

At the same time, databases are built up for RDs with input from the same FSR teams. These would provide biophysical and socioeconomic profiles of the sites where techniques and policy measures had proved successful. Socioeconomic profiles might include information on market access, population pressure and 'calendars' for the successful technologies. As such databases gain recognition, their use by FSR teams would widen and the process would evolve.

This route forward would have two short-term benefits:

- It would diffuse the issue of incorporating socioeconomic parameters into biophysical databases.
- It would outflank the problem of weak data on existing farming systems as an argument against their adoption as primary development units.

Once *in situ*, with research with farmers as their core role, the mandate of local FSR teams could be widened:

- Teams could help pursue greater system understanding, including measurement of key parameters for modelling of the dynamics of the development path for example, or the needs for sustainable farming, or the sequencing of interventions.
- Teams can channel local information to levels in the decision hierarchy where plans for, for example tenure arrangements, community regulation structures, pricing of water or remuneration for downstream costs – all influence local activities.
- Teams can contribute details of locally successful practices in production and land management to databases.

As the second of these supplementary functions implies, the processes of social, as well as technical innovation, can be facilitated by local research[31]. Providing local FSR teams with access to relevant global research requires a major effort in database development. Once such a database framework is operational, local field teams themselves would be well placed to enter, as well as to use, information.

It is particularly encouraging that 'hard core' disciplines are now reaching out towards the human side of the development equation. We have moved some way from Henry Nix's assertion, made over 10 years ago, when he said: 'I have argued that if it were possible to predict the performance of any crop at any location given a specified minimum set of site, soil, crop, weather and management data, we could indeed prescribe appropriate and relevant technologies at the farm or even the field level'[32]. Perhaps we will progress further down the path if all institutions and professionals with a role to play adopt a common hierarchical framework, and if the will is found to adjust both institutional mandates and structures to play these roles in partnership.

REFERENCES

1. **FAO, 1978–80.** Reports of the Agroecological Zones Project. World Soil Resources Report no. 48. FAO, Rome.
2. **Blagburn, C.H., 1961.** Farm Planning and Management. Longmans, London.
3. **FAO, 1984.** Land, Food and People. FAO Economic and Social Development Series no. 30. FAO, Rome.
4. **Collinson, M.P., 1988.** FSR in Evolution: Past and Future. Keynote address. 8th Annual Farming Systems Research/Extension Symposium. Winrock International, University of Arkansas, USA.

5. (Ibid.).
6. **FSSP, 1987**. Diagnosis in Farming Systems Research and Extension. Farming Systems Support Project, Volume 1 FSR/E Training Units, Participant Manual. University of Florida, Gainesville, USA.
7. **Upton, M., 1973**. Farm Management in Africa: the Principles of Production and Planning. OUP, London.
8. **Collinson, M.P., 1972**. Farm Management in Peasant Agriculture. Praeger, New York.
9. **Collinson, M.P., 1996**. Social and economic considerations in developing Resource Management Domains. Proceedings of the International Workshop on Resource Management Domains, August, Kuala Lumpur, Malaysia. IBSRAM, Bangkok.
10. **Byerlee, D., 1987**. From Adaptive Research to Farmer Recommendations and Extension Advice. Agricultural Administration and Extension, 27, 231–44.
11. **Perrot, C. & E. Landais, 1993**. Exploitations agricoles; pourquoi poursuivre la recherche sur les methods typologiques? Les Cahiers de la Recherche-Developpements, 33, 13–23. Reprinted in English in Systems Studies in Agricultural and Rural Development, J. Brossier, L. de Bonneval & E. Landais (Eds). INRA Editions, INRA, Paris.
12. **Collinson, M.P., 1996**. (Op. cit.).
13. **Byerlee, D., M. Collinson, *et al.*, 1980**. Planning technologies appropriate to farmers: concepts and procedures. Economics Program, CIMMYT, Mexico.
14. **Fresco, L.O., 1995**. Agroecological knowledge at different scales. In: Bouma, J. *et al.* (Eds) Ecoregional Approaches for Sustainable Land Use and Food Production. Kluwer and CIP, The Netherlands.
15. **Maxwell, S., 1986**. Farming systems research: hitting a moving target. World Development, 14(1), 65–77.
16. **Baker, D., 1993**. Inability of Farming Systems Research to deal with Agricultural Policy. Journal for Farming Systems Research-Extension, 4(1), 67–86.
17. **Bunting, A.H., Ed., 1987**. Agricultural Environments, Characterisation, Classification and Mapping. CABI, UK.
18. **Scholz, U., 1983**. The Natural Regions of Sumatra and their Agricultural Production Patterns – a Regional Analysis. Central Research Institute for Food Crops, Bogor, Indonesia.
19. **Carter, S.E., 1987**. Collecting and organizing data on the agro-socioeconomic environment if the cassava crop – a case study of a method. In: Bunting (Op. cit.).
20. **Weber, G., J. Smith & M.V. Manyong, 1996**. Systems dynamics and the definition of research domains for the northern Guinea Savanna of West Africa. Agriculture, Ecosystems and Environment, 57, 133–48.
21. **Fresco, L.O., 1995**. (Op. cit.).
22. **Conway, G.R., 1987**. The properties of agroecosystems. Agricultural Systems, 24 (2), 95–117.
23. **Collinson, M.P., 1996**. (Op. cit.).
24. **Esawaran, H., 1996**. Biophysical considerations in developing resource management domains. Proceedings of the International Workshop on Resource Management Domains, August, Kuala Lumpur, Malaysia. IBSRAM, Bangkok.
25. **Naisbett, J., 1994**. Global Paradox. William Morrow, New York.
26. **Byerlee, D., M. Collinson, *et al.*, 1980**. (Op. cit.).
27. **Janssen, W.G., 1995**. Aggregating economic knowledge for use at national, regional, sector and farm level. In: Bouma, J. *et al.* (eds) Ecoregional Approaches for Sustainable Land Use and Food Production: Proceedings of a Symposium on Ecoregional Approaches in Agricultural Research. Kluwer Academic Publishers, Dordrecht, pp. 143–166.
28. **Tims, W., 1995**. Responses to Sections A & B. In: J. Bouma *et al.* (Eds) (Op. cit.).
29. **Carter, S., P. Bradley, S. Franzel & J. Lynam, 1994**. Spatial Variation as a Factor in Natural Resource Management Research. International Institute for Environmental Technology and Management, Stockholm Environmental Institute.
30. **Delehanty, J., 1993**. Spatial projection of socioeconomic data using GIS: results from a Kenya study in the strategic implementation of a livestock disease control intervention. In: Dvorak, K. (Ed.) Social Science Research for Agricultural Technology Development: Spatial and Temporal Dimensions. CABI & IITA, Oxford.
31. **Carter, S., P. Bradley, S. Franzel & J. Lynam, 1994**. (Op. cit.).
32. **Nix, H., 1987**. The role of crop modelling, minimum data sets and geographic information systems in the transfer of agricultural technology. In: Bunting, A.H., 1987. (Op. cit.).

3.3 THE DEVELOPMENT OF DIAGNOSTIC METHODS IN FSR

John Farrington

PRA has ... generated a sense of community ownership of development projects and processes, and a recognition among administrators that farmer participation enhances the prospects of success. However, it is increasingly being seen as a 'new orthodoxy', and, like all orthodoxy, it is attracting challenges of diverse kinds. One has to do with intellectual property, viz. the argument that a number of the methods it embraces predate the term 'PRA'.

3.3.1 Introduction

A glance at the programmes for the international FSR symposia held since 1981 reveals a number of sea-shifts in major areas of concern, many of them having implications for the development and application of diagnostic methods. The principal concern over the first five symposia (1981–85) was to increase understanding of small-farm systems, and to develop methodologies for diagnosis, implementation of research and monitoring. By the seventh symposium, concerns over the link between FSR, macro-policy and communication had begun to emerge, but the emphasis was still on 'how systems work'. By the eighth symposium, the agenda had broadened to include major sessions on gender and intra-household analysis, farmer experimentation and natural resource management. This set the tone for subsequent meetings: by the 12th symposium in 1992 these new themes were joined by a major session on different types of institution. The 13th symposium in 1994 went further, considering the role of agriculture in the generation of rural livelihoods.

This contribution looks back at these changes and forwards to others which may arise for diagnostic methods in the future.

3.3.2 Methods, policy context and organizational change

This contribution argues that changes in methods have to be viewed in the context of major changes in the mandates and structure of research organizations, which, in turn, are influenced by overarching policy imperatives. The main interactions between policies, organizations and methods are presented on a broad canvas in Table 3.3.1. From the beginnings of agriculture to the middle of the 19th century, the imperative was to increase and secure household food production. Farmer experimentation was the sole means of achieving technical change in pursuit of this objective, and informal interaction amongst farmers the means of spreading technical change.

Agricultural science first became institutionalized with the establishment of Rothamsted Experiment Station in the UK in 1843, but many scientists retained their roots in rural communities, and farmer influence on the research agenda was strong. There was little difficulty in diagnosing the priorities for research in a 'whole farm' context. The first half of the twentieth century saw two important shifts. First, advances in plant and animal genetics and in the understanding of pests and diseases, plus the arrival of the internal combustion engine and agrochemicals, opened the door to the creation of specialist disciplines. Second, in Europe this took place against a growing policy imperative to increase the productivity of land (this pressure was less intense in the USA, with its lower population densities). Publicly funded research expanded rapidly during this period, but much of the expansion was in the form of specialist institutes and departments to house newly emerging disciplines. Their primary approach was reductionist in the sense that they assessed changes in productivity attributable to specific technical changes by isolating these from the remainder of the farming system. In this context, priorities for research were determined as much by the widening opportunities offered by science as by careful diagnosis of farmers' needs.

Policies to enhance the productivity of export crops led the European colonial powers to replicate these organizational structures and research approaches in the south. With political independence came a major drive towards research on food crops. However, by this time, many of the public sector research organizations in developing countries had been structured

Table 3.3.1. The evolution of FSR diagnostic methods: policy context and organizational change.

	1500 BC to 1850 AD	1850–1950 (S. Asia) 1850–1965 (Sub-Saharan Africa)	1950–1985 1965–1985	1985 to present
Policy imperatives	Increase food production	• Produce colonial export crops • Increase food production in north • Increase land productivity especially in Europe	• Increase food production in south • Increase land productivity especially in Europe	• Reduce food surpluses in north • Increase food production in south • Do above in ways compatible with sustainable natural resource management
Response				
Organization and structure	Informal interaction among farmers	• Shift from farm → station → laboratory-based research • Shift towards commodity institutes and commodity/discipline-based departments in north; these exported to south initially for export crop production	• Multidisciplinary teams, often led by social scientists, formed within national public sector research institutes to diagnose farmers' problems and opportunities and prioritize research • Efforts within national systems to link field-oriented with disciplinary researchers, and to create research-extension links • IARCs search for role in FSR, and for appropriate working relationship with national research services	• Reduced emphasis on diagnostic 'teams'; more responsibility on individual scientists or groups of scientists to identify clients' needs and design their work in response • Growing recognition that even with rapid methods, public sector alone does not have the resources for widescale diagnosis of the diverse needs arising from agroecological and socioeconomic (including gender and household) complexities • Hence, growing emphasis on multi-agency approaches and on having NGOs and farmers' groups diagnose needs and create demands on public sector • Efforts to increase responsiveness and accountability of public sector enhanced by above, but also by innovations in research funding, reward systems and decentralization
Methods/ disciplines	Farmer experimentation	• Generalist researchers rooted in rural communities gradually giving way to specialist disciplines • Increasing reliance on reductionist approaches • Research agenda increasingly driven by science • Few comprehensive efforts at diagnosis: to increase yields assumed to be overriding objective • No social science skills deployed	• 'Increase yield' diagnosis correct in high potential areas leading to success of Green Revolution deriving from reductionist approaches • Widespread failure of attempts to apply similar diagnosis and approaches to complex, diverse and risk-prone areas • To address above, economists (subsequently other social scientists) conduct extensive surveys to identify complexity, yet rationality of farming systems in CDR environments; 'rapid survey' techniques towards end of period	• Increasing reliance on rapid assessment and participatory diagnostic methods • These methods extended to understanding of complementarities between on- and off-farm resource management and their implications for research agenda • Increased efforts to understand implications of intra-household decision taking and gender issues • Increased efforts to locate agriculture's contribution within the multiple sources of livelihoods pursued by low-income farmers • Growing interest in systems approaches to address sustainability concerns in south and north

along discipline and commodity lines – occasionally even with separate specialist institutes – in ways unrelated to the needs of small farmers operating under complex, diverse and risk-prone conditions. Researchers' conventional diagnosis in this setting – supported by only limited contact with, often, the better endowed farms – was that the need to enhance yields per hectare was paramount.

Research prioritized on this diagnosis generated widely adopted packages only where three sets of conditions held: where pressure on land was high, where production conditions were favourable and stable, and where the main components of farming systems could easily be replicated on research stations. The rapid spread of the Green Revolution in the irrigated rice and wheat systems of South Asia is one example. Over time, elements of such packages were adopted selectively elsewhere as farmers perceived them fitting into their systems[1], usually where extensive farming options had disappeared. Elsewhere, inadequate diagnosis of farmers' requirements in a 'whole farm' context led to low adoption levels, and to a widespread view that farmers were 'backward' and unresponsive to conventional economic stimuli because of their unwillingness to adopt 'superior' technologies.

It is against such prejudice that the early farm management investigations of Norman in Nigeria[2], Collinson[3] and Ruthenberg[4] in East Africa, and Mellor[5] in India have to be understood. By examining how farm households deployed their labour and other resources in order to secure food requirements under high-risk conditions, they gave a new dimension to the understanding of farmer rationality. The early review by Gilbert *et al.* in 1980[6] suggests that much of this work was multidisciplinary, but led by social scientists working with intensive sample surveys of farmers' practices, frequent recording techniques and questionnaires, supplemented by the particular insights that biological scientists were able to provide. Bringing biological scientists and social scientists together to investigate field realities in this way had rarely been tried in small-farm agriculture. It set the trend for farm surveys for the 1970s and early 1980s, and much of the early institutionalization of FSR involved the formation of 'farm management' units or special teams. It should be noted, however, that the traditions on which FSR drew were by no means homogeneous: work in Senegal, for instance, had led to the setting up of 'unités experimentales'[7].

Two exponents broke the methodological mould of early FSR institutionalization: Hildebrand[8] by developing a 'sondeo' (literally 'sounding') method in Central America, which many regarded as 'quick and clean', and Collinson[9] by experimenting with informal, qualitative 'pre-surveys' using semi-structured interviews and checklists. These had two purposes: to characterize 'general attributes' and to identify parameters vital to system improvement for subsequent quantification in a formal survey. Collinson also argued that, when conducted in this setting, infrequent visits gave as good data on critical parameters as frequent visit methods. These were the precursors of 'Rapid Rural Appraisal'[10] and its derivatives.

3.3.3 Development of diagnostic methods in the 1980s

Two further methodological developments were reported in the early 1980s. First, the Rhoades and Booth *farmer-back-to-farmer* model developed at the International Potato Center (CIP) and described in 1982[11], stressed interdisciplinary rather than multidisciplinary work. It was characterized by interaction between farmers and researchers in which the conventional project cycle of diagnosis, experimentation, assessment and dissemination can be replaced, for instance, by approaches which begin with an experiment and end with a survey. Practical experience of interdisciplinary work at CIP led to the following characterization of the main stages of the model:

- In diagnosis the problem is identified jointly by farmers and researchers.
- Interdisciplinary team research develops potential solutions to the problem.
- Solutions are adapted to farmers' conditions in on-farm testing.
- Farmers play a key role in evaluation and further adaptation.

Second, the *farmer-first-and-last* model proposed by Chambers and Ghildyal in 1985[12] entails 'fundamental reversals of location and learning' and is characterized by:

- A diagnostic procedure involving learning with farmers.
- Technology generation on-farm and with-farmer.
- Using the level of farmer adoption as a criterion for evaluating research.

Chambers and Ghildyal outline the conditions – including institutional conditions – necessary for the approach to succeed. These include methodological flexibility and innovation, full interdisciplinarity, adequate resources for field work, scientific rewards geared to practical achievements (not merely to publications) and training in the necessary techniques for learning from farmers.

These models have different strengths and weaknesses. The *farmer-first-and-last* model, for example, is stronger on rhetoric than on the practicalities of how on-farm research might link with on-station or laboratory-based research (indeed, the model scarcely admits these as legitimate). However, the importance of both lies in the early impetus they gave to participatory research way ahead of the development of now widely used participatory rural appraisal (PRA) techniques.

In organizational terms, the 1970s and 1980s saw a search for ways to re-focus science on *field* problems and opportunities. This was not to imply that there was no useful role for reductionist research, simply that the danger of having station-based and laboratory research driven by priorities of little interest to farmers had to be countered. Some observers, particularly Chambers and Ghildyal, and Chambers and Jiggins[13], were extreme in their criticism of what they saw as research which had been outmoded by a growing understanding of farmers' own capacity for experimentation and by new, rapid methods of diagnosis. In fact the new methods and perceptions are not a substitute for reductionist methods, but are complementary in the insights and focus they provide, as indicated by the subtitle of the 1987 workshop from which 'Farmer First'[14] originated – Farmers and Agricultural Research: Complementary Methods.

3.3.4 Institutionalization of new approaches

Apart from the introduction of farm management units, the principal means of this re-focusing was through the creation of multidisciplinary diagnostic teams[15]. There were also efforts to create institutional links in several dimensions: between field-oriented diagnostic teams and commodity or discipline-oriented researchers[16], between research and extension[17] and between national research organizations and the International Agricultural Research Centres (IARCs).

Writing in 1988, Collinson[18] reflected on the slow pace of institutionalization of FSR in East Africa, citing various causal factors. He included vacillations in government and donor policies, changes in key senior personnel, paternalistic, top-down attitudes and organizational ethos, inappropriate role models, outmoded curricula for the training of scientists and poor reward systems. Even where, as in Zambia, an institutional mechanism had been created to inject systems perspectives into commodity research, the necessary changes in attitudes and perspective among specialist researchers were found to take much longer than the simple mobilization of diagnostic teams.

There can be no doubt that if the numerous shortcomings identified by Collinson were set right, the prospects for introducing systems perspectives in public sector research would be improved. But would exclusively public sector models, such as those in Ethiopia, Kenya and Zambia, be institutionally sustainable even if all these conditions were met? The dwindling core funds for public sector research institutes today makes them increasingly dependent on donors. Further, if staff and operating budgets do have to be cut, the greatest threat is to new initiatives such as these, and not to the more established commodity or discipline-based research.

Demands on agricultural research have been growing over the last decade, at the same time a core of public research budgets have declined. These are not simply demands generated by farmers and those, such as NGOs, who work with them in the context of a strengthened civil society. They also include the demand for 'sustainability' of natural resource management, now invoked at every turn. Within cultivated land this generates new demands for systems-based understanding of such externalities as salinization and chemical pollution, and of how interventions relying on joint action, such as integrated pest management, can best be designed. However, demands for sustainability embrace not merely cultivated land but, increasingly, the management of common resources such as water, trees and pasture on land adjoining cultivated areas – again, often demanding group approaches. Agricultural researchers have long been aware of the need to draw on such resources in order to ensure, for example, adequate supplies of fodder and so contribute to the sustainability of agriculture in such areas as the mid-hills of Nepal. But they now have the added burden of devising management practices and technologies to ensure sustainable exploitation of this wider natural resource base. Inevitably, as demands of this kind broaden, the complexities of diagnosis increase, giving added impetus to the search for cost-effective methods.

This growing disparity between expectations and resources has created an environment receptive to innovations of two broad, related types:

- The growing confidence in farmers' own capacity to identify their needs, has generated an array of 'farmer participatory research' (FPR) methods[19], the diagnostic components of which have recently developed in ways which are often both rapid and participatory[20].
- It is becoming clear to governments that they no longer have the resources to meet the wide range of potential research demands from complex, diverse and risk-prone areas. There is therefore much talk, and the beginnings of action, on the creation of partnerships between government and private sector organizations, both commercial and non-profit.

3.3.5 FPR and PRA

As the early review by Farrington and Martin in 1988[21] notes, FPR has its intellectual origins in the traditions of action research. In practical terms, participatory methods were first applied to technical change in agriculture by NGOs and to special projects to support farmers in identifying the opportunities and constraints they faced in agricultural development, in meeting these needs themselves if possible, and with help from government services if not. FPR was initially conceived and applied by NGOs in this 'empowering' mode and many early examples drew on such conventional anthropological techniques as participant observation. Few, however, could be classed as 'rapid'.

At the same time, a parallel mode of enquiry was developing in the form of rapid rural appraisal[22], drawing on efforts to diagnose farmers' needs without conventional questionnaire surveys or frequent recording techniques, avoiding their costs and rigidities and the risk of results so old they miss a 'moving target'[23]. To reduce any tendency to be purely extractive, this model evolved into PRA, the origins of which were reviewed by Chambers in 1994[24].

Two general observations need to be made, and the first concerns the relationship between PRA and FPR. The large number of case studies generated by the two has led some to equate PRA with FPR and Extension (FPR-E). There is, however, a basic distinction. FPR-E is an *approach* to the development of technologies, embracing diagnosis, screening, testing and verification that meet farmers' needs. As such, it is equivalent to FSR-E, but utilizes a wider range of methods and relies on a wider range of institutional linkages. PRA is one set of methods, and has been used primarily as a diagnostic tool. While it undoubtedly has potential at the evaluation stage of the research cycle, this remains largely unexploited. Importantly, it has little to offer at the experimentation stage. Efforts to ensure stronger farmer control over the experimentation process have, for instance, led to innovations such as the participatory varietal selection (PVS) highlighted by Witcombe and Joshi in 1995[25]. Techniques such as PVS are based on semi-structured interaction with farmers over one or more seasons in which their views on the design and management of trials, and their criteria for assessment of the results, are elicited. Much of the same kind of interaction had long been used in the variety of roles played by researchers and farmers in the joint conduct of on-farm research (OFR). A common theme of the numerous manuals on OFR, many of which predate PRA, has been the search for ways of increasing farmers' control over experimentation. A wide range of techniques for farmer-to-farmer extension are being discussed which, again, are distinct from PRA. These examples demonstrate that, while PRA has an important contribution to make to the understanding of farmers' needs, the concept and practice of FPR-E within a systems context goes beyond the group of methods embraced by PRA.

The second observation concerns the differing objectives, ethos and capacities of NGOs and government research and extension services. As noted, early approaches to participation were introduced by NGOs as part of a broad aim to *empower* rural communities. By contrast, government research and extension services view participation in a *functional* context – as a means of enhancing efficiency in the design and uptake of new technologies. There are, of course, overlaps and intermediate positions: the better-resourced government services may engage in more empowering approaches, some 'technology-focused' NGOs may be concerned more with functional than empowering participation, and increases in income generated by functional participation may lead to empowerment. But the broad differences in philosophy and mandate remain and set important limits on the types and depth of participatory approach that might reasonably be expected in government services.

3.3.6 Issues and prospects

For this reviewer, there are four issues in diagnostic methods that will occupy centre stage in the next decade:

- The role of PRA methods.
- The changing role of researchers as FPR gains ground.
- Multi-agency approaches.
- Expansion pathways.

The role of PRA

PRA has powerfully demonstrated the ability of village households to contribute to rural development planning. It has also generated a sense of community ownership of development projects and processes, and a recognition among administrators that farmer participation enhances the prospects of success. However, it is increasingly being seen as a 'new orthodoxy', and, like all orthodoxy, it is attracting challenges of diverse kinds. One has to do with intellectual property, viz. the argument that a number of the methods it embraces predate the term 'PRA'. Another is that enthusiasm for methods has led many to ignore differences in objectives and in the comparative advantage of different kinds of organization. It should hardly be surprising that departments of agriculture are unwilling to become involved in the more empowering forms of participation, but much criticism for failing to do so is implicit in the calls for 'new professionalism'[26]. There is increasing concern that enthusiasm for photogenic – perhaps gimmicky – techniques is not being matched by adequate care in the basics of unbiased sampling and questioning skills. Mosse, writing in 1995[27], sees the need to develop understanding at several levels of the organizational and political contexts in which PRA is conducted. 'At every level', he says, 'knowledge building, need definition, prioritisation ... are shaped by social relations, not just within rural society, but within project teams, the organizational interests which they are constrained to serve and the political environments in which they work'.

There is certainly scope for further refining PRA techniques in monitoring and evaluation and, at the diagnosis stage, in combining them with techniques for consensus building[28]. But a growing concern is the understanding of roles and process, and, as Alsop *et al.* argued in 1996[29], user-friendly techniques have yet to be developed for these purposes.

The changing role of researchers

The greater involvement of farmers in both the diagnosis of needs and in decisions over which technical options to test raises questions over the future role of researchers. Is it sufficient, as some would argue, for them merely to act as facilitators in processes led by farmers and, by implication, to observe with satisfaction the successful adoption of new technologies? Or do they retain some wider role? Many would argue that the mandate of researchers remains national (in some cases, provincial) and that an important role for them in FPR should be to identify not simply *what* works, but *why* and *how* so that 'baskets of choices' potentially relevant to other areas can be assembled. Thus, there is both continuity and change in the role of researchers: they continue to be concerned with identifying what is potentially relevant to wider 'research domains', but now assemble information to feed into these more from on-farm observations than from on-station experiments.

Multi-agency approaches

It is argued above that governments will tend to adopt functional approaches to participation, in contrast with the empowering approaches of NGOs. Nowhere is the contrast starker than in the different approaches to technologies requiring 'joint action' by the two types of organization. Some farm activities, including various IPM techniques, require joint action if they are to be effective and the management of common pool resources can only be performed by farmer groups. However, the record of government staff in forming anything more than very temporary groups is poor[30].

NGO approaches are much smaller in scale, more time-consuming and more intensive, relying on the development of a sense of community identity, and of leadership, participation and conflict resolution skills. The groups formed with NGO support are intended to become self-sustaining, to address their own community needs where possible and to make demands on government services where necessary. These objectives are not always achieved, of course, but the needs in agriculture and natural resource management are identified and addressed in this context. A particular problem facing NGOs is their limited awareness of the range of technical options available to meet farmers' needs, and limited access to such options. Proposals now abound for multi-agency approaches seeking to combine the strengths of NGOs in needs assessment and group formation with those of government in developing technical solutions.

The views and mandates of government and NGOs diverge to some degree, and the area of shared aspirations is limited. This does not imply, however, that NGO ideologies will necessarily be 'tainted' if they work with government. A very wide range of interaction has been identified[31] and even the most empowerment-oriented NGOs need not find it difficult to make technical demands on government. A widespread problem is that government services have historically been driven more by an ethos of delivering programmes than of responding to demands. Techniques based on process documentation and process monitoring are now being developed to assess how far joint activities proceed in line with expectations, and to bring them back on track where necessary[32]. The further development of multi-agency approaches and methods of monitoring them will clearly be a key influence on the type and extent of future multi-agency approaches, including the articulation by NGOs of farmers' needs to government services.

Expansion pathways

At the risk of caricature, NGO approaches towards diagnostic methods may be characterized as 'deep but small-scale', and those of government services as 'large-scale but superficial'. The intensive, face-to-face nature of NGO approaches has made it difficult to spread the approaches, but the strengths of both types of institution might be combined. Multi-agency approaches are not the sole solution. Access to powerful new media (television, video, local radio) is expanding rapidly in many rural areas. With a little imagination, improved use of these could lead to advances in diagnosis in two ways:

- The capacity of these media to enhance farmers' awareness of technologies that have worked well in areas similar to their own. This may stimulate 'self diagnosis', visits to other areas and testing and adoption at minimal cost to government or NGO.
- Their capacity to convey, in farmers' own words, how they set about forming groups and pressed their demands on NGOs or government services could have important demonstration effects.

As noted, changes are needed on both sides before such approaches will become widespread. Many regard direct interaction between researchers and farmers' associations as the most sustainable arrangement for the long term. However, only in commercial crops have associations (usually commodity associations) proven strong enough to influence research agenda[33] and there are grave doubts over whether small farmer groups can be created and sustained in semi-subsistence contexts without long-term external support, possibly by NGOs[34]. Moreover, even in the north, the extent to which farmers' organizations have been willing to finance research in the face of the withdrawal of public funding has been limited[35]. Further options for change include the following:

- Where public sector research and extension services are strong, they can be expected to introduce participatory appraisal on a wide scale, providing that appropriate reward structures and funding mechanisms are introduced.
- In some settings, farmers' associations may be able to articulate their members' requirements to research organizations.
- In others, NGOs may support local membership organizations in identifying or expressing needs, or may do so on their behalf.

Overall, what is clear is that there will be no single pattern for the introduction and wide implementation of improved diagnostic methods in the future. Local agroecological and socioeconomic settings will increasingly determine the choice of methods, and institutional configurations will have to be tailored to local settings.

REFERENCES

1. **Byerlee, D. & A. Siddiq, 1994.** Has the Green Revolution been sustained? The quantitative impact of the seed–fertiliser revolution in Pakistan revisited. World Development, 22(9), 1345–61.
2. **Norman, D.W., 1974.** Rationalising mixed cropping under indigenous conditions: the example of northern Nigeria. Journal of Development Studies, 11, 3–21.
3. **Collinson, M.P., 1972.** Farm Management in Peasant Agriculture. Praeger, New York.
4. **Ruthenberg, H., Ed., 1968.** Smallholder Farming and Smallholder Development in Tanzania. Weltforum Verlag, Munich.
5. **Mellor, J.W., 1966.** The Economics of Agricultural Development. Cornell University Press, Ithaca, NY.
6. **Gilbert, E.H., D.W. Norman & F.E. Winch, 1980.** Farming Systems Research: A Critical Appraisal. MSU Rural Development Paper no. 6. Michigan State University, East Lansing.
7. **ISRA, 1977.** Recherche et Développement Agricole: les Unités Experimentales du Sénégal. Institut Sénégalais du Recherche Agronomique, Dakar.
8. **Hildebrand, P.E., 1981.** Combining disciplines in rapid rural appraisal: the 'Sondeo' approach. Agricultural Administration, 8(6), 423–32.
9. **Collinson, M.P.,1972.** (Op. cit.).
10. **IDS, 1979.** Rapid Rural Appraisal. A conference held at the Institute of Development Studies, December. Mimeo Proceedings. Institute of Development Studies, Brighton, UK.
11. **Rhoades, R.E. & R.H. Booth, 1982.** Farmer-back-to-farmer: a model for generating acceptable agricultural technology. Agricultural Administration, 1, 127–37.
12. **Chambers, R. & B.P. Ghildyal, 1985.** Agricultural research for resource-poor farmers: the farmer-first-and-last model. Agricultural Administration, 20, 1–30.
13. **Chambers, R. & J. Jiggins, 1986.** Agricultural research for resource-poor farmers: a parsimonious paradigm. IDS Discussion Paper 220. IDS, University of Sussex.
14. **Chambers, R., A. Pacey & L.A. Thrupp, 1989.** Farmer First: Farmer Innovation and Agricultural Research. IT Publications, London.
15. **Galt, D.L. & S.B. Mathema, 1987.** Farmer Participation in Farming Systems Research. FSSP Networking Paper no. 15.
16. **Kean, S. & L.P. Singogo, 1968.** Zambia: Organisation and Management of the Adaptive Research Planning Team (ARPT), Research Branch, Ministry of Agriculture and Water Development. OFCOR Case Study no. 1. ISNAR, The Hague.
17. **ICAR, 1988.** Report of the ICAR Review Committee. Indian Council for Agricultural Research, New Delhi.
18. **Collinson, M.P., 1988.** The development of African farming systems: some personal views. Agricultural Administration and Extension, 29(1), 7–22.
19. **Farrington, J. & A.M. Martin, 1988.** Farmer Participation in Agricultural Research: a review of concepts and practices. Agricultural Administration Unit Occasional Paper no. 9. Overseas Development Institute, London.
   Okali, C., J. Sumberg & J. Farrington, 1994. Farmer Participatory Research: Rhetoric and Reality. IT Publications, London.
20. **Chambers, R., *et al.*, 1989.** (Op. cit.).
   Scoones, I. & J. Thompson, 1994. Beyond Farmer First. IT Publications, London.
21. **Farrington, J. & A.M. Martin, 1988.** (Op. cit.).

22. **IDS, 1979**. (Op. cit.).
23. **Maxwell, S., 1986**. Farming systems research: hitting a moving target. World Development, 14(1), 65–77.
24. **Chambers, R., 1994**. The origins and practice of participatory rural appraisal. World Development, 22(7), 953–69.
25. **Witcombe, J. & A. Joshi, 1995**. Farmer Participatory Approaches for Varietal Breeding and Selection, and Linkages to the Formal Seed Sector. Paper presented at the IDRC Workshop on Participatory Plant Breeding at the International Agriculture Centre, Wageningen, 24–30 July, 1995.
26. **Pretty, J.N. & R. Chambers, 1993**. Towards a Learning Paradigm: New Professionalism and Institutions for Agriculture. IDS Discussion Paper no. 334. Institute of Development Studies, Brighton.
27. **Mosse, D., 1995**. 'People's Knowledge' in Project Planning: The Limits and Social Conditions of Participation in Planning Agricultural Development. Agricultural Research and Extension Network Paper no. 58. Overseas Development Institute, London.
28. **Warner, M., 1997**. 'Consensus' Participation: an Example for Protected Areas Planning. Public Administration and Development, 17(4), 413–32.
29. **Alsop, R.G., R. Khandelwal, E.H. Gilbert & J. Farrington, 1996**. The human capital dimension of collaboration among government, NGOs and farm families: comparative advantage, complications and observations from an Indian case. Agriculture and Human Values, 13(12), 3–12.
30. **Chowdhury, M.K. & E.H. Gilbert, 1996**. Reforming Agricultural Extension in Bangladesh: Blending Greater Participation and Sustainability with Institutional Strengthening. Agricultural Research and Extension Network Paper no. 61. Overseas Development Institute, London.
31. **Farrington, J. & A.M. Martin, 1988**. (Op. cit.).
32. **Alsop, R.G. *et al.*, 1996**. (Op. cit.).
33. **Bebbington, A., D. Merrill-Sands & J. Farrington, 1994**. Farmer and Community Organisations in Agricultural Research and Extension: Functions, Impacts and Questions. Agricultural Research and Extension Network Paper no. 47. Overseas Development Institute, London.
    Carney, D., 1996. Formal Farmers' Organisations in the Agricultural Technology System: Current Roles and Future Challenges. Natural Resources Perspectives, no. 14. Overseas Development Institute, London.
34. **Mosse, D., 1996**. Local Institutions and Farming Systems Development: Thoughts from a Project in Tribal Western India. Agricultural Research and Extension Network Paper no. 64. ODI, London.
35. **Pollock, C., 1995**. Reconciling the Irreconcilable? The Effective Management of Agricultural Research. Paper presented to the ICAR/ODA Workshop on Research for Rainfed Farming, September 1995 at Central Research Institute for Dryland Agriculture, Hyderabad, India.

3.4 GENDER ANALYSIS: MAKING WOMEN VISIBLE AND IMPROVING SOCIAL ANALYSIS

Hilary Sims Feldstein

Development specialists have realized that ignoring women in the development process has meant ignoring a positive force for change. Increasing women's productivity has substantial payoffs for family livelihoods, including the better health and education of the next generation.

3.4.1 Introduction

The impact of FSR on agricultural research has been twofold. It showed that:

- Agricultural production was one part of a complex system guiding farmers' choices.
- Farmers had a sophisticated understanding about their current system and the costs and benefits of new technology, including specific preferences which governed its acceptability.

Initially, researchers saw technology as 'neutral'. They came to understand that acceptance could be size biased and that the constraints of low resource farmers differed from those of larger landholders. FSR focused attention on farmers and their diverse circumstances. However, a blind spot remained. While technology was seen as size biased, it was not seen as gendered. Too frequently, learning about and from farmers meant learning about and from male farmers. Women farmers and their knowledge and preferences about agricultural production were virtually invisible. Early guidelines to FSR did not include the analysis of gender-disaggregated data as a source of useful insights for scientists. Most FSR practitioners considered the household a 'black box' in which resources, responsibilities and benefits are equally (or optimally) shared. They assumed that one individual, usually male, could speak for all.

3.4.2 The importance of gender differentiation

In the 1930s and 1940s, a number of anthropologists, trained in careful observation of the daily lives of those they studied, detailed the productive activities of women and men[1]. But this kind of social analysis did not, in general, carry forward into FSR. Boserup's 1970 study, 'Women's Role in Economic Development'[2], reopened the case. She stressed the important contribution made by women to agricultural production, drawing attention to their different role according to population density, the intensity of production and the availability of hired labour. Modernization and intensification of agriculture would mean increased labour by both men and women. She pointed out that shared labour did not necessarily mean shared benefits; men were more likely to predominate with their access to markets and new technologies. Indeed, women's income and position could be at risk.

Beginning in 1975, the launch of the Decade for Women, researchers associated with US land grant universities working in developing countries provided abundant examples of women's roles in agriculture, including Horn and Nkambule-Kanyima in 1984[3]. Women often had exclusive responsibility for particular crops, livestock or operations. Their roles and preferences were sometimes complementary, sometimes parallel and sometimes in conflict with those of men[4]. Case studies from Indonesia showed how improved technologies could actually harm women's opportunities and welfare. The introduction of rice mills operated generally by men, replaced the hand pounding of rice carried out by women. The result was the loss of employment and income by poor women labourers.

In an early study of irrigated rice production in Gambia, Jennie Dey described how the assignment of irrigated plots to men transferred the ownership of rice production and its income from women to men[5]. In 1986, Christine Jones's widely circulated study in Cameroon showed how failing to take women's interests into account led to serious miscalculations in expected productivity and economic returns[6]. Both demonstrated that efficiency and equity were at risk when research or development failed to take account of women's roles in agriculture. There was increasing evidence that households were complex decision-making entities[7] and this sparked several efforts to combine improved understanding of household dynamics with FSR.

3.4.3 Strategies in the introduction of gender analysis into FSR

The inception and development of attention to gender issues within FSR are like a braid, made up of several strands that overlap and strengthen each other. These strategies were frequently opportunistic, building on existing openings and developing momentum. They fall largely into four categories:

- Conferences bringing together those studying intra-household decision making (principally anthropologists), and social scientists engaged in FSR (mainly economists)[8].
- The development of methods for use by FSR-E researchers to gain a better understanding of gender roles, and incorporation of these methods into the conduct of FSR-E[9].
- Specific studies to show how FSR could incorporate an intra-household perspective with useful results.
- Conferences and workshops that gave visibility to individual projects where knowledge of women's role had led to improved project outcomes[10].

These legitimized gender analysis and became the vehicles for the discussion and development of reliable gender-sensitive research methods. Three of the main training programmes for FSR were: CIMMYT's East African Economics Program based in Nairobi, Farming System Support Project's (FSSP) assistance to FSR projects funded by USAID, and the IRRI Farming Systems Training Program. Each of these came to accept the legitimacy of gender analysis and incorporated it into its training programmes. Looking back, the perseverance of Kate McKee of the Ford Foundation in identifying and funding different opportunities for furthering this issue was crucial. The Ford Foundation provided long-term support to the Intra-Household and FSRE Case Studies Project (IHH/FSR-E) from 1984 to 1994 and the Women in Rice Farming Systems (WIRFS) network from the mid 1980s to the mid 1990s – the two most persistent efforts to develop methods and engage FSR researchers.

The award of a large technical assistance and training contract to the FSSP to the University of Florida (UF)

This provided the means for stronger links between US land grant universities and FSR-E. Research by the Population Council of New York City showed that where women had a stronger bargaining position, fertility appeared to drop and women's own productive activities significantly enhanced their positions. Judith Bruce, with the support of the Ford Foundation, began a series of studies to help development planners understand women as producers. Constantina Safilios-Rothschild conducted studies comparing women's involvement in agricultural production with their invisibility in official figures. And a group from Harvard constructed a conceptual framework for a set of case studies for the World Bank and USAID, now known as the Harvard framework[11].

WIRFS Conference at IRRI

The WIRFS Conference at IRRI, supported by the Ford Foundation, was an early effort to explore the links between women and FSR. The meeting gave birth to the IHH/FSR-E Case Studies Project. Shortly after the conference, IRRI created the WIRFS Network, also funded by Ford and led by Gelia Castillo, Professor at the University of the Philippines at Los Baños, and Virgilio Carangal, head of the Asian Rice Farming Systems Network (ARFSN). The Women in Agricultural Development (WIAD) programme began to meet on an informal basis at the UF[12].

3.4.4 Milestones and achievements

1984

Joyce Moock at the Rockefeller Foundation and Kate McKee from Ford organized the Bellagio Conference on Understanding Africa's Rural Households and Farming Systems to address gender issues with the leading farming systems practitioners[13]. This conference provided the impetus to find practical tools for incorporating intra-household and inter-household analysis into FSR. The IHH/FSR-E Case Studies Project was initiated, led by Hilary Sims Feldstein and Susan Poats and, again, funded by the Ford Foundation. The project developed seven training case studies integrating intra-household considerations and FSR[14] and a handbook on gender-sensitive research methods[15]. The project also organized 'gender methodologies panels' at the annual FSR-E meetings.

1985

The Bellagio Seminar on Women and Agricultural Technology: Relevance for Research, sponsored by the Rockefeller Foundation and the International Service for National Agricultural Research (ISNAR), brought together the 12 international centres of the CGIAR[16]. The Conference agreed that gender was an important variable in agricultural research and that centres should recognize this by linking gender directly to technology generation and use.

The WIRFS Network was formed. At its first workshop, the priority was to build some regional experience. The network leader, Thelma Paris, an associate scientist at IRRI, used a judicious mixture of conferences, small research grants, individual mentoring and training, to encourage scientists from national programmes to incorporate gender analysis into their FSR.

1986

The monograph by Janice Jiggins in the CGIAR study on gender-related impacts and the work of the IARCs highlighted the accumulating evidence of women's substantial contribution to agricultural production and to household nutrition[17]. Jiggins was critical of the Centres' persistent gender bias that contributed to the invisibility of women in FSR programmes. The

UF/WIAD Conference on Gender Issues in Farming Systems Research and Extension took place with the support of three departments at UF, and the Ford and Rockefeller Foundations. Over 350 participants from all regions of the globe attended[18].

1987

The CIMMYT Networkshop on Household Issues and FSR was organized by Allen Low, Alistair Sutherland, Feldstein and Poats in Lusaka, Zambia, in April, 1987. Zambia was already a leader among national programmes in taking steps to incorporate a gender perspective in its research process. Biological and social scientists from nine national programmes spent 5 days discussing the IHH/FSR-E Zambia case, presenting papers and visiting the field[19]. This laid the groundwork for the incorporation of gender concerns into CIMMYT's East and Southern Africa FSR training and a wider appreciation of these concerns in general.

1988

The WIRFS conference, IRRI brought together national researchers from seven south-east Asian countries. Some, with small grants from the 1985 meeting, had applied a gender perspective to their ongoing farming systems work. For others, this workshop was their first exposure to gender analysis, widening the reach of the WIRFS programme in south-east Asia. In September, Feldstein and Flora provided a day of training on gender and FSR-E for the International Development Research Center (IDRC), reaching another major set of FSR actors. Tangible evidence that the argument for gender analysis had won recognition as good social analysis came when the very FSR practitioners who had baulked at using the words 'women' and 'gender' at the inception of the Case Study Project, suggested that gender issues be a subtheme of the 1988 conference.

1989

IDRC published the monograph 'The Gender Variable in Agricultural Research' by Feldstein, Flora and Poats in English, French and Spanish and gave it wide free distribution[20]. Later in the year 'Working Together: Gender Analysis in Agriculture', by Feldstein and Poats, was published[21]. Gender became a subtheme at the annual FSR-E conference, and sessions on intra-household analysis became the best attended.

1990

Attending the WIRFS Conference in Bogor, Indonesia, were senior agricultural administrators from India, Indonesia, Nepal, the Philippines and Thailand. Using the GAP conceptual framework, the researchers who had received small grants and training from WIRFS in 1988 presented their analyses of the sites in which they were working. The administrators' reaction was very positive and they endorsed a continuation of the WIRFS project and a request that IRRI do more to train their national scientists in gender analysis. IRRI subsequently incorporated gender analysis into its farming systems training as well as providing a separate course on gender analysis. Through the sustained commitment of IRRI and the Ford Foundation, the leading donor, the programme flourished for 10 years. During this period WIRFS worked with 23 organizations from National Agricultural Research Systems (NARS) in nine south and south-east Asian countries. WIRFS also began to have an impact on technical research conducted at IRRI[22].

By 1990 FSR was losing favour with donors but gender as a development variable was receiving more attention. In 1991, CGIAR launched its Gender Program. Beyond this, Agenda 21 explicitly discussed women's reliance on the environment for their livelihoods and the importance of involving women in sustainable natural resource management (NRM). The focus on NRM has provided a new venue for applying the principles of FSR-E and with it gender analysis, but also raises new issues.

3.4.5 Evolution of methods for collecting gender-sensitive and disaggregated data

The rubric, gender analysis, is often used to refer to both the conceptual framework, used for analysis and planning, and a range of methods for the collection of gender disaggregated data from both men and women farmers. The conceptual framework for gender analysis in FSR-E is an aid to understanding the pattern of roles and responsibilities in the farming system. The framework focuses the gathering of data and its analysis on four key areas: activities, resources, benefits and inclusion. The central question is 'Who does what?' Data is gender disaggregated in its collection and its analysis and information is gathered, arranged and compared on the activities of women and men farmers.

One of the major contributions of the framework is the use of a gender-disaggregated farming systems calendar. This shows when the labour constraint is most severe, and what and whose tasks would be affected by any changes in the farming system[23]. In Burkina Faso, for example, a linear programming of sex-specific labour inputs explained the limited extent of adoption of tied ridges, a technology for moisture conservation. Women's labour was constrained by their household responsibilities or involvement in 'own account' enterprises such as beer making or working in their own fields[24].

To predict the availability of other resources, researchers use the framework to analyse data on men's and women's access to, and control of, resources and the benefits from production. By comparing the costs and benefits of technologies by gender, researchers can anticipate the likely constraints to uptake. An analysis of benefits also includes an understanding of the shared and differing preferences of men and women for specific traits in a crop. Finally, monitoring the inclusion of women and men in the different stages of FSR enhances researchers' awareness of points at which gender-specific knowledge can usefully be brought to bear. While the focus of the framework is on gender and age, other important variables – ethnicity, class, life cycle stage – could also be incorporated into the analysis.

When the IHH/FSR-E Case Studies Project began, it followed the current methodology of FSR-E – diagnosis through informal and formal surveys, planning and design, on-farm experimentation, evaluation and dissemination. Each of these stages needed an expansion of methodology to ensure the collection of data about women as well as men. Attention to gender in informal diagnostic surveys means widening both the kinds of informants (more women) and the key questions. In Botswana and Zambia, researchers included female heads of household among those surveyed. In Sta Barbara in the Philippines, women members of joint households were interviewed separately in a resurvey that provided a better description of the farming system and respective roles. By interviewing women not included in the original surveys, researchers discovered that the processing of glutinous rice was an important source of household income. This predominantly women's activity became a fertile area for experimentation with improved technologies, from new varieties grown by men, to new post-harvest machinery used by women[25]. In Peru, work in the community revealed, unexpectedly, that women, rather than men, were responsible for the care of animals. Through discussion with the women, scientists learned that they did not use chemical dipping because of its high cost, and that there were local treatments for parasites. Scientists helped to organize a production research group to conduct trials based on local plants and chemical dipping. A local leaf proved to be as effective as the chemical, but it was scarce. This prompted the women to begin plant multiplication to overcome this limitation[26].

On-farm trials often required changes better to capture the practices and assessments of both women and men[27]. In Botswana and Zambia, female heads of households were also collaborators in on-farm trials and researchers held separate field days to ensure they heard the candid opinions of women farmers. In Zambia[28], researchers collected and analysed gender disaggregated data which proved particularly useful in technology surveys, as Sutherland wrote in 1994: 'For example, in an assessment of intercropping in Lusaka Province, male and female respondents saw different advantages and disadvantages. ... Certain treatments were ranked high only by the women respondents. This helped in the design and targeting of subsequent on-farm experiments'[29].

Including women as sources of information and partners in FSR continues to present methodological challenges. In many societies, men dominate the 'public domain'; women may not be present and when they are present they do not speak. Local custom may prevent men from questioning them individually. Group interviews and the use of more women field researchers have helped to overcome this constraint. Researchers may have to shell beans with women in the kitchen[30] or engage with separate focus groups[31] to hear women's voices. Understanding the general pattern of the roles of women and men may require very specific questions about the details of particular production activities.

In the late 1980s, FSR researchers and others began to expand the tool kit for learning from farmers to include mapping, focus groups, transepts and matrix ranking. Though still uneven in practice, these techniques prompted a shift away from extracting data to design solutions back at the research station towards more fully incorporating farmers in decision-making about what should be tried and how to test it. To use these tools, scientists concerned with gender and other variables have made adaptations to ensure that the voices of women and men are heard. This is especially true in the work of the Ecology, Community Organization, and Gender (ECOGEN) project at Clark University, the Sustainable Agriculture and Natural Resource and Environment Management Collaborative Research Support Project (SANREM CRSP), and UF's project on Managing Ecologies and Resources with a Gender Emphasis (MERGE).

As the participatory tools proliferate, adaptation to include gender analysis has two important elements. The first, when undertaking community mapping or transepts, is to ask questions about each element of the landscape, its different uses, ownership, and who has responsibility for each enterprise[32]. Activities, resources and benefit analyses can be conducted using cards with drawings of men, women and children and different enterprises. Using such cards, a group (all men, all women, or mixed) can collectively sort out men's and women's roles and responsibilities for each enterprise[33]. Participants can explain benefits by breaking down plants by their many products and the ownership, responsibility and beneficiary of each[34].

Second, to ensure that women's opinions are registered, researchers should conduct interviews with sensitivity. They must be careful about how questions are worded and what questions are asked, focusing on the areas where they know the women to be expert, for example, or conduct separate interviews keeping in mind that all women are not the same. These same principles may apply to groups differentiated by other variables such as class or ethnicity or age. 'Questions of Difference', a training video produced by Irene Guijt of the International Institute for Environment and Development (IIED) shows adaptations of participatory methods and the value of information from women to NRM[35].

3.4.6 Evidence on making a difference: efficiency, welfare, equity and empowerment

A question often asked is 'Does it make a difference?' Does understanding the gender dimension in division of labour, access and control of resources, responsibility and decision-making make a difference? The application of gender analysis to agricultural research is new, but there are now many examples of where it has made a difference. The answer to this question depends on the purpose to which the analysis is put. Because women's and men's different roles and interests may weigh heavily in the adoption of improved technologies, FSR leaders now recognize that gender differentiation is 'good' FSR.

Efficiency

One reason for its use is efficiency. In Rwanda, for example, scientists learned from on-farm trials that it was the women who were the predominant bean growers. In an experiment on farmer evaluation, women bean experts came on-station to discuss 21 varieties in the final stages of selection with breeders. This was much earlier in the breeding cycles than farmer evaluation of the few varieties moved forward into on-farm trials. The women experts selected for very specific traits; for planting where there is wind, for planting near bananas, and three or four varieties to plant in their own on-farm experiments. Their selections outperformed their own check varieties more frequently and by a greater margin than had breeders' selections in similar trials[36].

Gender analysis is used for an *ex ante* assessment of the fit of the technology into the farming system. It helps identify the appropriate collaborators for specific operations or enterprises. Gender analysis also helps identify who has local and practical knowledge about different parts of the farming system and of the natural resources upon which it draws. As in the

Sta Barbara case, this approach can lead to identifying overlooked areas where researchers' knowledge and experimentation can make a real contribution.

Welfare and equity

Gender analysis makes a difference in addressing issues of welfare and equity. It can be used *ex ante* to anticipate whether women's conditions are worsened or improved by a new technology. Concerns about displacing female hired labour have featured in the research by the International Crops Research Institute for the Semi-Arid Tropics (ICRISAT) on herbicides and have influenced the decision by the International Center for Agricultural Research in the Dry Areas (ICARDA) in developing a lentil harvester[37]. In Zambia, women's concerns about losing control of their crop negated a proposed intercropping of women's beans with men's maize[38]. Used more proactively, gender analysis provides a means for identifying those specific activities by women where research or extension can produce tangible benefits. For subsistence farmers, research to reduce weed populations would reduce women's drudgery and IRRI has engineered several machines scaled for women and women's tasks in production and post-harvest processing. In both cases, what is drudgery reducing in the case of subsistence farming may be labour displacing on larger farms. Such situations require sensitive analysis and exploration of options. The equity orientation is currently receiving increased attention. Development specialists have realized that ignoring women in the development process has meant ignoring a positive force for change. Increasing women's productivity has substantial pay-offs for family livelihoods, including the better health and education of the next generation[39].

Empowerment

A third area is women's empowerment, strengthening their ability to make choices about their own livelihoods. Researchers working with sheep in the Andes, where livestock were women's responsibility, helped form a women's sheep association to discuss research needs and to design trials. This forum became the vehicle for discussing many other issues and to formalization of the group as a place to discuss women's and community affairs[40]. In Guimba in the Philippines, an improved rice mill reduced costs and time for processing rice. The arrangements by women to run the mill became the basis for a women's association which eventually gave them a stronger voice in community affairs[41].

Whatever the intention – efficiency, equity or empowerment – gender analysis provides a 'map' for identifying men's and women's roles and helps improve the choice and design of improved technologies.

3.4.7 Continuing challenges

The application of FSR to NRM

NRM shifts the emphasis from the farm household to the community or communities in a larger landscape. Several elements may limit attention to gender: more complexity; more and different kinds of stakeholders, and, at the public community or intercommunity level, women may be invisible. While gender may not be the dominant variable in differentiating stakeholders, it is still important.

The role of women researchers

Male researchers and administrators often claim that adding women to the team will 'take care of the gender question'. A woman researcher may find it easier to reach rural women and her own experiences may help her understand their circumstances. However, being a woman does not automatically qualify one to carry out gender analysis. This is a skill that must be learned, by women as well as men.

The circumstances for women professionals

More professional women come to gender analysis workshops than to any other kind of FSR forum. For many, such a workshop is the first event of its kind that they have attended. Away from the main proceedings, they discuss the difficulties of their workplaces. Some are wholly accepted as members of the team, but many are ignored, belittled, sexually harassed and find their chances of career advancement extremely limited. The workplace rarely takes account of the greater time constraints on women than men as women continue to bear the greater responsibility for household production. There is a disturbing parallel between women's treatment in an organization and the seriousness with which the organization addresses the needs of women as clients.

The capacity to implement gender-sensitive research

The comparative advantage for gender analysis in FSR is with national programmes and NGOs, but with decreasing funds for agricultural research and development, the capacity for FSR is also shrinking. Some national programmes are now making a concerted effort to build gender analysis into their research, but many lack the resources or the commitment.

Resistance from men

However reasoned the argument for using gender analysis to improve the efficiency of research, many male scientists are still inclined to not listen, or to laugh or to ignore. Often the question 'Will it make a difference to what I am working on?' reflects their reluctance even to read the basic texts and consider their application to their own work. Their own socialization on women's roles at home often makes it difficult for them to take women seriously as colleagues or as sources of information. There are now a number of men who speak out forcefully for gender analysis and the importance of addressing women's needs, including the editor of this book. But in the long run it is practice and interventions at the field level which make a difference[42]. Given entrenched habits and attitudes, changes are slow.

3.4.8 Conclusion

Fifteen years of work in gender analysis has paid off in the increasing legitimacy and use of gender analysis in FSR-E and other agricultural research. Furthermore, gender analysis, by its focus on different kinds of farmers and their interactions, has made anthropological insights more visible in FSR, and has done so in a way which adds clarity to the presentation. A number of national programmes and NGOs in Africa and Asia are giving more attention to women as farmers and as fellow scientists. Women themselves are increasingly speaking out on their specific needs for technology, and the credit and extension services that enhance their productivity. However, those committed to excellence and equity in FSR will need to be persistent in ensuring that gender is fully considered in technology development.

REFERENCES

1. **Richards, A.I., 1961.** (Reprint; original 1939) Land, Labour and Diet in Northern Rhodesia. Oxford University Press, London.
2. **Boserup, E., 1970.** Women's Role in Economic Development. St Martin's Press, New York.
3. **Horn, N. & B. Nkambule-Kanyima, 1984.** Resource Guide: Women in Agriculture, Botswana. The Bean/Cowpea Collaborative Research Support Program, 1 October 1984. Michigan State University, Michigan.
4. **Cloud, K., 1985.** Women's Productivity in Agricultural Systems: Considerations for Project Design. In Overholt *et al.*, (Eds) Gender Roles in Development Projects. Kumarian Press, West Hartford, Connecticut.
5. **Dey, J., 1981.** Gambian Women: Unequal Partners in Rice Development Projects? In: Nelson, N. (Ed.) African Women in the Development Process. Frank Cass, London.
6. **Jones, C.W., 1986.** Intra-household bargaining in response to the introduction of new crops: a casestudy from North Cameroon. In: Moock (Op. cit.).
7. **Guyer, J.I., 1980.** Household budgets and women's incomes. Paper presented at the Symposium on Women in the Work Force, American Anthropological Association Meetings, 1979.
8. **Cloud, K., 1985.** (Op. cit.).
9. (Ibid.)
   Feldstein, H. Sims & S.V. Poats (Eds), 1989. Working Together: Gender Analysis in Agriculture, Volume 1. Kumarian Press, West Hartford, CT.
   Feldstein, H. Sims & J. Jiggins (Eds), 1994. Tools for the Field: Methodologies Handbook for Gender Analysis in Agriculture. Kumarian Press, West Hartford, CT.
   Thomas-Slayter, B., A. Lee Esser & M. Dale Shields, 1993. Tools of Gender Analysis: A Guide to Field Methods for Bringing Gender into Sustainable Resource Management. Clark University, Worcester, MA.

10. **Ashby, J., 1989**. Production and Consumption Aspects of Technology Testing in Pescador. In H. Sims Feldstein & S.V. Poats (Eds) Working Together: Gender Analysis in Agriculture, Volume 1. Kumarian Press, West Hartford, CT.
Paris, T.R., 1989. Philippines: Women in Rice Farming Systems, Crop–Livestock Project, Sta Barbara, Pangasinan. In: H. Sims Feldstein & S.V. Poats (Eds) Working Together. (Op. cit.).
Shinawatra, B., 1988. The incorporation of women's concerns in the design and impact assessment of a FSR/E project: A case of Northern Thailand. Paper presented at the 1988 Farming Systems Research and Extension Symposium, Fayetteville, Arkansas, 9–12 October, 1988.
11. **Overholt, C., M.B. Anderson, K. Cloud & J.E. Austin (Eds), 1985**. Gender Roles in Development Projects. Kumarian Press, West Hartford, CT,
12. **Poats, S.V., M. Schmink & A. Spring (Eds), 1988**. Gender Issues in Farming Systems Research and Extension. Westview Press, Boulder, CO.
13. **Moock, J. Lewinger (Ed.), 1986**. Understanding Africa's Rural Households and Farming Systems. Westview Press, Boulder, CO.
14. **Feldstein, H. Sims & S.V. Poats (Eds), 1989**. (Op. cit.).
15. **Feldstein, H. Sims & J. Jiggins (Eds), 1994**. (Op. cit.).
16. **Rockefeller Foundation and ISNAR, 1985**. Women and Agricultural Technology: Relevance for Research. Volume 1: Analyses and Conclusions. Volume 2: Experiencies in International and National Research. Report for the CGIAR Inter-Center Seminar on Women and Agricultural Technology. Bellagio, Italy, 25–29 March, 1985. The Hague, Netherlands.
17. **Jiggins, J., 1986**. Gender-related impacts and the work of the international agricultural research centers. CGIAR Study Paper No. 17. The World Bank, Washington, DC.
18. **Poats, S.V., M. Schmink & A. Spring (Eds), 1988**. (Op. cit.).
19. **Sutherland, A. (Ed.), 1987**. Report on a Networkshop on Household Issues and FSR. Lusaka, Zambia. Networking Workshop Report, No. 10. CIMMYT, Harare.
20. **Feldstein, H. Sims & S.V. Poats (Eds), 1989**. (Op. cit.).
21. (Ibid.).
22. **CGIAR Gender Program, 1994**. Partners in Selection: Women Bean Experts and Bean Breeders. Consultative Group on International Agricultural Research, Washington, DC.
Chater, S. & V. Carangal, 1996. On Farmers' Fields: Portrait of a Network. Manila: IRRI.
Paris, T., 1995. Women in Rice Farming Systems Network: Progress Report, 1991–95 (mimeo).
23. **Burfisher, M.E. & N.R. Horenstein, 1985**. Sex Roles in the Nigerian TIV Farm Household. Women's Role and Gender Differences in Development. Kumarian Press, Hartford, CT.
Feldstein, H. Sims & S.V. Poats (Eds), 1989. (Op. cit.).
24. **Nagy, J.G., H.W. Ohm & S. Sawadogo, 1989**. Burkina Faso: a Case study of the Purdue University Farming Systems Project. In: H. Sims Feldstein & S.V. Poats (Eds) Working Together. (Op. cit.).
25. **Paris, T.R., 1989**. (Op. cit.).
26. **Fernandez, M.E., 1994**. Women's Agricultural Production Committees and the Participative–Research–Action Approach. In: H. Sims Feldstein & J. Jiggins (Eds) Tools For the Field. (Op. cit.).
27. **Poats, S.V., H.S. Feldstein & D. Rocheleau, 1989**. Gender and Intra-Household Analysis in On-Farm Research and Experimentation. In: R. Wilk (Ed) The Household Economy, Reconsidering the Domestic Mode of Production. Westview Press, Boulder, CO.
28. **Baker, D., 1994**. Women and Trials Management in Botswana: Experiences with Farmer Groups. In: H. Sims Feldstein & J. Jiggins (Eds) Tools For the Field: Methodologies Handbook for Gender Analysis in Agriculture. Kumarian Press, West Hartford, CT.
29. **Sutherland, A., 1994**. Incorporating gender into a national programme of FSR. In: Feldstein, H. Sims & J. Jiggens (Eds) Tools for the Field. (Op. cit.).
30. **Ashby, J,. 1989**. (Op. cit.).
31. **Flora, C. Butler, 1994**. Using focus groups with rural women. In: H. Sims Feldstein & J. Jiggins (Eds) Tools For the Field. (Op. cit.).
Rocheleau, D.E., 1994. Investigating contradictions and mysteries. In: Feldstein & Jiggens (Op. cit.).
32. **Thomas-Slayter, B., A. Lee Esser & M. Dale Shields, 1993**. (Op. cit.).
Lightfoot, C., S. Feldman & M. Zainul Abedin, 1991. Households, Agroecosystems and Rural Resources Management. Manila: Bangladesh Agricultural Research Institute and International Center for Living Aquatic Resources Management.

33. **Narayan, D. & L. Srinivasan, 1995**. Participatory Tool Kit. The World Bank, Washington, DC.
34. **Buenavista, G. & C. Butler Flora, 1993**. Surviving Natural Resource Decline: Explaining Intersections of Class, Gender and Social Networks in Agbanga, Leyte, Philippines. An ECOGEN Case Study. Virginia Polytechnic Institute and State University, Blacksburg, VA.
Thomas-Slayter, B., A. Lee Esser & M. Dale Shields, 1993. (Op. cit.).
35. **Guijt, I., 1995**. Questions of Difference: PRA, Gender, and Environment. A Training Video. International Institute for Environment and Development, London.
36. **Sperling, L., M. Loevinsohn & B. Ntabomvura, 1993**. Rethinking the farmer's role in plant breeding: local bean experts and on-station selection in Rwanda. Experimental Agriculture, 29, 509–19.
37. **Rassam, A., 1985**. Farm labor by age and sex in north-western Syria: implications for two proposed technologies. Paper presented at the Farming Systems Symposium, Kansas State University, Manhattan, Kansas, 13–16 October, 1985.
Nour, M.A., 1985. In The Rockefeller Foundation and International Service for National Agricultural Research. Women and Agricultural Technology: Relevance for Research. Volume 2: Experiencies in International and National Research. Report for the CGIAR Inter-Center Seminar on Women and Agricultural Technology. Bellagio, Italy, 25–29 March, 1985. The Hague, Netherlands.
38. **Chabala, C. & R. Nguiru, 1989**. Intrahousehold Dynamics and FSR/E in Zambia, a Case Study of Traditional Recommendation Domain 3 in Central Province. In Working Together (Op. cit.).
39. **Kennedy, E. & P. Peters, 1992**. Household food security and child nutrition: the interaction of income and gender of household head. World Development, 20(8), 1077–85.
Hopkins, J., C. Levin & L. Haddad, 1994. Women's income and expenditure patterns: gender or flow? Evidence from Niger. Paper prepared for 1994 Annual Meetings of the American Agricultural Economics Association, August.
Kennedy, E. & L. Haddad, 1994. Are pre-schoolers from female-headed households less malnourished? A comparative analysis of results from Ghana and Kenya. Journal of Development Studies, 30(3).
40. **Kennedy, E. & L. Haddad, 1994**. Are pre-schoolers from female-headed households less malnourished? A comparative analysis of results from Ghana and Kenya. Journal of Development Studies, 30(3).
41. **Paris, T.R., C.P. Diaz, M. Hossain & A.B. Vasallo, 1994**. The process of technology development and transfer to women: a case of the microrice mill in Guimba, Nueva Ecija, Philippines. Unpublished paper.
42. **Goetz, A.M., 1996**. Local heroes: patterns of field worker discretion in implementing GAD (Gender and Development) Policy in Bangladesh. IDS Discussion Paper 358. Institute for Development Studies, Sussex.

3.5 RELATING PROBLEMS AND CAUSES IN FSR PLANNING

Robert Tripp

One of the values of causal analysis in FSR is that it forces participants to take a step back, consider a wider range of options than their own disciplines might suggest, and then focus on interventions that are relevant to the given circumstances.

3.5.1 Introduction

It is difficult to provide any precise definition for a movement as broad as FSR, but two elements are certainly characteristic. One is an insistence that priorities for research and extension programmes must be based on an understanding of farming practices within a holistic framework. The second is the conviction that technological innovations need to be developed and tested under farming conditions and management representative of target farmers. It is the expectation of FSR that these two elements should be directly linked; an understanding of farming system performance and constraints should naturally lead to the identification of appropriate technologies to be tested on farm.

But the translation of farming system diagnosis to relevant technology testing has not been straightforward. One technique has been used to make a more effective connection between diagnosis and experimentation – the management of causal analysis in FSR planning. This is a brief overview of the techniques that are used, providing an evaluation of the strengths and limitations of these techniques, and placing causal analysis in the broader context of the other adaptive and participatory agricultural research activities currently in use.

3.5.2 The conduct of causal analysis

Most FSR has been directed at public sector agricultural research and extension institutions. The aim has been to provide methods and techniques that are replicable and that can be incorporated into the operating procedures of these institutions. But the challenge has not been merely methodological. A concomitant purpose has been to impress upon researchers and extensionists the complexity and rationality of local farming systems and to convince them that textbook solutions to farmers' problems are often inappropriate. One of the most difficult aspects of conducting FSR, and in managing related training activities, has been precisely the point at which a diagnosis of the farming system is to be used to identify possible interventions.

Techniques for diagnosis, including various types of surveys, interviews, group meetings and observations, are relatively well described. Similarly, a range of sources provide guidance for the design of on-farm experimentation. But there are no standard methods for connecting diagnostic information to experimental design. In the absence of a robust linkage process, one unfortunate tendency has been to follow FSR diagnosis with proposals for testing and promoting the very technologies to which researchers were already committed before the diagnosis.

This challenge of stimulating an innovative approach to experimentation, consistent with the insights of farming systems analysis, was the principal motivation for the use of causal analysis in FSR. Causal analysis is by no means a universal component of FSR, however. The description in this section is based in particular on the experience of the CIMMYT and the International Center for Tropical Agriculture (CIAT), which used causal analysis in their training activities in FSR[1]. Causal analysis is part of a simple analytical sequence that includes:

- An identification of problems that restrict the productivity of a farming system.
- An analysis of the causes of those problems.
- Proposals for possible solutions based on an understanding of the causes.

The process is best illustrated with the example cited by Krisdiana *et al.* in 1991[2]. An adaptive research programme in Indonesia managed by the Malang Research Institute for Food Crops (MARIF) found that one problem in local maize production was uneven plant populations with high interplant competition. The immediate cause of this problem was the high planting density used by farmers. An initial response might have been a recommendation to change planting practices, but a farming systems perspective indicates the value of looking for the rationale (i.e. the causes) behind the practice. Three hypotheses were proposed (Fig. 3.5.1).

- Farmers may have overplanted as a reaction to seed quality problems.
- Farmers may have placed priority on maize thinnings as a source of animal feed.
- Farmers may have been trying to counteract the effects of early season pest damage.

The identification of effective and thus appropriate solutions obviously depends on assessing the relevance of these causal hypotheses. Possible solutions for seed quality, such as modifying sources, treatment or storage of maize seed, would be very different from those addressing the constraints of animal feeding, such as the identification of alternative fodder sources. In this example, the third causal hypothesis was the relevant one, and further research uncovered considerable shootfly damage that had led farmers to overcompensate in their planting practices.

The example illustrates several aspects of causal analysis. First, chains of causes are common; the immediate cause of the problem was overplanting, but this in turn was susceptible to causal analysis. Second, causes are not always immediately clear, and hypotheses need

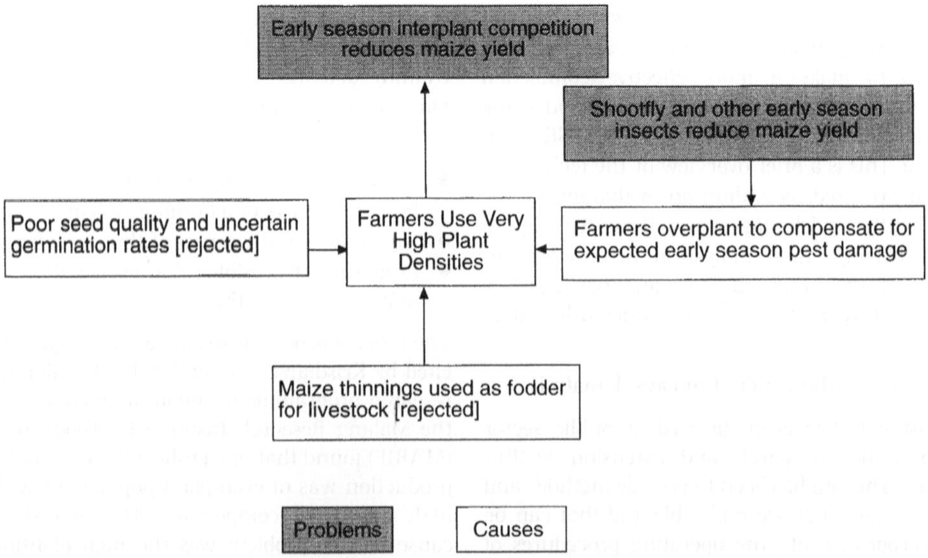

Fig. 3.5.1. Hypotheses on the causes of interplant competition in maize, Malang District, Indonesia. (Source: Krisdiana *et al.*, 1991.)

to be investigated through further conversations with farmers, observations, surveys or experiments. In addition, several layers of causes may have to be investigated before arriving at one that is appropriate for suggesting interventions. Once it was understood that shootfly should be addressed, simple seed treatments were tested. Further causal analysis examined delays in planting and reliance on maize–maize rotations as contributors to the shootfly problem, with the aim of possible modifications in rotation or planting date to deal with the pest problem (Fig. 3.5.2).

The complexity of causal analysis is such that diagramming is often helpful, as illustrated in Figs 3.5.1 and 3.5.2. This complexity is a serious concern; there are, after all, almost no limits to the breadth and depth of causal chains. The technique is only useful if it is managed as a pragmatic tool forcing researchers to understand as much as they can about the context of particular problems that have been identified before they start proposing interventions.

The causes of production problems can include natural conditions (e.g. soil type, rainfall pattern), socioeconomic conditions (e.g. food preferences, market requirements, current policies) and management practices. The use of non-practices as causes (e.g. farmers' unfamiliarity with chemical pest control in Fig. 3.5.2) should be limited, otherwise causal analysis becomes simply an inventory of the technologies that researchers or extensionists wish to recommend without considering a systematic diagnosis. The breadth of causal analysis should correspond to the capacities and mandate of the FSR programme. One common debate, for example, is the degree to which policy factors should be included in a causal analysis. To the extent that policies are susceptible to the actions and recommendations of the FSR programme, they are appropriately considered in a causal analysis.

The examples discussed above involve single problems, but the combination of the causal analysis of several problems often reveals interrelations that help to identify further priorities for research or intervention. It is not uncommon to find that several causal factors contribute to a particular problem and that they must be considered together in testing possible interventions.

Causal analysis should not be confused with two other common techniques in farming systems analysis[3]. It is not the same as problem ranking, carried out after an initial set of problems have been identified and involving prioritization on the basis of such factors as the importance of the problems and their susceptibility to amelioration. It is not the equivalent to analysis of solution feasibility, where possible interventions are ranked on the basis of parameters such as cost of research. Causal analysis stands in between problem identification and solution screening.

It is also important to emphasize that the sequence of problem–cause–solution should not be confounded with a methodological sequence in FSR. Although it is helpful to present FSR as a progression from diagnosis to planning to experimentation to assessment and feedback, the consideration of problems, causes and solutions goes on continually during the conduct of FSR. Causal analysis should not be seen as an isolated planning procedure but rather as a guiding principle for the conversations, surveys, experiments and data analysis of an entire FSR programme. The process of identifying and interrelating problems and causes begins with the first activities in an FSR programme and continues over each season with further diagnosis and experimentation. Hypotheses about problems and causes are discussed, tested and refined; the understanding of these relationships should progress in parallel with advances towards identifying useful interventions for the farming system.

3.5.3 Contributions of causal analysis

One of the values of causal analysis in FSR is that it forces participants to take a step back, consider a wider range of options than their own disciplines might suggest, and then focus on interventions that are relevant to the given circumstances. In the Indonesian example, extensionists might have recommended an educational programme to encourage lower planting densities; post-harvest specialists might have embarked on a seed storage campaign; socioeconomists may have encouraged more attention to livestock management in the farming system. In the context of the immediate problem and its actual cause, all of these would

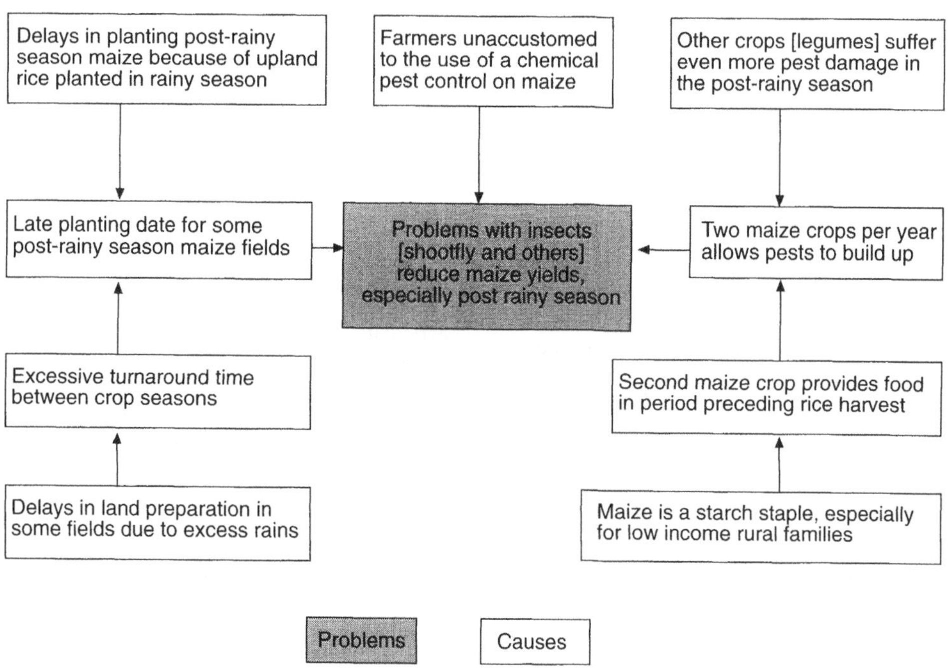

Fig. 3.5.2. Hypotheses on the causes of pest damage in maize, Malang District, Indonesia. (Source: Krisdiana *et al.*, 1991.)

have been misguided. The search for causes is an implicit part of most problem-solving techniques, but finding ways to make it an explicit and continuous part of FSR analysis stimulates participants to seek more information and to think carefully about the precise points of the farming system where innovation would be most productive.

It is important to realize that until the causes of a problem have been identified it is often difficult to proceed with interventions. One of the contributions of causal analysis to FSR planning is to encourage participants to articulate whether they are designing research and experimentation to test solutions for problems that are understood, or to seek further information to clarify what direction should be taken. In the Indonesian example, research was done on the quality of farm-stored maize seed, not as an intervention but rather to test the possible role of seed quality in the causal sequence.

Causal analysis helps provide a structure for FSR planning that limits the tendency to promote preconceived 'solutions looking for problems'. It offers a framework in which to discuss the rationale and interrelations of the farming system that can be reviewed and refined. As long as causal analysis is managed in a pragmatic fashion, it helps researchers to focus on creative solutions and to appreciate the degree of understanding required to identify useful innovations. Not enough attention to causes leads to inappropriate research and recommendations, but excessive pursuit of causal relationships diverts resources from the adaptive aims of FSR.

3.5.4 The limitations of causal analysis

Causal analysis has proven to be a useful tool for FSR planning, and for helping to communicate the basic premises of FSR. Its use does not, however, guarantee relevant research. As with any aspect of research methodology, warnings about the dangers of mechanical and unimaginative application hold true for causal analysis. The identification of useful interventions in FSR depends more on the skill and the experience of the participants than on any planning technique. Causal analysis helps to keep disciplinary bias at bay, but it certainly does not eliminate the possibility that the most influential members of the team will commandeer the direction of an FSR programme.

One concern with techniques such as causal analysis is that they may contribute to an excessively positivist conception of FSR. The sequence of problems, causes and solutions gives the image of a completely objective project aimed at discovering scientific truth. In fact, adaptive agricultural research is a much more subjective, iterative and political endeavour than is often acknowledged. Any tendencies to place such techniques on a pedestal of 'pure' science should be counteracted. One way of doing this is to admit that it is simply one more technique to increase the chances that the farmers, researchers and extension agents brought together for FSR will interact with tolerance, creativity and understanding.

Indeed, it is worth briefly deflating the position of causality in scientific endeavours. Science is sometimes envisaged as the search for single determining causes, but the reality of multiple causation and the influence of a researcher's frame of reference, and the scope of the enquiry being conducted, challenge such simple conceptions[4]. The structure of causality is not objectively determined, but is rather dependent on the context of the research and the purposes of the human actors[5]. In social science analysis, causality is only one of several ways of describing relationships among variables[6].

Another possible limitation to causal analysis (as described here) is the fact that it was conceived primarily as a technique for improving the FSR planning procedures used by public sector researchers and extension agents. Although FSR involves considerable interaction with, and participation from, farmers, the use of causal analysis in the planning of an experimental programme does not envision significant direct farmer involvement. As current movements in adaptive agricultural research include considerably more emphasis on farmer participation, and often less of a presence of public sector technical staff, the broader relevance of this technique is worth exploring.

3.5.5 Causal analysis in a broader context

Various types of causal analysis are common in project planning and design. Delp *et al.*, in 1977, described several planning techniques that involve causal analysis[7]. A number of donors have adopted the GTZ 'objective oriented project planning method', ZOPP[7], which begins by identifying a core problem and then analyses its causes and effects as a prelude to specifying possible alternatives for project attention. Examples of the use of causal analysis in agricultural research design include the method of farming systems diagnosis described by Lightfoot *et al.* in 1990[7] for training researchers. It uses systems diagrams, where farmers' problems are placed in the centre and causes, divided into primary, secondary, biophysical and socioeconomic causes, are arranged in circles around the problem.

The conventional concept of FSR planning is that of researchers and extension agents in an office or meeting room, debating the diagnostic and experimental data available to them. There is a need to involve farmers more in the planning process, and to consider the implications for causal analysis. In a technique described by Bunch in 1982 for World Neighbors activities[8], farmers first brainstorm to produce an initial list of problems, then work to refine the list, group similar concepts together and arrange them in order of priority. The iterative nature of planning would seem to indicate significant possibilities for shifting the locus of much planning to the field, taking advantage of innovations such as the 'regular research field hearings' described by Baker in 1988[9] or the farmer groups described by Norman *et al.*[10], that same year.

It is certainly the case that recent interest in participatory agricultural research methods provides few examples of what could be described as causal analysis. Perhaps this should not be surprising. First, there is a strong reaction by many against the 'empiricism' of FSR and indeed against what Roades termed 'the bankruptcy of social science methods used by FSR teams' in 1994[11]. The feeling here is that less structure and a wider range of methods would be helpful. In addition, farmer participation often emphasizes the importance of indigenous knowledge and explanatory concepts that may not be compatible with conventional causal analysis. Finally, the importance of empowerment[12] in farmer participatory methods directs attention away from the details of formal planning and towards the importance of farmers assuming control and responsibility for technology generation.

There is some debate regarding the degree to which FPR should be used to refine FSR[13], or whether it should complement[14], or indeed replace it[15]. In any case, it can be argued that the role of causal analysis in FSR can be usefully considered for FPR as well. Work in FPR has been responsible for a considerable expansion in the range of field techniques available for diagnosis, and impressive lists of alternative methods for engaging farmer experience and commitment for problem identification and prioritization are often presented[16]. However, this methodological diversification has not solved the problem of how to convert a description of conditions and problems into a plan of work. As Mosse said in 1996: 'The concept of "people's knowledge" misrepresents information production in the planning process and gives a deceptively participatory gloss to the more complex social dynamics of knowledge and the process of negotiation involved. ... Even where sophisticated methods of participatory appraisal are sensitively and effectively used, the local knowledge which they help to generate does not in any straightforward way translate into programme decision-making and action'[17]. The richness of participatory appraisal has yet to be matched by an adequate strategy for participatory planning.

In this sense, FPR shares much in common with FSR. Both have an unfortunate tendency to focus excessively on their diagnostic techniques; neither are very certain about how to proceed from diagnosis to planning; and neither can guarantee that their most powerful participants will not appropriate their course. Techniques such as causal analysis make a modest contribution to addressing these deficiencies in FSR. Political control is not going to be countered by methodological innovation. But planning and priority setting in a framework which encourages participants to examine their premises, acknowledge their biases and commit themselves to exploring alternative explanations can contribute to a more open and responsive programme of technology generation. Causal analysis is simply one way of promoting a state of mind in which puzzlement and respect for the complexities of local farming systems encourage an understanding that leads to meaningful improvement.

REFERENCES

1. **Tripp, R. & J. Woolley, 1989**. The Planning Stage of On-Farm Research. Identifying Factors for Experimentation. CIMMYT and CIAT, Mexico, D.F. and Cali.
2. **Krisdiana, R., M. Dahlan, Herianto, C. van Santen & L. Harrington, 1991**. From diagnosis to farmer adoption: MARIF's maize on-farm research program in East Java, Indonesia. In: R.Tripp (Ed.) Planned Change in Farming Systems. John Wiley, Chichester.
3. **Woolley, J. & R. Tripp, 1994**. Priority setting tools for research groups. In: J.D. Meindertsma (Ed.) Setting Research Priorities. Royal Tropical Institute, Amsterdam.
4. **Susser, M., 1973**. Causal Thinking in the Health Sciences. Oxford University Press, New York.
5. **Humphreys, P., 1989**. The Chances of Explanation. Princeton University Press, Princeton.
6. **Rosenburg, M., 1968**. The Logic of Survey Analysis. Basic Books, New York.
7. **Delp, P., A. Thesen, J. Motiwalla & N. Shesadri, 1977**. Systems Tools for Project Planning. Program of Advanced Studies in Institution Building and Technical Assistance Methodology, Bloomington, Indiana.
   GTZ, 1988. ZOPP: an Introduction to the Method. GTZ, Eschborn, Germany.
   Lightfoot, C., V.P. Singh, T. Paris, P. Mishra & A. Salmon, 1990. Training Resource Book for Farming Systems Diagnosis. IRRI, Manila.
8. **Bunch, R., 1995**. Two Ears of Corn: a Guide to People Centered Agricultural Improvement. World Neighbors, Oklahoma City.
9. **Baker, D., 1993**. Inability of Farming Systems Research to deal with Agricultural Policy. Journal of Farming Systems Research-Extension, 4(1), 67–86.
10. **Norman, D., D. Baker, G. Heinrich & F. Worman, 1988**. Technology development and farmer groups: experience from Botswana. Experimental Agriculture, 24, 321–31.
11. **Rhoades, R., 1994**. Farmer participation in priority setting. In: J.D. Meindertsma (Ed.) Setting Research Priorities. Royal Tropical Institute, Amsterdam.
12. **Okali, C., J. Sumberg & J. Farrington, 1994**. Farmer Participatory Research: Rhetoric and Reality. IT Publications, London.
13. **Ashby, J. & L. Sperling, 1994**. Institutionalizing participatory, client driven research and technology development in agriculture. Agricultural Administration (Research and Extension) Network Paper no. 49. Overseas Development Institute, London.
14. **Okali, C., J. Sumberg & J. Farrington, 1994**. (Op. cit.).
15. **Pretty, J., 1995**. Regenerating Agriculture. Earthscan, London.
16. (Ibid.).
17. **Mosse, D., 1996**. Local Institutions and Farming Systems Development: Thoughts from a Project in Tribal Western India. Agricultural Research and Extension Network Paper no. 64. ODI, London.

Part II

The Applications of Farming Systems Research

EDITORIAL INTRODUCTION

Mike Collinson

The introduction to this book adopted a definition of farming systems reseach (FSR) as 'understanding farmers' livelihood systems'. I believe this narrow definition puts FSR in its proper place without exaggerating its scope; as an 'aide' to research and development in traditional agriculture. Applications of FSR use this understanding in technology choice and generation, in development programming, and in policy formulation. I use the phrase development programming deliberately to include a widening range of activities, including its application to the introduction of new enterprises, particularly cash crops for the market, and to adding value through both group action and local processing. Farm system understanding offers a good basis for juxtaposing the cash, labour and land demands of new development opportunities on to the resource endowments and requirements of the current system, identifying potential clashes in resource use and formulating strategy for innovation. Understanding the farming system is also a potential asset for the successful operation of most enabling organizations, for project development in storage, marketing, processing and credit, as well as for capital works within the community.

Part II of this book describes the evolution of FSR applications. The emphasis is on technology generation, the origin of FSR, and the term on-farm research (OFR) is used to embrace both FSR and its application in on-farm experimentation (OFE), forming a useful shorthand when discussing the full R & D process. Two examples of its application in extension are followed by a discussion of its potential role in policy formulation.

THE CONTRIBUTIONS

The main contribution to Chapter 4 focuses on the application of FSR to technology development and was written by Ann Stroud and Roger Kirkby who have over 40 years of African field experience between them. They look at the history of OFE, review the evolution of concepts and methods on the where, what, who and how of OFE. They go on to assess the impact of FSR on research process in three thematic research areas; varietal improvement, agronomy and natural resource management, and livestock; they also examine the impact of 25 years of FSR on perspectives on technology development, and on the roles of farmers, researchers and institutions. Finally, they examine recurring issues in OFR before offering their view of the future.

Chapter 4.2 is a case study of technology development in action, focusing on a project seen as fairly typical of the way in which OFR has been implemented. Alistair Sutherland and J.N. Kang'ara describe a UK-funded OFR project in eastern Kenya, reaching some half a million people in 70,000 households. Its operating strategy was derived from an assessment of local institutional circumstances, aiming to sustain the processes it introduced after its completion. It showed results, in terms of benefits to farmers, over a 4-year period.

Chapter 5 has three contributions. Cornelia Flora and Charles Francis examine farming systems extension in the USA, drawing attention to two cycles of FSR in extension in the USA, first in the 1920s and 1930s when extension agents first sought to address the farm as a whole and latterly, since the 1980s with participatory approaches growing stronger and farmers, increasingly organized in groups, becoming more concerned about the sustainability of their systems. The role of FSR in the evolution of the extension services in Chile is tracked by Julio Berdegué, who outlines the radical changes in agricultural extension organization following privatization in 1978, and the parallel interest in FSR, largely in the voluntary sector. Farm systems development for policy formulation is described by John Dixon. He lays out the conceptual base and an operational process for a contribution from FSR to policy formulation and analyses both methodological and institutional problems inhibiting implementation.

A COMMENTARY: TECHNOLOGY CHOICE AND ADAPTATION

Over the 25-year history of FSR there have been dramatic developments in the processes for applying an understanding of farming systems to both technology choice and to technology development through OFR.

Despite early scepticism from the traditional research establishment, it is increasingly accepted that small farmers need different technologies to those of large farmers. Small farmers' limited power sources – hand hoes and, at best, ox-ploughs – have enormous implications for tillage capacity, both in terms of soil movement and for the timing of operations. Their limited and uncertain access to markets and institutions for credit and cash support requires their direct management of both climatic and market uncertainty through their farm practices, and restricts their choice of purchased inputs. Their criteria for the evaluation of new technology are set by cash scarcity, lack of institutional access, limited labour availability and family food preferences. Winning the battle for technologies unique to small farmers remains vital to the improvement of the culture of agricultural research establishments, and to their success.

Traditionally, station-based research identified technically ideal practices to grow the variety of a commodity identified as the best performer for an agroclimatic zone. At its most extreme, scientists' recommendations from station research failed to realize that a two-component recommendation on variety and fertilizer level, developed under optimum agronomy, has many 'hidden' components for the small farmer. These include the high management levels of the non-experimental variables (NEV), for example an optimal time of planting, spacing, weeding frequency and timing and pest control – all included to maximize the expressions of response to the two experimental variables and to attain statistical significance in the results. All of these NEVs may be new management practices, or new levels of practice, as far as the farmer is concerned, and without them the results from the new variety and fertilizer application may be considerably, sometimes wholly, diluted. Research prescriptions of these ideal technical models have traditionally been couched as extension recommendations and disseminated by staff, often with very limited education and training, virtually by rote.

In the 1980s indigenous technical knowledge (IK), new methods developed by farmers themselves, was promoted through networks such as Honey Bee with Anil Gupta, and Professor Mike Warren's group at Iowa State University. While extreme advocates saw IK as the only credible source of appropriate technologies, it is clear to most of those concerned with technology for smallholders that IK alone will not win the battle for better living standards. IK processes of farmer-led innovation and dissemination are currently being partnered with formal research institutions and networks for faster, more cost-effective innovation in smallholder agriculture[1].

The 1980s also saw Biggs and Clay[2] introduce the visionary concept of multiple sources of technology, potentially revolutionary for the FSR process. This concept held that new technologies appropriate for local farmers might be identified from research stations operating worldwide in similar agroecologies, and from IK of farmers elsewhere operating in similar circumstances. The concept effectively removes the monopoly of the local research station in dictating extension recommendations. Despite the insight of Biggs and Clay nearly two decades ago, multiple sourcing is still in its infancy. Where used it is usually operated on a personal level by an enlightened professional. Thorough implementation requires the development of databases of prototype, adopted and indigenous technologies for each major agroecology. Through the information highway such databases could be directly accessible by FSR teams to draw down a range of appropriate options for trial and adaptation by local farmers. NARS adopting the multiple source strategy would allocate staff and resources to contribute to, and to draw down from, these databases as part of their routine research programming and budgeting.

A further breakthrough in the wider area of agricultural development has been better understanding of its dynamics. Because the exclusive evaluation criterion in agricultural research has been yield – physical production per unit area – all research results, and therefore all recommendations, have been essentially land intensifying. In most of Africa, however, land has, until very recently, been farmers' most abundant resource. Recommendations to intensify its use often contradicted the local economic circumstances. Certainly there have been other problems – poor access to inputs, scarce cash and bad institutional management – but these were overshadowed, and even aggravated, by the fact that for much of Africa the cheapest way to increase incomes was to extend the area under cultivation by hiring machinery or labour. Simple sums in the 1960s demonstrated the futility of advocating the use of $20 for fertilizer when the farmer could gain twice as much by hiring power to cultivate an extra acre. Thus government extension and donor projects, built around research recommendations for land intensification, were frequently flying in the face of local economics, sometimes despite subsidies and credit for the purchase of inputs, with inevitable repercussions for the standing of extension staff in the eyes of rural communities. There is no universal watershed between extension of area and land intensification as means of income expansion, the appropriate strategy varies with the local situation and with the wider economics of the market. Better understanding of farming systems development, building on the work of Boserup[3] and Binswanger[4], has begun to clarify how to relate local situations to the extension/intensification watershed through the analysis of population density and market opportunity.[5] The upside to rising population densities and increasing land scarcity is first increasing relevance of the historical stock of technical knowledge accumulated by the formal research sector, and second, the expansion of local markets and widening opportunities for specialization – the clearest manifestation being the rapid growth of urban areas and their demand for food.

Progress in improving the OFR process has not yet had widespread impact on the adoption of new technologies by small farmers. There have been important successes, inevitably local, usually within a farming system. Many are recorded in the *Journal of Farming Systems Research and Extension* and in a growing volume of literature, including Tripp in 1991[6] and Franzel and van Houten in 1992[7]. The recording of both success and failure is critical, both for building and sustaining credibility with the process, as well as for its continual improvement. There is increasing impact on the R & D process in smallholder agriculture. Better articulation of their research needs and their system constraints by farmers is creating greater relevance in research priorities and in the design of new technological options by scientists. Two outcomes of an improved understanding of small farmers are proving particularly powerful: participatory breeding, addressed by Louise Sperling and Jacqueline Ashby in Chapter 11.4, and agroecosystems analysis, the focus of Clive Lightfoot's contribution in Chapter 11.5.

Problems clearly remain in the application of FSR to technology choice and development. Not least is the inertia in research institutions, much of it created by the general rundown of management in public institutions in many developing countries. The institutional dimension of FSR is the focus for Part III of this book. Suffice to say that one weapon in a 'phoney war' with the research establishment has been the attribution of high costs to FSR/OFR as a whole. The war has been 'phoney' in the sense that FSR/OFR is a stage added to an existing research and development process, not an alternative to it. The new stage needs evaluation on the benefits it brings for the costs it incurs. The cost of travel has been a particular *bête noire* of FSR, perhaps more in terms of its high demand for transport in institutions where, in the 1980s and beyond, working vehicles have been as scarce as hens' teeth. Meanwhile issues remain on the most appropriate process for OFR. Stroud and Kirkby discuss several of these in their contribution and I raise two here; first, the varied uses of the term 'scaling up' in contemporary systems and participatory literature.

Scaling Up

The term 'scaling up' is used, sometimes indiscriminately, in relation to six interlinked but discrete concepts – aggregation, coverage, diffusion, externality, extrapolation and hierarchy. More specific use of these terms in an FSR context will add clarity, help debate and boost the credibility of the approach:

- Aggregation – the cumulative impact of actions by many at the household level on the higher levels in the economic and ecological hierarchies.
- Coverage – the size of population which can be reached with a particular approach to agricultural development, given limited professional resources.
- Diffusion – the spread of changes among households.
- Externality – the impact of actions in one household, or one community, on others.
- Extrapolation – the use of FSR findings from one system, in other systems with a similar profile.
- Hierarchy – to initiate research at a higher level in the systems hierarchy, often on impact.

In an FSR context, diffusion is best defined as the spread of a change within the farming system for which it was selected. A change is identified in collaboration with selected groups of farmers at representative sites within the system, and diffusion is farmer to farmer, sometimes aided by the extension services or equivalent non-governmental or private sector agencies. Extrapolation takes that same change beyond the system for which it was developed. It may do so through selection by researchers and farmers in other systems who identify congruity with the natural, economic and cultural characteristics of the system in which it was a success. Externalities have long been identified by economists; hiring by local innovators may provide employment opportunities for other families, new production techniques adopted by local innovators may reduce prices in the market and influence the incomes of other families. More recently environmental externalities – the effects of downstream pollution or downhill soil erosion for example – on the activities of other farmers, have similarly engaged ecologists and widened the application of the term. Economists have long been active at the higher levels in the hierarchy, primarily for policy formulation. However, the aggregation of farm-level impacts remains an unsolved problem, one which is now also exercising ecologists, newly interested in understanding and measuring the environmental effects of farm management at the landscape and watershed levels. The expression 'scaling up' is sometimes used in reference to initiating research at higher levels of the systems hierarchy. It is important to emphasize that, although there are strong pressures to 'scale up' activities to higher hierarchical levels, it is largely an interest in the impact of actions which result from local decisions. Action still lies with farmers and the decisions they take, both in their households and in their communities. The issue of how much of this version of 'scaling up' falls within the aegis of FSR again resurrects the scope issue which is returned to in the final chapter of this book.

However, the critical 'scaling up' question nagging practitioners at the beginning of a new millennium, particularly participation and empowerment advocates, is that of coverage. Participation and empowerment absorbs high levels of professional input for each community addressed. In this it suffers the same replicability problem as intensive, quantitative data collection. The FSR emphasis on qualitative methods was born, *inter alia*, of the realization that the intensity of professional effort required dictated the coverage possible by any particular approach to R & D. Informal methods were, in general, a major effort to extend coverage of the rural population with a given level of professional and budgetary resources. A classic development battle of the 1960s was improvement versus transformation. For me, the conclusive transformation experience was that of the World Bank-supported Village Settlement Schemes in Tanzania in the early 1960s. Seventy such schemes were planned, each with 250 families (some 2000 people). These were resettled into new villages, with water and facilities; machinery was provided for cultivation of a new farming system. The aim was to reduce drudgery, improve housing and sanitation, increase incomes – in short to transform rural living. After six schemes had been implemented it had become clear that the sheer number of professionals required to manage the settlement and the farming was draining the country of agricultural graduates, focusing a scarce professional resource on a minute proportion of the rural population. FSR evolved the way it did; targeting farming systems rather than individual farmers, using rapid rural appraisal (RRA) as well as participatory rural appraisal (PRA) techniques, at least in part to manage the coverage issue. Coverage is taken up again in Chapter 12.

A holistic view of process

My second issue is that experts with particular disciplinary, methodological or institutional loyalties have become engrossed with particular stages of OFR, pursuing problems there which can be readily resolved elsewhere in the process. I must stress the need, in the systems tradition, to maintain an overview of the OFR process, and indeed the R & D process as a whole, when seeking improvement in any particular stage. Two examples are offered of the need for better balance between the stages in the adaptive research process:

1. Geographical information systems (GIS) and recommendation domain (RD) definition – FSR supports a typology of farming systems or RDs as a planning framework for agricultural development. Extensive efforts are being made to develop a viable planning framework by international institutions at a global level. These seek proxy parameters for which global data sets are available, to use through GIS. Yet there is no need to seek a final solution through global GIS. The sequential definition of domains is an alternative to building new, and perhaps expensive data sets at the macro level, and offers a way forward that can build close working partnerships across institutions, and the acknowledgement that each has a role in the success of the R & D process. For example, within GIS-defined resource management domains (RMDs), rapid rural surveys can identify spatially differentiated farming systems and broad resource endowment differences. In a third iteration, monitoring farmers' choices from among technological options offered within these preliminary RDs will allow closer profiling of the household circumstances for which particular options are chosen. This will finally refine domains and guide dissemination through extension.
2. Arguments are again being raised in favour of the measurement of farming system parameters through data-intensive surveys, by Baker in 1998, for example[8], and by Colin in 1994[9], among others. The quantification issue has been partly addressed in the context of modelling in the editorial to Part I of this book, but the issue also illustrates the importance of a balanced process. Take the example of the quantification of labour inputs at seasonal peak work periods to improve the *ex ante* and *ex post* evaluation of potential technologies. Detailed measurement is expensive. It is also unnecessary for a sound understanding of the main parameters involved, when such peak periods occur and what operations create pressures at these peaks. Initial estimates of the labour requirements of these operations can readily come from experience elsewhere. In the OFR process, once a range of innovative options is identified, the staff required in the field for experimentation can also undertake intensive data collection. Furthermore, the earlier qualitative diagnostic experience and the nature of the technologies identified with farmers, will have pinpointed those labour parameters that should be measured with precision to evaluate the options introduced. Such an iterative sequence is both cheaper and more accurate than an upfront intensive collection effort, as only a fraction of such an effort will prove useful. On the other hand, there is no guarantee that the appropriate fraction will even be part of a survey which is not founded on a sound understanding of the farming system.

FSR AND EXTENSION

Both the cases offered in the application of FSR to extension, the USA and Chile, are rather special. For the USA, with a capacity to provide intensive help in most local situations, even to the use of professionals in the planning of individual farms, the case demonstrates how the FSR approach can compete with others, even in the most sophisticated agriculture. For Chile, with a small peasant sector relative to its commercial agriculture, and the wealth to help this residual group on equity grounds, the case shows how the farming systems perspective, held by senior managers, helped develop innovative ways to address the small-farm question. Neither, however, provides evidence that farming systems understanding is a strong foundation for extension programming in countries dependent on a dominant small-farm sector for economic development. Such cases do not exist as yet. The potential applications of FSR in extension among smallholders are threefold: first, the use of a farming system typology for the organization of agricultural extension; second, the use of OFR-identified technologies as extension programme content; and third, and uncommon to date, the use of FSR understanding to formulate extension strategy for the introduction of innovations, whether new enterprises or new technologies.

There is, as yet, no record of any country formally organizing extension on the basis of farming systems. A few countries have developed a typology of farming systems for use in OFR itself, and Tanzania is one example, as shown by Ann Stroud in Chapter 7.1. One important barrier has been the common use of the same administrative units across the range of executive ministries, allowing national budgetary allocations to be made and compared. This has been aggravated by professional rivalries in the promotion of alternatives, particularly between land use scientists firmly wedded to physical parameters, and social scientists wedded to human ones. My contribution on typology in Chapter 3 attempts to reconcile these perspectives. Many countries have made progress in the use of FSR-based technologies for extension programme content. But there have been two major barriers: first, the rivalry with the existing research establishment as a source of extension recommendations, often reinforced by questions raised by scientists about the rigour of FSR methods. There have been particular difficulties with new plant material. In some countries it has taken years of effort to modify the formal varietal release procedures operated at the national level, and sometimes jealously guarded by senior plant breeders. Kean and Singogo reported on the struggle to get procedures modified in Zambia in 1988[10]. Other countries, and Kenya is one example as described by Sutherland and Kang'ara in Chapter 4.2, have eventually accepted that new plant material need not necessarily go through years of national testing prior to local release.

A second and perhaps the key barrier has been weak ownership of the OFR process in the extension services. All three major stakeholders in the development, dissemination and adoption process farmers, researchers and extensionists, need to feel ownership of the candidate technologies and other proposed changes in the farming system. OFR has successfully brought together farmers and adaptive researchers in an increasingly close partnership, but has too often excluded extension staff. The issue perhaps revolves around the question of how far research should seek a finished product – a technology proven acceptable to farmers, or should it be part of the extension function to 'finish' the product? Historically the link between research and extension has been a problem area. I have, along with many others, insisted that the root of the problem has not been poor communication between extension and research, but poor communication between farmers and both extension and research[11]. More recently a wide variety of devices have been used to draw extension into the adaptive research process[12] and some of these are examined more closely in the editorial to Part III on institutions.

FSR AND POLICY

Both practitioners and commentators have debated the need for FSR to focus on policy issues. Perhaps the most vigorous discussion of the historical failure to address policy, and the need to correct this in future comes from Baker[13]. He recognizes that even the early conceptual models by Norman, and by Byerlee *et al.*[14], noted the role played by the policy environment in influencing household decisions and acknowledged that FSR could inform policy. His main critique is that the practice of FSR has, in the main, treated policy as a given, not as something to be adjusted. He captures the argument in a nice phrase; FSR-E has had a 'commitment to a strategy of modifying technologies rather than modifying farmers' circumstances'.

In pursuing a policy orientation Baker calls for a return to in-depth household and village studies. For me this considerably complicates the issue. Effective use of such data on a national scale begs the solution of the aggregation question and perhaps oversimplifies the routine process of policy making. Dent[15] and others have put considerable effort into nesting modelling techniques for aggregation up the hierarchy of systems. Such nesting, if achieved, would base macro-economic modelling firmly on a foundation of micro-economic data. Netherlands professionals, including De Wit[16] and Van Keulen[17], are among those vigorously pursuing this goal. Baker, Dent and others[18] have discussed how far such methods should be embraced by FSR, or whether quantative micro-economic research to enhance both policy and development theory needs its own institutional niche in the establishment.

Experiences

Early FSR practitioners were usually the new boys on the block, working within a research service from a local station with local farmers. Their view of the world was clearly bounded by the existing policy environment, they had no hope of changing it. Gradually this perspective changed and attempts to address policy issues increased. In the early 1980s the International Maize and Wheat Improvement Center (CIMMYT) sought to demonstrate the application of FSR to policy formulation by generating case studies, for example on fertilizer pricing and credit in Panama and Haiti[19]. Tripp[20] in 1991 highlighted policy implications in six of the nine case studies reported there, and in 1992 Franzel and van Houten[21] offered five policy-oriented case studies in their record of FSR applications in Ethiopia. Beyond this Finan, in 1993[22], makes a plea for the wider use of FSR (phrased as FSIP – farming systems institutions and policy after Norman 1980[23]), to evaluate and adapt policy themes introduced under the structural adjustment reforms of the middle and late 1980s. So there has been experience, albeit in an *ad hoc* manner.

During the late 1980s the Food and Agriculture Organization of the United Nations (FAO) drew up a conceptual framework for an initiative in farming systems development. Training materials were prepared and field tested in Africa, the Carribean and Asia. From the early 1990s FAO promoted the idea of applying farm system understanding more widely within this FSD framework[24]. In partnership with bilateral donors it established regional programmes to support national governments in applying an FSD approach to the agricultural sector. Central to these initiatives were the broader questions of the institutionalization of systems approaches and applications to agricultural policy analysis.

Recent evidence is demonstrating that the decentralization of authority from central to local government creates opportunities for FSR to contribute to policy formulation at the municipal level. The Consortium for the Sustainable Development of the Andean Ecoregion (CONDESAN), a consortium of R & D institutions in the Andean countries hosted by the International Potato Center (CIP), has developed an explicit strategy to impact local rather than national policy. Where it has field sites CONDESAN seeks to persuade mayors to extend their interest beyond their municipalities into the hinterland which provides food and water to the towns. Consortium economists will help identify policy and legal instruments which mayors can use to improve natural resource management by communities and households and bring these to bear through the 'Mesas de Concertacion', a forum common to the Andean countries to bring municipal authorities, NGOs, universities and farmer associations into open discussion of problems and interests in their communities. The 'Mesas' work at many levels and, by reaching a horizontal consensus at one level, add weight to the message moving up to higher levels. CONDESAN hopes that convinced mayors will eventually be heard at national level. A similar initiative is underway in Colombia supported by CIAT. A serendipitous element to the emerging success of this strategy is that, urged on by the international community, Latin American/Caribbean (LAC) governments are decentralizing responsibilities and budgets to the municipal level. CONDESAN's timing has been such that municipalities have the funds to pursue a wider vision than in the past. Despite this serendipity the principle of directly engaging local policy makers when working with the communities surrounding OFR sites seems a useful one. The devolution of decision making empowers local stakeholders, increases the awareness of local issues, and reduces the aggregation problem. As environmental considerations multiply, local government authorities, particularly urban authorities, will necessarily increase their interest in the activities of their hinterlands. FSR offers a process for better understanding these.

Process in policy making

John Dixon's contribution on policy applications for FSR focuses on process. He describes the variety of information sources available for policy formulation in most developing countries, and, at the same time, the haphazard way in which, if at all, these sources are used. Policy formulation is more usually an informal balancing of stakeholder perspectives rather than rigorous research and analysis, with the outcome often decided by the weight of stakeholder influence. This is often true whether formal studies are available or not, and questions the cost-effectiveness of in-depth research for routine policy formulation. *Ad hoc* formal policy studies are typically precipitated by strong political involvement or by issues which demand an organized response to prevent policy shifts against the national interest. Clearly, intensive research cannot be commissioned for most national-level policy decisions. The costs of the studies on such a wide scale, the technical problems of aggregating local results, and the time lag between study and decision, make such research a blunt instrument for influencing other than longer term policy trends.

CONCLUSIONS

Despite isolated experiences of applying the FSR approach to policy formulation, Baker's critique carries weight: applications of FSR to policy formulation have never gathered the momentum they deserve. This situation is complicated by three factors:

- The complexity and 'informality' of the policy formulation process, particularly at the national level.
- The cost of the in-depth data collection techniques advocated, and the weakness of the analytical tools for aggregation.
- An overriding imperative to sustain the institutionalization of FSR which demands that the needs of its various applications are not confounded.

Baker, in 1993, was perhaps being too naive about the process of policy formulation, expecting too much from a research input. In practice it is an informal often political process, usually a vague one. Even the in-depth detailed studies Baker calls for would usually do no more than add weight to the case of one or another stakeholder engaged in the formulation process. This, and the data-intensive techniques for the modelling and analysis required, must query policy research as a central application for the limited FSR capacity on the ground at present. Information useful to policy analysis is perhaps best treated as a by-product of their basic farm improvement role.

Herdt in 1987[25] and Tripp in 1990[26] favour a separation of functions. Micro-level policy research deserves its own capacity, readily mobilized when local information can illuminate national issues. Such capacity might often first turn to work completed by FSR cadres. Those cadres in place, given their closeness to the community, might well serve as 'field facilitators' for outside professional resources with policy and/or theory mandates, in the same way that they can bring natural scientists into contact with farmers. However, if too much weight is given to FSR for policy purposes the danger is that social scientists in an 'FSR establishment', because of its skills and field location, are overloaded with *ad hoc* studies and become exclusively a vehicle for policy research. It is too early, and existing FSR establishments seem too fragile, to serve as more than an information source for would-be policy analysts. Added to this, the skills mix and the institutional partnerships required will be different for policy formulation as a core FSR application.

In my view, care is needed in adding greater weight to a policy analysis role for existing FSR cadres. It can threaten the progress made in introducing FSR capacity into existing institutions by confounding the role of, and skills needed by, FSR professionals, as well as by widening the organizational link required. Such institutional linking has so far proved the Achilles' heel of FSR, and it remains the most formidable barrier to success in its goal of farm improvement. The topic is pursued further in discussing promotional strategies in Chapter 12.

NEW FARMING SYSTEMS DEVELOPMENT – WAS IT A CANARD?

An FSR application popularized by the 1985 World Bank-sponsored review of FSR by Norman Simmonds in 1985[27] was New Farming Systems Development (NFSD). It was Simmonds own terminology, and he argued that 'NFSD contrasts with the preceding (OFR/FSP) in seeking to generate revolution rather than evolution, to build radically new systems *ab initio* ... in the real world not all changes should or can be stepwise; many farmers' circumstances cry out for radical alterations.' NFSD has received little attention over the last 15 years, both adoption theory and on-the-ground experience reject it as a canard. However, Simmond's discussion did shed early light on widening the scope of FSR from its origins in technology adaptation.

History demonstrates the futility of attempting to introduce new farming systems to farmers. Jolly's work in Trinidad[28] showed how the reconciliation of what was planned and what farmers could cope with took over 5 years of trial and error. World Bank experience with the Village Settlement schemes in Tanzania in the early 1960s showed that the overheads in terms of professional supervision to pursue a new farm plan were too high to ever be cost-effective[29]. The smallholder resettlement of the White Highlands in Kenya after independence, in which farm plans drawn up to maximize family income were largely ignored by resettled farmers in the interests of securing family food supplies[30]. On the academic scene in the 1960s, despite the tenet from Schultz in 1963 that small farmers effectively allocate their resources to satisfy their priorities[31], optimizing models were widely used to demonstrate the incomes that small farmers might aspire to. As I pointed out then[32], the management revolution implied is inevitably too radical for small farmers to absorb. The real challenge is plotting a path from the existing system towards an optimal system, and pursuing this at a pace that reflects farmers' resource base, management capabilities and risk preferences. At the same time the dynamics of the market place as well as technological innovation ensure that the optimal system itself is constantly changing. Both factors make it clear why stepwise improvement, rather than transformation, is the only way forward for the smallholder sector in the large. Our knowledge of small-farm adoption has improved; it works component by component, or via small packages of easily absorbed, complementary components, initially on a small scale, then more widely as the farmer understands how to adapt the system to the needs of the new components.

When developers have promoted new systems they have ignored what we have learned about adoption among resource-poor farmers, vulnerable to risks and producing food for their families. A new system is anathema in a constantly changing market place, out of date before it can be operational. The need is not for new farming systems but for new technical systems, for rotation, for storey, intercropping and relay cropping which provide high and sustainable physical productivity – this is where new vision is required. FSR has the capacity to assess which of these systems fit identified farming systems and which sequencing of technical components will be attractive to the farmers operating them.

CONCLUSIONS ON FSR APPLICATIONS

Across agricultural research as a whole, balancing systems with component research is best seen and promoted as a reorientation of a reductionist process historically too isolated from the small-farmer constituency, in terms of both priorities and relevance. New and often complex technical systems aiming at sustainable productivity remain the appropriate focus for both strategic and applied research. Their successful introduction will depend on their being broken down into components introduced in an appropriate sequence that immediately brings a sustainable increase in income to the farm families concerned. The sequence will vary in different production circumstances. 'Good' contemporary agricultural research identifies and designs sustainably productive technical systems, components of which can be extracted and shaped to integrate with farmers' existing systems. Those components that really make a difference will win a place in farmers strategies and themselves become drivers of change in the system.

Within FSR itself, process and scope have evolved dramatically. In the early days OFR developed as an adaptive research step, modifying technologies developed on the local research station, to make them more compatible with the circumstances of local farmers. Currently still evolving, best practice now sees FSR as a process for understanding a farming system and then identifying and shaping development opportunities from a wide range of sources. There is a confusion in current debate. A significant part of the FSR constituency itself has difficulty relating to issues which appear to be far removed from what concerns them. There remains a need to clarify the scope of FSR and the implications of the different applications in order to maintain interest, commitment and momentum.

On the ground progress towards this best practice remains slow for a variety of reasons, the best understood are poor sourcing of information on opportunities, poor skills from inadequate training and education, and weak institutional commitment. To date, 'revolutions' in the process and in the culture of research establishments have only happened here and there, most often in the voluntary sector. Part III of this book focuses on the institutional experience in the public sector. Poor management and inertia there has been exacerbated by political apathy on research, and has encouraged the rundown of institutions. This is proving a tough nut to crack.

REFERENCES

1. **Ashby, J.A., 1990**. Evaluating Technology with Farmers: a Handbook. CIAT, Cali, Columbia.
   CIAT, 1991. Institutionalizing Local Leadership for Farmer Participation in Agricultural Technology Generation and Transfer in Rural Communities. Report of a Special Project. CIAT, Cali, Columbia.
2. **Biggs, S.D. & Clay, E.J., 1981**. Sources of Innovation in Agricultural Technology. World Development, 94, 321–36.
3. **Boserup, E., 1965**. The conditions of agricultural growth; the economics of agrarian change under population pressure. Allen & Unwin, London.
4. **Binswanger, H.P., 1986**. Evaluating research system performance and targeting research in land abundant areas of sub-saharan Africa. *World Development*, 14(4), 469–75.
5. **Smith, J. & G. Weber, 1994**. Strategic research in heterogeneous mandate areas: an example from the West Africa savannah. In: J.R. Anderson (Ed.) Agricultural Technology: Policy Issues for the International Community. CAB International, Wallingford.
6. **Tripp, R. (Ed.), 1991**. Planned Changes in Farming Systems: Progress in On-Farm Research. Wiley, Chichester.
7. **Franzel, S. & H. van Houten (Eds), 1992**. Research with Farmers: Lessons from Ethiopia. CAB International, Wallingford.
8. **Baker, D., 1998**. Micro Strategies for Macro Factors. Invited Paper presented at the 15th International Symposium for Farming Systems Research-Extension, Pretoria, South Africa, Nov/Dec 1998.
9. **Colin, J.P., 1994**. Systems of production and production of knowledge: reflections on the basis of Ivorian and Mexican experiences. Journal of Farming Systems Research-Extension, 4(3), 31–46.
10. **Kean, S. & L. Singogo, 1988**. Zambia, Organisation and Management of the Adaptive Research Planning Team (ARPT), Research Branch, Ministry of Agriculture and Water Development. OFCOR Case Study no. 1. ISNAR, The Hague.
11. **Collinson, M., 1988**. The development of African farming systems; some personal views. Agricultural Administration and Extension, 29, 1–16.
12. **Kean, S. & L. Singogo, 1988**. (Op. cit.).
    Tripp, R., 1991. (Op. cit.).
    Almy, S.W., M. Besong, T. Woldetatios, C. Pouborn, C. Ateh & M. Mboussi, 1991. The Extension Agent in the Research Process. Working across Ministries in Cameroon. Journal of FSR-E, 2(3) 69–84.
    Galt, D.L. & A.F. Al-Khadi, 1991. Farmers meet research and extension staff. JFSR-E, 2(3) 31–50.

Shumba, E.M., 1994. Institutional linkages that enhance the value of OFR for smallholder farmers: the Zimbabwe experience. Journal of Farming Systems Research-Extension, 4(3), 137–146.

Anandajayasekeram, P. & Stilwell, T., 1998. Institutionalisation of Farming Systems Approach: Lessons from Eastern and South Africa. In: P. Anandajayasekeram & T. Stilwell (Eds) Institutionalisation of Farming Systems Approach in Eastern and Southern Africa. SAAFSR-E, Swaziland, and FARMESA, Zimbabwe.

13. **Baker, D., 1993**. Inability of farming systems research to deal with agricultural policy. Journal of FSR-E, 4(1), 67–86.
14. **Norman, D., 1980**. The Farming Systems Approach: Relevancy for the Small Farmer. Rural Development paper no.5. Michigan State University, East Lansing, Michigan, USA.

    Byerlee, D., M. Collinson *et al.*, 1980. Planning Technologies Appropriate to Farmers. Concepts and Procedures. Economics Program, CIMMYT, Mexico.
15. **Dent, J.B., 1993**. Potential for systems simulation in farming systems research? In: F.W.T. Penning de Vries, P.S. Teng & K. Metselaar (Eds) Systems Approaches for Agricultural Development. Kluwer and IRRI, Netherlands.
16. **De Wit, C.T., H. Van Keulen, N.G. Seligman & I. Spharim, 1988**. Application of interactive multiple goal programme techniques for analysis and planning of regional agricultural development. Agricultural Systems, 26, 211–30.
17. **Van Keulen, H., 1993**. Options for agricultural development: a new quantitative approach. In: F.W.T. Penning de Vries *et al.* (Eds) Systems Approaches for Agricultural Development, pp. 355–61. Kluwer Academic Publishers, The Netherlands.
18. **Herdt, R., 1987**. Whither Farming Systems? Paper presented at the 7th Annual Farming Systems Research and Extension Symposium, Little Rock, University of Arkansas.

    Tripp, R., *et al.*, 1990. Farming Systems Research Revisited. In: C. Eicher & J. Staatz (Eds) Agricultural Development in the Third World, 2nd edn. John Hopkins University Press, Baltimore, MD.
19. **Martinez, J.C. & J.R. Arauz, 1984**. Developing appropriate technologies through on-farm research: the lessons from Caisan, Panama. Agricultural Administration, 17, 93–114.

    Yates, M., *et al.*, 1988. Fertiliser in Les Cayes, Haiti: Addressing Market Imperfections with Farm Based Policy Analysis. CIMMYT Economics Working Paper 88/01. CIMMYT, Mexico, DF.
20. **Tripp, R. (Ed.), 1991**. (Op. cit.).
21. **Franzel, S. & S. van Houten (Eds), 1992**. (Op. cit.).
22. **Finan, T., 1993**. Structural Adjustment, Policy Reform and the Resource-Scarce Farm Family: the Waning Role of FSR. JFSR-E, 4(1), 125–38.
23. **Norman, D., 1980**. (Op. cit.).
24. **FAO, 1989**. Farming Systems Development: Concepts, Methods, Applications. FAO, Rome, Italy.

    FAO, 1994. Farming Systems Development: A Participatory Approach to Helping Small-scale Farmers. FAO, Rome, Italy.
25. **Herdt, R., 1987**. (Op. cit.).
26. **Tripp, R., *et al.*, 1990**. (Op. cit.).
27. **Simmonds, N., 1985**. Farming Systems Research: a Review. World Bank Technical Paper no. 43. The World Bank, Washington, DC.
28. **Jolly, A.L., 1953**. A Report on Unit Farms to the Caribbean Secretariat, Trinidad. Imperial College of Tropical Agriculture (mimeo).
29. **Collinson, M., 1972**. Farm Management in Peasant Agriculture: a Handbook for Rural Development Planning. Praegar, New York.
30. **Nottidge, P. & J. Goldsack, 1968**. The Million Acre Settlement Scheme. Routledge, UK.
31. **Schultz, T., 1963**. Transforming Traditional Agriculture. University of Chicago Press, Chicago.
32. **Collinson, M., 1972**. (Op. cit.).

Chapter 4

FSR in Technology Choice and Development

4.1 THE APPLICATION OF FSR TO TECHNOLOGY DEVELOPMENT

Ann Stroud and Roger Kirkby

This growing understanding of small farmers and their circumstances was the beginning of a farmer-oriented rather than a 'top-down' approach to technology development, a trend that continues to evolve.

4.1.1 Introduction

On-farm experimentation (OFE), as a major component of FSR, came into being primarily in response to the change in 'clientele'; from large to small-scale farmers with colonial independence (in Africa) and a growing awareness that established research approaches were not achieving the expected results with this constituency worldwide. Bentley's 1994 statement on farmer participatory research – 'The current strong interest in FPR (farmer participatory research) was conditioned more by dissatisfaction with formal sector agricultural research than by new information about the value of traditional farmer knowledge'[1] – equally well reflects the central issue driving the birth of FSR in the early 1970s, and its subsequent evolution.

Through the 1960s the realization grew that small-farmer circumstances were complex; they had multiple objectives, differing levels of access to productive resources, preference for short-term benefits over longer term ones when survival was at stake, specific farm practices to manage uncertainty and risk, and unanticipated opportunity costs associated with resource transfers between enterprises.[2] This growing understanding of small farmers and their circumstances was the beginning of a farmer-oriented rather than a 'top-down' approach to technology development, a trend that continues to evolve.

FSR focused on interdependencies and interrelationships between technical and human elements. It borrowed theory from rural development, farm management economics, systems thinking and agronomists' on-farm trials. It blended economists' perceptions with agronomic concerns to identify small-farmers' unique qualities and build on these by working with them, not only in identifying research priorities, but also in developing potential solutions to their problems. OFE became an important initiative to generate technologies useful to small farmers in an efficient manner and as a complement to on-station research.

As early as the 1940s, researchers were seeing the value of moving experiments onto farms: 'The simplest and indeed the only sure means of assessing the value to agriculture of a new discovery is to test it by the field experiment method in fields which are under ordinary crop production. Experiments in the cultivator's fields are, therefore, essential links in the chain connecting discovery and its application to practice'[3]. However, only a few researchers did move on to farmers' fields, this sited experiments in more representative ecologies but researchers continued to direct and manage them. Evolution has continued in response to limited progress in improving adoption, to tightening research budgets, and to new preoccupations. The new methods of farmer participatory research (FPR) have modified, indeed in some authors' minds replaced, FSR. New preoccupations, particularly the burgeoning concern with natural resource management and sustainability[4], and also the inclusion of the resource-poor[5] and rural women[6] have increased the complexity of technology development. Similarly, it is increasingly appreciated that change at the farm level may be stimulated by community action or policy shifts and may impact not only on the farm, but also on the community, landscape and agroecosystem as a whole. The operational scale of FSR is expanding to embrace these widening implications.

Throughout their history FSR and OFE have benefited from reviews, discussion, analysis and refinement[7]. Past evolution has been strongly influenced by three major factors; a more overt inclusion of farmer participation, the realization that options are important for varying circumstances and new thematic areas which have emerged – sustainability, natural resource management and gender among others. Projecting into the future, we see farmers becoming more sophisticated in demanding services, with modern communications, GIS and decision support models becoming increasingly useful in setting priorities for technology development and in its dissemination. We anticipate development of more *micro* or *inward* approaches to resource management at both household and community levels, including empowerment of communities; as well as more *macro* or *outward* uses of modelling, GIS and databases for analysis of diversity and for extrapolation. Several trends beyond the traditional borders of the farm will affect OFE including natural resource management, livelihood analysis, the role of NARIs in agricultural transformation and the integration of the production–distribution–consumption sequence[8].

OFE approaches have been greatly influenced by changing philosophy and better conceptualization of what interventions, at what hierarchical level, benefit whom. Much of the evolution discussed here has been stimulated by iteration between learning from practical experience and concepts and philosophy. Our contribution offers more detail on the reasons for moving experimentation onto farms in partnership with farmers. It looks at the development of concepts and methods which changed ideas on where OFE is done, who it is done with, how its priorities are identified, and finally how it is done: the 'where', 'who', 'what' and 'how' of OFE. We review developments in systems perspectives, how these have influenced OFE, and how changes in the roles of farmers, researchers and institutions have also driven its evolution. We also look at the development of OFE in three major research areas: crop varieties, natural resource management (including trees) and livestock with their special problems. Finally, our contribution reviews two recurring issues in experimenting on farmers' fields and revisits expectations for the future.

4.1.2 Why OFE?

In Africa during the 1930s, colonial governments expanded their use of research stations, mostly selected to emphasize cash crops of interest to their metropolitan economies. These structures and priorities were inherited (mostly in the 1960s) by governments which added food crops to better address the needs of the mass of small farmers, and added substations to expand coverage of a more diverse set of agroecological zones. Subsequent rates of uptake of results by the small-farmer constituency were disappointing. Essentially FSR, including OFE, evolved to address this issue.

Movement towards using off-station sites was intended in the first instance to capture a more realistic sample of soil and climatic conditions, particularly for research related to soil fertility and pest and disease complexes[9]. A second motive was to test for transferability of technologies developed under the artificially favoured management of most stations: optimal fertility, seedbed preparation, pest/weed/disease controls and easy access to inputs, tractors and timely operations. In a few early cases, 'traditional' management systems and farmer experimentation were seen as logical and important to build upon[10]. Even these few scientists, however, carefully selected sites and imposed experimental design as used on research stations. They did collect a more complex set of data to allow statistical analysis to understand variability. Multilocational testing, often at sites contracted from village schools or others to ensure researcher control, was adopted by crop breeders and others as they came to realize that on-station environments were not representative of the range of farmer conditions, an idea given credence by developed-country researchers' successful interactions with farmers, for example in the land grant system in the USA (see for example Flora and Francis in Chapter 5.1).

As the 'place' for doing research evolved, so did the methods and assessment of results. Technical factors identified as limiting yield began to modify experiment design, usually as simulated experimental treatments and, less commonly, as suboptimal treatments[11]. OFEs were designed by some practitioners not to optimize yield but to identify *stable* cropping systems relevant to small farmers operating under serious labour and capital constraints[12]. Thus, the selection of experimental treatments, such as varietal types and cultural practices, gradually became more relevant to small-farmer conditions and interests. Likewise, non-experimental variables, traditionally kept at non-limiting levels to allow expression of yield differences and to maximize the probability of measuring treatment differences statistically, were modified to be more representative of farmers' conditions, usually suboptimal from a technical standpoint. Assessment methods have changed from the purely statistical to economic analysis and more recently to farmer assessment, which many feel is a more meaningful way to integrate biophysical and socioeconomic considerations. The focus on maximizing yield has not vanished, but rather is augmented by consideration of farmer and consumer preferences for varietal selection, feasibility of labour requirements for agronomic practices, systems sustainability, equity or gender considerations and others. Eventually, a number of reasons for conducting research on farmers' fields emerged that are value-laden. These values permeated the evolution of the 'way' to conduct experiments.

4.1.3 The original expectations for OFE

Initially FSR, much in line with the Green Revolution philosophy, addressed productivity. In the early 1980s systems sustainability and social equity entered the picture as reactions to early outcomes and to the restricted vision inherent in 'improving productivity'. New goals of sustainable agriculture sought to maintain or increase biological and economic productivity, enhance efficiency in use of inputs, increase production stability, increase resilience to environmental change, minimize adverse environmental impacts and ensure social compatibility[13].

The broad goal for OFE in the context of FSR has not changed over time: to enable the cost-effective identification of technologies that are more quickly adopted by more small farmers. The use of a systems perspective provides an improved understanding of farmers and their circumstances, and in turn aids the diagnosis of priority problems, the identification of research opportunities and the improvement of planning in applied and strategic research programmes. The process is interactive and dynamic, and the focus for improvement is not necessarily predetermined. Its application calls for an appreciation of farmer knowledge, of location specificity, of the influence of exogenous factors, and an understanding of how and why farmers compromise on optimal technical practice[14].

It was within the FSR context that OFE was originally applied as a means for better mobilizing the results from the local research station to reach farming communities. The emphasis, early on and still today for many programmes,

Box 4.1.1. Reasons for OFE.

- To rigorously test technology developed on-station or elsewhere under a more representative range of environmental and human conditions
- To understand new technical relationships relevant to clients' conditions, for example interactions between crop varieties and low soil fertility
- To enable farmers to evaluate technology under their conditions using their criteria, including socio-economic factors
- To identify new researchable problems perceived by farmers, to understand their compromises on technically optimal management and to use these in refining other diagnostic information
- To seek and apply farmers' knowledge in all stages of the research process
- To improve farmers' capacities for experimenting

was on the adaptation of recommended technologies to suit the circumstances of a greater number of local farmers, and then planning more relevant research on the station. We shall see how these expectations have changed over time, and how OFE enhanced links between on-station and on-farm researchers and farmers, brought disciplines purposefully together for the first time, and changed disciplinary roles and relationships. It also broadened the types of disciplines involved to include sociologists, policy makers and others. Methods and approaches for technology development evolved to embrace the values inherent in the expectations of OFE, first and foremost among them being the opportunity to get closer to farmers and to enhance their 'voice' in the research process.

4.1.4 Where to do OFE, what to work on, who with and how

The plethora of decisions to be taken in conducting OFEs is one reason why researchers have found FSR, including OFE, difficult to implement. Most FSR has been, and unfortunately often still remains, in a pilot mode, outside the nationally recognized processes for R & D, and consequently has been much more complex to implement than the annual rolling programmes of station-based research in which broader decisions are taken by managers. Several contributions in this book address these 'overhead' issues; many related to the institutionalization of FSR as a whole. Much of the early OFE was funded through donor programmes in which, at worst, funds and personnel were parachuted in for 3–5-year periods. These 'external' projects often answered the 'where, what, who and how' questions with little regard for the subsequent transfer of the decision making into a national programme. Neverthless, experiences outlined in this section illustrate that there are a number of areas in planning and implementing OFE which practitioners have grappled with over time; within each subsection progress and evolution is discussed.

Where to do OFE: RDs

The evolution of thinking on typologies for agricultural development is discussed in Chapter 3.2. RDs which reflect discrete farming systems and therefore integrate environmental and human circumstances, are widely used for stratifying farmers into groups. The underlying principle is that farmers managing similar farming systems use similar technologies and will often adopt the same improvements. RDs typically have a spatial dimension, for example differentiating among crop combinations across land types and soil toposequences and a resource endowment dimension that recognizes household wealth differences which result in different farming systems. The priority setting process through farmer consultation has often failed to take into account this resource endowment dimension and unwittingly missed the 'target' of the very resource poor, at both international and national levels[15]. Farmers, even within small geographical areas, do not necessarily share the same experiences, access similar resources or share the same sense of sustainability[16].

Recent international preoccupation with managing the environment has stimulated the concept of RMDs: areas with a similar pattern of climate, land types and broad social and economic circumstances with similar implications for resource management[17]. RDs and RMDs are essentially different levels in a hierarchy of systems. For example, while striga (witchweed) may be an important pest on specific soil types throughout an RMD, farmers operating different systems within the RMD, and grouped in different RDs by wealth, might need different solutions to the striga problem – the purchase of fertilizer perhaps being feasible for the better-endowed farmers while tolerant varieties are a more feasible solution for the poor[18].

Delineation of RDs needs to remain flexible, because farmer typologies are as dynamic as farmers themselves and some environments, mountainous areas for example, are highly heterogeneous. OFE is slower and generally more costly than diagnosis, the selection criteria for both RDs and experimental sites demanding careful thought. Tanzania is one country which now has a national-level farming systems zone map which can be linked to databases containing farming systems and farmer characteristics and constraints. With the advent of GIS it is becoming increasingly feasible to map RDs, and a good deal of methodology is under development, an example of progress in Brazil is reported in Chapter 11.2. However, over-reliance upon GIS, a tool for describing spatially definable dimensions (a major soil type, or distance to market), also risks under-emphasizing, even omitting, non-contiguous determinants such as microenvironments and wealth classes.

What to work on: setting priorities for OFE

Priority setting for OFE has swung from one extreme to another. In its early days FSR, generally acting in an 'extractive' mode, allowed researchers to gain an understanding of the system and to discuss key problem areas and possible solutions with farmers. More recently the participatory movement has taken great pains to establish farmer and community priorities, often without much understanding of the farming system as a whole. Both extremes have usually ignored broader policy issues when establishing an OFE agenda.

Two levels of priorities need consideration, particularly where resources are limited. Where a national map of RDs is available, OFE programme focus is likely to be guided by national or local government priorities related to domestic food needs, export opportunities, reaching the rural poor, etc. From a national perspective, resources need to be deployed to those RDs where such priorities can best be addressed and which show potential for extrapolation to other areas. While national and local priorities need to be reconciled in any OFE programme, meeting farmers' own priorities will be of vital importance to its success. This requires flexibility and may lead to the adjustment of researcher-determined priorities.

The three-way balance in an OFE programme to meet farmer, community and government priorities is in a second iteration of controversy. In its early days, scientists questioned FSR's heavy reliance on farmers' views. They perceived small farmers as having limited knowledge and likely to unduly confine the scope for change by influencing the research agenda. The second iteration of this controversy is between promoters of participation for empowerment and practitioners of FSR. Many participatory practitioners argue for an OFE agenda based wholly on local farmer and community priorities. However, opportunities for solutions usually go beyond local knowledge. The combination of outsiders, who understand the breadth of opportunities available in technology, market access or policy change, and local farmers, can together identify opportunities relevant to local change. Local farmers' knowledge cannot provide all the answers, neither do farmers necessarily recognize all the problems, particularly those with longer term consequences such as the potential for soil mining through nutrient-extractive technology in a low-input system.

The pragmatic answer seems to be that both participation and systems understanding are important in a partnership between farmers and researchers. Eliciting and acting on local priorities gains the trust of the community, and creates an environment for local learning or 'empowerment' that can have longer term benefits[19]. It is this trust which allows deeper system problems to be confronted more easily. In this sense participation and empowerment really are prerequisites to the creation of partnerships that increase the success rate and reduce the costs of traditional on-station research. Achieving this usually means concentrating limited research resources in a few communities. The challenge then is extrapolation: to understand in the course of planning just how site-specific the 'local' priorities are and to what extent the solutions identified can be disseminated to a larger mandate area.

Putting priority setting into practice remains potentially controversial among collaborating research partners. Kirkby *et al.*[20] used meetings with individuals and communities to design the research agenda, to identify individuals interested in participating in experiments on particular components of the agenda and to obtain rapid feedback during the first year of on-farm trials. In this way the research team refined their understanding of farmer priorities and preferences derived from surveys, quickly eliminated some experimental factors, identified traditional practices and explained the influence of socioeconomic factors. Most successful OFE programmes have used this sort of iterative process.

The participation of farmers as active stakeholders complicates the setting of priorities, particularly if gender and wealth differences are to be taken into account. Occasionally, priorities established jointly with farmer research groups (FRGs) confirm a research agenda already developed by researchers themselves, as in a Zambian case[21]. More commonly, the participatory process leads to a wider agenda and requires the drawing in of new expertise to address the unexpected. This diversity and, at the same time the local specificity of an OFE agenda, requires careful consideration. At the same time agendas are likely to be heavily influenced by the selection of stakeholders. For example two innovations favoured by villager committees in the Aga Khan Rural Support Programme in the northern areas in Pakistan were improved bulls and forest trees, although only the wealthy farmers had the land to spare for fodder and forest tree planting[22]. In itself this is not a problem unless programme priorities among the RDs involved are distorted. Being aware of who benefits, and participant's relationships to the target community, is essential and may lead to complementary OFE targeting of other RDs. New skills in facilitation and old skills little tapped, especially from social science, can help in this process.

Who to work with: identifying representative farmers

The question of 'who' are the farmers that one is working with has two main aspects: first, the identification of representative groups to discuss priorities, experimental design and evaluation, and second, who should join in experimentation. Within an identified RD, working with representative farmers raises no theoretical problems – reconciling the social and ecological profile of the RD with individual farmers should be straightforward. Inability or neglect to do this in practice has been one of the downfalls of FSR programmes. There are many practical issues involved, particularly where RDs are differentiated by wealth and gender and are not geographically contiguous. Both women and the poor tend to remain invisible, and interactions with outsiders are often dominated by élites, particularly within mixed communities that include poorer households.

The pros and cons of working with volunteers versus a purposively stratified random sample have been widely discussed. In the early years OFE farmers tended to be 'faceless', chosen by researchers for ease of communication and readiness to collaborate, their accessibility and responsiveness presaging easier and more demonstrable impact. Selecting 'master' progressive and 'model' farmers deemed to be easier to work with, more ready to follow protocols and likely to influence others subsequently, was an approach rooted in many pre-FSR extension programmes. In general most were male, unrepresentative farmers who tended to be the most accessible, but not necessarily those who are the innovative experimenters best equipped to participate in technology choice and development for the community.

Researchers have developed tools for differentiating by gender and wealth. Feldstein and Poats[23] developed a user-friendly gender analysis framework now widely applied to understand differentiated decision making in the household and to distinguish the resource base of men and women householders. Grandin's[24] wealth ranking tool to establish resource endowment levels is only now gaining popularity in practical applications. Over time, common sense and prodding from social scientists and donors have posed provocative questions:

- Who are these farmers and what do they represent?
- How are women farmers being targeted?
- If our project goal states we are targeting resource-poor farmers, who are they and are we addressing their needs with our technologies?

Such questions have undoubtedly stimulated better differentiation among target groups. The urgency for this has been reinforced by the growing international concern with poverty and recognition of its link with environmental degradation. There has, however, been a lag in incorporating such understanding into the agenda for technology development and into the criteria applied when involving farmers in experimentation. Granted the principle involved working with women and the poor has proven more difficult than anticipated. Women usually do have different objectives and interests, are often inaccessible: are too busy to attend village meetings; it may be socially unacceptable for them to have a visible, verbal presence or to be approached by male researchers; and they may be shy about working with officials from outside. Women and poorer farmers may not have the necessary resources to participate easily: their land area may be too small, labour not enough and their ability to accept risks from trying new technology too limited.

Researchers have developed strategies to reach these groups. For example, holding separate interviews with women and men; timing meetings so that women can attend; and working on specific issues of interest to women[25]. Practical experience has been varied. Norman and his team[26] used FRGs in Botswana to elicit participation by farmers other than male-headed, richer households. Using farmers as experimenters, Sumberg and Okali[27] were able to observe how technology was used in the context of the household and developed a better understanding of household decisions governing tree foliage use by men, and the feeding of small ruminants by women. Researchers in Zambia initially tried to stratify their selection by socioeconomic criteria, but the process was too labour intensive and the resource-poor often performed disappointingly in OFE; they had other priorities, did not want to assume the risk, the individual trial farmers operated in isolation without community control. The Zambia programme later switched to the use of FRGs[28].

A major conclusion for any of the methods of farmer selection is the importance of *knowing* who you are working with. This should be clear at a sufficiently early stage to permit adjustment to minimize it as a source of bias and to solicit participation in a sensitive way. In addition, OFE requires different management scenarios depending on the degree of researcher involvement, and the level of risk and innovation implied[29].

Trial site choice

When establishing a trials programme there are two levels of decision on site choice: first, where to establish the field effort within the RD, and subsequently the choice of site in the fields of collaborating farmers.

Prior analysis of the extent to which a site represents an RD has been uncommon. Considerations of cost have led much on-farm research to be located conveniently for researchers, usually clustered in a relatively small area of the RD to aid logistics and access. Nevertheless the advantages and disadvantages of dispersing, clustering and stratifying experiment locations need to be explicitly weighed. Other criteria advocated by practitioners include: the state of community organization, the willingness and enthusiasm of the community and/or individuals to host research, reliability of access during the rainy season, the presence of an extension or development worker, environmental or management risks, logistics and distance between trial sites, amount of research supervision required, and the number, size and complexity of trials necessary to adequately sample variability. Cost considerations related to logistics, data reliability and farmer willingness and interest that may lead to further innovation and spread of technology, tends to be more important in OFE than in other FSR activities; these aspects may indeed warrant greater weight than representativeness. The compromises involved in these choices need to be described in drawing conclusions and the compromises themselves kept in perspective by the knowledge that an RD site is a lot closer to reality than the research station.

Site choices can be researcher driven at the community and village level, and farmer driven at the local and field level. Even when the choice is driven by farmer interest, researchers need to question what wider group is represented by the field site or level of farmer management applied; both strongly affect the interpretation and extrapolation of the data. Tools useful in site selection include maps, survey information and locally made transect diagrams. More use should be made of sample frames, coupled with criteria for making hierarchical choices on region, subregion, village, group or set of farmers, and fields within farms. Such a framework at least offers the researcher greater opportunity for relevant analysis. Carter[30] and others have used map overlays, databases and correlation (now easy to accomplish using GIS) to visualize both environmental and socioeconomic heterogeneity.

Stepping down in scale, within-field site choice involves random or deliberate selection as the two main options. Stewart[31] put forward such considerations as trial objectives coupled with knowledge already available in making the choice. Deliberate selection implies some knowledge of the effects of individual factors being tested and presupposes a rationale for the selection criteria (e.g. crop nutrient treatments at a low fertility site). With increasing farmer participation in OFE, trial site choice has, to a greater extent, been devolved to farmers who understand the objectives. This has spurred the use of different and appropriate data collection and statistical methods to handle interpretation of information derived from a wider range of circumstances.

Identifying technologies for problem solving

An early assumption in OFE held that component technologies, usually in the form of 'packages' (e.g. crop variety, spacing, time of planting and fertilizer level), would be developed on-station and passed to an on-farm research team for validation, with information on successes or problems being fed back to the on-station 'designers'. Historically, high rates of farmer rejection were due in part to the on-station origin of technologies. Poor feedback mechanisms between on-station and on-farm research teams and the questionable reliability of on-farm testing by researchers, who often distanced themselves from their clients by the more extractive methods employed, penalized successful adaptation and brought FSR a poor reputation.

Five factors are in the process of changing this situation:

- Better understanding of the adoption process of small farmers.
- Improved problem identification and causal analysis.
- Increased farmer participation in the OFE process.
- Increased exposure of on-station researchers to farmers.
- Offering multiple choices for new technology.

Causal analysis, increased farmer participation and changing roles of researchers are covered in their own right in section 4.1.6 below. Numerous studies, such as that by Byerlee and Hesse de Polanco[32] with barley producers in Mexico, have shown that farmers normally adopt technology components sequentially; and that complex packaged technologies with multiple products, such as alley cropping, cannot accommodate wide variation and uncertainty in farmer conditions[33]. However, even technologies that traditionally have been 'packaged' can usually be broken down into components for sequential introduction. It is generally more efficient to leave farmers, responsive to their own circumstances, to incorporate components into their system and for researchers to monitor this incorporation for lessons that may require follow-up by the formal research sector.

Participatory techniques proving their worth in problem identification include visualization through mapping and diagramming, identifying and using indigenous knowledge and causal analysis. The drawing of problem trees is a common tool in causal analysis. Tripp and Woolley[34] made explicit the fact that problem identification does not lead directly to solutions but that causal chains can be identified, and when necessary better understood, through experimentation. One important (but as yet little exploited) use of causal chains is to create multiple entry points that widen the options for solution. Better understanding of farmer differences has also widened awareness of the need to provide more options. In essence, 'multiple options' rather than 'single best bets' have begun to revolutionize the approach to technology identification and development. The former presupposes client-driven choice while too often the latter has been researcher driven.

In a seminal article in World Development, Biggs and Clay[35] put forward the idea – obvious with hindsight – that the local research station is only one potential source of useful technology for a farming system. Other potential sources include local farmers who have already addressed the problem (an aspect of indigenous technical knowledge), research stations worldwide operating in similar conditions of climate and soil, and farmers worldwide who have already faced and overcome the problem. Until now only limited efforts have been made to organize information in ways that allow local professionals to draw on such extensive sources, but clearly GIS and the new information technologies now make this practical. As an example demonstrates, it is creeping into OFE practice. Within Tanzania's FSR programme, while most technologies still come from the research station, the Lake Zone team tested a number that had been developed elsewhere.

Wider African experience brought into the project through technical support from The Netherlands affected the selection of ox weeders, a milk churner, a wheelbarrow, grafting of cassava and cottonseed cake for feeding oxen. Livestock researchers examined the use of indigenous feed and pest-control options.

Using a participatory approach a better set of potential solutions are likely if farmers' local knowledge and researcher 'world knowledge' are brought together[36]. While farmer interest is now generally better captured and the station research agenda is more relevant, the site specificity of the agenda, particularly with heavily participatory approaches, are still controversial in terms of the effort invested. Extrapolation and therefore linkage with regional and national priority setting remain a challenge.

Types of trials

Many early papers about OFR organized typologies of trials by function (state of knowledge and type of technology being tested) or the consequences for implementation[37]. Most typologies are embodied in the categories shown in Box 4.1.2. Developing an operational framework was a necessary and important step, and for the first time served to relate trial function to management and design choices (see 'Trial design' and 'Trial implementation'). This helped individual researchers and programmes move onto farms, and assisted more experienced practitioners with the multitude of decisions demanded by dynamic OFE. It also offered a terminology that practitioners could relate to. The framework refined thinking about trial functions beyond the research station, but may have initially been too rigid.

Evolution here has more to do with 'how' the different types of trials have been used than with the the categories themselves. Some early typologies were couched as sequential research stages through which technology was expected to pass before being 'ready' for farmers. Regardless, most researchers now feel that OFE has an integrated extension function and that

Box 4.1.2. Types of trials.

- *Exploratory and diagnostic trials:* used to confirm or clarify problems and causes, to assist in making a tentative diagnosis, or where 'exploration' is necessary. This includes what IRRI called 'constraints research', in which experiments are used to understand reasons for non-adoption of apparently promising technologies[38]. Exploratory trials are usually relatively simple with one or two treatments added or removed from the farmer's practice, or a simple factorial combination at two levels so as to determine which are most important or interacting. Finally, such trials may be aimed at better understanding causes or exploring the chain of causes
- *Refinement trials:* to provide information on technology performance or the refining or fine-tuning of potential solutions. These trials tend to be larger, as a range of options may need screening under farmer conditions
- *Verification or validation trials:* on the basis of strong evidence of success from limited sites the technology needs to be verified across a diversity of circumstances before it can be recommended with confidence. Results are used to formulate recommendations, identify their limitations, modify technology and this may lead to new questions that require further research. Researchers have argued to keep these trials simple, with few treatments so that farmers can understand them. However, there has been a tendency to underestimate the capacity even of extremely resource-poor farmers to compare large numbers of treatments if they specialize in the crops concerned; Sperling *et al.*[39] (1993) provide an example for beans in Rwanda
- *Demonstrations:* usually having one or two treatments placed in relatively large plots for ease of visual comparison. Although by this stage most researchers and extensionists feel that the technology is ready for transfer, monitoring can be useful in new environments and can encourage a more dynamic and participatory approach to extension
- *Farmers' experiments:* typically not listed in the early days, the experiments are designed, implemented and interpreted by farmers. Farmers' management of technology that they have sourced from research, extension or neighbours can be monitored jointly. Researchers too are likely to learn from the innovative ways that farmers incorporate new ideas into their systems[40]. Ashby[41] reviewed the comparative advantage of farmer experiments; they are particularly cost-effective in speeding up the process of adaptation and testing

the 'stages' are really a continuum. In addition, it was initially understood that farmers entered the process at its later stages, but now farmers are being brought in at all stages (see 'Changing roles: farmers' below).

Trial design

Researchers engaging in OFE were forced to revisit decisions on trial design. They found it challenging to maintain on-station experimental standards when working on-farm, due to irregular topography, non-uniform management practices (e.g. sowing-date differences), heterogeneity of cropping patterns and small farm sizes. Numerous publications offered guidelines, Mutsaers and Walker[42] systematically address numerous experimental design considerations: number of factors/treatments and field sites, replication, experiment size, management of non-experimental variables, degree of farmer management and control/check treatments, with recommendations for various trial types. Many others have also provided guidelines[43] to assist decisions on interpretation, error terms, measuring of interactions, confounding, degree of accuracy needed, etc. Ultimately, each case needs to be considered according to its objectives, the variability at the site, and the need to explain differences reliably. It is these choices that make OFE difficult to implement by standard protocols.

Selection of control or check plots has been controversial, with numerous choices for different circumstances. Superimposed trials offer a more realistic farmer control. Farmer controls or farmer's practice are potentially variable within the experimental area, but may not be carefully specified in the experimental write-up. Farmers most certainly use controls or comparisons that differ from those of researchers when making their assessments. It is easier for them to compare the performance of a treatment with what normally happens over time and space in their own circumstances; whereas a researcher relies on a control plot which is in close proximity to the treatment. Thus farmers, at their local level, have a wider frame of reference[44]. The use of gender analysis to understand the division of labour, control and access to resources, and the specific constraints faced by different groups can help research design, for example by appreciating a situation where women make decisions on seed selection and men determine where the fertilizer is used[45]. New techniques have increased the confidence and therefore the use of farmer evaluations by previously sceptical researchers, although there is more to be done in understanding farmers' frames of reference.

In the early days of OFE, researchers sought specific guidelines on the number of replications to use for each kind of trial and kind of conditions; now however, researchers worry less about prescriptions. Like other design aspects, the choice of the number of replications is dependent on several factors, making it difficult for the novice to feel confident: precision, labour and space required, type of treatments, the level of prior knowledge of site and area variability, among others. Replicating treatments across farms rather than within a farm became a logical and statistically acceptable option when a crucial aspect of evaluation was performance across the range of variation represented by an RD.

The level of design complexity was another area often debated. In general, simplicity has been the rule. A simple experiment (e.g. a variety comparison or paired-plot fertilizer test) at many sites can yield a large amount of useful information, provided that a wide range of variables are monitored: farmer preference, soil, cropping history, weeds, pests and diseases, and other environment and management data[46]. On the other hand, infrequently used designs were also suggested to avoid having large experiments: trials with non-factorial arrangement of treatments, confounded factorials, stepwise designs, incomplete blocks and others. More sophisticated analysis techniques may be required (see 'Issues of analysis and interpretation of results' below). Although Box[47] found that more complex designs may be necessary to safeguard statistical relevance, such trials may be more difficult to interpret for many agronomists and social scientists. Sperling *et al.*[48] maintain that farmers can handle large and more complex experiments if they are 'experts' in the thematic area or the subject matter; the key may be to bring in or acquire the skills and take the time to identify local experts. New efforts in monitoring participatory research should yield further clues as to what is appropriate to various circumstances.

Trial implementation

In the late 1970s and in the 1980s the degree of farmer involvement in OFE were categorized in a number of ways, either sequencing their involvement over time or identifying their role in experiments with different purposes. On-farm research teams in east and southern Africa developed categories to help distinguish various management strategies:

- 'Researcher managed, researcher implemented' (RM/RI) where the researcher maintains responsibility for the trial, controlling variation and management, but enters into a contractual arrangement for the farmer's land and labour services.
- 'Researcher managed, farmer implemented' (RM/FI) trial where the researcher plans the trial but works together with the farmer in implementation so that some factors (experimental and non-experimental) are controlled, but the farmer is encouraged to manage things 'normally' so that the researcher can see how the test options perform under farmer management.
- 'Farmer managed, farmer implemented' (FM/FI) trial where the researcher and farmer may consult on design but the farmer makes all decisions on how to implement, including the non-experimental variables.

As farmer participation in OFE increased, various practical arrangements were arrived at including new ways to manage trials. Researchers have had to deal with farmers' expectations and the consensus in the literature is that researchers must make a conscious effort to foster farmer ownership. Several reports have indicated that decentralized trial management is necessary to devolve experimentation largely to farmers, and that the collection of feedback on a regular basis is essential[49]. Researcher time allocation across tasks changes. While FRGs on the one hand increased the ability to manage large numbers of farmers, they also brought new questions; how large a group could be handled, what types of farmers are included, how to handle the dominance of some types, etc.

Efforts to categorize OFEs continue, often for training purposes, with an increasing emphasis on the need for flexibility. Table 4.1.1 highlights points to consider in decisions on trial management.

The management of non-experimental variables (NEV) has undergone large changes in philosophy; from being standardized to being allowed to vary with the situation. Even in researcher-managed trials, NEVs cannot be controlled to the same extent as on-station. Rules concerning standardization generally align themselves with the degree of control required as determined by the overall trial objectives and the subject matter. Where interactions are expected, such as between fertilizer and weeding, the researcher may want to control a critical NEV[51]. Allowing farmers to intervene has meant that data collected on non-treatment conditions must increase if there is a need to explain or understand variation

Table 4.1.1. Points to consider when deciding on level of farmer involvement[50].

Points to consider	More researcher Involvement	More farmer involvement
Type of precision wanted	Control of on-site variability and detailed measurement of treatments	Measure and understand variability in the target group Identify resource conflicts
Confidence in the technology	Less	High
Risk to the farmer	High	Low
Type of experiment	Refinement, exploratory	Validation, demonstration
Trial objectives	Require tighter control of the experiment	Farmer conditions are needed for implementation and assessment
Trial complexity	Greater	Less
Number of experimental sites	Smaller	Larger

between the treatments or the sites. Farmer management is an increasingly attractive option for controlling experimentation costs, which can be surprisingly low, affordable in just about any programme[52], while also reaping the benefits from farmer participation. Moving to this mode reduces concerns about timing of operations, but may increase concerns about biases in management. These and other aspects are discussed in the literature and guidelines are offered.

Trial evaluation

Trial evaluation has evolved from considering traditional yield and biophysical performance data to considering economic performance by enterprise and/or by system, the effect on the natural resource base and system sustainability, stability analysis and finally to relying heavily on farmer assessment, which can integrate numerous factors from farmers' perspectives. The tendency has been for researchers to rely on the latter when farmers have greater involvement (usually at later stages of testing), and to continue to rely on traditional performance criteria and classical statistical significance when researcher managed (usually at earlier stages). This has often pre-empted farmers' choices and reduced their options.

Farmer assessment tools, preference and matrix ranking and carefully constructed open-ended evaluations, have provided researchers with frameworks for quantifying preferences, understanding and ranking useful farmer criteria, as well as offering more formal fora for farmer involvement in evaluation. Information solicited has proved extremely useful as feedback into variety selection, farm implement design and other research programmes. It also demonstrates how farmers can integrate and weight a number of factors, hitherto a challenge for researchers using their conventional statistical and economic methods.

Participatory monitoring and evaluation are being introduced as a means of measuring longer term effects, such as in soil fertility and in soil and water conservation programmes. This elicits farmer-derived indicators as well as researcher indicators of progress and change, and can be at the hierarchical level of a field trial, a watershed or a farming system[53]. This set of tools, derived from the log frame approach to project monitoring, offers researchers a 'larger' picture, in terms of both scale and time, and provides input into impact studies. A potential 'trickle-down' effect into OFE operates where trial data parameters are linked to, or are the same as, the indicators.

4.1.5 A review of the impact of OFE in three thematic research areas

Variety development, crop improvement and seed systems

Although collaboration between International Agricultural Research Centres (IARCs) and national programmes has resulted in substantially increasing crop yields on farms in high-input situations, plant breeders have not displaced the pre-eminent position of landrace varieties grown in many low-input systems[54]. Farmers' variety preferences were distinctive, but were initially poorly understood by national and international researchers involved in variety development programmes. The 'Green Revolution' approach of many IARCs that supplied standard regional or global germplasm nurseries to national programmes dependent upon imported fixed lines, initially did little to improve adoption performance unless accompanied by well-oriented local selection. An example is the rejection of the International Rice Research Institutes (IRRI's) high yielding, early maturing rice varieties by the majority of small farmers and consumers in Tanzania who preferred medium to long strawed, aromatic types.

Many OFE practitioners feel that crop variety testing is so straightforward as to offer few lessons in generating the more complex technologies that are increasingly needed to sustain systems under pressure. However, experience suggests that even successful variety development is not so simple in many environments. 'Where to test', 'what to select for' and 'how to evaluate varieties under farmer conditions', three major issues in the early 1980s, are still debated today. Before the advent of FSR, multi-locational testing was a common practice to select germplasm adapted for heterogeneous ecological conditions. Multilocational tests for less productive environments were often under-represented and farmers were involved in a contractual mode, providing only land and labour but not advice nor evaluation.

A surprisingly persistent issue is whether suboptimal, even low-input, environments should be used in plant breeding. As early as the mid 1980s groups of breeders involved in a range of crops in Africa recommended ultimate testing under a relevant range of farmer conditions[55], but wider progress has been slow. Feedback from OFE has been critical in moving from the traditional situation where breeding criteria were set from the plant breeder's point of view to one where farmer-based criteria for evaluating crop varieties are increasingly used. At the same time the conventions of experimental precision had dictated the use of high-input regimes to force the expression of differences between varieties. Many breeding programmes were disinclined to factor in the constraints faced by farmers in improving crop management practices. This has been exacerbated by poor communication with farmers and/or on-farm researchers and has inhibited setting NEVs at more appropriate levels in selection trials.

While breeders still debate these issues, a consensus is emerging among the profession that adaptation of varieties to poor and difficult environments is most likely to be achieved by selecting in those environments, and there is increasing acceptance that farmers, no matter how poor they may be, are critical selectors who make sensible decisions. In the case of Tanzania rice varieties (cited above) it took 10 years before feedback was systematically collected from farmers, through OFE, and used in adjusting selection criteria. Once local selection criteria and farmers' trade-offs are understood and responsibility for variety development is decentralized to local research stations with breeding staff capable of making adequate numbers of crosses, exotic germplasm often remains indispensable as parental material for introducing new characters desired by farmers.

Today farmers are increasingly being brought into the picture, acting as 'advisors' to breeders on what may work in their micro-environments and taking unfinished materials home for selection under their own conditions (see Sperling and Ashby, Chapter 11.4). Experiments to compare breeding by farmers and trained plant breeders are underway, and no doubt much is yet to be learned on their respective roles. Small farmers producing beans in eastern Africa are adept at managing large numbers of varieties to address a range of needs related to home consumption, risk avoidance, exploitation of micro-environments and generation of cash income; yet the single most widely grown bean variety is one that is acceptable at a modest price to low-income urban consumers despite not being preferred for consumption by farm families. Useful tools have been developed[56] to solicit from farmers the effects of gender, resource level and differences in utilization; an understanding of trader influence is an ongoing issue for marketed materials.

Agronomy, crop management and natural resource management

In traditional commodity driven station-based research, agronomy really had no life of its own. Crop management practices were a by-product of the effort of agronomists to give breeders recommended packages for spacing, time of planting, fertilizer and weeding regimes for their materials. In the continuing search for improved varieties, successive iterations followed as new materials were identified. Best practice was aimed at the maximum exploitation of yield potential of the commodity. It is these agronomic components of packages that have been the least adopted.

Over the 20 years between 1970 and 1990 agronomy and crop management experienced two revolutions: first the 'systems' revolution and second the 'sustainability' revolution. In conventional agronomy perhaps rotation experiments come closest to embracing the new dimensions these brought.

THE 'SYSTEMS' REVOLUTION. The impact upon OFE of the systems revolution in agronomy had two dimensions: the breakdown of traditional packages into components, and the shaping of components to constraints identified in farming systems.

Research station recommendations were usually a package designed to give optimum yields for the commodity concerned, and at first OFE or adaptive trials drew on these recommendations for testing under farmer conditions.

However, small farmers could rarely absorb such packages as a whole, and then OFE was used to identify those components that offered a 'main effect' which provided an attractive return in the farming system. The strategy held that introducing a sequence of components, each profitable in its own right, would accumulate interactions and provide extra benefits.

Soon it was realized that farmers' priorities (to produce a diversity of crop and livestock products in order to satisfy food preferences, manage risk and exploit market opportunities) and the resource constraints under which the farmer operated (particularly seasonal labour supply) required the reshaping of interventions to fit these patterns. The role of OFE was to identify whether recommendations fitted the system, or were still viable given the degree of modification needed to make them fit, and to provide feedback to the on-station commodity teams.

Diagnostic work enabled researchers to understand systems interactions and led to the definition of linked problems and solutions, often by subsystem or commodity, and these became the content of the OFE programme. However, it remains difficult to reintegrate solutions or to really apply systems thinking when designing agronomic experiments, even when working in multidisciplinary teams. The nature of the experimental design encourages a breakdown into small parts or pieces. Looking for an intervention that addresses multiple issues or problems remains difficult, though intercropping and agroforestry are two areas where this has had some success.

Intercropping and cropping systems research were two early predecessors of the fuller systems revolution. Cropping systems work began in the 1960s and formed a strong thrust in agronomy in the irrigated rice systems in Asia. As Richard Harwood notes in Chapter 2.3 the early goal here was to maximize the use of water and incoming radiation, usually through relay cropping. Most intercropping research initially followed the usual research station pattern of seeking an 'ideal' in terms of yield, and land equivalent ratios were used as a measure of performance. The expectation in the 1970s was that intercropping would lead to more relevant technology and indeed, as FSR gained momentum, economists were able to analyse the rationality and flexibility of farmers' intercropping systems as a base for improvement. Biophysical scientists later concurred after testing these hypotheses experimentally, but still tended to emphasize optimizing yields rather than reducing risk and lowering labour requirements[57]. Waddington[58] later emphasized the need for a high level of farmer involvement and use of a multidisciplinary approach. He concluded that there had been too much intercropping research on crop densities and spatial arrangements, with a dearth of research on fertility, weed and pest control, and interactions with labour. For minimizing land requirements in OFE, he recommended keeping experiments simple with use of incomplete factorials or confounding designs.

From the mid 1980s OFE received a boost from agroforestry which, although agroforestry practices had been used by farmers for hundreds of years, was suddenly captured for research. Agroforestry is a 'composite' technology. Similarly to livestock: there are multiple products and diagnosis is needed to identify the combination of interests to farmers. Trees take a long time to mature, and other factors contributing to trial complexity include asynchronic production, complex ownership rights and the need for large areas for spatial arrangements. While non-conventional analytical methods such as non-parametric statistics are useful, early experience showed that consulting farmers at a very early stage helps ensure that years of work are not invested in blind alleys (*sic*)[59]. Given that agroforestry is a relatively new and complex area of research, agroforesters have been relatively quick to go on-farm and developed innovative sampling to estimate yields. The 'tree' is seen both as a commodity and as a source of services for users and systems (e.g. soil fertility replenishment).

Moving to OFE in agronomy created challenges, many of which remain. One is the specificity of farmers' conditions with dramatic differences in moisture and fertility interwoven with social (farmer preferences) and economic (labour, risk, land, markets) differences determining the use of cultivar types, and intercropping patterns. Complex conditions gave rise to more complex designs and analytical tools. Concepts developed to sample and understand variable and dynamic situations. Stoop[60] used a toposequence to understand land use and plan experimentation. Models are another way of dealing with the large variations involved in soil fertility[61] and help better descriptions of agroecological zones by mapping the variation in yield potentials. In solving pest, disease and weed problems it is difficult to find the consistent infestation levels (in space and time) that provide definitive answers: OFE was used early on as experimenters gravitated to situations where the problems were found. As such problems arise from a complex interaction of management and environmental factors, 'prescriptive' solutions were rarely successful for small farmers. Nevertheless, systems interactions with labour, and considerations of cost and safety, were not always given due weight when researchers chose their treatments. Gradually, the involvement of economists and increased awareness of health and environmental hazards have increased 'feasibility testing' to choose realistic management options. In addition, farmers have become more involved and, for management of some pests, have been successfully organized into integrated pest management (IPM) groups[62]. Principles are being taught to farmers so that they can apply them in their own contexts, while researchers take on more of a monitoring and advisory role.

THE 'SUSTAINABILITY' REVOLUTION. As the 1980s progressed, the importance of sustainability, and natural resource management for the long term, was increasingly highlighted by both ecologists and economists. Sustainability permeated the international community following the Brundtland report 'Our Common Future' in 1987. Concern about systems sustainability paralleled the rising environmental movement on other fronts. Previous concentration on productivity was and still is largely due to farmers' demands for more income and more food, as well as concern with alleviating poverty and creating impact, with little initial concern for the 'extractive' nature of some technologies or the possible degradation of agrobiodiversity. The sustainability question has stimulated a widening partnership between agronomy and ecology. One result has been the birth of agroecosystems analysis (see Lightfoot in Chapter 11.5) which has brought three new dimensions to OFE:

- Lifted attention from the commodity level up the systems hierarchy to the whole farm and watershed or community, providing a full systems context for the identification of potential improvements.
- Shifted attention from rates and dates to the management of soil and water resources.
- Added long-term goals of sustainability to those of shorter term productivity gains.

Lightfoot and Noble[63] noted that most OFE dealt with commodity components or factors. They felt that a holistic treatment of farming problems was required to respond to the rising concerns about natural resource management (NRM) and systems sustainability. They argued that the 'enterprise' focus must change to a 'livelihood' focus. The vision of the farm as a collection of enterprises in a continuous space should be transformed into that of relationships between land and water resources, which are often non-contiguous and may be communally owned. Lightfoot and Noble developed an innovative protocol to enable farmers to appreciate their environment and the potential integration of its elements, to capture IK, and incorporate new elements to enhance sustainability.

A number of approaches to address sustainability issues have been suggested and tried in recent years, usually employing holistic analysis techniques preceding experimentation. Tools such as resource mapping with material flows at different scales can assist farmers to classify and perceive their environment and envision change by identifying technological or management options. These can be incorporated into their systems using diagramming in the design phase and encourage discussion about new ways to reallocate and integrate resources and enterprises. Farmer experimentation on various processes follows, and then farmer evaluation and participatory monitoring of the system. The overall aim is to diversify links within the farm system in a way that both stabilizes and improves farm production.

Alternatives include land use system analysis and nutrient analysis. These can be used as diagnostic tools to help pinpoint areas for experimentation, and as evaluation tools once a new practice has been employed, in order to measure effects at various scales of influence. Land use systems analysis allows the quantitative analysis of effects of cropping practices on production and environment, and includes (quantified) integration of biophysical production and socioeconomic feasibility. Inputs, outputs and transformation processes are defined using a hierarchical approach[64], that is defining a land use system which is made up of subsystems, which in turn are made up of biophysical components each with its own related operational sequences, outcomes and influence on other scales. Reference is increasingly made to spatial units in which the unit of management is not necessarily the plot or farm level but large units such as a watershed or a geographical region.

Nutrient flow analysis, a measure of sustainability, seeks to understand the nutrient status of parts of a system at different scales. Budelman and Smaling and Braun[65] are among many researchers who are experimenting with this tool, increasing its participatory nature so that it can be used by farmers and extensionists to evaluate the efficiencies of a given system, make appropriate adjustments and monitor system 'health'. Farm and farmer typologies and land use analyses are being brought into a process which is rapidly evolving.

Although OFE in agronomy has yielded disappointingly few examples of adopted technology, feedback on the understanding of variability and risk, and of labour issues and economic feasibility has been invaluable in redirecting on-station research. Second-generation technology development should be more productive, and certainly the emphasis has moved to soil fertility and IPM work in the 1990s. More programmes are leaving seeding rates and dates to farmers. However, institutionalization and application of these processes has been slow (see Part III) and books recently published on the subject carry much the same messages as in the early days.

The challenges of livestock OFE

Livestock researchers tend to be more familiar with systems thinking than crop scientists due to the complexities of providing for feed and fodder. Neverthless they face a formidable list of difficulties and OFR with livestock has lagged behind progress in other research areas. Difficulties include:

- The small number of observations due to few livestock units on farms.
- Large variation in experimental units by sex and age and high statistical variability in measurements.
- Inability to control the lactation stage.
- Need for greater emphasis on the choice of farmer-manager than site.
- Mobility including transhumance.
- Higher costs and wider risks (disease, weight and production loss), including expensive compensation to farmers for animal losses.
- Longer duration experiments.
- Farmers' emotional ties to livestock (with motivational and social consequences) and the difficulty in valuing complex, non-market inputs and outputs.

In addition, grazing trials can place heavy financial demands on national programmes (fencing, transport and technical staff) which are seldom met without project assistance[66]. Experiences recording these difficulties began to appear in the literature in the late 1980s[67], together with concerns shared with crop researchers on how to zone and identify target groups, livestock types and production systems.

Progress has been made in several areas. In conventional research large samples and longer term trials (2 years) were used to test effects, as with trypanosomiasis control and anthelmintic treatments. OFE on disease control has become increasingly common due to the immediate effects of treatment attracting livestock keepers (e.g. see Chapter 4.2). Nutrition improvement and feed resource management are two other areas in livestock research in which OFE has advanced.

Some of the unique challenges associated with on-farm livestock work are briefly elaborated here. Farmer selection is guided by more restrictive criteria: ownership of a minimum number of animals, minimum size grazing area. Ownership situations are often linked with a management style that complicates selection and confounds the outcome. In southern Ethiopia there are six ownership arrangements for draft animals depending upon who purchases them, who houses and feeds them, and who has the rights to use the various livestock products. On sample size, observing too few animals per farm can seriously affect average farm performance and cause large variability between farms. Solutions include keeping both treated and control animals within a herd, and using control and treated flocks when it is difficult to keep treatments separate. In forage-feeding trials, one replication per site may be reasonable given the size of 'plot' needed to supply sufficient forage. Various types of records are needed along with performance (reproduction, growth) and each may require a different time scale or recording frequency.

Trials can be augmented by long-term monitoring, where many variables affecting productivity are recorded[68]. Farmers have been successfully taught to monitor their own livestock performance in some trials. Farmer assessment and intermediate indicators can avoid waiting to record an increased calving ratio: rather, farmers judge the condition of the pregnant females at a certain point in the season against feed regimes and can anticipate a higher ratio.

Baker *et al.*[69] experimented with 'regular research field hearings' (RRFH) to see if farmer participation improved in small ruminant trials. The objectives were to increase communication between research, extension and farmers to build an understanding of the farming system and production problems; to impart knowledge; and to increase farmers' understanding of how they could apply the technological aspects of the trial. Repeated contact with farmers did increase understanding and trust between the parties, and improved the application of the technological package. Farmer assessment is most important, multiple objectives in livestock keeping are often difficult to weigh and anyway require farmer input.

Despite the inherent difficulty of OFE with livestock, useful design considerations and guidelines have slowly emerged, for example, guidelines for successful participatory forage research are now available[70].

Conclusions

To conclude, each thematic area has evolved. Farmer participation greatly changed the direction of plant breeding with farmers' input making it potentially more efficient. Recognition of the farmer's assets; their local knowledge, their germplasm and their expertise continues to impact the breeding process. In agronomy, variation still poses the major challenge; however, the idea of providing a range of options and allowing farmers the leeway to adapt potential solutions to their conditions is slowly gaining credence. Again it is increased farmer participation which has pushed change in this direction.

Conceptual development continues to deal with systems and subsystems, the interactions, the linkages and their complex and dynamic nature. The use of decision trees and decision support systems[71] are evolving as ways to manage variation using experimental results as input. Conceptual change has been driven by increased interaction with farmers, by the increased understanding of the whole range of farmer circumstances, and by experimentation itself. A relatively slow move towards OFE in a true systems context is occurring, for example in intercropping, agroforestry and sustainable management of the resource base.

Thematic evolution has also occurred. For example, pest management work has become more systems oriented, soil fertility work has taken new directions (inorganic and organic combinations, biomass transfers from hedgerows to cropped areas), much weed management work now embraces the labour and gender implications. What remains challenging is to integrate the components into the system, to work as a multidisciplinary team and not merely within one, and to function effectively as a part of a revitalized research and development continuum.

4.1.6 The major dynamics 1970–98

Changing perspectives; components, systems and system hierarchies

Over the last 25 years OFE has come under the influence of new perspectives on agricultural R & D which stem from systems theory and systems thinking. On-station research driven by experimental design has sought statistical precision in the development and assessment of best-bet components and yet its methods provided the foundation for early OFE. The reductionist tools have been increasingly supplemented by methods based on systems thinking and the concept of linked systems – as nested subsystems or hierarchies. This has led to more holistic experimentation on more complex subsystems and considerations of landscape, rather than just plot-level effects.

Cropping systems research and FSR/OFE were initiated from increasing concern over stagnant productivity in the smallholder sector and were developed by practitioners in the field. Both preceded the more recent, more formal, efforts to apply systems theory to agricultural R & D. Collinson[72] saw the existing farming system as the essential starting point for development, the base on which to graft productivity improvements. OFR with a farming systems perspective in CIMMYT parlance was seen as a problem-oriented subset of FSR, in which specific subsystem linkages were recognized while introducing stepwise and usually component-related changes to selected enterprises. A major impact from practice was to focus OFE on problems of the farmers and their systems, rather than on the hypotheses of researchers seeking yield maximization. Also the realization that components could not be experimented on in isolation from the system as there were inevitably ramifications on resource use and on other enterprises as well as interactions beyond the farm boundaries (markets, input suppliers, off-farm employment, among others). Solutions could be 'indirect', e.g. weed problems in beans might be solved by introducing a herbicide for maize, freeing labour for bean weeding. However, even with these systems insights, FSR practitioners have continued to experiment more frequently on component changes than on more complex subsystems.

In the early 1980s, systems theory and thinking began to penetrate agricultural R & D at many levels. Plant, crop and later, soil and water modelling addressed systems at different levels of a biophysical or ecological hierarchy, while farm and more aggregative economic modelling addressed the human dimension. Soft systems modelling (e.g. farmer diagrams) offered a user-friendly approach allowing better integration of the human and biophysical dimensions, and was perceived as particularly appropriate to small-farmer circumstances. In the more formal models the objective functions still cannot adequately balance the diversity of small-farmers' goals, and the accurate measurement of a farm system's multiple parameters, particularly labour and output, remains very expensive due to the intensive manpower input needed to collect the data required.

Several examples support the advantages of using a systems approach. In West Africa, a team of international scientists decided that the main constraint in a small ruminant production system was the system itself and went on to design and test a series of new ones. Sumberg and Okali[73] feel that new farming systems conceived by researchers are too often based on a single principle such as profit maximization, and they advocate a more flexible farmer-designed and farmer-adjusted approach. Lightfoot *et al.*[74] concluded that the cropping pattern research, used successfully in Asian lowland irrigated rice systems, was not useful for upland farmers: the patterns were too resource demanding and did not address the need for responsiveness to risky conditions and changing opportunities.

Conventional OFEs are limited in scope (number of trials and farmers reached) due to the need for researchers' control, supply of inputs and data collection. Lightfoot *et al.* also said that 'farmers' problems are systems problems as they usually involve many components of the whole farm system'. They believed that many of the tools used in FSR, such as rapid diagnostic methods, did not adequately link biology and socioeconomics. They tried other participatory tools which incorporated indigenous technical knowledge (ITK) and systems logic, to link both biological and socioeconomic concerns to the design and testing of options. Thus, a systems approach to design can be taken: for example, legume-enriched fallows should both enhance soil fertility and reduce labour costs in coconut fallow rotation systems in the Philippines. Treatments can consider multiple benefits: mucuna as a weed control/soil cover crop rather than for soil fertility improvement alone. Lightfoot and Minnick[75] further argued that farmers must assist in designing new farming systems to cope with complicated environmental issues and with multiple product enterprises such as agroforestry.

New research procedures, beyond questionnaires and component technology trials, are needed to accommodate the ideas of larger numbers of farmers and communities to permit testing at a farm systems or landscape level. In particular, diagramming has enhanced communication on system changes, including visualizing steps in testing a technology experimentally, in integrating enterprises (nutrient, labour, material management), teasing out farmers' knowledge and reaching agreement on what might be feasible and expected changes.

The early conceptual models in FSR implicitly acknowledged the importance of system hierarchies. The CIMMYT approach emphasized the importance of information on local rainfall, soils, policy and markets in understanding farmers' decisions. However, early models ignored the output side of the equation; the impact of OFE generated changes at farm level on the economy and the ecology at large. Two new dimensions emerged strongly in the late 1980s to change this: first, what we earlier described as the 'sustainability revolution' raised the profile of impact through farm-level changes on the watershed and the ecology beyond, and vice versa. It stimulated much of the thinking of practitioners such as Conway and Lightfoot and placed environmental criteria on a par with productivity in the evaluation of opportunities for change on the farm. Second, policies, infrastructure and service programmes were seen to be linked with opportunities for change on the farm, and could be influenced by the appropriate flows of information. This strong interaction widened the scope for OFE to highlight 'potential' rather than 'immediate' on-farm opportunities, and to offer information to policy makers and service providers to help mobilize this potential.

Current thinking stretches the 'system' concept in a further dimension: 'food systems' linking production storage, transport, marketing and transformation into consumable, saleable, value-added products. It brings in stakeholders well beyond the farm gate: processors, lending agencies, market entrepreneurs and the like. At the very least, the researcher is asked to consider preferences from a variety of potential users, in potentially dynamic and volatile markets[76]. When starting from this wider system, the emphasis shifts from packaged technology to an analysis which identifies potential innovation paths; for example, looking at credit and marketing systems available to support technology.

An application which remains controversial is the development of new farming systems (or subsystems). Kirkby *et al.*[77] worked at both levels: evaluating patterns of crop and livestock species in relation to land use, environmental risks and self-sufficiency as well as simultaneously studying components (fertilizer response, varieties and crop management practices). Zandstra[78], writing from his Asian cropping systems perspective at IRRI, regarded OFE partly as a process for comparing an 'experimental' with an existing system. Indeed their strategic orientation and the desire to avoid site-specific limitations has encouraged several IARCs to focus on designing new farming systems.

It has been claimed that new farming systems would encourage more radical, and therefore more rapid, change[79]. Optimizing farming systems became a popular modelling exercise as mathematical programming blossomed during the 1960s and 1970s. However, a better understanding of small farmers' stepwise adoption behaviour sheds light on the difficulties of introducing new farming systems. First, the optimum itself will not be stable for long, influenced by the dynamics of the market, shifts in policy and ever newer technology. Second, the path farmers might take from their existing system towards the optimum is itself subject to the vagaries of the weather and the same market policy and technology dynamics. Given stepwise farmers' adoption behaviour, development is necessarily a step-by-step path which never ends[80]. Illustrating this perspective with an agroforestry example, Sumberg and Okali[81] recommend that researchers carefully examine their objectives: do we really need a narrowly defined optimum alley width for a given situation and a series of precise alley farming packages for specific crop combinations, or is it better to identify a generalized model that can be extended flexibly on an ecological or even regional basis? They suggest that it is best to develop a broad understanding of the possible ways to fulfil local objectives as the basis for identifying a range of management options for farmers in various circumstances. In this context, new technical options developed in the form of decision guides with their accompanying principles, offer more useful goals for research than creating new farming systems. There remain major challenges in optimizing interventions to achieve change balanced across productivity, sustainability and equity from within the resource endowments of the system.

Changing roles: farmers

The need to understand how technology performs under farmer conditions is only satisfied if a partnership with farmers provides real interaction. Farmer involvement has become a major feature of OFE. There are now many reports of farmers' positive contributions which have led to increased appreciation of them as 'experts' in their own right. So far the appreciation is coming mainly from the NGO and project sectors; in national agricultural research scientists are still reluctant to give equal status to small farmers.

EARLY DAYS. Agricultural scientists have wanted to maintain precision and control, using the research process they were trained in, until the technology was 'ready' for farmers to 'verify' and use. Critics have long highlighted the gap between the 'mind-set' of scientists and farmers, and the scientist's attitude that has kept farmers at a distance[82]. Neverthless, even in the early days some advocated the need for increased farmer involvement in selecting technology, modifying treatments, sharing in evaluation and as a source of technology[83]. Without farmer involvement research remained top-down, even if conducted on farm. Researchers have subsequently tried to address the issue of social distance by increasing the investment in interaction time with farmers[84]. The root cause remains irrelevant higher agricultural education.

INDIGENOUS KNOWLEDGE. Although anthropologists recognized that farmers do experiment, they had traditionally published in specialized disciplinary outlets. Before the 1970s development theorists presented IK as inefficient, inferior and an obstacle to development. An accumulating understanding of small farmers, and the increasing urgency for improving their situation, generated a surge of interest in IK and in farmers' own experimental processes. Advocates began underscoring the promise of IK for improving agricultural production and sustainable development. Today IK has become a legitimate alternative source of technology, and beyond this, better knowledge of farmers' research methods is encouraging their use in tandem or combined with more formal processes.

There is abundant documentation on farmers' knowledge of micro-environments and the variable conditions under which they live and operate. Examples include the highly complex indigenous land use systems in the Philippines[85]; farmers in a small project area in the Ecuadorian Andes who used 100 crop associations in managing their agroecological diversity and risks; Rwandan bean farmers actively manage mixtures, averaging 12 varieties each, which are adapted to soil and seasonal climatic differences.

In 1987 Rhoades[86] presented modern agricultural science as basically an afterthought to the great technological breakthroughs made by farmers over the centuries, 500 generations during which they had an evolutionary impact on plants, animals and the land. He argued that there was still a great gap in our scientific knowledge about farmers, regardless of the hundreds of studies conducted: 'We the non-farmers lack the basic understanding of farmers' own research methods, their schemes of information exchange, their informal farmer-to-farmer extension methods, and their approaches to generating new technology or designing new farming systems'. Rhoades was emphasizing indigenous processes rather than indigenous technology. Farmers have a rational concept of experimentation. They modify, develop and adapt technologies encountered through interaction with researchers and extension, or acquired from other farmers. In their informal experimentation, their observations and intuition guide them in making technological choices and in developing production strategies. Rhoades found from case study work in Peru that farmers tended to experiment with small quantities before moving to a larger investment; they chose and adapted elements or parts of a package or a principle; social factors, such as theft prevention, privacy and gender were as important as technical and economic factors, new materials were often added to but in general did not replace traditional materials; and farmers used their own criteria to evaluate technology, weighted according to their own preferences. Rhoades concluded that farmers use similar methodological steps to researchers:

problem formulation, formulation of a testable hypothesis, testing the hypothesis empirically and validating or invalidating the hypothesis. The difference is that farmers have very specific goals and the results of the experiment must be practical.

Many reports have catalogued examples of farmer experimentation and IK in variety selection, adoption of new crops, modification of cultivation implements, development of cultural practices for weed control, soil and land-use system classification, and adjustment of a cropping system and practices to spread labour[87]. However, while the use of IK by NGOs and in projects has increased, use has been much more limited in the formal research sector. Underutilization of IK may be due to lack of methodologies and training in how to identify IK, lack of awareness or general non-appreciation of IK, or farmers not being empowered to develop and improve technologies. Box[88] suggested that researchers could copy farmer experiments, and make them amenable to statistical analysis; and likewise researchers' experiments could be reconstructed by cultivators in their fields. Lightfoot[89] felt the use of indigenous experimentation was worth the effort but slow and in need of help from researchers. He cautioned that soliciting farmer knowledge would be difficult without researchers learning new skills. Examples are the farmer field schools for educating farmers on principles needed in IPM. Potts *et al.*[90] purposely studied and documented farmers' experimentation techniques to incorporate lessons into their own research process and increase the probability that new technologies would be appropriate and acceptable. Like Rhoades, Potts' group found that farmers wanted quick results because time, land and cash were limited. They worked on factors rather than changing the whole system and exhibited systems thinking of their own by considering production to marketing aspects. They first worked at a small scale to minimize risk and to obtain a 'feel' for the technology, and used either a control or several reference points. Farmers evaluated concepts rather than collecting quantitative data usually relying on impressions. When experience was limited they collected precise data and initiated their own recording systems. Farmers used their own criteria; for example, the flexibility of the system was more important than immediate economic returns. All such findings are useful for the better implementation of OFE.

While indigenous technology itself is a valuable additional source for farmers in other systems to draw on, the use of indigenous processes of experimentation and dissemination may be the ultimate vehicle for close, cost-effective partnership between local and more formal researchers, and for the cost-effective spread of information across villages and communities.

CLARIFYING FARMERS' ROLES IN RESEARCH. The need to clarify roles has pushed researchers closer to farmers. Farrington, Chambers and Jiggins, and Sumberg and Okali[91] explained the contribution expected from researchers and farmers: the researcher's roles are to widen the range of technology available to the farmer by identifying system constraints and opportunities for the further exploitation of natural potential; through observation to identify technical problems for further investigation; and to provide counselling to farmer groups. The farmer should contribute to the definition of the research agenda, provide specific local knowledge, conduct and evaluate research and disseminate the results.

Biggs[92] categorized relationships with farmers into four types: 'contractual' where the researcher merely contracts the use of the farmer's resources; 'consultative' where the farmer is consulted about the farming system used; 'collaborative' where researchers and farmers are partners in the research process; and 'collegial' where farmers draw on research within the development context. Conventional practice is gradually moving into a collaborative relationship. Farmers are currently more appreciated as experimenters in their own right; more research programmes see the value of farmer designed and managed trials and of the increasing involvement of farmers and organized farmer group representatives as stakeholders in research planning and priority setting[93].

Farmers are becoming more fully involved in research through the use of participatory methods, and the increased trust these engender. Their input is gradually being accepted, not just in the later stages for verification but up-front in technology design. There is great potential for monitoring farmers' actions and results: for the use of IK and indigenous experimentation processes, both in systems understanding and in technology design[94]. There is also great potential in giving more weight to farmer criteria and assessment in trial implementation and evaluation. Lightfoot *et al.*[95] say that practitioners must make the jump from information gathering to building farmer skills in experimentation, decision making and impact assessment as part of the research process. Across large areas of Latin America, CIAT[96] has shown that small farmers can be successfully trained to carry out and interpret more formal experiments. While improving their effectiveness in research should increase participation and shift some of the costs of OFE to the beneficiaries, a potential danger is that some researchers perceive these farmers as better assistants than partners.

The 1994 review by Bentley suggests that farmer–research partnership is easier said than done and there are few documented examples of farmer–researcher generated technologies; there are many more success stories of researcher-generated (hybrid seed, agrochemicals) and farmer-generated technologies (domestication, organic manures, local varieties and breeds, basic farm tools). Numerous factors work against FPR: farmers are difficult for researchers to meet; they have different observation skills, experimental styles and scales of operation; social distance is an obstacle, and many researchers are not full time in farmers' fields anyway. Even with these challenges, many are attempting it and believe it is the way forward.

Changing roles: researchers

As section 4.1.5 suggests, agronomists and soil and crop improvement scientists have had to cope with a growing volume of OFE with implications for their traditional ways of doing business. Their introduction to participatory methods has been through short courses and field exercises, whereas educational systems need to change before most agriculturalists can fully embrace participatory concepts and skills.

A second change has been accommodating social scientists, as members of FSR teams, into the research process. The introduction of social scientists into research establishments was largely a reaction to poor adoption amongst small-farmer communities, partially attributed to natural scientists placing too much emphasis on biological potential and yield. It was hoped that an economist's perspective would bring an understanding of how farmers allocate scarce resources and why this often compromises the technically optimal. Collinson[97] was one of the earlier proponents of including an economist in the research team. He believed in synergy and division of labour; the biological scientist would understand how to exploit biological potential, and the economist would understand what would enhance the farmer's goals. He saw this interaction as crucial to making experiments consistent with farmer objectives.

The perspective of economists initially served to legitimize farmer practices: maize–bean intercropping in Latin America was shown to reduce risk and produce a consistent income with lower investment at the yield levels achieved by small farmers. Economic analysis also enhanced the understanding of options: monocropped beans could give greater returns if farmers were able to employ an intensive management package and reduce costs by using family labour and non-purchased local materials[98]. Economic analysis combined with agronomic parameters enabled technical packages to be evaluated for profitability and risk and sometimes explained non-adoption[99]. The economic perspective has also increased consideration of labour productivity, a major concern in many African farming systems, where seasonal labour peaks constrain incomes. Detailed guidelines have been developed for the collection of labour and other farm data, both in terms of understanding the problem and in trying to find solutions[100]. This has made experimentation more sensitive to the labour issue which has dogged acceptance of many station-derived research recommendations. An example is the case of early planting dates throughout much of eastern and southern Africa. Early dates interact strongly with recommended crop densities and fertilizer levels, but many farmers using hoes cannot achieve them for more than a tiny area of land. Although economic analysis is currently routine for some researchers, including agronomists, to analyse experiments and alternative cropping systems, it is still generally underutilized.

Economists have played a role in FSR and OFE for years and are often members of FSR teams. In Ethiopia, economists are charged with carrying out most on-farm trials as a 'service' for the agronomists and other technical scientists who generate technologies on station. In the literature, reference is often made to 'socioeconomics', which presumes that economists and sociologists are one and the same, which they are not. Rhoades[101] complained that economists have reduced the farm household to the model of a business firm with its outputs, inputs and reactions to profit. Rural sociologists have been members of FSR teams so rarely[102] yet their recent conceptual contributions have created an escalating demand for better understanding of gender, household resource allocation, farmer preference, partnership, local institutional dynamics and community communication systems, information that is useful in all stages of OFE. Social constraints to adoption have been 'discovered' after years of participatory experimentation with farmers[103]. Examples involve conflict over land use, organization of animal traction for row seeding, returns to women on fertilizer investments, and erosion control, all of which need to be worked out by the community. With hindsight, sociological input, more often available with NGOs, could have speeded adoption and avoided some expensive and unproductive OFE initiatives. Increased emphasis on collaboration with farmer groups requires communication skills not easily obtained from the ranks of biophysical scientists and economists and promises an important future for anthropology and rural sociology in OFE. In Chapter 10 David Norman, a farm economist, Constance McKorkle, a rural sociologist and Peter Hildebrand with Dennis Keeney, speaking as a pair of agronomists, review the impact of their respective disciplines on FSR, and the impact of FSR on their disciplines.

Changing roles: institutions

OFR, the combination of FSR and OFE, is considered an innovative process of technology development that complements station-based research. Adoption requires either institutional reorganization or partnerships across institutions. In the 1970s OFR was relatively new but was seen by some (especially in Africa), as competing for staff and resources in a situation where commodity programmes were still relatively weak and understaffed. Where OFR was adopted it usually operated independently of the on-station process and their complementarity was not always clear. Yet the two are necessarily interdependent: commodity programmes need to be strong enough to generate options and absorb farmer-level information from on-farm research, and OFR needs to be strong enough to test options with farmers and provide meaningful feedback to commodity groups. It is becoming increasingly accepted that a complementary programme of on-station and on-farm experimentation is more efficient.

FSR and OFE demanded changes in orientation and practice of NARS institutions. Various formats were tested: special OFR teams, commodity teams using an FSR approach, extension programmes with OFE responsibilities or 'collaborative' teams in which the FSR team included at least one commodity team member. Nevertheless, links between commodity research and OFR teams tended to be weak and the incorporation of feedback with a farming systems perspective difficult. Castillo[104] maintained that FSR as a concept should be directly incorporated into agronomic, livestock and forestry research, not maintained as a separate unit, but noted that it is usually easier to create a parallel programme. A commodity orientation is still apparent in most research institutes and serves to marginalize those systems perspectives that aim beyond the commodity. On the whole, reorganization of the research process in public sector institutions to use OFR has been minimal.

Nevertheless, there are examples where public commodity programmes have successfully used OFE. Zeigler[105] reported how a late-maturing maize variety selected on-station primarily as high yielding was rejected by Burundi farmers. When analysed in the farmers' context it offered no advantages, while reducing stability and increasing risk. This feedback was utilized by the maize programme to develop a compatible, early-maturing maize variety. Thomson *et al.*[106] changed the orientation of on-station researchers by demonstrating the usefulness and 'systems fit' of feed legumes in barley fallow systems.

Extension involvement in on-farm testing has been widely accepted as a cost-efficient link between research and extension; much OFE would not have been possible without extension inputs. However, existing organizational structures often inhibit communication between research and extension. Limited qualifications and lack of training opportunities make quality work difficult to achieve; extension staff too are thin on the ground; and top-down, centralized approaches (such as the original Training and Visit promoted by the World Bank) are contrary to the farmer-orientation of FSR. Extension workers in future need to be conversant with participatory methods, be more concerned with offering options, observe how farmers react, become more consultative and strategic in giving advice, and provide feedback to research on constraints and needs.

Institutional partnership is a new watchword, certainly among the donors. The value of partnership must remain at the forefront; where benefits arise from the inputs of both farmers and researchers. Managing such collaborative alliances is another new skill required in on-farm research. The burgeoning presence of NGOs at the local level and their often successful engagement with the community is seen as bringing complementary skills to the science and formality characterizing the public research and extension establishment. In Latin America it is predominantly the locally based NGOs that have engaged communities in a participatory process for technology identification and development. This is also increasingly true for Africa and some parts of Asia. Few countries in Africa are, as yet, sufficiently comfortable with the operations of NGOs to encourage partnerships with their public research and extension establishments, although the number of exceptions is growing. Partnership and the reorganization of institutions to enable a more effective R & D process is perhaps the greatest ongoing challenge faced by the FSR movement. The issue is discussed at length in the contributions to Part III of this book.

4.1.7 Two recurring implementation issues in OFE

Introduction

Many manuals and guidelines have been published in the past 20 years in an effort to adjust conventional experimental methods for use on-farm[107]. Many experimental designs and many methods of analysis have been described including regression, modified stability analysis, diamond trials[108], superimposed trials, paired comparisons, dispersed randomized blocks and incomplete factorials. All have their uses. Nevertheless, OFE implementation problems still remain. Dealing with variation and choosing methods of analysis are two recurring issues throughout the history of OFE that symbolize the tension between being 'scientific' and being 'realistic'.

Understanding and dealing with variation

Various techniques have evolved, the use of zoning, land use typologies and RDs to limit sources of variation and help researchers place trials to 'control', sample or understand heterogeneity be it spatial or temporal. Neverthless OFE practitioners have continuously been challenged by variation at two scales:

- Between farms and farmers, that is across trial sites.
- Within farms caused by physical diversity, by diverse farmer management or within the trial itself by experimental variation.

Off-station multi-locational testing, used prior to OFE, sought to differentiate optimum crop management according to the diversity of climate and soil conditions. For this researchers had to understand variability in greater detail to decide where to place their trials ecologically, how many trials to put out, and what data to collect to link conditions to results. A particular feature, usually physical (rainfall or soil type), would be chosen to determine placement of trials. For example, Kirkby *et al.*[109] chose a soil fertility gradient in their Andean site as the major factor that explained much of the variation and used this to repeat a series of trials along other ecological gradients (slope, elevation, rainfall and wind), even though these zones in themselves were very heterogeneous.

Variability within an experiment was initially managed by the researcher and was originally one of the most critical challenges in moving from the protected environment of the research station into the field conditions that farmers have to manage. It has been a contentious issue since the early days of FSR, and is still perceived as a threat by scientists rooted in the analysis of variance (ANOVA) culture. Researchers moving into farmers' fields felt more comfortable controlling variation (for the same reasons as in on-station trials) by the design used, by setting uniform management of the NEVs and by managing the trials themselves. Yet they still found themselves faced with high coefficients of variation (CVs), high levels of data loss and lack of significant differences between treatments. They started to realize that conventional trial designs did not necessarily work. For example, high treatment numbers can raise the CV because a randomized block design (RBD), the one most often used, does not adequately control heterogeneity[110]. Thus, OFE practitioners have found it challenging to maintain the same experimental standards expected for on-station research, due to irregular topography, non-uniform management practices (e.g. sowing-date differences), heterogeneity of cropping patterns and small farm sizes. The semi-arid areas show the problem at its most extreme, as rainfall and runoff variation are high and cause varying or poor plant stands and large differences within small areas.

Various statistical techniques have emerged as being useful to help scientists manage variation. Hildebrand[111] developed a modified stability analysis both to help explain variation and to allow for prediction across differing environments. Thus experiments repeated across an environmental and/or management gradient could be usefully analysed together and permitted prediction to other sets of conditions at relatively low cost. However, it has been found to be important to characterize carefully the gradients and their variable aspects to be able to make reliable predictions. Historically, many collected data sets would have been more useful in interpreting responses to varied conditions if more time had been invested in fully profiling sites. Such analyses are useful, particularly when applied in conjunction with modelling to allow interpolation and (less surely) extrapolation, along gradients of one type or another.

Multivariate analysis or *ex post* stratification can make sense out of a large range of observations and has been used to characterize micro-environments, as well as to provide insights into adoption or performance patterns[112]. Along these lines, Cady[113] proposed a three-stage process:

1. Yield–environment correlation analysis to determine which environmental variables should be included in the combined analysis.
2. Analysis of variance to look at treatment differences and trial differences separately or combined.
3. Regression analysis and prediction where yield is a function of factors managed by farmers and of environmental variables.

One drawback has been that some methods to explain how technologies perform under varying conditions, such as multiple regression of yield against site variables, have not been easy for the 'ordinary' field agronomist to use. Further, methods developed empirically in the field often remain unpublished[114].

In agronomy a major shift has been made towards the measurement of variability as an essential part of the trial rather than trying to control it. This new concept caused changes in the implementation and design of trials. New techniques such as single replications, stratification of site and farmer selection to sample variability were tested, among others, to deal with the high variability, small farm size and limited research budgets. Interestingly, the idea of using single replicates on large numbers of farms was made by Stewart as early as 1949 and later picked up by other researchers. A useful decision guide to the efficiency of one-replicate yield testing under any set of location and replication costs is provided by Dofing and Francis[115].

To add to the complexity of viewpoints on this subject, economists too have helped address variability. To deal with seasonal variability, for example in rainfall, Nagy and Sanders[116] calculated a production risk by using a simple (but crude) percentage of farmers involved in farmer-managed trials who lost cash. This overcame difficulties with more formal ways of calculating risk when there were data deficiencies. Staff at Michigan State University[117] developed the 'MSTAT' programme including a computerized risk analysis which calculates an 'index of variation' to rank interventions. Beyond this a variety of social aspects (gender, wealth and local organizations) began to demand attention and added a further dimension to variability. This is another area still in its infancy.

Some researchers have overcome concern with site specificity and variation by relying on farmers to fine tune and adapt to a wide range of conditions. Farmer participation and an understanding of local processes for technology development and diffusion have been helpful. Farmers have been allowed to manage more, particularly setting their own NEVs. At first heavier farmer involvement generated much unusable data, and on-farm researchers have begun to put more emphasis on monitoring farmer experiments to learn from a variety of experience. Increasing demands for rapid impact requires deeper knowledge on what does well where, with whom and why. Tensions remain between allowing farmers to take charge and adapt, and the evolution of more rigorous approaches to classify and understand variation.

One of the most recent challenges is of working at the wider hierarchical levels; the landscape and catchment area, particularly in NRM with the increasing importance of measuring environmental impact. A recent challenge here is the integration of 'new' disciplines – geographers, land use planners and ecologists familiar with working at wider scales. New terms are entering the vocabulary of agronomists – for example, looking for niches that can be improved through agroecosystems analysis (see Lightfoot in Chapter 11). Dealing with variability in this context means understanding and exploiting it, rather than controlling it. Monitoring or modelling variable processes at these wider scales is used to project impact, and variability is built in, sometimes through fuzzy logic that takes into account a multitude of possibilities.

Issues of analysis and interpretation of results

Analysis and interpretation of results, like dealing with variation, has exhibited a 'tension' between science and a developmental orientation, particularly in two areas:

- Who decides what is best?
- Do the methods used have to be acceptable to all stakeholders?

Is the research output intended for the scientist or the farmer has become a common question. 'Does the researcher contribute to the science of agriculture or make a contribution to agriculture?' Many would agree with Lockeretz[118] that the objective of most research is not a professional research publication but an improved agricultural system. Some feel that the aim is to generate technology that farmers like and use, while others still feel that the main contribution is to understand technical relationships and make a contribution to science. Both adaptive and strategic objectives have their place (sometimes in different institutional settings), and require appropriate approaches to analysis and interpretation.

Assessment methods for OFE have diversified since 1970 as OFE evolved. The incorporation of other disciplinary perspectives in the research process and the participation of farmers both had major impacts on the analytical tools used and the interpretation of results. The controversy has ranged between entrusting farmers with assessments, seen to integrate economic, labour, social and technical factors, urged by professionals such as Ashby, and the development of increasingly sophisticated statistical tools and rules, to deal with variable conditions and farmers' adjustment of treatments. Mutsaers and Walker take the view that farmer assessment should be complemented by data to verify farmers' assertions[119].

Over time, as multidisciplinary teams have worked together and as farmers have participated in experiments, parameters used to assess trials became more complex, embracing wider dimensions and new criteria. For example, measuring the benefits of potential new varieties, apparently simple technologies, has extended beyond yield and disease resistance to include other farmer objectives and preferences. Where stability is important then analysis should account for levels of stress tolerance; when home consumption is the end use the material must be acceptable to the farm families concerned; and if the end use is the market, then consumer preference or monetary value is important. Evaluation is complicated for more complex technologies such as livestock management, agroforestry or intercropping, though for intercrops total outputs can be converted to a common 'currency'; nutrition, monetary value or land equivalent ratios according to farmer circumstances.

A radical change was the introduction of economic analyses of experiments[120]. Once this sort of analysis was appreciated, a contribution from economists was in increasing demand from research teams. Later, agronomists applied economic assessments: response functions based on profitability and risk considerations (e.g. in fertilizer trials), assessments of labour demands and returns to labour (e.g. weed control options), returns over variable costs and marginal benefit cost ratios. Zandstra[121], himself an agronomist, suggested that it was necessary to check for conflicts with other enterprises, and minimally to evaluate for effects on other subsystems beyond the plot scale. These were important steps towards the acceptance of whole-farm analyses, already a well-established analytical tool with farm economists.

Analytical tools that were considered a more scientific way to understand complex environmental interactions, such as multiple regression analysis, have been used to a greater extent than in more controlled experimentation. Although various user-friendly statistical packages were developed for these analyses, use has been limited in developing countries where statistics is not a strong discipline and access to computers and suitable software is poor (although rapidly improving).

Out of the expansion of methods and viewpoints, arose a concern – how to combine, weight and/or integrate various methods of analysis. Whole-farm modelling developed in the early 1960s to integrate farm activities and resources had offered one way forward. Production systems research facilitated development of simulation models like Decision Support Systems for Agricultural Technology (DSSAT) in the 1980s to attempt this type of assessment. However, considerations of sociocultural acceptance, equity, access to and control of resources, decision-making patterns, possible implications of gender roles, and production and consumption preferences are not easily weighed in whole farm modelling[122]. For many researchers, farmer assessment gradually became an important tool for integrating and weighting the measures of different types of benefits.

Researchers began to learn from farmers' own evaluations, on labour inputs for example. But there was continued concern about biased behaviour and subjective assessments: can we really believe what farmers tell us? Are they really seeing an improvement or just trying to please us to get inputs again next season? As assessment methods improved, became more quantitative and had greater reliability, trust in the use of a variety of farmer assessment tools spread for eliciting quantified farmer responses: Bao board (based on a traditional game), matrix rankings, preference rankings, the measurement of productivity, labour calendars and discussion through interviews[123].

Another concern was whether the plethora of analytical methods could be assessed and guidelines developed about who should evaluate, using what method under which set of circumstances. For example, for a variety verification trial, statistical analysis (demanded by many national variety release procedures) and a full farmer assessment could be used; for an exploratory or diagnostic trial on agroforestry, one might use nutrient analysis, yield and biomass measurements, and farmer ranking. Generally, the greater the degree of farmer involvement the more emphasis is placed on farmer comments and opinions rather than on 'hard' data. Sumberg and Okali[124] contended that, whatever basis farmers have for making decisions, their weighting and combining of factors goes well beyond the analysis and synthesis that formal experiments use. The resulting assessment, while not being statistically based, is held as valid within the context of the particular production system.

Although there has been a long experience in FSR of examining physical performance over a range of biophysical environments, interests have largely stayed at the level of the farm. Recently, evaluating the effects of technologies at wider geographic scales and socially at the community level has gained in importance. With the burgeoning interest in sustainability and NRM, environmental impact has come to the fore to balance the previous interest in productivity gains. The scaling up of changes at the farm level to landscape levels has recently become a preoccupation for both agronomists and ecologists. Methods development is addressing the valuation of natural resources, how to monitor their status, how to measure off-site costs and impacts, and how to measure the short- and long-term conservation effects of changes in farmers' practices. Methods are also needed to handle considerations of impact on equity and society, and economic effects of scaling up that are reflected by price changes through the market. An awareness is gaining ground of the potential interactions, extending beyond the farm gate, between social arrangements (e.g. sharing of a resource) and technology adoption.

Associated with heightened interest in environmental issues is a search for analytical indicators to anticipate wider impact over time. This is raising difficult methodological questions: can the same quantifiable indicators be used over different scales; how does one integrate multiple indicators that are valuable at different scales; how are indicators valued by different stakeholders – issues that are taking us into new realms of analysis. Participatory monitoring is taking this idea to the level of the farmer or community and allowing for the expression of their interpretation and interest. The choice of indicators then becomes an indicator in itself; what are the farmers' interests and how do they value change?

Lightfoot *et al.*[125] started to address farmers' use of indicators to evaluate farm system change. In their monitoring system, farmers in the Philippines used diagramming, recorded inputs and outputs, and measured indicators for economic efficiency, biological material recycling, species diversity and resource system capacity. Some of the initial issues confronted in this initiative were the integration of indicators for measuring impact on sustainability, the causal relationship between the indicators, and assigning values to resources and flows (see Chapter 11.5). Researchers are now interested in taking this to wider, more complex scales, including looking at landscapes as a whole.

Thus analytical methods continue to evolve: seeking balance between the experts and the users; covering temporal and spatial concerns; integrating different perspectives, and relating all these to the various objectives. The driving force, however, is to know or better understand the real difference and to link change to action and learning; learning to analyse and analysing to learn.

4.1.8 The future

Initially, 'improved' technologies were developed on research stations, often, even in developing countries having huge smallholder sectors, with commercial farmers in mind. Results from early on-farm testing demonstrated why many of these technologies were not suitable for small farmers. The broad goal for OFE has not changed over time: the cost-effective identification of technologies that are more quickly adopted by more small farmers. Early expectations saw OFE first as a means of adapting local station results to the needs of local farmers as understood by researchers; and second, as a help in more relevant programming of station research. These remain the strategies in many current OFE programmes, but adoption levels in many programmes remain modest. The reasons are numerous, and some are related to the OFE process itself.

Waterworth and Muwamba[126], in a review of southern Africa research programmes, reported that only about a third of the technologies made available were adopted by farmers, largely due to input supply constraints. Other causes of non-adoption were related to the research process itself: superficial problem diagnosis, poor implementation of field trials, inadequacies in analysis and interpretation of trial results, and high turnover of research staff[127]. Similarly, Fujisaka's[128] studies on innovations intended to improve sustainability of Asian upland agriculture showed many factors inhibiting adoption despite the use of participatory approaches. These factors included problems incorrectly identified, farmers' practice proving better than the innovation, innovations not working under some local circumstances and extension agents failing to demonstrate correctly and/or targeting the wrong farmers.

Poor implementation has often been due to institutional constraints, including the lack of training for researchers and extension staff, both in OFE methods and in participation with farmers. Problems also come from incentive systems related more to doing the research than to seeing research results being used. Rhoades[129] recounts a pertinent statement made by a colleague: 'The main obstacle in providing farmer participatory research is the research workers themselves – both social and biological scientists. It is my general experience that a vast majority of research workers prefer to do research about a problem rather than research to solve a problem. ... Solving problems is much more difficult than doing research about a problem, so why get too close to this deeper area by including farmers with real problems in your team! I think those who have ventured into this high risk area have enjoyed the risks and have seen farmers are not only good research workers but excellent and efficient extension workers'. Now that research has entered into an era of accountability for achieving impact from its results, will the thinking of individual researchers change in time to restore institutional confidence?

Despite these problems, flexibility in the process and trial and error in implementation has brought evolution in OFR, especially if one accepts that participatory research is an important refinement rather than a separate, new process. Evolution has been continuous: many articles and books analyse and expound on changes and innovations. Those FSR programmes that have been supported for years were able to experiment with options, and those that underwent continual self-assessment were able to evolve most successfully. Process development has been dramatic. Particularly useful in achieving FSR's broad goals have been integrated research planning between FSR and other researchers, training extension workers to use a problem-oriented approach, considering policy makers as clients to solve input supply problems[130], improving links between research and extension, increasing the level of farmer participation and orienting the research process to be able to address the issues of women and the poor[131].

A result of this 25-year evolution is that OFE is generating renewed expectations for research efficiency. Better understanding of small farmers and their circumstances and the potential from new information technology are fundamental contributing factors, and four innovative concepts underpin the methodological innovations:

- The perception of the farmer as an empowered partner rather than a receiver of generated wisdom.
- The concept of multiple sources of technology that has destroyed the hegemony of the local research station.
- The concept of a range of solutions as options from which farmers can choose and adapt (rather than a technical package as a final product).
- The understanding of farmers' own experimental abilities, and of the high levels of spatial diversity and local variability, has usurped the need for final, precise recommendations from applied research.

Even though these changes are embodied in the best current practice, many programmes do not yet adhere to them. Impact often remains weak and research efficiency low. Further evolution is needed, or else the persistent problems of institutional reorganization and cultural distance between scientists and small farmers will continue to inhibit the effective mobilization of new technology. Increased reliance on farmer experimentation appears to be a particularly significant development. Small farmers have a clear ability to experiment rationally, their ability improves through interaction with researchers, and the interfacing of farmers with the formal sector can bring in options and methods from a wide variety of sources. Since farmers have been testing options since early times, should not, as Ashby[132] suggests, adaptive testing be devolved to farmers who would identify and transmit local recommendations to other farmers?

To achieve this, the formal research sector would need to decentralize and use a more interactive operational model, anticipate diverse needs, incorporate feedback sooner, modify research quickly to meet client requests and assure a range of options to address client needs. Less time would be spent on producing final, precise recommendations and in finishing varieties and testing these at the national level. Researchers would need tools, including GIS and decision support systems, to help them choose widespread strategic issues and identify where different solutions are likely to be widely adopted by farmers. Field research would evolve into monitoring and evaluating farmer assessments and the systems compatibility of optional solutions. However, the necessary changes in policy and procedures in the formal sector, for example research station planning and farmer recommendation committees, as well as in effective communication between farmer representatives and scientists, imply long-term institutional and cultural change and continued evolution in roles and skills.

The factors likely to play a role in directing future evolution are the increasing numbers of scientists wanting research to be more farmer and impact-driven; global trends towards liberalization, decentralization and democracy that favour a client orientation; and the demands of donor funding for increased accountability (most examples cited in this chapter come from projects that have been externally supported). Whether or not one likes all the implications, innovative institutional arrangements are being forged as a result of declining resources allocated to research (both national and international) coupled with a burning need for impact on food and welfare while maintaining a fragile resource base. But will change be fast enough to encourage investment in research partnerships, or will governments despair of their smallholder constituency?

REFERENCES

1. **Bentley, J.W., 1994**. Facts, fantasies and failures of farmer participatory research. Agriculture and Human Values, 11(2&3), 140–50.
2. **Belshaw, D.G.R. & M. Hall, 1972**. The analysis and use of agricultural experimental data in tropical Africa. East African Journal of Rural Development, 5, 39–71.
3. **Stewart, A.B., 1949**. Memorandum of colonial fertilizer experiments. 1. Planning and conduct of fertilizer experiments. HMSO, London.
4. **Ranweera, N.D., M. Dixon, N.S. Jodha, 1993**. Sustainability and agricultural development: a farming systems perspective. Special Issue; Sustainability farming systems. Journal of Asian Farming Systems Association, 2(1).
   Stroud, A., 1995a. An evaluation of on-farm research methodologies used in the Ministry of Agriculture, Tanzania. Paper presented in Design implementation and analysis of on-farm trials. Royal Tropical Institute, Amsterdam.
5. **Ravnborg, H.M., 1992b**. Resource poor farmers: finding them and diagnosing their problems and opportunities. Paper presented at the 12th Annual Farming Systems Symposium, Michigan State University.
6. **Feldstein, H. Sims & S. Poats, 1990**. Working together: Gender Analysis in Agriculture, Volume 1. Case Studies. Kumarian Press, West Hartford, CT.
7. **Tripp, R. (Ed.), 1991**. Planned Change in Farming Systems: Progress in On-farm Research. John Wiley, New York.
   Okali, C., J. Sumberg & J. Farrington, 1994. Farmer Participatory Research: Rhetoric or Reality. Intermediate Technology Publications, London.
8. **Li, H. & Z. Wang, 1993**. Concepts of Chinese ecological agricultural research. Journal of Asian Farming System Association, 2, 133–50.
9. **Stewart, A.B., 1949**. (Op. cit.).
   Boughton, D., E. Crawford, J. Howard, J. Oehmke, J. Shaffer & J. Staaz, 1995. A strategic approach to agricultural research program planning in sub-Saharan Africa. MSU International Dvmt. Working Paper no. 49. Michigan State University, Lansing, MI.
10. **Johnson, A.W., 1972**. Individuality and experimentation in traditional agriculture. Human Ecology, 1(2), 149–59.
11. **Kirkby, R.A., 1981**. The study of agronomic practices and maize varieties appropriate to the circumstances of small farmers in highland Ecuador. Cornell University, PhD thesis.
12. **Stoop, W.A., 1986**. Agronomic management of cereal/cowpea cropping systems for major toposequence land types in the West African savanna. Field Crops Research, 14, 301–19.
13. **Lightfoot, C. & R. Noble, 1992**. Sustainability and on-farm experiments: ways to exploit participatory and systems concepts. Paper presented at the 12th Annual Farming Systems Symposium, Association for FSR/E. 13–18 Sept. Michigan State University, East Lansing, MI.
    Li & Wang. (Op. cit.).
14. **Norman, D.W., 1980**. The farming systems approach: relevancy for small farmers. MSU Rural Development Paper no. 5. Department of Agriculture Economics, Michigan State University, East Lansing, MI.

Collinson, M.P., 1982. Farming systems research in Eastern Africa: the experience of CIMMYT and some NARS, 1976–81. MSU International Development Paper no. 3. Department of Agriculture Economics, Michigan State University, East Lansing, MI.

15. **Ravnborg, H.M., 1993**. Targeting international agricultural research towards the rural poor. CDR Working paper 93.4. Centre for Development Research, Copenhagen, Denmark.
16. **Ravnborg, H.M., 1992**. Sensing sustainability: farmers as soil resource managers. Centre for Development Research Working Paper 92.6. Copenhagen, Denmark.
17. **Dumanski, J. & E. Craswell, 1998**. Resource Management Domains for evaluation and management of agro-ecological systems. In: IBSRAM Proceedings no. 16 International Workshop on resource Management Domains. Kuala Lumpur, 1996.
18. **Stroud, A., 1989**. Weed Management in Ethiopia: an Extension and Training Manual. FAO, Technical Cooperative Program, Rome, Italy.
19. **Lightfoot, C., J.P. Dalsgoood, M. Bimbao & F. Fermin, 1993**. Farmer participatory procedures for managing and monitoring sustainable farming systems. Journal of Asian Farming System Association, 2, 67–87.
20. **Kirkby, R.A., *et al.*, 1981**. (Op. cit.).
21. **Kean, S.A., 1988**. Developing a partnership between farmers and scientists: the example of Zambia's adaptive research planning team. Experimental Agriculture, 24(3), 289–99.
22. **World Bank, 1996**. The Aga Khan Rural Support Programme: a third evaluation. A World Bank Operations Evaluation Study. The World Bank, Washington, DC.
23. **Feldstein & Poats** (Op. cit.).
24. **Grandin, B., 1988**. Wealth Ranking in Smallholder Communities: a Field Manual. Intermediate Technology Publication, Nottingham, UK.
25. **Penninkhoff, P. (Ed.), 1995**. Design, implementation and analysis of on-farm trials: an assessment of field project experience. Tanzania Workshop Proceedings. Royal Tropical Institute, The Netherlands.
26. **Norman, D., D. Baker, G. Heinrich & F. Worman, 1988**. Technology development and farmer groups: experience from Botswana. Experimental Agriculture, 24(3), 321–31.
27. **Sumberg, J. & C. Okali, 1988**. Farmers, on-farm research and the development of new technology. Experimental Agriculture, 24(3), 333–42.
28. **Ndiyoi, M.C. & W. Heemskerk, 1995**. On-farm experimentation: the case of bulrush millet in W. Province, Zambia. In: Design, Implementation and Analysis of On-farm Trials. Workshop Proceedings. Royal Tropical Institute.
29. **Stroud, A., 1995**. Revisiting expectations of FSR: experiences and challenges. Paper presented at SAAFSR/E Conference, 2–4 Oct., 1995. Harare, Zimbabwe.
30. **Carter, S.E., 1987**. Collecting and organizing data on the agro-socio-economic environment of the cassava crop: case study of a method. In: Bunting, A.H. (Ed.) Agricultural Environments; Characterization, Classification and Mapping. CAB International, Wallingford.
31. **Stewart, 1949**. (Op. cit.).
32. **Byerlee, D. & E. Hesse de Polanco, 1982**. The rate and sequence of adoption of improved cereal technologies: the case of rainfed barley in the Mexican Altiplano. Economics Program Working Paper 1982–4. CIMMYT, Londres, Mexico.
33. **Sumberg & Okali, 1988**. (Op.cit)
34. **Tripp, R. & J. Woolley, 1989**. The Planning Stage of On-farm Research: Identifying Factors for Experimentation. CIMMYT, Mexico and CIAT, Colombia.
35. **Biggs, S.D. & E.J. Clay, 1981**. Sources of innovation in agricultural technology. World Development, 9(4), 321–36.
36. **Okali, Sumberg & Farrington, 1994**. (Op. cit.).
37. **Christianson, P., 1981**. Observations on the major classifications of field trials used in FSR. SAFGRAD, Farming systems unit. BP 1783, Ouagadougou, Upper Volta.

Zandstra *et al.*, 1981. (Op. cit.).

Matlon, P., R. Cantrell, D. King & M. Benoit-Cattin (Eds), 1984. Coming Full Circle: Farmers' Participation in the Development of Technology. IDRC-189e. IDRC Ottawa, Canada.

Hildebrand, P.E. & F. Poey, 1985. On-farm Agronomic Trials in FSR/E. Lynne Reinner, Boulder, CO.

Steiner, K.G., 1987. On-farm Experimentation Handbook for Rural Development Projects: Guidelines for the Development of Ecologically and Socio-economically Sound Extension Messages for Small Farmers. Sonderpublikation der GTZ no. 203. Technical Centre for Agricultural and Rural Cooperation (CTA), Eschborn, Germany.

Lightfoot, C. & R. Barker, 1988. On-farm trials: a survey of methods. Agriculture Administration and Extension, 30, 15–23.

38. **De Datta, S.K., K.A. Gomez, R.W. Herdt & R. Barker, 1978**. A Handbook on the Methodology for an Integrated Experiment-Survey on Rice Yield Constraints. IRRI, Los Baños.
39. **Sperling, L., M.E. Loevinsohn & B. Ntaborvura, 1993**. Rethinking the farmers role in plant breeding: local bean experts and on-station selection in Rwanda. Experimental Agriculture, 29, 509–19.
40. **Budelman, A., 1996**. The search for sustainability: nutrients, trees and farmer experimentation in Sukumaland agriculture. Working paper 16. Tanzania/Netherlands FSR Project, Ukiriguru Agricultural and Research and Training Institute/Royal Tropical Institute, Amsterdam, The Netherlands.

    Havekort, B., J. van der Kamp & A. Waters-Bayer, 1991. Joining Farmers' Experiments. Experiences in Participatory Technology Development. Intermediate Technology Publishers, London.

    Sumberg & Okali, 1988. (Op. cit.).
41. **Ashby, J.A., 1987**. The effects of different types of farmer participation on the management of on-farm trials. Agriculture Administration Extension, 25(4), 235–52.
42. **Mutsaers, H.J.W. & P. Walker (Eds), 1991**. On-farm research in theory and practice. Proceedings of a workshop on design and analysis of on-farm trials. 27 Feb. to 3 Mar., 1989. IITA, Ibadan, Nigeria.
43. **Stroud, A., 1985**. On-farm Experimentation: Guidelines for Using OFE Methodology in Crops, Livestock and Agroforestry Experimentation. Teaching Notes no. 13. CIMMYT East African Economics Programme, Nairobi.

    Matlon, P., *et al.*, 1984. (Op. cit.).

    Hildebrand, P.E. & F. Poey, 1985. (Op. cit.).

    Steiner, K.G., 1987. (Op. cit.).

    Mutsaers, H.J.W., G.K. Weber, P. Walker & N.M. Fisher, 1997. A Field Guide for On-farm Experimentation. IITA/CDA/ISNAR.
44. **Sperling, L., 1996**. Personal communication.
45. **Sperling, L., *et al.*, 1993**. (Op. cit.).
46. **Kajiru, G.J., R.O. Kileo, A. Stroud & A. Budleman, 1998**. Validating a fertilizer recommendation across a diverse cropping environment. Nutrient Cycling in Agroecosystems, 51, 163–73.
47. **Box, L., 1987**. Experimenting cultivators: a methodology for adaptive agricultural research. Agriculture Administration Network, Discussion paper no. 23. ODI, Regents Park, London.
48. **Sperling, *et al.*, 1993**. (Op. cit.).
49. **Potts, M.J., G.A. Watson, R. Sinung-Basuki & N. Gunadi, 1992**. Farmer experimentation as a basis for cropping systems research: a case study involving true potato seed. Experimental Agriculture, 28, 19–29.

    Farrington, J., 1988. Farmer participatory research: editorial introduction. Experimental Agriculture, 24(3), 269–79.

    Kean, S., 1988. (Op. cit.).
50. **Stroud, A., 1993a**. Conducting On-farm Experiments. CIAT, Cali, Colombia.
51. **Mutsaers & Walker, 1991**. (Op. cit.).
52. **CIAT, 1996**. Bean Programme Annual Report. CIAT, Columbia.
53. **Farrington, J. & J. Nelson, 1997**. Using logframes to monitor and review farmer participatory research. AGREN network, Network Paper no. 73. ODI, London.

    Lightfoot, C., *et al.*, 1993. (Op. cit.).
54. **Simmonds, N.W. & M. Talbot, 1992**. Analysis of on-farm rice yield data from India. Experimental Agriculture, 28, 325–29.
55. **Kirkby, R.A., 1984**. Crop Improvement in Eastern and Southern Africa: Research Objectives and On-farm Testing. IDRC Regional Workshop, Nairobi, Kenya, 20–22 July, 1983, IDRC-218e.

    ter Horst, K. & E.R. Watts (Eds), 1983. Plant Breeding for Low Input Conditions. Proceedings of a Workshop. Department of Agriculture, Zambia.
56. **Ashby, J.A., 1990**. Evaluating Technology with Farmers: A Handbook. CIAT, Cali, Colombia.
57. **Stoop, W.A., 1986**. (Op. cit.).
58. **Waddington, S.R., 1990**. On-farm experimentation with intercrops: a summary of issues from a workshop held in Malawi. Farming Systems Bulletin no. 7. (Sept–Dec). CIMMYT, Nairobi.
59. **Franzel, S., C. Holding, J.K. Ndufa, O.C. Obonyo & S.M. Weru, 1996**. Farmers and trees – linking research and extension. Agroforestry Today, 8, 19–21.

ICRAF, 1991. Methods for participation: on-farm agroforestry research. Summary proceedings of an international workshop. Compiled by S. Scherr. ICRAF, Nairobi.
60. **Stoop, W.A., 1986**. (Op. cit.).
61. **Smaling, E.M.A. & A.R. Braun, 1996**. Soil fertility in sub-Saharan Africa: new dimensions, new challenges. Communications in Soil Science and Plant Analysis, 27 (2&4), 365–84.
62. **Thrupp, L.A., 1996**. New Partnerships for Sustainable Agriculture. World Resources Institute, Washington, DC.
63. **Lightfoot, C. & R. Noble, 1992**. (Op. cit.).
Lightfoot, C., *et al.*, 1993. (Op. cit.).
64. **Stomph, T.J., L.O. Fresco & H. van Keulen, 1994**. Land use systems evaluation: concepts and methodology. Agricultural Systems, 44, 243–55.
65. **Budelman, A., 1996**. (Op. cit.).
Smaling, E.M.A. & A.R. Braun, 1996. (Op. cit.).
66. **Thomson, E.F., R. Jaubert & M. Oglah, 1992**. Using on-farm trials to study the benefits of feed legumes in barley based rotations of NW Syria. Experimental Agriculture, 28, 143–54.
67. **Nordblom, T.L., A.E.K.H. Ahmed & G.R. Potts, 1985**. Research methods for livestock on-farm trials. Proceedings of a workshop held at Aleppo, Syria. 25–28 March, 1985. ICARDA and IDRC, Ottawa.
68. **de Leeuw, P.N., 1990**. On-farm livestock research in sub-Saharan Africa: a review of results, problems and issues. Farming Systems Bulletin, no. 7. CIMMYT, Nairobi, pp. 1–12.
69. **Baker, G., H.C. Knipscheer & J. de Souza Neto, 1988**. The impact of regular research field hearings (RRFH) in on-farm trials in north-east Brazil. Experimental Agriculture, 24(3), 281–8.
70. **Cheng, Y. & P.M. Home, 1997**. Field Experiments with Forages and Crops: Getting It Right The First Time. Forages for smallholders project. ACIAR, Laos.
71. **Fischler, M. & C.S. Wortmann, 1999**. Green manure research in eastern Uganda – a participatory approach. Agroforestry Systems (in press).
72. **Collinson, M., 1981**. A low cost approach to understanding small farmers. Agriculture Administration, 8, 433–50.
73. **Sumberg, J. & C. Okali, 1998**. (Op. cit.).
74. **Lightfoot, C., O. De Guia & F. Ocado, 1988**. A participatory method for systems-problem research: rehabilitating marginal uplands in the Philippines. Experimental Agriculture, 24(3), 301–9.
75. **Lightfoot, C. & D.R. Minnick, 1991**. Farmer-first qualitative methods: farmers diagrams for improving methods of experimental design in integrated farming systems. Journal for FSR/E, 2(1), 57–69.
76. **Boughton, D., *et al.*, 1995**. (Op. cit.).
77. **Kirkby, R.A., et al., 1981**. (Op. cit.).
78. **Zandstra, H.G., 1982**. An Overview of Farming Systems Research. Keynote address, AFSRE Symposium, Kansas State University, Manhattan, Kansas.
79. **Simmonds, N.W., 1985**. Farming Systems Research: a Review. World Bank Technical Paper no. 43, World Bank, Washington DC.
80. **Collinson, M.P., 1972**. Farm Management in Peasant Agriculture. Praegar, New York.
81. **Sumberg, J. & C. Okali, 1988**. (Op. cit.).
82. **Chambers, R., 1980**. Understanding professionals: small farmers and scientists. IADS Occasional Paper. New York.
83. **Kirkby, R.A., 1983**. On-farm experimentation: new methods for an old concept? ICRAF Newsletter, 9.
84. **Box, L., 1987**. (Op. cit.)
Potts, M.J., *et al.*, 1992. (Op. cit.).
85. **Conklin, H.C., 1957**. Hanunoo agriculture, a report on an integral system of shifting cultivation in the Philippines. Forestry Development paper no. 12. FAO, Rome.
Voss, J., 1992. Farmer management of varietal bean mixtures in central Africa: implications for a technology development strategy. In: J. Moock and R. Rhoades (Eds) Diversity, Farmer Knowledge and Sustainability. Cornell University Press, New York.
Kirkby, R.A., 1981. (Op. cit.).
86. **Rhoades, R.E., 1987**. Farmers and experimentation. Discussion Paper 21. ODI, London.
87. **Havekort, B., J. van der Kamp & A. Waters-Bayer, 1991**. Joining Farmers' Experiments. Experiences in Participatory Technology Development. Intermediate Technology Publishers, London.

88. **Box, L., 1987**. (Op. cit.).
89. **Lightfoot, C., 1987**. Indigenous research and on-farm trials. Agriculture Administration, 24, 79–89.
90. **Potts, M.J., *et al*., 1992**. (Op. cit.).
91. **Chambers, R. & J. Jiggins, 1987**. Agricultural research for resource poor farmers. Part I. Transfer of technology and farming systems research. Agriculture Administration, 27, 35–52.
Farrington, J., 1988. (Op. cit.).
Okali, C., *et al*., 1994. (Op. cit.).
92. **Biggs, S.D., 1989**. Resource-poor Farmer Participation in Research: Political, Administrative and Methodological Considerations. OFCOR Comparative Study no. 3. ISNAR, The Hague, The Netherlands.
93. **Ashby, J.A., 1986**. Methodology for the participation of small farmers in the design of on-farm trials. Agriculture Administration, 22(1), 1–19.
Ashby, J. & L. Sperling, 1995. Institutionalizing participatory client driven research and technology development. Development and Change, 26, 753–70.
Merrill-Sands, D. & J. McAllister, 1988. (Op. cit.).
94. **Farley, C., 1996**. Smallholder knowledge, soil resource management and land use change in the highlands of south-west Uganda. PhD dissertation, University of Florida, Gainesville.
Ashby, J.A., 1986. (Op. cit.).
95. **Lightfoot, C., *et al*., 1993**. (Op. cit.).
96. **CIAT, 1995**. CIAL (local agricultural research committees) Handbooks nos 1–9. Centro Internacional de Agricultura Tropical (CIAT), Cali, Colombia.
97. **Collinson, M.P., 1974**. Transferring technology to developing economies: the example of applying farm management economics in traditional African agriculture. World Development, 2(2), 93–7.
98. **Francis, C.A. & J.H. Sanders, 1978**. Economic analysis of bean and maize systems: monoculture versus associated cropping. Field Crops Research, 1, 319–35.
99. **McIntire, J. & L.K. Fussell, 1989**. On-farm experiments with millet in Niger: crop establishment, yield loss factors and economic analysis. Experimental Agriculture, 25, 217–33.
100. **Collinson, M.P., 1986**. Collecting information on rates of work to evaluate the labour productivity of possible innovations in OFR/FSP. Farming Systems Newsletter, 24 (Jan.–Mar. 1986). CIMMYT, Nairobi.
Spencer, D.S.C., 1991. Collecting meaningful data on labor use and farm size for economic analysis associated with on-farm trials in sub-Saharan Africa. In: Mutsaers, H.J.W. & P. Walker (Eds) (Op. cit.).
101. **Rhoades, R., 1987**. (Op. cit.).
102. **IRRI, 1982**. The Role of Anthropologists and Other Social Scientists in Interdisciplinary Teams Developing Food Production Technologies. IRRI, Los Baños.
103. **Posner, J.L., E.W. Crawford & M. Kamuanga, 1990**. Farmer participation in on-farm trials: the case of lowland rice in Southern Senegal. Journal for FSR/E, 2(2), 125–34.
104. **Castillo, G.T., 1994**. Keynote address: sustainable agriculture: in concept and in deed. Journal of Asian Farming System Association, 2, 159–85.
105. **Zeigler, R.S., 1986**. Application of a systems approach in a commodity programme: evaluating Burundi highland maize. Experimental Agriculture, 22, 319–28.
106. **Thomson, E.F., R. Jaubert & M. Oglah, 1992**. Using on-farm trials to study the benefits of feed legumes in barley based rotations of NW Syria. Experimental Agriculture, 28, 143–54.
107. **Zandstra, H., *et al*., 1981**. (Op. cit.).
Shaner, W.W., P.F. Phillips & W.R. Schmel, 1982. Farming Systems Research and Development: a Guideline for Developing Countries. Westview Press, Boulder, CO.
Hildebrand, P.E. & F. Poey, 1987. (Op. cit.).
FSSP, 1987. Analysis and interpretation of on-farm experimentation. FRS/E training units, Volume 3.
Steiner, K.G., 1987. (Op. cit.).
Neeley, Matata & Kysiombe, 1991. Statistical Techniques for the Analysis of On-farm Crop Trials. A Manual. CIMMYT, Nairobi.
Stroud, A., 1993a. (Op. cit.).
Mutsaers, *et al*., 1997. (Op. cit.).
108. **Allan, A.Y., 1969**. Maize diamond demonstration plots. Contributions: Third East African Cereals Research Conference. East African Agriculture and Forestry Research Organization, Nairobi.
109. **Kirkby, *et al*., 1981**. (Op. cit.).

110. **Chiduza, C. & P. Nyamudeza, 1990**. Implementation of On-farm Experimentation in Marginal Environments: Issues and Challenges. Farming Systems Newsletter no. 7. CIMMYT, Nairobi, pp. 113–19.
111. **Hildebrand, P., 1984**. Modified stability analysis of farmer managed on-farm trials. Agronomics Journal, 76, 241–74.
112. **Jongman, R.H.G., C.T.F. Ter Braak & O.F.R. van Tongeran (Eds), 1995**. Data analysis in community and landscape ecology. Cornell University Press, New York.
113. **Cady, F.B., 1991**. Extrapolation of experimental results from a series of on-farm trials. In: Mutsaers, H.J.W. & P. Walker (Eds), 1991. (Op. cit.).
114. **Penninkhoff, P., 1995**. (Op. cit.).
Kajiru, J.G., *et al.*, 1998. (Op. cit.).
115. **Dofing, S.M. & C.A. Francis, 1990**. Efficiency of one-replicate yield testing. Journal of Produce Agriculture, 3, 399–402.
116. **Nagy, J.G. & J. Sanders, 1990**. Agricultural technology development and dissemination within a farming systems perspective. Agricultural Systems, 32, 305–20.
117. **Michigan State University, 1986**. MSTAT: a Microcomputer Statistical Program. MSU, USA.
118. **Lockeretz, W., 1995**. Removing applied agricultural research from the academy. American Journal of Alternative Agriculture, 10(1), 19–24.
119. **Ashby, J.A., 1990**. (Op. cit.), **Mutsaers, H.J.W. & P. Walker (Eds), 1991**. (Op. cit.).
120. **CIMMYT Economics Program, 1976**. From Agronomic Data to Farmer Recommendations: an Economics Training Manual. CIMMYT, Mexico.
121. **Zandstra, H.G., 1982**. (Op. cit.).
122. **Cloud, K., 1984**. Women's productivity in agricultural systems: considerations for project design. In: C. Overholt, M.B. Anderson, K. Cloud, & J.E. Austin (Eds) Gender Roles in Development Projects. Kumarin Press, West Hartford, CN.
123. **Penninkhoff, P., 1995**. (Op. cit.).
Franzel, S., *et al.*, 1996. (Op. cit.).
124. **Sumberg, J. & C. Okali, 1988**. (Op. cit.).
125. **Lightfoot, C., *et al.*, 1993**. (Op. cit.).
126. **Waterworth, J.V. & J. Muwamba, 1989**. Report of a Study on the Use of On-farm Research Results by Extension in Zambia and Swaziland. Mimeo. CIMMYT, Harare.
127. **Waddington, S. & A. Low, 1988**. CIMMYT's on-farm research support program for eastern and central Africa. In: B. Gelaw (Ed.) Towards Self-sufficiency. Proceedings of the Second Eastern, Central and Southern African Maize Workshop. CIMMYT, Harare, Zimbabwe, pp. 160–7.
128. **Fujisaka, S., 1994**. Learning from six reasons why farmer do not adopt innovations intended to improve sustainability of upland agriculture. Agricultural Systems, 46, 409–25.
129. **Rhoades, R., 1987**. (Op. cit.).
130. **Low, A. & S. Waddington, 1991**. Farming systems adaptive research: achievements and prospects in Southern Africa. Experimental Agriculture, 27, 115–125.
131. **Harrington, L., 1995**. Sustainability in perspective: strengths and limitations of farming systems research in contributing to a sustainable agriculture. Journal of Sustainable Agriculture, 5(1/2), 41–59.
132. **Ashby, J.A., 1993**. (Op. cit.).

4.2 EXPERIENCES IN APPLYING FSR IN SEMI-ARID KENYA

A.J. Sutherland and J.N. Kang'ara

It was not difficult to recruit farmers for the experiment. As soon as the first farmer found the concoctions effective, the word spread quickly to neighbours with similar problems who asked to be included in the experiment. This made it possible to treat enough animals to allow for statistical analysis of results.

4.2.1 Introduction

Based at Embu in the semi-arid Mount Kenya region of eastern Kenya, the Dryland Applied Research and Extension Project (DAREP) is exploring practical ways to improve the quality of life for approximately 70,000 smallholder farming families. DAREP was created in 1993 to develop sustainable agricultural technologies and participatory methodologies within an FSR framework. It operates from the Kenya Agricultural Research Institute's (KARI) regional research centre, which covers eight districts of almost 20,000 km^2 in Embu, Thoraka-Nithi and central Isolo. DAREP makes good use of the decentralized local research and demonstration sites inherited from an earlier project funded by the UK's Overseas Development Administration, which also included a component of applied research. With KARI taking the lead on coordination and implementation, DAREP also involves the Kenya Forestry Research Institute (KEFRI), the local extension arm of the Ministry of Agriculture, Livestock, Development and Marketing (MOALDM) and the Natural Resources Institute (NRI[1]), UK, which manages technical cooperation input on behalf of ODA's natural resources research strategy programme.

Activities are implemented through a core interdisciplinary team including an agronomist, a livestock specialist, an agroforester, an agricultural engineer, an agricultural economist, a social anthropologist and, more recently, an entomologist. Other specialists from the implementing national institutions are brought in whenever necessary.

DAREP is, above all, practical. The two cases highlighted in this chapter show how the FSR process has been applied, through a novel methodological and project framework, to produce useful technical interventions within a short time period.

4.2.2 Applying the FSR sequence

DAREP followed the FSR process during implementation. Farming systems characterization, diagnosis of priority constraints, trial planning, technology testing, evaluation and extension/dissemination have all been undertaken with farmers and in close collaboration with extension at all stages. The various activities have overlapped somewhat and have been compressed into a relatively short time-frame; characterization and diagnosis started in mid 1993, followed almost immediately by trials and a dissemination process which began after about 2 years.

Various modification helped reduce the amount of time between diagnosis and dissemination, as follows.

Systems characterization and targeting

The project framework had a predefined target group and target area and so characterization activities were limited to describing important production system and household differences within the project area.

Use of existing infrastructure

The inherited research infrastructure was used to implement research activities in which scientists had confidence, without having to wait for results from diagnostic surveys.

Planning interventions with farmers

This actually increased the time needed for the diagnostic surveys, as did further consultation of professional experts and literature reviews. However, both activities greatly increased the confidence of researchers in the interventions proposed.

Continued dialogue

Continuing dialogue between researchers and farmer experts included informal surveys, PRA, farm visits and farmer open days at the sites.

Interdisciplinary team planning

Matching research plans with time and resources meant that activities received support. Only those activities that could be accomplished within the agreed budget and time frame were initiated.

After the final round of research planning meetings in March 1995, 30 research activities had been endorsed and a number of these, including the two documented here, were underway.

4.2.3 Animal health

A livestock reconnaissance survey conducted in September 1993 provided an overview of production constraints[2]. The first broad-based diagnostic survey conducted in Tharaka in November 1993[3] found that animal health, rather than nutrition and breeds, was seen by farmers as their priority constraint. A similar picture was found during the Mbeere PRA conducted in May 1994[4], although nutrition and breed were also seen as major issues.

Farmers were asked to list those animals that were of importance to them. Cattle came first, followed by goats. Goats, rather than cattle, were selected as the primary target for research intervention for two reasons. First, the main problem with cattle was tick-borne diseases, and there were few researchable interventions available at the time to address this problem. Second, while cattle were generally owned by older men, the ownership of goats was more widespread, including women and younger men. So successful interventions with goats stood to benefit a larger group of farmers.

Four main goat health problems were identified by farmers and local veterinarians during the surveys; contagious caprine pleuropneumonia (*lvuri*), worms (*njoka*), gall sickness (*nduru*) and mange (*mung'uru*). Because the first was already being addressed through vaccination campaigns, interventions on worms and mange were suggested as they were endemic in the area and the existing local knowledge on their control held out the hope of finding a sustainable solution in a short time.

Methods

Mange is a wasting skin disease of goats that is particularly feared because, once it gets into the herd, it results in high animal mortality. Infected animals are not fit to eat or sell. Some farmers had found commercial products for mange too expensive and had started looking for locally available alternatives. A focused PRA using group discussions and visits to local herbalists came up with a list of about eight local concoctions already being tried by farmers. This list was further screened through discussion with farmers, and a trial was designed comparing three of the local concoctions with which farmers and the researcher felt comfortable with two of the recommended commercial medicines and one herbal treatment of neem solution. The three local concoctions were tamarind and castor oil, old engine oil and *mwarwa* (*Albezia anthelmintica*), and old engine oil plus mothballs and salt. The two recommended commercial medicines were an organophosphorus acaricide (Supa dip) and Ivormectin's Ivomex which were used as experimental controls.

The mange control trial was conducted on farm with infected herds in two phases. In the pilot phase, from September to December 1994, 41 animals from three different farmers were treated using seven of the treatments. As a result of this, two of the less promising treatments were dropped and neem was added as a treatment after a tour to ICIPE. The verification phase was conducted during a fresh outbreak of mange in April and May 1995 in Kamayaki and Kaamwa, respectively, and a further 120 animals were treated. The trial required close collaboration with the local veterinary officers, one of whom reported the first outbreak of mange in Chikariga locality. The researcher launched the pilot phase almost immediately, and he and the local vet treated the infected animals together.

Further modifications in the trial took place during implementation. Although the trial was originally designed for goats, it was found that sheep and even calves were infected, and these were also included to establish good relations with the farmer collaborators.

Treatment of the animals involved one application per week of the local concoctions by the researcher together with the farmer owning the animals for a period of 4 weeks. To maintain experimental standards for comparison, the local concoctions were supplied and prepared by the researcher. Data collected included information on body condition, weight change, intensity of infection and the farmer's opinion. Some of the animals were also photographed before and after the treatment.

It was not difficult to recruit farmers for the experiment. As soon as the first farmer found the concoctions effective, the word spread quickly to neighbours with similar problems who asked to be included in the experiment. This made it possible to treat enough animals to allow for statistical analysis of results.

Results

All the local treatments were effective against mange (Table 4.2.1). However, neem was less effective than the other treatments and some farmers asked to switch from neem to castor oil plus tamarind. With castor oil plus tamarind and old engine oil plus *mwarwa*, treated animals recovered faster than those being treated with the commercial drugs, as shown by the greater reduction in intensity of infection. The visual results, however, were so impressive, and farmers were so enthusiastic, it was not necessary to wait for statistical analysis before reaching a conclusion on the efficacy of the treatments.

A problem-driven dissemination process began shortly after the experiment was concluded. In August 1995, when a mange outbreak occurred in Gategi, another locality about 100 km south, the researcher organized a farmer-to-farmer tour. Farmers with infected animals from Gategi visited the Kaamwa farmers who showed them how they treated their animals using local concoctions. The Gategi farmers went home and applied the tamarind mixture with castor oil with good results.

4.2.4 Household food shortages

Problem diagnosis

During the diagnostic PRAs, household food shortages emerged as a significant problem for most households in the project area. Using food availability calendars constructed by farmers, two hunger periods were identified. The most serious began in mid September and continued through to early January, while the second started in late April and continued until early June. Diagnosis showed that household food security was a main indicator of wealth, while inability to feed one's family signified poverty. The poorer households were more affected by the hunger period in late October to December than the richer households as this is when they hire out their labour for weeding in exchange for food. Poorer households, in effect, serviced richer ones with larger fields and better weed management of millet, the main cash crop of this season. As a result poorer households did less weeding in their own fields, or had smaller fields producing less grain, and became locked into a cycle of dependency and periodic hunger. Women were particularly affected by this situation due to their multiple roles in the household. Poorer women not only had to weed for other farmers, but were also expected to weed their husband's fields, cook, fetch water and firewood, take care of sick children and even herd the livestock – all in the course of the day.

Diagnosis showed that farmers had a range of strategies for coping with food insecurity. The most direct strategies were storage of grains (particularly millet which could be stored without chemicals), eating early maturing crops such as pearl millet and cowpeas in the field,

Table 4.2.1. The effects of mange treatments on goats after 4 weeks.

Treatment	Fall in intensity of infection	Change in body condition	Change in body weight	Number of animals
Ivomex	−2.631	0.808	2.350	31
Supa-dip	−2.811	0.840	1.407	17
Castor seed oil + tamarind	−3.269	1.258	1.980	25
Old engine oil + *mwarwa*	−3.688	1.154	1.746	17
Old engine oil + mothballs + salt	−2.946	0.940	1.988	10
Neem	−2.338	0.747	1.647	20

Notes:
1. The reduction in intensity of mange infection observed in castor seed oil plus tamarind and old engine oil plus *mwarwa* treatments were significantly higher than neem and Ivomex ($P < 0.05$) but not the others.
2. The body condition and body weight changes were not significantly different.

Table 4.2.2. Technology basket displayed for addressing household food security issues.

Type of technology	Choice
Early maturing pearl millet	5 varieties
Early maturing cowpea	11 varieties
Early maturing sorghum	10 varieties
Cassava	6 varieties
Sweet potato	6 varieties
Dwarf early pigeon pea	10 varieties
Sweet potato planting methods	3 methods
Dry season vegetable preservation methods	3 methods
Groundnut introductions	9 varieties
Proso and foxtail millet comparison	2 varieties

and planting drought-resistant late maturing food crops such as two season sorghum and pigeon peas. Less direct strategies were sale or barter of assets (labour, handicrafts and livestock) in exchange for food, and relying on food relief programmes. The poorer households had fewer coping strategies, and those without livestock were most acutely affected.

Planning interventions with farmers

The researchers identified a range of researchable options relevant to the hunger problem through the diagnostic process. Direct options included reducing the span of the hunger period by introducing earlier maturing varieties of existing food crops and drought-tolerant cultivars of cassava and sweet potatoes as supplementary starch staples. Indirect options included the introduction of alternative low-input cash crops and high value horticultural crops, reduction in livestock risks to provide a more reliable buffer in times of drought, and labour-saving technologies to reduce drudgery and improve the timeliness of weeding.

Providing a basket of choices

Through the local research sites, the project introduced a larger range of cropping technologies than that requested initially by farmers, as an alternative to selecting one or two interventions that might have a big impact. The technologies listed in Table 4.2.2 were displayed on the local research sites, either as replicated trials or as unreplicated observation plots.

Experimental evaluation

The technologies displayed at the local sites were visited by farmers during open days and evaluated using participatory matrix ranking and guided group discussions. In addition to farmer assessment, agronomic data and researcher observations were recorded. During the open days farmers were invited to choose which technologies they would like to try out on their own farms. One item in particular, a new pearl millet variety from the International Crops Research Institute for the Semi-Arid Tropics (ICRISAT) called ICMV 221 aroused a lot of interest and many farmers requested its seed to test on their farms. ICRISAT assisted by bulking ICMV 221 in a dry season nursery to speed distribution to a larger number of farmers so that larger plots could be planted (Table 4.2.3).

For the three seasons April 1994, October 1994 and April 1995, 90 farmers ran on-farm trials comparing the performance of ICMV 221

Table 4.2.3. On-station yields and days to maturity for pearl millet varieties.

	ICMV 221		Kiraka local		KPM1	
Season	Yield kg ha^{-1}	Days to maturity	Yield kg ha^{-1}	Days to maturity	Yield kg ha^{-1}	Days to maturity
October 1993, 8 sites	1937	76	513*	90	1627	88
April 1994, 6 sites	1220	80	1413	92	1487	89

* Data from two sites only.

Table 4.2.4. On-farm yields for pearl millet varieties.

	ICMV 221		Kiraka local		KPM1	
Season	Yield of kg ha^{-1}	Number farmers	Yield of kg ha^{-1}	Number farmers	Yield of kg ha^{-1}	Number farmers
April 1994	937	19	827	6	865	17
October 1994	830	52	972	31	823	28
April 1995	655	19	540	10	616	8

with other improved and local varieties (Table 4.2.4).

Although on-farm yields across the three varieties were similar, farmers' evaluations clearly indicate that ICMV 221 has the potential to make an impact on the hunger problem. ICMV 221 was popular because it matured earlier than the local millet varieties. In terms of taste and ease of separation of grain from the head, ICMV 221 was similar to the local variety which can be eaten fresh in the field. Farmers also liked its compact head, high yield and its softness, which make it easier to grind than the local varieties. A further advantage for household food security is that ICMV 221 is not as marketable as the local millet and so less likely to be sold in order to raise cash. This could reduce domestic tensions where men control millet sales and marketing leading to arguments and wife beating when wives complain about men selling food stocks or vice versa.

ICMV 221 does, however, have disadvantages which underline the importance of diversity. Farmers consider its storage qualities inferior to the local varieties and note that because it matures earlier, it attracts birds. If both varieties are grown the bird-scaring period is longer. Some farmers have turned this to an advantage by planting ICMV 221 later, spreading their labour for planting and weeding, so that it matures at the same time as the local variety and shortens the bird-scaring period. This is a particular advantage for poorer households, giving them a potentially longer window for planting and weeding. To avoid excessive loss from bird damage, a compromise management strategy is to plant a small portion of ICMV 221 for early food, and plant the rest later in order to reduce bird scaring and spread out labour.

While ICMV 221 was the main 'success story' in addressing food insecurity, farmers have also adopted early maturing varieties of sorghum and cowpeas, cassava and proso millet as a new food crop and are currently exploring groundnuts as a potential cash crop. There has been less uptake of dwarf pigeon peas and sweet potatoes.

Dissemination

Dissemination of the new technologies has been effected through the local site network and the farmer collaborators involved. They have passed on popular new varieties and crops to neighbours and friends. Local farmers have been able to purchase seed of the new varieties from the bulking plots at the research sites that are now managed by local committees. This has worked well for all of the crops, with the exception of pearl millet because of the difficulty of maintaining its seed purity. Some training of the field staff has been undertaken on this issue, but more needs to be done in terms of educating local farmers, extension staff and NGOs.

4.2.5 Lessons learned

A wide range of sources for the accessing of new technology is vital if good FSR diagnosis is to be swiftly followed by the introduction of new technology options to farmers. This was a weak area in early farming systems projects which relied heavily on the technical knowledge of individual agronomists and/or on the availability of 'on the shelf' technologies in the national research systems. These cases show that both international research centres and local farmer experts are important sources of new technologies, in addition to anything the national research system has to offer.

Using international and local technology sources can produce results quickly and with minimal cost to the national research system. ICMV 221 was an introduction from India, which had not completed the usual prolonged series of national variety trials before coming to the project. It was made available through the close collaboration that existed between ICRISAT and the national research system which recognized that adaptive research is not just a conduit for the products of strategic research, but should also access technologies from other sources, including existing local knowledge. The case of mange in goats shows that farmers' own knowledge, tempered through a formal on-farm experimental approach, provided a quick and effective technical solution to an important problem. Use of a scientific approach meant the results could be presented with confidence not only to other famers but also to professionals in extension and research. Knowledge that had been fragmented and individual was brought together into a more coherent whole and shared for the benefit of a larger constituency.

Strong linkages, both with extension and strategic research, were essential components in the process. In the mange case the full involvement of livestock and veterinary extension officers in diagnosis, planning and experimentation provided ownership and commitment to the new technology, even to the point of continued collection of information on local remedies and collaborative planning of new research ventures. Strategic research provided important support with the identification of the parasite, and will facilitate identification of the active ingredients in the local concoctions.

The inclusion of extension and dissemination in the project design encouraged swift uptake of the technologies. Some FSR programmes have seen the role of the researcher as stopping at the point where a technology has been proven on farm. Technical recommendations have been passed on to extension only to be hindered by bottlenecks in extension and input supply systems[5]. The local and rather unique site infrastructure inherited at the start of the project provided an invaluable capacity for pilot extension and dissemination activities, but at the same time raises questions of replicability.

Even though the site infrastructure inherited at the start of the project was unique, it did contain replicable elements.

- Field implementation of the research and dissemination activities did not require highly trained staff, but utilized unemployed school leavers who were competent in local languages and easily trained on the job. As they were trained they effectively trained others with similar backgrounds.
- The low level of external inputs used in running the research activities made them relatively sustainable with low capital costs; the main operating cost being the labour input. While employing local labour is not cost-free, it has the combined benefits of injecting cash into the rural economy, providing an informal means for extending the new technology and making worthwhile use of government or donor funds.
- The research was conducted within the community, on sites belonging to local community members or schools. Many rural communities have such areas available and would be happy for them to be used for agricultural research and demonstration activities.
- The sites were managed by a local committee elected by the local community and played an important role in organizing extension and dissemination activities, particularly seed bulking. In the sites where project funding of research and dissemination activities has stopped, the committees have continued to organize seed bulking activities on their own, without external inputs, demonstrating the sustainability of the approach.

Providing a basket of choices rather than narrowing down to best-bet solutions, has proved to be a workable approach in a farming system which is both risk-prone and diverse. For the dryland cropping programme, the local site infrastructure was ideal for introducing the basket, allowing the project, rather than farmers, to carry the initial risk. The experience with a large range of crop varieties has shown that farmer preferences vary significantly from one local site to another and, if resources allow, farmers prefer to exercise choice in technology selection.

Their participation in narrowing down a larger number of options increases their ownership of the technologies and their interest in dissemination. More conventional procedures in FSR have involved 'pre-screening' by researchers, followed by a minimum period for running trials to produce statistically valid results before making 'recommendations'. These might have to pass through a committee structure before they are 'released'. When farmers participate more actively in the technology screening process they also formulate their own recommendations in the process. Farmer-to-farmer visits provided a fast and effective means of extending a technical message from one community to another, and avoided a long and expensive process of scientific validation, bureaucratic approval and publication of technical recommendations.

It is important for national researchers to have the confidence for extensive dialogue with knowledgeable farmers and go on-farm at an early stage. In the earlier days of FSR, young national researchers were often operating in the shadow of older expatriates who had a limited capacity to dialogue with farmers, and also a tendency to use on-farm trials for verification, rather than for exploration. In the case of mange the nature of the disease further facilitated the dialogue process as farmers were willing to carry the risk of experimental failure rather than lose their animals.

4.2.6 The future – a vision statement

To what extent could the DAREP experience be replicated in other parts of Kenya, and what would be the resource implications? These questions need to be addressed by looking at institutional settings and at the resource situation (human, training, equipment and operational funds).

The institutional setting is broadly favourable for replication of the DAREP approach. A memorandum of understanding encourages close collaboration between researchers and extension in OFR and demonstration activities. A comprehensive network of research centres covering the country has the mandate to conduct adaptive research within a particular region. Regional research programmes are coordinated through an assistant director, providing an opportunity for both coordination and support to exchange information and ideas between centres. Communications between centres are improving through fax and email facilities and there are opportunities to share research results between centres in order to reduce unnecessary duplication. An ODA-funded adaptive research project is proposing to hold joint meetings on a pilot basis to enable the sharing of ideas between three different regional research centres. Other donors are also supporting adaptive research and encouraging sharing of experiences.

However, at the level of the centre and the individual scientist, there are some potential bottlenecks concerning the formation of effective interdisciplinary teams. The culture of research has tended towards promotion of a 'hard science' orientation with individualism, disciplinary and commodity approaches prevailing, and also with a degree of hierarchy in research management and implementation. This culture, while changing slowly, could seriously hamper greater decentralization through small interdisciplinary teams focusing on particular local areas of recommendation domains. A major shift in thinking and operational practice at some of the centres would be required to replicate the DAREP approach.

Turning to human resources, these are relatively abundant. At the field level, most districts in Kenya have many more trained agricultural extension staff than the DAREP project area and there is an abundance of very capable school leavers with good results who are looking for any kind of employment opportunities. Professional manpower requirements are slightly more limiting. In the DAREP case, for example, about four to five full-time researchers covered two districts, but most of them also had other responsibilities. This represented about 20% of the research manpower at a regional centre covering eight districts. In other words, the project absorbed the full quota of available professional manpower, with none to spare, particularly as a further 20% were out on training. The other nine regional research centres, and four national centres which also hold a regional mandate, have comparatively fewer districts to cover and smaller numbers of scientists allocated to regional research duties per district. This means that the DAREP approach is broadly replicable in terms of professional researcher input, but it would mean scientists and centres operating at stretched capacity.

Training is a somewhat limiting resource factor. A number of the researchers have some experience with on-farm trials, surveys and participatory methods, but there should be more sharing of experiences and approaches in order to build up confidence. Specific training may be helpful, particularly as the training is field based and experiential, and can be quickly followed up by implementation activities. More theoretical training which deals with the concepts and methods of on-farm and participatory research, and also basic social science training would also benefit individuals who have a strong interest in publishing the results and experiences of OFR. Projects like DAREP also form a useful pool of expertise from which others can be trained.

The main equipment required for such an approach is a reliable means of transport for field work. Within DAREP, over-reliance on transport and frequent field visits has been reduced by the delegation of many responsibilities to field staff and farmers. However, researcher mobility remains crucial to effective OFR. Having an interdisciplinary team using common research sites and farmers allows for rationalization of vehicle use and also field staff use.

In common with most national research systems in the region, operating costs are the major resource constraint and it seems unlikely that funding will increase substantially in the near future. The effects of structural adjustment and other economic reform programmes, and the rather low priority given to agricultural research by national governments, make it likely that government operational funding may remain low and that donor projects will continue to be one of the main sources of operational funds. The only other hope would seem to lie in seeking more collaboration with local NGOs and other projects with an agricultural technology component. Such projects often have funds and effective extension networks, but are lacking in technical expertise. They may be prepared to fund OFR operational costs, particularly if they are convinced that the results will benefit their constituency of farmers.

REFERENCES

1. **NRI, 1993**. Final Report EMI Dryland Farming and Dryland Applied Research Project 1988–1993. Natural Resources Institute, Chatham, UK.
2. **Kang'ara, J.N. & A.J. Sutherland, 1993**. Report of a Livestock Reconnaissance Survey in Lower Embu and Tharaka-Nithi Districts. Dryland Applied Research and Extension Project, KARI, Embu, Kenya.
3. **DAREP, 1994**. Farming in Tharaka: Report of Informal Diagnostic Survey of Farming Systems in Tharaka North and Central Divisions of Tharaka-Nithi District. Kenya Dryland Applied Research and Extension Project, KARI, Embu, Kenya.
4. **DAREP, 1995**. Farming in Mbeere: Report of Informal Diagnostic Survey of the Farming Systems of Lower Embu (2 Volumes). Dryland Applied Research and Extension Project, KARI, Embu, Kenya.
5. **Waterworth, J.V. & J. Muwamba, 1989**. Report of a Study of the Use of On-farm Research Results by Extension in Zambia and Swaziland. CIMMYT Regional Office, Harare.

Chapter 5
FSR in Extension and Policy Formulation

5.1 FARMING SYSTEMS EXTENSION IN THE USA

Cornelia Butler Flora and Charles Francis

There was an increasing gap between commodity-based research and the commodity interest groups that focused on the short-term economic bottom line needed for farm survival. Different actors on the agricultural stage appeared to be playing different roles, or were appearing in completely different dramatic productions.

5.1.1 Introduction

The farming systems approach in agricultural extension in the USA has had two distinct periods. The first was farm management extension in the 1920s and 1930s when extension agents, as they were called at the time, considered agronomic, animal husbandry and economic elements of the farm in concert. In general, the approach was prescriptive in nature, stressing what should be done rather than the logic of what was being done[1].

The second period began in the early 1980s at the time of a major economic crisis on farms. This coincided with the return to the USA of a number of scientists who had worked overseas on long-term assignment, bringing with them new ideas and approaches already being practised in the developing world. The revival of interest was catalysed by publications such as 'Farming Systems Research and Development' by Shaner *et al.* in 1981[2], and by the annual meetings that began at Kansas State University, University of Arkansas, Michigan State University and the University of Florida[3]. The initial novelty was the focus on the system as a whole, generally from an expert perspective. Between the early 1980s and the 1990s the participatory aspects of farming systems became more prominent in extension, coupled with the growth of farmer-led concerns about the sustainability of agricultural systems. The participatory methods that grew from these experiences have now become an integral part of many extension programmes in the USA and elsewhere in the temperate zone[4].

5.1.2 An early focus on whole farm systems

The early multidimensional approach to farm planning was exemplified by practical programmes in Missouri called 'Balanced Farming', back in the 'dirty thirties'. Agents took a farm and family focus and looked at production, economics and what today we would call family quality of life. The extension agent, who was concerned with soil conservation as well as preserving the family farm, sat down with farmers across the kitchen table and helped them plan their total farming systems. This resulted in whole-farm plans that were consistent with the soil and other resources on the farm and that met the unique goals of each farm family. It was a labour-intensive system and was only possible because of the structure and orientation of extension in the USA. Extension was institutionalized as a federal system in the USA with the 1914 passage of the Smith-Level Act. This created outreach programmes of the state agricultural colleges in a cooperative network called the Federal State Extension Service. All of the extensionists were, therefore, employees of the state and of the Federal Government. Since local extension was also partially funded at the local level, many were county employees as well. This tripartite funding meant close attention to the needs of farmers in each county. Those who used the system most effectively were often the decision makers, members of the county boards, and the powerful farmers with access to resources. This early Cooperative Extension Service (now called the Cooperative Extension System) was conceived as a technology transfer system, in which knowledge was derived from the land grant researchers and Department of Agriculture scientists, and transferred through extension to local users. From the start, it was an education programme.

Beginning in the 1930s, a great deal of extension agent's time was spent explaining to farmers the ever more complex federal farm programmes, with an increasing emphasis on productivity and economics. This continues to the present day. The changing nature of agriculture, however, with the farm crisis blamed to some extent on the advice received from land grant universities, led to a reappraisal of the role of some of the representatives of the land grant system in the field. By this time, extensionists in many states were involved not only in technology transfer and education, mostly on complex farm programmes, but technology adaptation as well. While most scientists continued to work on a commodity-component basis, with a growing proportion of research dollars going to fundamental research at the microbiological level, agents and specialists were being forced to deal with whole-farm systems. There was an increasing gap between commodity-based research and the commodity interest groups that focused on the short-term economic bottom line needed for farm survival. Different actors on the agricultural stage appeared to be playing different roles, or were appearing in completely different dramatic productions.

5.1.3 A second systems iteration

A number of American scientists who had worked in research and education in developing countries returned to the USA in the late 1970s and early 1980s. The overseas activities of agricultural economists and agronomists such as David Norman in Africa, Perry Phillips in Asia, Charles Francis in South America and Peter Hildebrand in Central America were supported by bilateral development dollars. These funds aimed to replicate the American experience and institutions in other countries in order to increase Third World productivity. While overseas these scientists observed that heavy investment in institution building and commodity research was not increasing the productivity or standard of living of limited resource farmers. Each of them had helped to develop farming systems research to improve technology choice and design, based on the premise that technologies should focus on the situation of the farmer. On their return they found an American research and extension system that was extremely rigid, discipline oriented and narrowly bound, and at times even territorial. It seemed that the USA was lagging behind the progress already being made, and the lessons being learned, in developing countries. Alongside a growing number of scientists, many of whom had shared common experiences in developing countries, they explained the advantages of a farming systems approach that included multidisciplinary teams and farmer participation to their own institutions and to their colleagues.

The first easily identifiable efforts include field research with some American students and foreign nationals who were being trained in the USA in the early 1980s. Much of the support came from USAID investments in land grant universities designed to increase their capacity to meet the needs of developing countries and serve as contractors to, and ambassadors for, American foreign assistance efforts.

The strengthening grant at Kansas State University, where David Norman was located, was based on a farming systems approach to agricultural development. The first Farming Systems Symposium, held in the autumn of 1981 at Kansas State University[5], aimed to rethink the relation of extension and research in the light of farmers' circumstances. The main thrust was international, on the assumption that the current research and extension system in the USA was already meeting the needs of the agricultural sector.

Soon after this, USAID implemented the Farming Systems Support Project through the University of Florida, which continued to support the symposium and included a domestic farming systems project for limited resource farmers in Virginia. That first project was motivated by the need to move small farmers from dependence on a single cash crop – tobacco – and diversify their agricultural income. The assumption was that when resources are limited, and the income and area farmed is relatively small, their circumstances would be similar to those faced by farmers in the developing world. These small farmers were being missed by the current extension system, and particularly by the Land Grant System. Its structure meant that extension met the needs of powerful and vocal farmers and agricultural processors, rather than those of the politically disempowered. The only whole-farm focus in the USA had been institutionalized into the farm management specialist whose training was in agricultural economics. The resulting focus on short-term economics ignored a number of complex realities on farms and in communities[6].

As the farming systems approach came from developing countries to the USA, the extension system seemed more open to the changes than the agricultural research establishment. The American farming systems projects in the early 1980s in Florida, Hawaii and Virginia stressed multidisciplinary teams and local participation. Small programmes were established domestically with USAID funding in the early 1980s at the University of Florida and Virginia Polytechnic Institute. The Winthrop Rockefeller Foundation funded similar efforts in West Virginia.

The University of Hawaii began to institutionalize a farming systems programme within the College of Tropical Agriculture and Human Resources. Once again, returning scientists with international experience in farming systems, among them Perry Phillips, insisted that this approach be implemented in Hawaii as well as in their international programmes. The team began working with small-scale sugar growers, using methodologies that had proved effective in developing countries. These early approaches followed the farming systems process of diagnosis, design, on-farm trials and analysis of the results. The basic assumption was that to get the technology right it was critical to implement an interdisciplinary approach and include farmers in the diagnosis of problems, and in the design, conduct and interpretation of on-farm trials. An additional assumption was that the most appropriate technology would be demonstrated and disseminated through the testing phase, and that adoption would multiply as more farmers and specialists became aware of the success of technology and the effectiveness of the process. The ideas developed by Robert Rhoades[7] and Robert Chambers[8] were critical in informing this approach, as they demonstrated that farmers could understand the principles of the technology itself and adapt it to their particular situations, serving as extensionists in the process.

Extension in the USA has traditionally used field demonstrations as a form of technology transfer. Demonstrations were located on experiment stations or in the fields of farmers willing to lend land to the researcher or extension agent. They would often carefully replicate a credible experiment station design to assure statistical validity and thus prove the superiority of a new technology. One of the major difficulties in implementing a farming systems approach in extension in the USA was moving the extensionist – now referred to as an extension educator rather than agent – from a mentality of demonstrating the 'right' technology to an open-ended approach using on-farm trials of alternative and farmer-selected technologies. Farming systems stressed alternatives; the demonstration approach stressed *the* answer.

On-farm trials, in contrast, are based on the actual production conditions of farmers on commercial farms. As in developing country situations, scientists working on farming systems research and extension in the USA had to grapple with the needs and complexities of whole-farm systems compared to the need for more component research. Well into the 1980s, and today, most continue to do the kind of research that is publishable and rewarded; many perceive this as only possible through component research. Some scientists and their students persisted in pursuing the on-farm research (OFR) paradigm, and results of both the methods and the results from experiments were published in refereed journals, such as Franzleubbers and Francis in 1992 and 1994[9].

5.1.4 Sustainable agriculture and farming systems

While scientists were testing and demonstrating an alternative model of a systems-based approach to research and extension, farmers were getting organized as well. They were protesting against the lack of knowledge with extension about alternative ways to produce and market crops that were both farmer- and environment-friendly[10]. Farmers found they were getting more information from each other than from the extension system. As their organizations became more sophisticated, farmers were determined to do their own research on their own farms. The techniques of the FSR approach for on-farm trials proved particularly useful to them.

Many of the early efforts to utilize a farming systems approach to sustainable agriculture were made more in spite of the land grant system than because of it, although 'deviant' scientists participated with the farmers as colleagues to help with the design and implementation of the trials[11]. Because of the limited numbers of participating scientists, these efforts often lost the interdisciplinarity of the FSR models, losing the focus on technology developed for very specific ends, based on the farmer's analysis of his or her most pressing problem.

A major shift in the willingness of some scientists to plan and operate in a systems context occurred with the availability of federal funding to support sustainable agriculture. A farming systems approach proved particularly amenable to analysing multiple outcomes, including productivity, economic return, energy efficiency, quality of life or the well-being of all members of the farming household, such as intra-household issues and environmental outcomes. Encouraged by the research money available through the Low Input Sustainable Agriculture Program (LISA), later the Sustainable Agriculture Research and Education Program (SARE), scientists found that working in teams changed the way they designed their research and the way they worked. Many found these new kinds of interactions too time-consuming, but others found them intellectually exciting and followed through over the long term. The first efforts of many researchers focused more on new ways to work together across disciplines, rather than on farmer involvement.

The SARE/ACE programme, or to give it its full title of Sustainable Agricultural Research and Extension and Agriculture in Concert with the Environment, included grants to farmers to conduct their own research, often with the support of land grant scientists. Funded by the American Department of Agriculture and the Environment Protection Agency, the availability of such grants was reviewed by a board that included farmers. This shifted the balance of power towards researchers and farmers interested in alternative systems. Stronger relationships were formed between groups of farmers, who became the source of diagnosis as well as design and on-farm trials. This mirrored the growing understanding in developing countries of the importance of working with farmer groups, forming new ones when necessary.

The Practical Farmers of Iowa, formed in 1985, is now one of the leaders in farmer-led research, focusing on the on-farm trial aspects of farming systems research (FSR) and extension. The farmers of the Land Stewardship Project (LSP) in Minnesota came together as a reaction against the poor stewardship by insurance companies on the land they acquired during the farm crisis. LSP soon turned to on-farm trials as a means of empowering farmers to find alternatives to the system that had resulted in many losing their land. Similar groups that focused on sustainability issues and the need to develop appropriate technology began to form across the country. These groups were organized at first in opposition to the land grant system, a group of academics they perceived as interested only in component research, and who ignored systems approaches and sustainable technologies. Initially, representatives of the land grant system tended to denigrate farmers' research and their audacity in positing alternative sources of knowledge. Purist researchers saw farmers' efforts as an attempt to retreat to the technologies of the 1920s with consequent dramatic declines in agricultural production.

5.1.5 Institutionalization of farming systems-based extension

Despite an initial frontal assault by the agricultural establishment, new partnerships between land grant institutions and farmer groups interested in sustainable agriculture began to form by the 1990s, adopting a systems focus for their work. In many parts of the country, the methodology for this OFR and monitoring was spurred by the holistic resource management (HRM) approach. First introduced by Alan Savory in 1989[12], this was another example of the north learning from the south, bringing experiences with animal agriculture from Africa to the USA. The HRM method demands analysis of a whole system and the optimization of multiple outcomes: quality of life, economic goals and environmental goals. HRM takes the household very seriously, and, whenever possible, training is carried out with farm families.

Many farmer groups such as the LSP and the Practical Farmers soon shifted their approach to include the household and community. Including women in the meetings meant moving the schedules to weekends so that the women who worked off the farm or had responsibility for school children on weekdays could attend. The groups provided not just child care but serious child development activities in the meetings, making the meetings a joy for children as well as adults. The focus on community linkages was spurred by the realization that we not only had to address on-farm issues but issues of linkages off the farm as well. This effort was spurred by a grant programme from the W.K. Kellogg Foundation, the Integrated Farming Systems Program that linked sustainable agriculture groups with land grant colleges in 18 states. Programmes designed to include community outreach and integration of rural and urban interests were implemented in the mid 1990s in Arkansas, California, Delaware, Georgia, Iowa, Kansas, Kentucky, Maryland, Massachusetts, Michigan, Minnesota, Montana, Nebraska, North Carolina, Ohio, Pennsylvania, Washington and Wisconsin.

FSR and extension in the USA has been built by farmer organization involvement with the land grant faculty, and with an emphasis on sustainability. Because of the relatively high level of education of farmers in the USA, they are more easily viewed as peers by at least some land grant scientists and extension educators. The participatory on-farm trial dimension also changes the relation of extension educators to the traditional researchers in the land grant system. There are a number of examples of serious research by extension-farm teams from California and Washington, throughout the Midwest, in the north-east and in the south. One such team has been organized by the LSP and works with scientists from the University of Minnesota, Iowa State University, the Minnesota Department of Natural Resources and other agencies. A graziers group working to improve the sustainability and profitability of their livestock–crop systems meets on a regular basis to design tools for monitoring impacts that are both environmentally sound and production oriented. In addition, participants are monitoring quality of life on a systematic basis. Changes in the farming system in this group can be justified in terms of quality of life issues, especially family relations, as well as through environmental or economic rationales.

The intertwining of farming systems research and extension with sustainability is perhaps most clear in the SARE/ACE. Under Title XII of the enabling legislation there is funding for training specialists and educators in Extension as well as in the Natural Resource Conservation Service (NRCS) in sustainable agriculture. Funds for this training have been allocated by region. The training includes in some sense a farming systems approach with a very strong emphasis on farmer participation, an elevation of the importance of on-farm trials and demonstrations, and the relation of technology to the environment. One notable example is the Chapter 3 training carried out in 1996 in the north central region, organized under the title 'Everyone is a Teacher, Everyone a Learner'[13]. Despite the recent increase in training, only some 25% of extension efforts use a systems approach. Most of those come from extensionists trained in human development by the traditional colleges of home economics. Individuals in those disciplines are trained in a systems approach, which is not part of the standard training in colleges of agriculture.

5.1.6 Conclusion

One important result of greater farmer involvement through farming organizations, including the Regional Sustainable Agriculture Working Groups, has been the impact on agricultural research in many land grant institutions. Technology is now being developed based on new farmer-derived criteria and a broader based discussion of priorities. The field extension educator is learning a new and important role, moving away from service provision to become an educator who can help local people solve their own problems. There is currently a huge debate over the focus, funding and structure of extension in the USA. The infrastructure of the current extension system has great potential to reach a large number of citizens throughout each state. Extension educators and specialists working together with farmer groups in sustainable agriculture, using an FSR and education approach, provide one important model for the future.

REFERENCES

1. **Gilbert, E.H., D.W. Norman & F.E. Winch, 1980**. Farming Systems Research: A Critical Appraisal. MSU Rural Development Paper, no. 6. Michigan State University, East Lansing.
2. **Shaner, W.W., P.F. Philipp & W.R. Schmehl, 1981**. Farming Systems Research and Development: Guidelines for Developing Countries. Westview Press, Boulder, CO.
3. **Sheppard, W.J. (Ed.), 1982**. Small Farms in a Changing World: Prospects for the Eighties. Proceedings of the Kansas State University, 1981. Farming Systems Research Symposium, Kansas State University, Manhattan, Kansas.
4. **Francis, C., C. Edwards, J. Gerber, R. Harwood, D. Keeney, W. Liebhardt & M. Liebman, 1995**. Impact of sustainable agriculture programs on US land grant universities. Journal of Sustainable Agriculture, 5, 19–33.
5. **Sheppard, W.J. (Ed.), 1982**. (Op. cit.).
6. **Flora, C.B., 1983**. Farming systems research and the land-grant system: transferring assumptions overseas. Proceedings of the 1982 FSR Symposium. C.B. Flora, Ed. Kansas State University, Manhattan, Kansas.
7. **Rhoades, R.E. & R.H. Booth, 1982**. Farmer-back-to-farmer: a model for generating acceptable agricultural technology. Agricultural Administration, 11, 127–37.
8. **Chambers, R., A. Pacey & L.A. Thrup, 1989**. Farmer First; Farmer Innovation and Agricultural Research. IT Publications, London.
9. **Franzleubbers, A.J. & C.A. Francis, 1992**. Farmer participation in research and extension: N fertilizer response in crop rotations. Journal of Sustainable Agriculture, 2, 9–30.

   Franzleubbers, A.J. & C.A. Francis, 1994. Energy output:input ratio of maize and sorghum management systems in eastern Nebraska. Agriculture, Ecosystem and Environment, 53, 271–8.
10. **Watkins, G., 1990**. Participatory research: a farmer's perspective. American Journal of Alternative Agriculture, 5, 161–2.
11. **Rzewnicki, P.E., R. Thompson, G.W. Lesoing, R.W. Elmore, C.A. Francis, A.M. Parkhurst & R.S. Moomaw, 1988**. American Journal of Alternative Agriculture, 3, 168–73.
12. **Savory, A., 1988**. Holistic Resource Management. Island Press, Washington, DC.
13. **Carter, H. & C. Francis (Eds), 1995**. Everyone a teacher, everyone a learner. North Central Region. Chapter 3 Training Program, SARE, Cooperative Extension Division, University of Nebraska, Lincoln, Nebraska.

5.2 THE EVOLUTION OF THE ADVISORY SERVICES IN CHILE AND THE ROLE OF FSR

J.A. Berdegué

Too often in Latin America, governments and institutions have started from scratch every time there is significant political or administrative change, or when new financing is obtained from a new bilateral or multilateral source. This tendency constrains institutions from accumulating knowledge and expertise, in a process of continuous improvement. In the case of Chile, even the change from the military to the democratic government in 1990, did not lead to an institutional rupture.

5.2.1 Introduction

Chile has a system of technology transfer services for small-scale farmers that is the product of at least 17 uninterrupted years of experimentation and evolution. This chapter will describe and assess this continuous process of change, the role played by FSR, and the product of this process, the Agricultural Advisory Service (SAA), founded in 1995.

The Agricultural Development Institute (INDAP) was founded in 1962, as a public service within the Ministry of Agriculture, to provide organizational, technical and financial assistance to small-scale farmers, in the context of the national debate that soon led to the Agrarian Reform programme. INDAP staff were involved in the creation of some of the first peasant organizations and in the implementation of many publicly funded productive and social projects, some of them large scale. By 1973, INDAP's 5000 staff members were involved in many projects and activities, from training women on 'home economics', to the import and distribution of agricultural inputs and machinery.

After the military coup of 1973, INDAP was deeply reduced in size and scope, so that by 1989 there were less than 1000 staff members and only two programmes: Technology Transfer and Credit. New laws limited the type of beneficiaries with whom INDAP could work, effectively excluding rural salaried workers and, in a strict interpretation of the law, rural women and youth, and restricting the types of services it could provide to them. The advent of a democratically elected government in 1990 brought about numerous changes and, by 1997, INDAP was reaching almost 120,000 households, a number which has remained more or less constant since then. A 1994 study, based on data collected for the Ministry of Planning, estimated that 64% of the households attended by INDAP were below the official poverty line, of which 34% were below the extreme poverty line. One of the programmes operated by INDAP is the SAA (formerly known as the Technology Transfer Program) which works with about 50,000 households, and includes specific services for 35,000 women. The programme has reached a total of around 85,000 people since its launch, almost half of them women.

5.2.2 FSR in Chile: its non-governmental origins

During the 1980s, the systems approach was basically restricted to two non-governmental organizations. One of these, AGRARIA, began to work with a systems approach in 1982, introducing the concepts and methods of the francophone Recherche-Developpement, maintaining close relationships with French specialists. The other, GIA (Agrarian Research Group), began implementing FSR-E projects in 1984. At the same time, a few more or less isolated individuals in the universities and the governmental Agricultural Research Institute (INIA) also began to conduct research based on a FSR-E approach and began to established informal collaborative agreements with AGRARIA and GIA.

Towards a new conceptual framework

The work of these NGOs, together with contributions from others, soon led to the development of a new conceptual framework to analyse Chile's small-farm subsector, and to design new development strategies, including institutional arrangements. Some of the key concepts were that small farms function according to rational criteria different to those of large, commercial enterprises; and that a top-down, linear and mechanistic approach was a serious constraint to a dynamic system of technological innovation. These new ideas remained more or less academic throughout the 1980s, but in 1990 they provided the foundations and much of the detail of the new government's policies concerning research, extension and rural development.

Testing, development and adaptation of methods

In parallel with the development of new concepts, new field methods were tested and adapted to Chilean conditions, and a large number of agronomists and agricultural technicians were trained. In the 1980s, these new methods had been applied only in NGO projects, and in the Secano (rainfed agriculture) programme of INIA. When political conditions changed, however, these methods were rapidly adopted by the government for agricultural research and development, widely adopted by INDAP and, to a smaller but still significant extent, by INIA. In the universities, however, the new system methods remained restricted to a few research groups.

Building human capacity in systems

The GIA and AGRARIA programmes included large training components from the very start. GIA's annual programme to train university-level and technical-school level staff from NGOs has been running for over 12 years and has trained around 350 people from most of the Chilean NGOs specialized in agriculture, and most of these have been working as INDAP consultants since 1990. For several years now, many of the 50 or so people who have taken part in GIA's 'Young Scientists Training Program' have held executive positions in the government or in NGOs.

AGRARIA, with the cooperation of French organizations and of national universities, has for several years organized a very intensive, month long, training workshop designed to disseminate the concepts and methods of Recherche-Developpement, training about 120 university-level people in total. GIA and AGRARIA have also developed a group of some 30 highly skilled experts who maintain close professional relationships.

5.2.3 The evolution of the technology transfer services

It is possible to distinguish three periods in the evolution of Chile's extension service since a privatized system was initiated in 1978, as follows.

The period of maximum liberalization: 1978–83

The Ministry of Agriculture implemented the Entrepreneur Technical Assistance (ATE) Program, giving small and medium farmers subsidized stamps with which to pay for the technical assistance provided by an independent agronomist or medical veterinarian. Government intervention was minimal and the individual farmer was responsible for selecting the professional who would provide the services. The farmer could terminate the contract at any time. This system failed completely, for one main reason – the false assumption that there is a market of technical assistance services in the rural areas of Chile. In most rural areas there was rarely one qualified agronomist. The farmer had to hire the one person who visited the farm, and this contract was established on the basis of the information that the service provider (the extensionist) wanted to transmit to the farmer. If the results were poor nothing happened, since the cost of the service was fully subsidized and there were no other competing professionals in the area.

The period of maximum uniformity: 1983–90

As a response to this failure, the system was replaced in 1983 by the strictly regulated Integral Technology Transfer Program (PTTI) which serviced 13,700 farmers, 9% of small-scale agricultural producers. In 1987 this was joined by the Basic Technology Transfer Program (PTTB) which covered some 13,000 minifundistas, very small-scale farmers, usually practising subsistence agriculture in poor areas of the countryside. The government determined which private Technology Transfer Consultant (CTT) firm would provide the service in a given area, using a system of public bidding. For political reasons, many qualified private organizations, such as NGOs and small farmers' organizations, were in effect barred from participating.

The improvement plans of 1990, 1992 and 1994

Three main changes were introduced in 1990, after the democratically elected government took control of INDAP. First, an increase in the number of farmers serviced by the system. By 1994, coverage had increased by 91% compared to 1989, with a 61% budget increment in real terms. Second, additional resources were used to increase the participation of very poor, minifundista farmers. Third, a number of changes were made to improve technical and methodological contents, and to reduce the per capita cost of the programme. To improve cost-effectiveness, programming would be based on progression and graduation. After about 9 years, farmers would 'graduate' from the two-stage programme.

Professionals exposed to systems approaches in NGOs had a significant influence on discussions about INDAP's institutional mission and basic conceptual approach. Some argued that several rural and social development programmes should be implemented, so that INDAP could become an institution promoting the integral development of all poor rural families. Others saw the key task for INDAP as supporting small-scale farmers to carve an effective niche in Chile's export-oriented market economy.

In 1992, INDAP and the Ministry of Agriculture adopted a 'Plan for the Improvement of the Technology Transfer Program'[1] which restricted INDAP's mission to the development of small-scale farmers as producers. This adjustment was further advanced in 1994, with the approval of a Strategic Management Program. This stated that the institutional mission was to 'consolidate the productive peasant sector as an economic and socially relevant agent, on the basis of competitive production systems and efficient functional organisations'. After 4 intensive years of cumulative changes, the PTT was given a new name to reflect its new orientation: the SAA.

5.2.4 Characteristics of the SAA

The SAA is the product of a process of cumulative innovations, gradually introduced over a period of at least 17 years. Too often in Latin America, governments and institutions have started from scratch every time there is significant political or administrative change, or when new financing is obtained from a new bilateral or multilateral source. This tendency constrains institutions from accumulating knowledge and expertise in a process of continuous improvement. In the case of Chile, even the change from the military to the democratic government in 1990 did not lead to an institutional rupture. The multilateral loans which contribute to INDAP's funding, including two large loans from the World Bank, have always been integrated within a common framework.

The reforms of 1992 were derived from two simple ideas:

- A single tool could be used only for a limited number of possible purposes.
- If resources were limiting, it was essential to establish priorities.

At the field level, the planning and evaluation mechanisms were, to a large extent, focused on the operational aspects of extension; for example the number, type and dates of the field activities. A planning system has since been created, which includes the definition of development strategies for each of the main agricultural systems within each microregion. It also features medium-term development plans for each local group (15–20 households) in the SAA, and, more recently, the identification of formal projects including technical assistance and loans to finance specific associative market-oriented enterprises at the farm or community level. In all cases, economic criteria to a large extent guide the definition of objectives, which must lead to improved productivity, market participation and profitability.

Diverse operational modalities

The SAA is a system of 17 specific operational options or 'modalities', offering different services to meet different needs. The most important modality is SAA-Local, which is destined to become the 'standard' option. A group of about 270 families within one microregion are serviced by a CTT firm, which has an average budget of about US$382 per household-year to cover the costs of providing the agricultural advice. The CTT, with varying degrees of participation from farmers, defines a local development project for each local group. INDAP pays the CTT according to the degree to which it has completed the annual plan of operations and the specific objectives defined in the project.

Complementary private and public roles

The system recognizes three key actors: INDAP, CTT and local farmers' organizations, and assigns specific responsibilities to each of them.

INDAP has the following responsibilities:

- To fund the system.
- To define the global policies and to assign accordingly the budget and other resources to the different regions, areas and rural social strata.
- To regulate and conduct the public bidding process.
- To define the general methodological and operational frameworks of the SAA.
- To supervise, evaluate and grade the CTT firms.
- To provide training to the extensionists of the CTT.

CTT and local farmers' groups are responsible for:

- Defining development strategies and projects, mid-term objectives and goals.
- Defining, programming and implementing activities for the achievement of those goals and objectives.
- Defining the specific extension approaches and methodologies to be used at the field level.

All participating farmers must contribute to the costs of the services they receive, on the basis that farmer participation will be greater if they have a direct financial commitment. By 1996, all of the households were paying between $13 and $65 each year. Payments are made directly to the CTT, and a system of penalties is in place to stimulate the CTT to recover this cost.

The function of the CTTs today is to assist farmers to detect and respond to market opportunities through local or microregional development projects organized around particular market-driven objectives. Each project will integrate the four basic services that INDAP makes available to its clientele: the SAA, the Financial Service (credit and financial subsidies), the Agribusiness Service (marketing and agroindustry) and the Organizational Development Service.

The new market orientation is also being strongly supported by INDAP's Extensionist Training Programme. In 1990, close to 100% of the training workshops were devoted to traditional agronomic issues: weed control, new crop varieties, etc. By 1995, 65% of the training resources were being invested in workshops directly related to key issues in the implementation of the new market approach, and the same trend is seen in the allocation of funds in the Specialised Training Program.

Linkage with agricultural research

In 1990, a Co-operation Agreement was signed by INDAP and the INIA, creating a formal and systematic mechanism for joint activities in favour of the small-scale farm sector, at the local, regional and national levels. INIA channelled important funds from its own budget to set up five Technology Validation and Transfer Centers (CATT). There are well over 50 joint local-level area commissions (which also include representatives from the CTT and from the farmers' groups), eight regional commissions and one national commission. The area commissions meet on a monthly basis to review the work in the CATTs, and to plan and conduct informal training events, field trips and short diagnostic studies with the participation of INIA researchers, INDAP staff and external supervisors, and the technical staff from the CTT firms.

Farmer participation

Farmer participation in this system is highly variable. INDAP requires that local groups must always formally approve the medium-term plan or the project and the annual work plan, before they are accepted as the guidelines for the activities of the CTT. However, the degree of effective participation is determined by the personal disposition and preference of the agronomist or agricultural technician in charge. INDAP does not intervene beyond a certain point, because this would run counter to the attitude of leaving most field-level decisions in the hands of the CTT and the farmers' groups. Institutionally, INDAP has formal advisory boards at the national and regional levels and in several of the local area offices, in which the farmers' organizations are very well represented. These boards discuss many of INDAP's policies and the most important specific measures, although they have no decision-making capacity. While the degree of farmer participation has improved enormously since 1990, proactive participation is still fairly rare. In 1995, INDAP adopted three specific measures that should lead to improved farmer participation:

- The increased financial contribution of the participating households.
- The participation of the households in the regional level pre-selection advisory committees that screen those CTTs that want to bid to provide the SAA services.
- The inclusion in the formal CTT evaluation and grading system of an item which reflects the opinion of the participating farmers about the efficiency, quality and relevance of the services provided by the CTT.

Farmers' organizations

As a result of SAA-Local, an effort has been made to support the development of 'common interests local groups', which are groups of usually five to 10 farmers who work together to implement an 'associative market-oriented enterprise'. This usually involves some sort of investment, at least partially financed by a loan from INDAP's credit lines. The groups are usually informal, in the sense that they lack a formal legal status. At the microregional level (SAA-Microregional), the emphasis has been placed on working with formally constituted farmers' organizations, such as provincial or regional cooperatives or trade unions. According to the definition of most of these microregional projects, the process of modernization that they are supposed to stimulate should affect not only the agricultural systems and the small-scale farmers, but the organizations themselves. The end result has been that it is usually impossible for a small-scale farmer to participate in the SAA as an isolated individual. He or she must be active either in a community-based group, a 'common interest local group' or a formal organization, which INDAP recognizes as its counterparts in the SAA and its other services.

Quality control

Quality control is implemented by INDAP through several different mechanisms, starting with the requirements established to register any organization or firm as a CTT. The system of public bidding used to allocate the public subsidies has been refined on several occasions. Today it involves a process of anonymous proposals, evaluated on the basis of purely quantitative criteria, and with a strong role of public notaries who certify most of the documentation submitted by each CTT in support of its bid. This has narrowed the likelihood of non-technical criteria playing a role in the allocation of the public subsidies in favour of any particular firm or organization. All consultant firms are evaluated and formally graded twice a year, through a fairly transparent mechanism based to a large extent on objective criteria. Individual extensionists, and even complete field-level technical teams of the consultant firms, can be removed from the system if they fail to meet a minimum annual grade. The reports of the Technical Assistance Consultants (CAT) carry a strong weight in this formal evaluation and grading process.

5.2.5 Assessment – the influence of FSR

Once the political conditions of the country allowed, the influence of systems approaches was rapidly felt in INDAP. After 1990 no fewer than 20 senior NGO staff with systems backgrounds came to work for INDAP, INIA or similar institutions, taking top management positions. Since 1990 individuals trained in the AGRARIA or GIA field projects have held the Chair of the most important technical unit of INDAP. The key INDAP advisors in defining the technology transfer improvement plans of 1990, 1992, 1995 and 1996 were all systems experts. There is, therefore, a solid 'systems' influence at the national, regional and local levels.

Systems perspective and methods

Since 1990, there has been a relatively consistent effort to introduce systems-based concepts and methods into the 'corporate culture' and the operational procedures of INDAP, in particular at the field level. The definition of extension was changed from a concept of transmission of information from professionals to farmers, to the idea of an education and communication process in which farmers and their families play an important and active role. The 'women's component' of the technology transfer service was radically changed from a home economics approach, to one which is fully based on the concept of gender and in the methodology of gender analysis. In 1991, a cooperation programme was established between INDAP and INIA, which implemented joint committees at the national and regional levels, and tasks forces at the local level involving researchers, INDAP staff, employees of the CTT and, in some cases, farmers. A linkage mechanism was put in place that now involves many hundreds of researchers and extensionists in most of the regions of the country.

Starting in 1990, much effort has gone into linking extension with other agricultural development services, such as credit, marketing, small-scale irrigation and the strengthening of local organizations[2]. Field methods have also been changed and an effort has been made to establish strong links with OFR. Large numbers of extensionists now conduct simple on-farm tests either on their own initiative or in collaboration with INIA specialists. Finally, a strong emphasis has been placed on developing participatory extension methods and in training extensionists and local groups of farmers in their use and applications. The system has been radically reoriented from one based on individual farm visits, to one that emphasizes group activities with local organizations and communities. All these changes have also resulted in a much more flexible system, in which the extensionists and the local groups of farmers can plan the objectives they will try to achieve in a 3–5 year period.

Training

Each year nearly US$400,000 is invested in training INDAP and CTT staff, with a focus on activities designed to promote the new systems-based orientations. These have included courses and workshops on systems principles, gender analysis, methodology workshops and so on. This effort is not, however, as effective as it should be. The technical staff coming out of the universities and schools are ill-prepared to deal with the reality of small farmers, and also because a high percentage of extensionists leave the system each year to earn higher wages in the private sector.

Cost and operational efficiency

The budget of the programme increased by 254% between 1983 and 1995. The largest increases took place in 1987 with the introduction of the Basic Technology Transfer Program, and in 1990–92 with the expansion of the programme after the election of the democratic government. This differs from the trend seen in most Latin American programmes, where the public funds for this type of programme have tended to decrease sharply. The number of participating households increased by 370% over the same period and the number of individuals reached by 625%. The cost per individual reached fell from $555 to just $226.

There is no doubt that the PTT operated through private CTTs is significantly more cost-effective than a full public alternative would be. A larger fraction of the total cost can be used to pay for field-level activities rather than for fixed office level expenses. The World Bank survey[4] of 420 households in a poor, rainfed region in Central Chile, found that for every dollar spent in INDAP's Technology Transfer Program, the participating households generated an additional income of US$3.33.

While many CTTs and farmers claim that the system is still 'too bureaucratic' and that INDAP requires 'an exaggerated amount of paperwork', there is no doubt that there has been an enormous improvement in agility and operative efficiency. Issues which are common to many extension systems in Latin America such as agricultural inputs that arrive weeks after the planting dates, cars that cannot operate due to lack of money to pay for gasoline and so on, have essentially vanished from the agenda in Chile.

Impact

Several studies have assessed the impact of the programme though none have fully weighed the income effects of falling agricultural prices. Monardes *et al.* in 1993[3] evaluated adoption processes and rates in small farms that participated in the PTT in the poor, dryland area of Region 6, concluding that those small farmers who adopted these recommendations obtained the best economic results. The study of a representative sample of 1000 farms, commissioned by the World Bank in 1995, found that the PTT had a significant impact on such variables as: productivity per unit of land, physical total output, proportion of total household income derived from on-farm activities and proportion of the total labour of the head of the household allocated to on-farm activities. Another World Bank study of 1995[4] concluded that the annual family income of those who participated in INDAP's programme was US$1200 greater than that of non-participating households – around 46% higher than the average family income in the region.

5.2.6 Conclusions

INDAP has introduced a large number of major innovations in its technology transfer systems for small-scale farmers in recent years, building on both its positive and negative experiences since 1978. The farming systems perspective has played a significant role in these changes without explicitly imposing the FSRE label on INDAP. The strong emphasis on local initiative and operational flexibility means that the system must leave room for the approaches that CTTs or farmers organizations may want to put in place. More is accomplished through training and by pushing for a permanent process of innovation and improvement using systems concepts, than by issuing an administrative decree that makes the systems approach 'official'. The main successes of the service have been:

- Achieving reasonable levels of cost-effectiveness, operational efficiency and flexibility.
- Stimulating the organization of farmers, particularly at the local level.
- Putting in place innovative mechanisms to assess market demands, and to respond to them.
- Involving the private sector in the provision of technology transfer services.
- Stimulating farmers to contribute to the funding of these services.
- Promoting significant improvements in productivity, production levels and income.
- Focusing a large share of its financial resources on poor households and on rural women.
- Experimenting, for the first time in Chile, with new ways to link agricultural research with small-scale producers and with the extension system.

On the other hand there are still important areas in which INDAP's SAA needs to improve. The issues being faced today by small-scale farmers and the technical staff working with them, are infinitely more complex that those of the past. The current system has limited capacity to deal with this complexity in an effective way.

While the market orientation of the SAA is very relevant to the small-scale commercial farmers, who account for around half of the participating households, it is much more difficult to operationalize the new concepts and approaches for the poorest minifundista households. There is intense debate within INDAP about effective ways to serve the poorest sectors of its clientele within a market-oriented conceptual framework. Farmer participation is clearly insufficient and the system is still not truly accountable to the participating households. INDAP and the CTT continue to be the active elements in the partnership with farmers. While there is resistance to change within INDAP, it is also true that farmers' organizations are often unwilling to get involved in the issues of agricultural and technological development. Formal evaluation methods are still weak, sporadic and unsystematic at all levels and a proper evaluation system has yet to be developed. Similarly, the extension approaches used are not yet exploiting the opportunities afforded by the new information technologies. This is not only a methodological problem, but an expression of a conceptual limitation. The actors in the system including the scientists, university academicians and intellectuals specializing in agrarian issues, have not really grasped the fact that development now means expansion of the relevant knowledge and information base available to rural communities, and of their capacity to manage this resource. The focus is still on the products rather than on the processes and forces for development.

REFERENCES

1. **Berdegué, J.A., P. Brevis, G. Escobar, R. de Kartzow, D. Ossandón, G. Quiroga & F. Rodríguez, 1992**. Perfeccionamiento del Programa de Transferencia Tecnológica. INDAP, Santiago, Chile.
2. **Berdegué, J. & G. Escobar, 1996**. New directions of the systems approach for the modernization of Latin American peasant agriculture. In: M. Sebillote (Ed.) Systems-Oriented Research in Agriculture and Rural Development. Proceedings of an International Symposium, Montpellier, France, 21–25 November 1994. CIRAD-SAR, Montpellier, France, pp. 73–99.
3. **Monardes, A.T., T. Cox, D. Narea, E. Laval & C. Revoredo, 1993**. Evaluación de Adopción de

4. **World Bank, 1995**. Chile: Strategy for rural areas. Enhancing agricultural competitiveness and alleviating rural poverty. Report N12776-CH of the Natural Resources and Rural Poverty Division. World Bank, Washington, DC.

5.3 A FARMING SYSTEMS CONTRIBUTION TO AGRICULTURAL POLICY ANALYSIS

John Dixon

From a policy analysis perspective, agricultural resource management decisions are concentrated at three levels: the sector, where government resources are allocated; the village or community level, where communal resources are managed; and the farm-household level, where the bulk of agricultural production decisions take place.

5.3.1 Introduction

The goals of agricultural policy are diverse, but typically include increased agricultural productivity, contribution to national economic growth, macro stability, improved distribution and sustainability. The forces shaping policies have always been complex[1] and many policies, but agricultural policies in particular, are an important determinant of farm household behaviour. They exert strong influences on technology adoption, enterprise choice and farm investment[2]. The success or failure of most agricultural policies is determined by the ways in which the many different types of farm-households respond to changes in the policy environment. Many policies persist which are significant impediments to sustainable, efficient agricultural development.

The development theories of the 1950s and 1960s emphasized the role of market failure. There was confidence in the role of government to correct market failure, and development planning was perceived as a mechanism for efficient resource allocation. In Africa especially, independence marked shifts in the focus of agricultural policies; from industrial export crops to food crops, and from estate agriculture to smallholder production. However, since the early 1980s, following the widespread acceptance by donors of the Berg Report's recommendation for economic liberalization and structural adjustment by the donor community[3], agricultural policies in many developing countries have been substantially modified. Moreover, the dramatic political reorientations in east Asia and eastern Europe have led to major shifts in agricultural policies in these hitherto centrally planned economies. Finally, the environment has assumed growing importance in policy making, especially following the publication of 'Our Common Future' (the Brundtland Report) in 1987, and the Earth Summit in 1992.

With the implementation of the Berg Report recommendations influenced by the international community, developing country government expenditures reduced and the roles of the private sector and of the NGOs expanded. This increase in the institutional stakeholders in agricultural development, coupled with the expanding emphasis on farmer empowerment, has complicated the institutional framework within which policies are analysed, formulated, implemented and evaluated. It is argued here that the better understanding of smallholder farming systems can contribute to the formulation of more effective agricultural policies; both by the *ex post* assessment of farm level impacts of policies; and by better *ex ante* analysis of the potential impact of policy alternatives. Evidence is drawn from case studies and the experiences of national and international research and planning organizations.

This contribution continues with an overview of the conceptual and operational frameworks linking farm-level information to policy making, a review of the process of farm systems based policy formulation and its evolution in practice. Operational prerequisites for effective information flows are identified, and lessons from ongoing experience discussed. Finally, the prospects for a farming systems contribution to agricultural policy adjustment are assessed.

5.3.2 Conceptual framework and operational strategy

Hart's contribution (Chapter 3.1) on a conceptual framework for FSR applications has demonstrated an evolution from the original narrow, farm-level perspective to a hierarchy of system levels. This embraces both farm and nation, acknowledging the important interactions between decisions made by the farm family and those made by national policy makers. In many ways this widening of the conceptual framework mirrors the widening applications, admittedly almost always *ad hoc* applications, of farming systems understanding over the last 25 years. Table 5.3.1 offers a calendaring of the widening functions for farm systems understanding in relation to agricultural policy.

The operational implications of the hierarchical conceptual framework are clear. From a policy analysis perspective, agricultural resource management decisions are concentrated at three levels: the sector, where government resources are allocated; the village or community level, where communal resources are managed; and the farm-household level, where the bulk of agricultural production decisions take place. For our purposes, a farm-household system can be considered a combined farm and family unit, or any group of such units. Such a grouping is reflected in one of the many uses of the term farming system and, in FSR, is commonly referred to as a 'Recommendation Domain'.

The local agricultural production and consumption processes of the farm-households within a community, together with their physical environment, sociocultural environment, and policy and support services environment, are collectively referred to as an agricultural system. Just as the crop, livestock and family components of a farm-household system interact, so each farm-household and agricultural system is influenced by the state of related agricultural systems, through trade, competition for resources, social relationships and other 'horizontal' links.

There are also numerous 'vertical' links between hierarchical levels; marketing channels, transportation systems and political processes which connect farm-households and farming systems to the local agricultural system and the agricultural sector. Government provides some of these vertical links, for example with the support of input distribution schemes. However, most of the impact on farm-household and farm system decisions is created through the myriad legislation, regulations and policies which create an external 'policy' environment. Farm-household or system-based policy analysis is complicated by the fact that many of these vertical links are reciprocal: for instance, agricultural price policy decisions induce changes in production and consumption patterns of agricultural commodities, and are themselves modified in the light of the resulting production or consumption response.

Making farming systems understanding fully operational in policy analysis would entail a process similar to that used in its application to technology generation and transfer, though the formal testing of promising policies in the real world is rarely possible:

- Assembly of secondary information.
- Agroecological zoning and agricultural or farm system characterization (if not already established).
- Exploratory diagnosis, using informal methods, with the identification of constraints and opportunities.

Table 5.3.1. Changing relevance of farming systems understanding to agricultural policy.

	1970s	Early 1980s	Late 1980s	Early 1990s	Late 1990s
Broadening functions					
Research	***	***	***	**	**
Incorporating extension	*	**	***	***	***
Incorporating support services		*	**	***	**
Incorporating agricultural sector policies			*	**	**
Expanding stakeholders (beyond farmers)					
Public organizations	***	***	***	**	**
Incorporation of civil society (NGOs)		*	**	***	***
Incorporation of commerce (private sector)				*	**

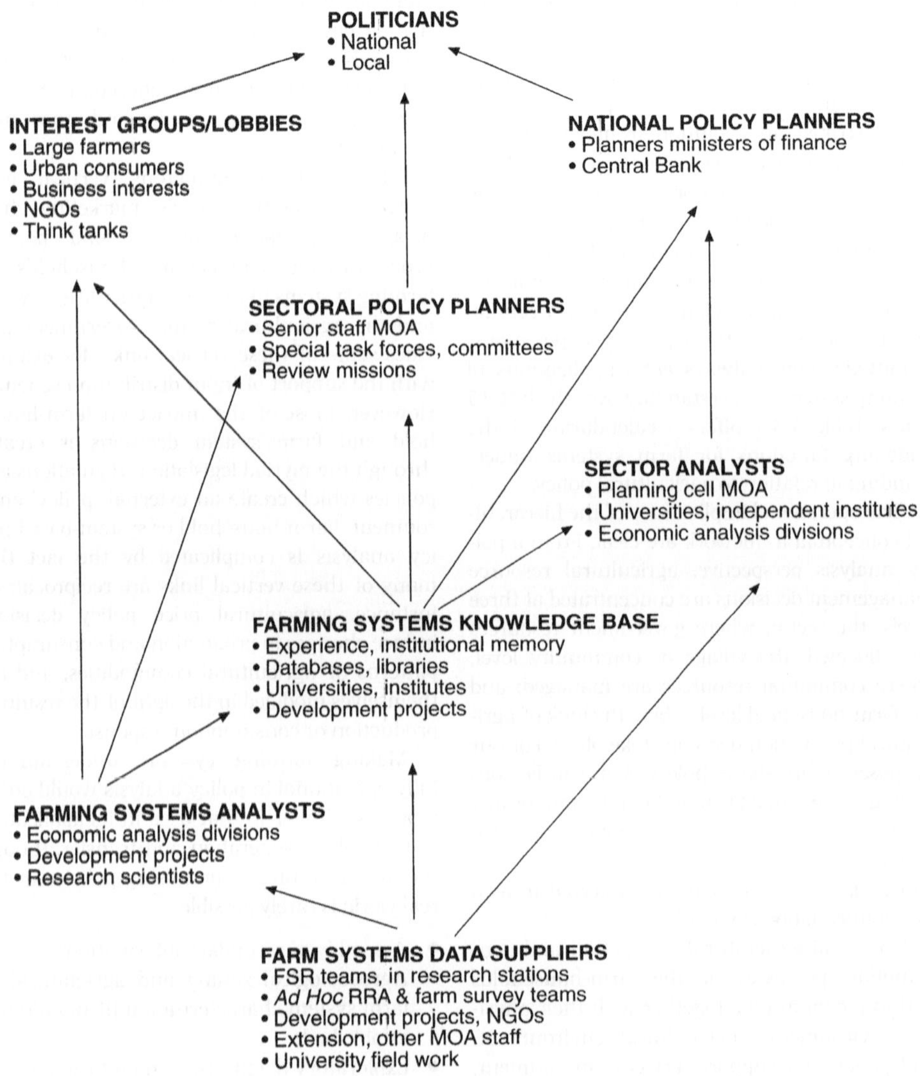

Fig. 5.3.1. Common generalized institutional structure.

- Verification survey (some circumstances only).
- Assessment of alternative policies (during implementation).
- Monitoring response to new policies.

The interested reader is referred to Friedrich and Hall (1989)[4] and Dixon (1993)[5] for a complete description of these steps.

Little progress has been made in using this as a comprehensive operational strategy for the public sector. Its operationalization is bedevilled by both technical and institutional difficulties. Technically, there are the well-known problems of integrating biophysical and socioeconomic information and, most importantly perhaps, the question of scaling up; the aggregation of farm-household information to a meaningful level for policy analysis. Beyond these technical issues many facets of the current public institutions infrastructure present barriers to its operation.

5.3.3 The existing institutional environment

Figure 5.3.1 displays a generalized but fairly typical constellation of institutions supplying and using farm-household data in developing countries. The model emphasizes the two-way flow of information that would theoretically be desirable and the flows represented can also be considered in a supply and demand framework, where the upper part of the diagram represents demand and the lower part the suppliers. While Fig. 5.3.1 is focused on the major areas of government administration which play, or might play, a role in the use of farm systems information in agricultural policy analysis, as mentioned earlier, the number of stakeholders and the diversity of interrelationships between them is growing, NGOs now play a significant role.

Although the current strength of links between these groups varies, the possibility of several different flows of information are evident. Senior planners, politicians and interest groups, dictate the demand for farm-household data in policy formulation. The numerous institutions working at the farm-household level, including government departments, projects and NGOs, generate a supply of farm system information which can flow to researchers, policy analysts and planners who represent an interface between demand and supply. However, in most developing countries the total policy analysis capacity at this interface is quite limited, often only a few individuals scattered among several institutions. It excludes senior decision makers who have little time to devote to such pursuits.

The principal users of farm-household and farm system data in a policy design context are the sector analysts and policy advisors, mostly located in ministries of agriculture and planning, or their equivalent. Their clients in turn are national planners and politicians. These analysts have a basic minimum requirement for generalized agricultural statistics, but usually need additional farm-household and system data. The nature and scope of this data depends on the topics needing analysis, the policy and political agendas, usually determined by the politicians and senior planners, with varying levels of influence from special interest groups, including large farmers, exporters, agro-processers and consumers, etc. The potential for the application of farming systems data in policy is probably greatest for these *ad hoc* analyses.

In developed countries researchers at universities and institutes are perhaps the major users and they are often also the major suppliers of farm-household data. Politicians and special interest groups are also sometimes important direct users of farm-household data.

5.3.4 Changing dimensions of data flows

The information flows from farm system understanding and results from farm-household analyses can be discussed in a supply and demand context. In many developing countries farm-household data supply is fragmented and often uncoordinated[6]. Typically there are one or more government departments which are actively engaged in routine farm data collection and analysis, the Directorate of Economics and Statistics in India is one strong example. Very often, agricultural (and other) censuses are the responsibility of a non-agricultural department, whose staff may not understand the intricacies of farming systems. In addition there are many *ad hoc* farm-household system studies. These are often conducted by research groups or contracters rather than by the line departments responsible for routine agricultural data collection. Their resources are usually fully committed to regular surveys and they often lack the flexibility needed for the organization of field programmes involving rapid rural appraisal (RRA) or farming system surveys at short notice.

A large volume of farm-household and system data is also collected separately by externally supported projects for benchmark, constraints analysis, and monitoring and evaluation purposes – one well-known example is the World Bank household survey in the Ivory Coast. Other major sources of farming systems information are the FSR teams in research stations, many of them trained by Consultative Group for International Agricultural Research (CGIAR) centres, notably the International Maize and Wheat Improvement Center (CIMMYT) and the International Rice Research Institute (IRRI). In some countries there have been attempts to assemble and systematically assess existing farm-household data for secondary uses; Food and Agriculture Organization of the United Nations/United Nations Development Progamme (FAO/UNDP) supported one such effort in Nepal, but, in general, governments have underinvested in the compilation and synthesis of farm-household and farm system information.

The available farm-household data are transformed by various methods of micro-analysis or farming systems analysis, very often by the collecting organizations. The results, occasionally accompanied by the original data, may flow directly to sector analysis exercises but more frequently become part of the farming systems knowledge bases. These are the accumulated data and understanding of farming system structure and function held individually and institutionally, and typically scattered around various institutions. These are rarely collected or compiled in a systematic way because of the lack of capacity for synthesis and the lack of a common purpose across the institutions organizing collection.

The 1950s and 1960s saw a growth in the number and size of government institutions, supported in part by donor policy, and many of the institutional and organizational features developed during that phase persist to this day, despite the downsizing resulting from structural adjustment programmes (SAPs). Narrow specialization was a common phenomenon of the rapid growth in the 1950s and 1960s including institutions created to focus on specific commodities or particular development problems. This led to isolation and blinkered information needs, restricting the demand for farming systems understanding and limiting the value of the transfer of information between organizations. Beyond this, most policy-making organizations remain highly centralized. Information flows tend to be top-down, rather than bottom-up, restricting the flow of information on farm-household and agricultural system circumstances which might illuminate policy design and adaptation, despite the expanding influence of farmer participation, particularly through the NGO sector.

5.3.5 Key areas for change

Farming systems understanding and its applications are now a common feature of the agricultural research and development programmes of a great many developing countries. Their characteristics are well known, as Simmonds illustrated in 1985[7], although their cost-effectiveness is debated (see, for example, Anderson 1990[8] and Chapter 7.3). The emphasis remains on technology generation and, perhaps increasingly, technology transfer. However, the FSR process is, in principle, directly applicable to agricultural market development, credit provision and, of particular relevance for this chapter, agricultural policy analysis. This wider role was clearly recognized by David Norman and Mike Collinson, two of the founding fathers of FSR, who considered that 'FSR ... consists of two thrusts towards increased productivity: (1) the development and dissemination of relevant improved technologies and practices; and (2) the implementation of appropriate policy and support systems to create opportunities for improved production systems and to provide conditions conducive to the adoption of technologies already available'[9]. Farming systems development (FSD), promoted and supported by FAO, addresses this second thrust[10].

As FSR has matured there are increasing applications to policy, both the *ex post* evaluation of micro-impacts and the *ex ante* analysis of alternatives. However, it remains true to say that applications remain largely *ad hoc*. More coherent use of farm-household system information for policy purposes would be stimulated by changes in four key areas, as follows.

Decentralization

In applying farm systems understanding to agricultural policy making, lessons can be drawn from the disappointing experience with institutionalization of FSR in technology development. Many such programmes seem to have lacked the resources or skills to involve farmers collaboratively in research planning and review[11]. The hierarchical structure of line ministries and quasi-government institutions both slows the horizontal flow of information, and discourages effective communication of significant findings of farming systems studies by junior staff to their seniors[12]. Efficient use of farming systems understanding in agricultural policy formulation would require change in these aspects of organizational culture.

In developed countries, farmers' organizations and various interest groups provide some of this information set. In developing countries, however, where such constituencies have evolved they still tend to be too weak to discharge this function. The organizational models proposed for decentralized planning seem the most conducive to the use of farming systems information in policy analysis. For example in the Andean CONDESAN consortium[13] (Consortium for the Sustainable Development of the Andean Ecoregion) made up of national research institutions and NGOs, convened by the International Potato Center, the parallel decentralization of central government functions and budgets to local governments has provided a strong momentum to the use of farming system information in municipal policy formulation. Information on the farming systems in municipal hinterlands has proved of great interest for planning in areas such as urban food supplies, water use and environmental conservation.

It seems that problems are less intractable at the level of local government aggregation, and there is less interinstitutional conflict than characterizes the narrow focus of traditional line ministries competing for budgets at the national level.

Zonation

National policies are too general to provide the fine tuning required to alter incentives in different ways in different farming systems. However, preceding the aggregation issue are more straightforward questions of zonation. More appropriate zoning is required to provide a baseline for better differentiating producer responses.

Information for policy formulation is traditionally and characteristically based on administrative regions, in line with existing statistics and implementation realities. Administrative regions are only serendipitously related to farming systems and understanding is confounded when data collection is not organized within a framework of differentiated farming systems. There is an ongoing trend towards the greater use of non-administrative zoning. Thailand, for example, has delineated agroeconomic zones for planning purposes[14]. The National Planning Commission of India defined 15 agroclimatic zones and 125 agricultural subzones which are based in both planning and research-extension coordination[15]. In a more participatory approach, Lightfoot, in 1990, reported on how farmers in Malawi delineated meaningful land use zones which provide a framework for agroecosystems analysis and the forecasting of farmers' responses to alternative policies[16]. However, such zonation is by no means universal; many developing countries lack agroecological zonation, and a further step beyond it is required to set up systematically defined farm system types (see Low, 1991[17] and Chapter 3.1). Moving the zoning issue forward requires that governments are convinced of the efficacy of the farming system as the basic unit for agricultural development in the smallholder sector.

Several recent developments are encouraging the zoning of agricultural systems and the characterizing of farm-household systems. First, the prominence of sustainability issues in policy formulation has led to greater attention to natural resource management, with an obvious link to agroecological zoning and to management practices in farming. Second, agroecological zones have been delineated in a growing number of countries, and provide a sound background framework for the characterization and typing of farm-household systems. For those programmes focused on production constraints, agricultural system zones offer an excellent framework for programme design and implementation.

Data integration and aggregation

A further technical challenge is the effective integration of biophysical and socioeconomic data in characterization, and its sourcing from methods as diverse as informal participation and remote sensing. Informal low-cost techniques are usually applied to develop a sound understanding of the constraints and potential of the different types of the farm-household systems designated through zoning. These techniques include RRA and participatory rural appraisal (PRA), both involving the participation of farm-household members in the identification of constraints and possible solutions. The sound understanding gained of the farming systems can also be used as a basis for simple models of representative farm-households. These models can also contribute to improved agricultural policy analysis, in the first place, by improving the mental images, or notional models, of farm-household systems – such notional models form the basis of many quick policy decisions. Ultimately, once the data integration problems are overcome, farm-household system models will be combined to provide assessments of aggregated impacts of policy changes, though the techniques for aggregation still remain a challenge.

In practice, the degree of detail and level of accuracy of notional models is still generally fairly poor and varies with the experience of the individual. The differences in depth and detail of existing knowledge of farming systems tend to follow a fairly predictable pattern[18]. Often the knowledge of production practices and yields of staple cereals and major export crops is reasonable, at least for the more accessible areas, major market-surplus regions and special rural development areas. Conversely, production processes of newly introduced, subsistence or minor activities are often not well understood by those charged with policy formulation. Above all, the knowledge of interactions between system components is typically very limited.

The spread of microcomputers will soon bring farm-household modelling within reach of most policy analysts. The user friendliness of 1990s spreadsheets, programming software and special-purpose farm simulation packages favours much wider use of computerized farm-household system analysis, though the high cost of formal data collection remains a problem.

Two aspects of modelling remain critical: first, methods for aggregation, second, estimates of the adjustments in the farm-household system to changes made in the policy instrument. In practice, it will be advisable for modelling to be kept relatively simple to maintain balance and a system-wide view, and so as not to unduly tax the limited resources available for policy analysis.

Data collection and analysis

These issues of data integration and aggregation cry out for a coherence in the organization of data sources, and in data collection. Since many policy changes are made at very short notice there is often insufficient time for deep analysis. In such a simple approach a major source of mistakes is the error of omission. Checklists represent one way to minimize this danger, particularly when the analysis is performed by a single individual or few individuals under pressure of time. The headings in a checklist correspond to the major social and private goals which are relevant to the decision to be taken, such as production, employment, income and stability. The checklist items under each heading refer to pertinent characteristics of the farm-household system under consideration.

The use of notional models is an aid to understanding how decision makers integrate informal observations with information about agricultural systems from more formal sources, based on a repertoire of experience which might include a few field visits and discussion with farmers. The checklist and notional models can be developed in staff workshops ideally, but not necessarily, supplemented by farm visits.

A formal household survey is sometimes conducted to verify the conclusions of the exploratory diagnosis, or to refine estimates of the prevalence or magnitude of particular farm-household characteristics. The results of such surveys might include the extent of resource degradation, frequency of different types of crop rotations, proportion of cultivated land under vegetables, or season feed, labour and cash-flow profiles. A small sample is preferred, in order to limit the expense, required staff and time, and the probable measurement error. The design of the sample should take into account the various policy analysis domains. The survey results also help to elaborate farm models which can be used to deepen the understanding of production constraints, e.g. by the estimation of resource opportunity costs.

The search for viable policy options is an important part of the policy analysis cycle, and farm systems understanding can contribute by identifying farmers' views of the feasibility of options. Each option needs to be assessed against a variety of criteria including policy objectives, and the expected responses from both farm-household systems and the involved institutions. As the last step of the policy analysis cycle, impact on different farm-household systems should be monitored and evaluated. Such assessments need to evaluate the adjustments made to the policy in different farm-household systems. Both informal and formal survey methods are useful.

5.3.6 Prospects for the use of farm systems understanding in policy analysis

The policy environment is an important determinant of the pattern and level of agricultural production, and the rate of technological change. That farmers respond to price changes or other modifications of incentives is no longer seriously doubted, but an understanding of the magnitude of these responses and the differences in the levels of response of farmers operating different systems are crucial for the successful design of many agricultural policies. The likely impact on incomes and production of particular input or product price policies will benefit greatly from a careful analysis of the farm-household and agricultural systems. It is for this reason that an understanding of farming systems represents a potentially powerful complementary approach to existing methods of agricultural policy analysis. Despite that potential, it has so far had little impact on policies beyond research policy.

Farm systems understanding has a role to play in policy analysis, as a complement to, and occasionally a replacement for, conventional methods, under the following circumstances:

- When aggregate data for time-series or other analyses might be suspect or even unavailable, particularly when dealing with minor activities, remote areas or very weak national data systems.
- When a diversity of heterogeneous farming systems hinders the estimation of response to alternative policies without appropriate disaggregation.
- Where equity considerations are important and systems are diverse, complementary farming systems information will be valuable.
- Structural change or the rapid flux of policies are common difficulties in conventional time series based policy analyses.
- Decentralized planning often entails in-depth analysis of farming systems.
- Minor crops or activities are rarely adequately measured in most agricultural statistical series.
- Small developing countries may not have the infrastructure or resources to maintain comprehensive current agricultural statistics, but may get value from field-based farming systems teams providing a number of outputs, including information for policy analysis.

Some FSR supporters advocate a narrow 'traditional' role for FSR teams. In 1990, for example, Posner and Gilbert[19] argued that such teams would be overloaded if other tasks related to, for instance, natural resource management, watershed management and policy, were to be added to their technology testing work. However, FSR teams would seem to have the responsibility to communicate findings with significant policy implications to policy analysts. Moreover, it is concluded here that other agencies, including policy-related institutions, should adopt a farming systems perspective, though this does not imply the establishment of separate farming systems teams in each agency, which would generally be undesirable.

The principal constraints to wider use of farming systems data and analyses are firstly institutional and secondly technical in the development and adoption of suitable methods. Key developments in methods are needed to help the profession realize the potential of farming systems understanding to complement existing methods of agricultural policy analysis:

- Integrated agroecological zonation and farm-household characterization.
- Notional farm-household system models.
- Policy-oriented RRA.
- Dynamic farming systems analyses.

Greatest progress would result from the adoption of a farming systems perspective throughout policy-related institutions, and improved communication within and between such institutions.

REFERENCES

1. **Bates, R.H., 1981**. Markets and States in Tropical Africa: the Political Bias of Agricultural Policies. University of California Press, California.
2. **Mellor, J.W. & R. Ahmed (Eds), 1988**. Agricultural Price Policy for Developing Countries. Johns Hopkins University Press, Baltimore.
3. **Caldwell, J.S. & E. Akobundu (Eds), 1997**. Agricultural systems and policy. Special edition of Journal of Farming Systems Research – Extension, 6 (3).
4. **Friedrich, K.H. & M. Hall, 1989**. Farming Systems Development (FSD). Farming Systems Research/Extension Symposium, Lafayette, Arkansas.
5. **Dixon, J., 1993**. Policy issues, farming systems development and sustainable agricultural and rural development. Twelfth Session of the Farm Management Commission for Asia and the Far East. Dacca, Bangladesh.

6. **Low, A., 1991**. The Utilization of Farming Systems Data in Agricultural Policy Analysis in Eastern and Southern Africa. Draft Working Paper, Farm Management and Production Economics Service, FAO, Rome.
7. **Simmonds, N.W., 1985**. Farming Systems Research – a Review. Technical paper 43. World Bank, Washington, DC.
8. **Anderson, J.R., 1990**. FSRE Impact Inquisition: Investor Issues. Asian Farming Systems Research and Extension Symposium, Asian Institute of Technology, Bangkok, Thailand.
9. **Norman, D. & M. Collinson, 1985**. Farming systems research in theory and practice. In: J.V. Remenyi (Ed.) Agricultural Systems Research for Developing Countries. ACIAR Proceedings of Workshop, Hawkesbury Agricultural College, Richmond, Australia.
10. **FAO, 1989**. Farming Systems Development: Concept, Methods, Applications. FAO, Rome.
11. **Merrill-Sands, D. & D. Kaimowitz, 1989**. The Technology Triangle: Linking Farmers, Technology Transfer Agents and Agricultural Researchers. ISNAR, The Hague, The Netherlands.
12. **Low, A., 1991**. (Op. cit.).
13. **Rueda, J.L., 1996**. CONDESAN: Un Neuvo Enfoque de Investigation para la Ecoregion Andina. CONDESAN, Lima, Peru.
14. **Katikarn, K., 1990**. The Utilization of Micro-data in Agricultural Policy Analysis in Thailand. Draft Working Paper, Farm Management and Production Economics Service, FAO, Rome.
15. **Vyas, V.S., D.S. Tyagi & M.S. Rathore, 1991**. The Utilization of Micro-data in Agricultural Policy Analysis in India. Draft Working Paper, Farm Management and Production Economics Service. FAO, Rome.
16. **Lightfoot, C., 1990**. Integration of aquaculture and agriculture: a route to sustainable farming systems. Naga: ICLARM Quarterly (Jan.), 9–12.
17. **Low, A., 1991**. (Op. cit.).
18. **Chambers, R., A. Pacey & L.A. Thrup (Eds), 1989**. Farmer First: Farmer Innovation and Agricultural Research. Intermediate Technology Publications, London.
19. **Posner, J.L. & E. Gilbert, 1990**. Sustainable agriculture and FSR teams in semi-arid West Africa: a fatal attraction. Farming Systems Research–Extension Symposium, Michigan State University, MI.

Part III

Institutional Commitment to Farming Systems Research

EDITORIAL INTRODUCTION

Mike Collinson

Part III of this history addresses institutional commitment to farm systems research (FSR), and consists of Chapters 6, 7 and 8. In the 1970s and early 1980s, public sector institutions were the main focus for promotional efforts, which came mainly from the international community. However, the regional histories of FSR in Part IV speak to the growing diversity of institutional involvement, with NGOs particularly strong in Latin America. This widening involvement of NGOs was also recorded by Farrington and Bebbington in 1993[1] when they expressed the view that NGOs are too divided, have their own constituencies and are too local to create the pressures required for government institutions to reorient and reorganize. This suggests that support from the international community will continue to be vital in promoting change in public institutions in many developing countries.

THE CONTRIBUTIONS

Chapter 6 includes country cases on the introduction of FSR into National Agricultural Research Services (NARS). The case of France is followed by an example from Senegal in west Africa and potted histories from eight east African countries. The contribution on France, by Jacques Brossier and his colleagues at the National Institute for Agricultural Research (INRA) focuses on the only developed country to have an institutionalized cadre of systems researchers, making France a world leader in the application of systems methods at the levels of the farm and the 'terroir' or broader landscape. Michel Benoit-Cattin provides a short history of the Unites Experimentales in Senegal, an early French initiative in the application of systems research at the farm level in developing countries. Willem Stoop and his colleagues outline the research programme of the Sikasso Production Systems and Natural Resource Management team (PSNRM) of the Institute d'Economie Rural (IER) in Mali. To conclude Chapter 6, Stuart Kean and Creasy Ndiyoi contribute potted histories of the development of FSR in eight countries in east and southern Africa and offer a synthesis on progress in the region as a whole.

Chapter 7 looks at three dimensions of the organization and management of FSR. First Ann Stroud traces the 20-year history of the institutionalization of FSR in Tanzania and concludes with an analysis of the reasons for the slow progress. Stuart Kean and Creasy Ndiyoi, both national coordinators for the Adaptive Research Planning Team (ARPT) in Zambia in their day, contribute an analysis of the influences of key stakeholders in the adoption and application of FSR in Zambia. Elon Gilbert, who participated in the International Service for National Agricultural Research (ISNAR) study which collected data on the costs of OFR in nine countries in the late 1980s, examines the 'real' costs of on-farm research (OFR) versus on-station research (OSR). Comparisons across countries proved difficult, and the data were not published at that time. Though now dated these are *still* the only comparative data available.

Chapter 8 includes three contributions on training in FSR; in east and southern Africa, at the International Centre for Development-Oriented Research in Agriculture (ICRA), and on the compilation of a text for FSR training. Ponniah Anandajayasekeram, initially in charge of training for the International Maize and Wheat Improvement Center's (CIMMYT) east and southern African network, and latterly with the Food and Agriculture Organization of the United Nations (FAO)-operated network under the South African Centre for Co-operation in Agricultural and Natural Resources Research (SACCAR) umbrella, looks back over 15 years of training for FSR in the region. Richard Hawkins, anglophone Training Coordinator at ICRA, an institution providing graduate training in 'development-oriented research in agriculture', examines ICRA's origin and goals and the evolution of its operational strategy. Finally, Bill Shaner gives an account of a global search to bring FSR best practice together and make it available as training material to new practitioners and to trainers in the early 1980s.

A COMMENTARY

Interestingly, FSR is a product; first of colonial independence, with new politicians responding to constituencies made up from the mass of small farmers, and second, of international action to fill the professional void as the old era came to a close. The accounts by some of the 'old dogs' of FSR in Chapter 2 reflect its predominantly expatriate origins. Among these Collinson at CIMMYT and Harwood at IRRI spent part of their careers at the two International Agricultural Research Centres (IARCs) that played major roles in the promotion of, and capacity building in, FSR from the early 1970s. Several IARCs established in the late 1960s and early 1970s, such as the International Institute of Tropical Agriculture (IITA) in 1967, the International Center for Tropical Agriculture (CIAT) in 1967, and the International Center for Agricultural Research in the Dry Areas (ICARDA) in 1977, had FSR explicit in their regional mandates. The Technical Advisory Committee (TAC) of the Consultative Group on International Agricultural Research (CGIAR), itself established in 1971, carried out a review of FSR across the IARCs in 1977. The conclusions of the review on the role of FSR remain surprisingly current some 20 years later. Ten years before Brundtland, they embrace environmental and policy linkages, both of which (re)surfaced as issues in the late 1980s. Some major conclusions of the 1977 review are summarized below[2], including the need to:

- Understand better the problems and needs of the farmer.
- Improve the efficiency of the agricultural research process by focusing priority setting on farmers' problems, and by designing technologies with farmers' circumstances in mind.
- Take into account both the interactions between technologies and between technologies and the environment, and thereby improve the appropriateness of the generated technologies.
- Ensure that these technologies contribute to the long-term maintenance and enhancement of agricultural productive capacity.
- Facilitate the linkages between research and extension, delivery systems and the farmer.
- Assist the formulation of development policies and methods which address the problems of the farmer.

Peculiarly, it was IRRI and CIMMYT, both globally oriented commodity-based IARCs with no farming systems mandates, that led the promotion of systems-based research to national agricultural research services. In Chapter 2, Richard Harwood tells the story of the early days in the Cropping Systems Programme at IRRI and the initiation of the Asian Cropping Systems Network in 1974. The network created partnerships with national research services for joint activities in and across countries, and pioneered on-farm studies, including both surveys and experiments. As Harwood explains, the Cropping Systems Programme embraced the social sciences in the early 1970s and became essentially a farming systems programme. It was renamed the Asian Rice Farming Systems Network in 1983. The network attracted 12 country members in its first 5 years and by 1993 had expanded to 16 countries. The network wound up in the mid 1990s as a result of changing priorities in both IRRI and the donor community.

In 1974 regional programmes were adopted as a new element in CGIAR strategy and, in 1975, CIMMYT began to deploy staff to regions ranked as a priority for research on wheat and/or maize, the Center's mandated crops. In east and southern Africa, the Andean region of Latin America, the Indian subcontinent and in south-east Asia, programmes were established to promote and build national capacity in OFR with a farming systems perspective (CIMMYT's FSR terminology). From CIMMYT's point of view there were two objectives: first, to mobilize its technologies more effectively by improving national understanding of small-farmer needs, and second, to channel effective demand from farmers, through the national research service, to better shape CIMMYT's own programmes and products. The CIMMYT east and southern African programme, established in 1975, networked formally until 1993 when it was wound up, though CIMMYT regional staff remain active in OFR.

Perhaps the largest promoter of FSR was the United States Agency for International Development (USAID) with some 70 bilateral FSR programmes round the world. USAID also established the Farming Systems Support Project (FSSP) in 1982. A consortium of 21 American universities, led by the University of Florida, Gainesville, FSSP responded to and supported national initiatives in FSR around the world, with Africa, particularly West Africa, as a dominant focus. The legacy of FSSP, in addition to human capacity in FSR in a number of developing countries, has been a valuable set of training materials and also the Association of Farming Systems Research and Extension (AFSRE), together with the regional FSR associations in Asia and Africa. A description of these takes up Chapter 9 of this history. Other important promotional and support networks include International Farming Systems Research Methodology Network (RIMISP) in Latin America, described by Julio Berdegué in Chapter 9. The Semi Arid Food Grains Research and Development (SAFGRAD) programme for the Sahel region encouraged FSR in west Africa, and more recently, the network operated by FAO under the SACCAR umbrella helped to build capacity in eastern and southern Africa.

All of these international networks were donor funded and were parallelled by large bilateral investments. While USAID was a major contributor through Title XII grants for country projects staffed by American universities, the International Development Research Center (IDRC), The Netherlands, the Canadian International Development Assistance (CIDA) and the Scandinavian countries were also enthusiastic investors. The Netherlands and IDRC in particular were appreciated for taking a longer-term view and their investment continues. On the multilateral front a survey of World Bank agricultural research and development programmes in the early 1990s showed that over two-thirds included FSR components. There were hopes that the international and regional professional associations which formed in Asia and Africa in the early 1990s, could take over some support functions, but sustaining them is proving difficult, as donor interest in FSR has declined.

INSTITUTIONALIZATION: AN ANALYSIS OF THE CONTEMPORARY ISSUES

Gaining institutional commitment to FSR as a new stage in the R & D process for agricultural development has proved difficult in the developing countries. To date FSR has rarely had a fair trial in the public domain and performance has often been weak. The reasons include:

- General institutional failure; public research organizations as a whole have degraded.
- The expansion of manpower in national agricultural research organizations was paralleled by declining programme funding as research budgets remained static or even fell. Operating funds per professional fell from 40–50% to 10–20% of recurrent budgets as salaries absorbed an increasing proportion. This trend particularly penalized research in farmers' fields with its heavy demands on transport.
- Early promotion of FSR by attacking the results of past research helped alienate the establishment within which FSR teams had to work.
- The curricula in university agricultural faculties, often a direct import from a metropolitan university, featured large-scale commercial agriculture, machinery and monocrops. Small farmers and their production strategies were ignored. Such undergraduate training distanced would-be agriculturalists from the realities of farming in their own countries.
- Newly graduated young social scientists had little credibility, particularly with the hard scientists dominating research establishments.

- A failure to analyse the institutional implications of FSR and identify organizational options for countries in different circumstances.
- The faddism and impatience of the donor community.

General institutional failure, including the weakness of research institutions, even in their classical on-station experimental role, remains the most pervasive of these reasons. Institutional stagnation has created an environment inimical to successful innovation in the research process. Added to this general weakness is the inherited orientation in institutional culture: institutions in the ex-colonial territories often had the characteristics of metropolitan institutions. After independence they were frequently captured by in-country élites with lifestyles reflecting their metropolitan counterparts. Research institutions were no exception. Where expatriates had settled in agriculture, as in many African and Latin American countries, research institutions were often programmed to their particular needs as articulate and preferred clients. Even when local scientists came to dominate the research establishments, their training in a metropolitan tradition led them to identify more easily with commercial farmers, both expatriate and local, than with the peasantry. Any shift in focus to small resource-poor farmers implied a loss of contact, influence and prestige for researchers who saw themselves as the peers of the larger commercial farmers. Beyond weak management and a culture alienated from the small-farm sector, its lack of political appeal has made research an early focus for budgetary savings. In the face of all this it would have been surprising if the introduction of FSR into research institutions had been easy to achieve, particularly as it required an expansion of the professional establishment to include the social sciences at a time when budgets were falling.

Researcher allegiance, mandate and organization

The need to understand natural processes as the source of new and better ways of management in agriculture has long been accepted as strategic research. Its wide relevance crosses country boundaries, for example, underpins the rationale for the international agricultural research centres. But it is the applied research paradigm of identifying prototype technologies on the basis of a technically ideal commodity management system which underpins the mandating and organization of most National Agricultural Research Institutions (NARIs). In such commodity-based institutions, scientists and managers give their allegiance to their commodity and their disciplines. The commodity-focused, disciplinary-based research approach, and the institutional organization to support it, have persisted.

An exercise used in training in east and southern Africa in the 1970s and 1980s illustrates the dilemma of the current organization with its disciplinary and commodity loyalties. In the course of FSR training, station-based scientists of diverse disciplines would visit a community of small farmers. They would be asked to look around a farm, ask questions of the family and identify priority problems. The results were easy to anticipate; the soil scientist pinpointed physical erosion or low soil fertility, the breeder poor varieties, the entomologist bugs, the pathologist diseases and the agronomist poor management practices. Subsequent discussion with the farmer and a group of his neighbours searched out some of their own priorities which rarely included those identified by the disciplinary specialists.

The top-down organization of research and extension has reinforced the prescriptive nature of recommendations emanating from the 'technically ideal commodity management system' as the product of the research approach. The 'have solution will travel' mentality fostered by commodity and disciplinary allegiance has been particularly distorting and damaging at the local level in pressing changes onto farmers which they cannot use. The inevitable reaction has been lowered morale in the extension and research cadres and lowered political appeal of the traditional process. It is a history which highlights allegiance to the beneficiaries as a more appropriate driver of research culture, mandate and organization.

As a process with strong beneficiary allegiance, FSR/OFR has been criticized by many research managers who feel it is too costly in vehicles and operational allowances. It is time-consuming and potentially expensive due to site specificity. However, much of the effort put into its development has been the quest for cheaper methods. Qualitative diagnosis from the mid 1970s became a hallmark of FSR. Early on-farm experimentation (OFE), often conducted with individual, widely separated farmers across recommendation domains (RDs) was risky: if operational funding failed, or roads became impassable, or researchers were called to a meeting at a critical time, large losses of data and efficiency followed. This was one of the major reasons for the move towards using farmer research groups[3], facilitating clusters and working with extension or NGO staff located closer to the farmers. Placing greater emphasis on farmer management reduced the need for researchers to be present at critical moments. As Gilbert shows in his contribution in Chapter 7, in the late 1980s the costs of OFR were cheaper than research on the station. With the use of participatory methods researchers have devolved an even greater responsibility for the adaptive testing and subsequent monitoring to farmers[4]. The costs of OFE have gradually been reduced. Based in Colombia, CIAT teaches farmers to do more formal experiments and they are successful in obtaining analysable results interpreted by the farmers themselves. A CIAT report[5] in 1996 estimated operational costs to be as low as $450 for a set of farmer-designed, farmer-implemented bean variety trials with 220 farmers across northern Tanzania.

The issue is not really one of costs however, but of cost-effectiveness and research efficiency. FSR is not a substitute for the traditional process but an additional stage to improve its relevance. The low research efficiency of the traditional process is well illustrated by an example of rice varieties in Tanzania. Researchers used their own and the International Rice Research Institute's (IRRI) criteria (yield, pest and disease resistant, short-strawed, and early maturing) when selecting varieties on-station; after 10 years' work potential varieties entered on-farm testing where most small farmers rejected them in favour of local aromatic, long-strawed varieties which are more marketable, easier to cut and produce more fodder and thatch.

To date most public institutions have not been penalized for producing products that cannot be used by farmers, though it is becoming more widely accepted that a complementary programme of on-station and on-farm experimentation is more efficient. Farmers must play a major part in the on-farm element of such a partnership. Ashby[6] suggested that farmers, as clients, should have a key role in evaluating performance as well as sharing in implementation and costs. To achieve impact farmer participation is not enough; who participates, how the participation is orchestrated in terms of group organization, and management and leadership are key elements for success. Managing such collaborative alliances is another new skill required in OFR.

FSR as a beneficiary-driven interface

My own preference is for regionally based FSR teams, each mandated to a designated group of farming systems. Only such regional deployment will allow convincing input into development programming and policy formulation. The need is for FSR teams with allegiance to their communities, diagnosing with farmers and with an open slate in seeking options for solving problems and for improving farms.

Where to locate such teams institutionally poses a more difficult question. While FSR was seen in its early, narrow role of adapting output from the local research station to the needs of local farmers it was almost universally located in research institutions. For its emerging wider role this poses a problem. The line management of departments in public institutions are notoriously blinkered. Partnerships, even across departments within the same ministry, can be difficult. The history of research and extension bears this out – linkage has proved problematic even between departments which operate different parts of what is best considered as a unified process. The problem has perhaps not been so much poor communication between extension and research as commonly diagnosed, but poor communication between farmers and both extension and research – essentially the one-way traffic so strongly condemned by Chambers[7].

On the face of it, particularly with the wider FSR role embraced by emerging best practice, the extension services do offer an alternative home. Extension is usually organized on an area basis, much more congruent with FSR deployment than the commodity and disciplinary structure of the research services. There is a need to give extension greater ownership of the messages it carries to farmers. And RDs or farming systems offer a better organizing paradigm for extension, and FSR offers a stronger base from which to influence research programming. Low *et al.* in 1992[8] summarized a number of observations on housing OFR in extension, saying: 'Logic, as well as experience from Asia[9], suggests that OFR should be as much an extension as a research activity'. They also warn that: 'Adoption of an OFR approach in extension implies moving away from packaged recipes towards providing farmers with options and advice on how to improve production. This change in the role of extension not only requires new skills on the part of the extension staff at all levels[10] but also needs sanction and support from directors of research and extension'. Finally, they summarize efforts to draw closer to the extension services in the adjacent countries of Malawi, Zambia and Zimbabwe. In Malawi the OFR teams were deployed under the control of area-based extension programmes. Extension managers had their own view of the role of the teams and were particularly insistent that diagnosis was superfluous because they themselves were well aware of farmers' problems. In Zambia the OFR teams included research/extension liaison officers to enhance information exchange between the teams and extension. In Zimbabwe after independence in 1980 the role of Agricultural Extension Officers (AEOs) at the district level was redefined to include the adaptation of technology and the production of appropriate extension messages for use by field staff. In 1986 a central committee was established, made up of research and extension managers, which sought to ensure that on-farm programmes in research and extension were complementary and that results were shared. Shumba, in 1994[11], reported that the committee in Zimbabwe has been a successful linkage device.

Al-Khadi and Galt in 1991[12] and Almy *et al.* in the same year[13], reported on initiatives to draw research, extension and farmers closer through FSR in an OFR context. Roy, in 1996, recorded a commitment in Bangladesh to mainstream FSR into the extension services to enhance location specificity and ensure that technologies developed are demand-driven and client-oriented[14]. Roy notes: 'The proposition that FSR be incorporated in the country's mainstream extension services is no longer debatable. But the practical approaches and innovations in management required to foster a truly FSR perspective are yet to be developed'.

Roy's point is significant. Very little thought has been given to the organizational and cultural changes which would effectively mobilize FSR anywhere but within research institutions. Both Ewell[15], as a part of the ISNAR-managed On Farm Client Oriented Research (OFCOR) study programme, and Low *et al.*[16] have drawn lessons on OFR and research extension linkage from earlier experiences but, given the widening role for FSR, it remains a yawning gap in our understanding. These problems have been compounded by the lack of professionalism in managing and advising institutions by donors. There has been little or no analysis of the forms of R & D organization suited to particular circumstances of different countries. Hayami and Ruttan, in 1984[17], extended their theory of induced innovation to institutions, and, though this is credible where farmers are able to articulate their needs and expect a response from both markets and institutions, it is difficult to accept where farmers have no voice and markets have largely been controlled through government. In recent years smallholder empowerment as promoted by the participation movement has become an important complement to the structural adjustment pressed on developing countries by the international community.

CONCLUSIONS

The traditional approach to applied agricultural experimentation, and the research organization to implement it, promotes allegiance to commodities. This has slowed progress in improving the relevance of research output to smallholders. Over the last 30 years other factors have certainly contributed to the persistence of outmoded research paradigms and organizational forms. The NARIs have faced crises in both management and finance. Sorely needed increases in professional manpower have had to be supported from falling funding. As salaries have increasingly dominated recurrent budgets the operational funds for experimentation have reduced. Beyond this generic weakening of public R & D institutions, university agricultural education, often based on

Western curricula, has failed to come to grips with the needs of professionals destined to work with small farmers.

Reforming the professional mind-set is particularly important and Anandajayasekeram's account of the shift from short-course, in-service training, to an emphasis on the introduction of FSR concepts and methods into university courses at the undergraduate and graduate levels, offers some encouragement. Other writers have emphasized the need for curricular changes in tertiary education in agriculture in the developing countries, including Doppler and Maurer, in 1993[18]. It is a role that associated Western universities could pursue more vigorously.

An increasing number of governments will likely sanction private sector sponsored research for subsectors where they have a propriety interest. But public services will continue to be needed for food crops, particularly those that do not come onto the market, for fragile areas, and for natural resource management more generally. I have to conclude that many developing country governments cannot afford to let go and depend on the private and informal sectors. My own convictions are that public sector agricultural research will remain vital for developing agricultural economies with vast numbers of very small farmers. I do not see an easy alternative to strategic and applied research continuing to be organized by disciplinary and commodity, but with a change in culture. I agree with Hall who, in 1993[19], highlighted complementarities between the local organization and community-based expertise and the broad influence and scientific and technical capacity of government as offering a natural partnership.

The public sector makes huge investments in agricultural R & D, and poor governments need a good return on their scarce revenues. By improving the effectiveness and efficiency of these investments, FSR can also improve the political credibility of the R & D process. FSR offers a natural interface between public research and extension in the R & D process for agriculture, and between public research and external agencies, NGOs, farmers' associations and the private sector, all increasingly involved in development. These could all draw from an FSR-based interface that:

- Helps to articulate the needs of farmers to applied research establishments, whether public or private.
- Draws down commodity and factor output proven in the conditions of the local ecology, relevant to farmers' needs and selected for its productivity and plasticity to accommodate to the local system.
- Moves away from prescription by providing a basket of choices, made with farmers, which farmers then assess for themselves, in their own fields and provides understanding of local farming as a resource for wider development programming and policy formulation.

At least three alternatives to FSR's location in the research services present themselves:

- That FSR is located in the extension service. Its area-based organization is much easier to reconcile with FSR than with the commodity and disciplinary organization of research institutions.
- The diffusion function is undertaken through enhanced traditional farmer-to-farmer processes in association with FSR which umbrellas multiple sources of innovation as well as OFE. Extension service responsibilities are revised to include support for farmer-based dissemination processes and the facilitation of input supply, credit and other services needed to mobilize innovations as they spread through farmer-to-farmer contact.
- A new community development paradigm, in which social scientists play a very wide role in community interfacing, linking back to a full range of services from public and private institutions.

A choice of institutional home will depend on the promotional strategy which seems most appropriate in the circumstances of the individual country. The issue merits further discussion in Chapter 12, but certainly the findings of the ISNAR-managed OFCOR study on the linkage question remain relevant[20]: the extension service should be included from the beginning, not as an afterthought once results need mobilizing; there are limits to informal cooperation. Extension staff resent collaboration with researchers as an additional, unrewarded burden. Researchers are put off by extension staff's low motivation and poor management; formal linkage methods are needed to underpin collaboration. These should be visible at each level of the administrative hierarchy.

REFERENCES

1. **Farrington, J. & A. Bebbington, 1993**. Are There Lessons from NGO–Government Links? In: Institutionalization of a Farming Systems Approach to Development, p. 219. Farms Systems Management Series No. 5. FAO, Rome.
2. **Dillon, J., D. Plucknett & G. Valaise, 1978**. Farming Systems Research at the International Agricultural Research Centers. Technical Advisory Committee, CGIAR. FAO, Rome.
3. **Norman, D., D. Baker, G. Heirich & F. Worman, 1988**. Technology development and farmer groups: experiences from Botswana. Experimental Agriculture, 24(3), 321–31.
4. **Sumberg, J. & C. Okali, 1988**. Farmers, on-farm research and the development of new technology. Experimental Agriculture, 24(3), 333–42.
5. **CIAT, 1996**. Annual Report of the Bean Programme. CIAT, Cali, Colombia.
6. **Ashby, J.A., 1991**. Adopters and adapters: the participation of farmers in on-farm research. In: Tripp, R. (Ed.) Planned Change in Farming Systems: Progress in On-Farm-Research. John Wiley, London.
7. **Chambers, R., 1989**. Reversals, institutions and change. In: Chambers, R., Pacey, A. & Thrupp, L.A. (Eds) Farmer First: Farmer Innovation and Agricultural Research. Intermediate Technology Publications, London, pp. 181–93.
8. **Low, A.R.C., S.R. Waddington & E.S. Shumba, 1991**. On-farm research in southern Africa: the prospects for achieving greater impact. In: Tripp, R. (Ed.) Planned Change in Farming Systems: Progress in On-farm Research. John Wiley, London.
9. **Pickering, D.C., 1985**. Sustaining the continuum. In: Cernea, M.M., Coulter, J.K. & Russell, J.F. (Eds) Research, Extension, Farmer. IBRD/UNDP Symposium, Washington, DC.
10. **Denning, G.L., 1985**. Integrating agricultural extension programs with farming systems research. (ibid.).
11. **Shumba, E.M., 1994**. Institutional linkages that enhance the value of on-farm research for smallholder farmers: the Zimbabwe experience. Journal of Farming Systems Research–Extension, 4(3), 137–45.
12. **Al-Khadi, A.F. & D.L. Galt, 1991**. The bottom up and the top down: area farmers meet regional and central research and extension staff in Jordan. Journal of Farming Systems Research–Extension, 2(3), 31–50.
13. **Almy, S.W., M. Besong, T. Woldetatios, C. Pouborn, C. Ateh & M. Mboussie, 1991**. Extension agents in the research process. Journal of Farming Systems Research–Extension, 2(3), 69–84.
14. **Roy, I., 1996**. Integrating FSR into the national extension system; a case of Bangladesh. Journal of Farming Systems Research–Extension, 6(2), 45–53.
15. **Ewell, P., 1988**. Organization and management of field activities in on-farm research: a review of experiences in nine countries. OFCOR Comparative Study Paper no. 2. ISNAR, The Hague.
16. **Low, A.R.C., S.R. Waddington & E.S. Shumba, 1991**. (Op. Cit.).
17. **Ruttan, V. & Y. Hayami, 1984**. Induced innovation model of agricultural development. In: C.K. Eicher & J.M. Staatz (Eds) Agricultural Development in the Third World. Johns Hopkins University Press.
18. **Doppler, W. & M. Maurer, 1993**. Farming Systems Development in Academic Training. Institutionalization of a farming systems approach to development, p. 91. Farms Systems Management Series no. 5. FAO, Rome.
19. **Hall, M., 1993**. Summary and Conclusions, (ibid.) p. 323.
20. **Merill-Sands, D.M., S.D. Biggs, R.J. Bingen, P.T. Ewell, J.L. McAllister & S.V. Poats, 1991**. Institutional considerations in strengthening on-farm client oriented research in national agricultural research systems: lessons from a nine country study. Experimental Agriculture, 27(4), 343–74.

Chapter 6

FSR: Some Institutional Experiences in National Agricultural Research

6.1 THE SYSTEMS RESEARCH DEPARTMENT AT INRA

J. Bonnemaire, J. Brossier and B. Hubert

Research took leave of its academic methods to keep a group of scientists in the field who gave priority to organizing for the group involved, rather than for disciplinary programme logic. Their conclusions were ultimately validated.

6.1.1 Introduction

France is often considered as 'separate' from other industrialized countries due to circumstances of history and culture and there are clear differences in a number of its institutions. One that is entirely unique in the industrialized world is the Department for Research on Agrarian Systems and Development (SAD). SAD is located within INRA. This contribution explains why and how SAD was created. We try to capture the particular French context, drawing some comparisons with the UK and the USA. SAD's achievements are outlined, and the way in which its multidisciplinary culture has created a new view of some disciplinary issues is examined. First, we give a brief description of the INRA-SAD department as it is today.

6.1.2 Defining SAD

INRA employs a total of 8500 staff. It has five scientific divisions, four of which are physical environment and agronomy, animal production, plant production and agrofood industries. The fifth consists of the Department of Rural Economics and Sociology, the Department of Biometry and Artificial Intelligence and SAD, which employs just 200 people. These include 100 scientific researchers, grouped in some 15 locations throughout France.

The overall objective of SAD is to study the practices and processes of development with a view to:

- Understanding local dynamics, clarifying which trends result from broad, wide-ranging forces and which from specific and unusual local situations.
- Analysing how processes of change combine vertical forces, such as macro-economic factors, subsector organization and available technology, and horizontal or spatial forces such as resource management, organization of activities and types of cooperative action and conflict.

Development is examined by studying the process of technical and organizational innovation, giving priority to two themes:

- Organized action – how knowledge is produced and shared, and how learning takes place.
- Territorial aspects – organization of social groups and networks as well as the technical systems, ecological structures and processes that exist within a defined geographical area.

Inductive approaches prevail based on observations, case studies, participatory approaches and modelling. SAD research contributes to the joint production of knowledge with groups of local actors so that they can improve their capacity to conceive and carry through their own projects. The research has a strong *ex ante* component; current events must be dealt with, but future events and adaptive behaviour must be anticipated. Research is organized around five themes that deal with the processes of technical change as they affect farms, economic structures, land use and society as a whole:

- Technical systems and innovations.
- Farm management, and technical and organizational learning.
- Means of coordination within subsectors.
- Spatial organization of activities and landscape dynamics.
- Territorial aspects and forms of social integration.

6.1.3 The particularly French context

The long-standing development of systems research in France and the expansion of multidisciplinary research, both of which led to the creation of SAD in 1979, are linked to the Higher Schools of Agriculture (Grandes Ecoles) peculiar to France. In these schools all the agronomists of the 1950s, 1960s and 1970s received a background in the three disciplines fundamental to the study of the transformation of production systems and agrarian systems, and which are also central to SAD; livestock research, agronomy and economics. Multidisciplinary research in SAD was facilitated by this common culture and language in education. This does not, however, mean that French educational methods should be replicated outside France to ensure the development of systems research. Indeed, the model is dated. Its multidisciplinarity is limited to the three disciplines mentioned – not sufficient for drawing up a systems science blueprint. Second, in France and elsewhere, there is insufficient training in multi- and interdisciplinary approaches. Systems science and constructivist epistemologies still find it difficult to penetrate academic circles.

The French public service tradition

Public service in France, initiated under Napoleon, is centralized, often technocratic and has expanded since the 1940s. This mission for public service, which other countries envy, still clearly prevails. Created as recently as 1946, INRA has become the largest public institution for agricultural research in industrialized countries. Its creation took place during the reconstruction phase of the economy, in which agriculture was seen as an essential component. France chose to create large public service organizations responsible for areas such as reconstruction, national security and economic development.

INRA staff still have a missionary drive for the cause of farming. This ethos of service for society has always been one of the fundamental reasons for INRA's existence. During a recent debate about the objectives contracted between INRA and its supervisory ministries, this remark was made: 'In France, those who work for the state feel a special responsibility. It may be true we can never be fired, but the main reason people want to work for the state in France is because there is a sense of honor about public service' (IHT, Tuesday 25 June, 1996).

What justifications are there today for such a large institution devoted to public service? Its mission for fundamental research is not challenged, yet INRA clearly positions itself as an institution involved in applied and objective-oriented research. For many years, the fragmentation of agriculture justified INRA, since it was not possible for farmers, as highly dispersed groups of actors, to finance *ad hoc* research. The justification today is found in the new demands made by society. Hence the social contract between agricultural research and the nation now embraces not only agriculture and agrofood industries but stretches to consumer needs for quality food products and for protection of the environment and natural resources. A certain amount of bureaucratic self-preservation is involved, contradicting the systems approach and representing something of a paradox. Since World War II, in particular, France has favoured centralized technocratic procedures for spreading the message on modern farming innovations. Research work on marginal areas and developing countries follows in this same tradition of public service and SAD's research on deprived areas in France has links with subsequent work undertaken in the developing world. The SAD department was created prior to similar research departments within the Centre International de Recherches Agronomiques pour le Developpement (CIRAD) and the Institut Français de Recherche Scientifique pour le Developpement en Cooperation (ORSTOM), the French institutions responsible for research overseas.

The same public service and bureaucratic tradition stimulated encouragement from the Ministry of Research through staff incentive programmes in the 1970s and 1980s. The same is true of requests from other technical ministries to set up research teams in order to solve sociopolitical problems, such as forest fires in south-eastern France and the economic development of Corsica.

6.1.4 Comparative history of French and British agriculture

Differing agrarian culture and dynamics in these two European countries resulted in different policy approaches to agriculture and different research options. In the UK, agriculture has long been more homogeneous, more professional and more élitist than in the Latin countries, including France. Small landholders were eliminated fairly early, leaving fewer farmers who were also better trained. And a community of thought and work has existed between great landowners and scientific circles and agricultural public servants since the time of Darwin, the first great cattle or sheep breeders, and perhaps even earlier. Enclosures had started in the 16th century under the Tudors, with the development of towns and the beginnings of the cloth industry as the driving force of sheep farming. While this phenomenon stalled during the 17th century following social unrest and the Enclosures Edict, the process of agrarian individualism soon accelerated with the arrival of the Industrial Revolution towards the end of the 18th century.

A class of great landowners progressively prospered in the UK. Very often they were not only farmers but also had interests in industry and other connections between town and country. These landowners, men of science and public service, were accustomed to meeting at the agroindustrial interface, understanding each other without difficulty. Similarly the message passed successfully from the early research station to the farms.

In France, the situation evolved in a different manner[1]. From the 16th until the 18th and 19th centuries, the conception of the rural economy, inherited from Latin agronomists, remained that of a global system very diverse in its components. Only then, under pressure from the physiocrats, the British agricultural model and developments in industry, was emphasis placed on the role of large farms in the diffusion of agricultural progress, of capital and profit in agriculture, or on the need to apply the sciences to agriculture as well as other sectors.

The agricultural crisis at the end of the 19th century and the protectionism that resulted brought about a radical change in approach. Small landowners were considered an element of social stability and many theories about the social, economic and political advantages of small farms were expounded. Throughout the first half of the 20th century, the policy was to institute development on a cooperative or mutual basis (credit, insurance, supply and sales) to give the same economic advantages to small farms and large farms alike. To give impetus to agriculture, the emphasis was on structural and organizational change as well as on techniques and science.

After World War II, French agriculture became aware that it needed to catch up, and the UK was held up as a model. Small French landowners were expected to contribute to the reconstruction of the country. The aim was increased production and reduced costs in order to satisfy increasing consumer demand and make manpower available for industry.

The farm community was, however, much more diverse, complex and far larger than that found in the UK. It was impossible for policy to cope with this diversity or to deal directly with the multiple situations involved. With the help of the agricultural union movement inspired by the Catholic Church, and a resurgence of prewar ideology on progress in agriculture, the prevailing rationale favoured setting up intermediate technostructures to help farmers organize themselves and help each other through cooperative groups. 'Development' occurred when farmers compared techniques and results with their neighbours. Technical progress and its diffusion, although fed by scientific knowledge, remained mainly endogenous to the farming sector due to the predominant dynamics of local groups of farmers. As a result, the farm has always been viewed as a holistic system.

Small landowners in France were more numerous and had less training than those in the UK, and most were imbued with Catholic ideology on productivity, which put the emphasis on mutual help and collective action for collective social betterment. In addition, France is the only country with a strongly centralized system among European countries possessing a 'farming' culture. It was not surprising, therefore, that as early as the 1950s, France generally gave priority to an agricultural development approach that was more 'technocratic' and institutionalized than in the UK. There, short-cuts for transferring technical knowledge already existed, and it was created or taught directly to a network of agricultural enterprises which were more homogeneous, less numerous, more technically up-to-date and better linked into the market economy. One UK example was the former Milk Marketing Board, which monopolized the distribution of small bottles of milk to houses by direct sale.

6.1.5 The USA: emergence of non-institutional systems research

Multidisciplinary systems research into complex farming systems is without doubt more recent in the USA. Beginning at the end of the 1980s, new areas for systems research are now shooting up everywhere[2]. This is not an organized phenomenon, but trends are being set and certain questions and approaches are repeatedly cropping up: erosion, pollution, sustainability, participatory approaches, the involvement of actor-groups in the research, and so on. The teams involved still have little contact with the AFSRE.

At present in the USA, several terms are used to define new ways of farming: sustainable agriculture, low input agriculture (LIA), low input sustainable agriculture (LISA), low external input sustainable agriculture (LEISA), alternative agriculture, regenerative agriculture, organic agriculture, profitable agriculture and clean environment (PACE), best management practices (BMP) and agroecology, to name just a few. There is some ambiguity and confusion concerning the concepts behind the various terms, some of which have political, even ideological, implications making them difficult to clarify. Local organizations increasingly request research and stimulate response by providing the necessary support.

In France the decision was to institutionalize interdisciplinary and holistic research as INRA-SAD. In the USA the choice was to create *ad hoc*, flexible centres in the land grant universities. Sometimes the impetus for such initiatives comes from outside the university. For example, the Leopold Center of Iowa State University was created by the State of Iowa, possibly against the better judgement of some at the university. The aim is to give administrative and logistic flexibility so that concrete problems can be solved more easily despite the rigidity in the disciplinary subdivisions of the university system.

6.1.6 The origins of, and reasons for, the SAD department within INRA

Holistic production systems between the 16th and the 19th century have always been the subject of traditional studies in rural economics and agronomic science in France. Brossier, in 1987[3], noted that, as far as Anglo-Americans are concerned, the development of the farming systems concept is more recent and is connected with research and development in developing countries. In the USA for example, the tendency has been for teaching on narrow disciplinary lines, creating a top-down chain from laboratories to users. The USA based its agricultural progress on considerable scientific investment, focused at first on genetic research such as the creation of strains and varieties that were highly productive – and highly consuming of inputs – and simply distributed to farmers.

In spite of Borlaug and his Nobel Prize, the repeated failure of attempts to develop agriculture in developing countries finally convinced Anglo-American researchers, a prevalent group in the international organizations, of the importance of production systems concepts and the use of FSR-E methods for rapid diagnostic assessment and adaptation of improvements for potential users.

Preliminary interdisciplinary research

At its inception in 1946, INRA provided back-up for the reconstruction and modernization of French agriculture with production growth as a clear objective, and, as per the American model, complete trust in research. Research would produce techniques to be diffused to an élite group of productivist smallholders, cooperatives and mutualists who would then set an example to the groups as a whole. There was, therefore, a need to synthesize the knowledge produced at the farm level, to make it accessible by the various types of regional agriculture. In 1959 INRA, as the research institution of a centralized state, decided this mission could be achieved by creating a special structure to handle the Application of Research to Extension (SARV), equipped with experimental farms for the purposes of synthesis and demonstration.

Soon afterwards the close relationship existing between the government and the farming profession led the latter to take responsibility for extension. Thanks to the green laws (1960 and 1962) which gave priority to economic organization and structural policy, this soon became 'Agricultural Development'. Financed by parafiscal taxes on products, it aimed at the redistribution of means among regions, production sectors and types of public. In 1964, the Experimentation and Information Service (SEI) took over SARV's organization of information transfer from INRA and research at the interfaces insufficiently addressed by disciplines.

Little by little, given the divergence between technical advice and agrarian situations[4], the increasing numbers of specialists at INRA felt the need for a broader approach. Researchers, many of whom retained a farming culture owing to their social origins and training, became progressively aware of the problem through experiences of marginal areas and teacher-researchers. Hence, the first research work undertaken by the SEI had three objectives:

- The diffusion and adoption of technical innovations.
- The pooling of new knowledge from individual disciplines at the farm level.
- To make appropriate responses to farmers' needs across the diversity of agrarian situations.

The Cooperative Research Programme carried out in 1963–65 on the high livestock farming plateaux of the Aubrac area in the Massif Central (RCP Aubrac), set up by CNRS, ethnologists and museologists, together with INRA livestock researchers and agronomists, was a critical experience[5]. Livestock research and ethnosocioeconomic research was confronted with the collapse of a traditional, very complex production system due to economic crisis and the irrelevance of modern techniques. Research took leave of its academic methods to keep a group of scientists in the field who gave precedence to organizing for the group involved, rather than for disciplinary programme logic. Their conclusions were ultimately validated.

The French rural economists, like the agronomists and livestock researchers close to the modernization process, quickly became aware of its other side; the difficulties of deprived areas, economic slippage of certain categories of smallholders or of certain production systems. The SEI researchers, upholders of both the traditional INRA research approach and of the novel SEI approach, undertook a series of research projects in the early 1960s to study 'regional potential' to identify how technique and environmental diversity affected yield[6]. Until 1970 these projects focused on the obstacles to technical progress, but were then redirected towards the production conditions and choices of farmers – individual farmers at first, and thereafter within continuous land areas using the notion of the local agrarian system[7]. Hence in Corsica the concept of 'system of practices' emerged to describe a technical and economic operation – a situation in which practices were stable and farm structures variable![8] In the Causses area, the limestone plateau where Roquefort cheese is produced, genetic and other researchers working at the Toulouse centre had to balance the constraints of system intensification, enhancing land areas of great diversity, and reconciling the logic of different disciplines and pluridisciplinary programmes.

The teacher-researchers of the Universities of Agronomy (Grandes Ecoles) played a decisive role in these initiatives, particularly as they wished, above all, to train for action, enjoying more freedom than the researchers at INRA with its academic needs in specialized scientific production. At the same time, under the pressure of teacher-researchers such as Henin and Sebillotte, agronomy itself diversified its approach, placing the emphasis on aspects such as quality, field observation, physics and, above all, a model integrating different branches of knowledge where the farmer and his decisions had their place. Henceforward, agronomists studied objective-orientated operating systems made up of plant community, soil, climate and human activities, together with their interactions. In 1980 the French government created two research teams for missions in south-eastern France and in Corsica to bring research closer to the action in areas where a dual problem existed – forest fires and the development of livestock farming. These teams, faced with the complexity of the local production systems, and also with action in a socially accepted FSR approach, found themselves in the newly formed SAD.

Because of their contact with economists and agronomists, but above all with field realities (particularly as a result of experience in developing countries), livestock researchers have progressively developed a certain number of approaches on applied livestock systems at different organizational levels. As for the economists who contributed to the creation of this research trend, they find themselves at the junction of several schools of thought:

- Company economics applied to the farm and its management.
- Economic modelling (using linear programming) firstly used to normative ends and thereafter to analytical ends (underpinned by the postulate of overall farmer rationality).
- Studies of agriculture compared on various regional levels in France and worldwide.
- Promotion of the agrarian system concept after Dumont.

A large number of farm typologies based on farmer projects and situations have been developed with a view to grasping more fully the diversity of agriculture. Beginning with a study of the obstacles to adopting technical progress, the teams of researchers analyse farmer decision making, farm functioning and the paths taken to implement change. The analysis of farmer practices (which, why, how, with what results) has proved to be essential in this analysis of farms. Increasingly this minute analysis of farms goes hand in hand with a more holistic approach to the small regions involved, the territorial and social dynamics (lands, farmers, landscapes or Pays, Paysans, Paysages).

Converging experiences, institutional incentives and systems analysis

Experiences in France and in developing countries were being refined and were converging to create a multidisciplinary systems approach to research in farming. The Ministry of Research encouraged this movement as early as 1968 with successive scientific committees working on a widening field of study. Exchanges were increasingly promoted with other research organizations, particularly the CIRAD and ORSTOM, active in developing countries, resulting in parallel study experiences.

By the beginning of the 1980s, this research work was gradually organized around the systems concepts which represent the three epistemological origins of the SAD: Simon and his theory on optimal decision and procedure rationality, Piaget and his constructivist epistemology as a basis of learning theory, and Morin and the paradigm of autoecoreorganization in modelling complex systems. The theoreticians of systems modelling, including Le Moigne in 1990[9], also made important contributions.

At the heart of our own scientific field, contributions by Sebillote in 1974 on the relationships between agronomic science and agriculture, by Petit in 1975 and 1981[10] on the theory of adaptive behaviour, by Osty in 1978 on the family–farm system[11] and by Teissier in 1979[12] on the relationships between practices and techniques, all demonstrate the interest in this approach for research into farming activity. Management science made its official debut in 1983 when a specific section was created at the CNRS.

As all this scientific thought and experience accumulated a need arose for an appropriate support structure, leading to the creation of SAD at the end of 1979, staffed by about 50 researchers. Its initial field of study was the implementation of a common research approach which saw farms as managed systems[13]. The approach had internal coherence, a specific objective, included descriptions of the way practices were used, insertion in a local environment and in a system subsector, homogeneous grouping using typologies and the inclusion of both spatial and temporal dimensions. Recourse to systems analysis and to modelling, and the similar scientific thought on this subject are elements of unity between research at different scales and between research by different teams.

The approach can be interpreted as an enlargement of agronomic science in which agricultural activity is analysed in its geographical, economic and social context. Study of agricultural activity, starting with the basic unit that one farm represents, led to the identification of guidelines for research into the systems used based on the 'technical facts' of farm practices and how these were implemented[14].

6.1.7 The SAD interdisciplinary culture and a new look at disciplinary issues

The science in which SAD is involved concerns the dynamics of farm activity, giving priority to the study of technical skills because of their biological, physical and human ramifications, their organizational implications and their social relevance. In research oriented to development we have become interested in technical systems for two reasons:

- They are action systems, elaborated by social groups to affect the world; which means that they are particularly relevant subjects for our research.
- For their theoretical side: technical systems are central to the relationships between social groups and the physical and biological environments in which these groups wish to act, and have at once social, symbolic and physical ramifications.

The study of technical systems enables these different dimensions of human activity to be examined simultaneously, especially the relationships between culture and nature.

Interdisciplinarity is necessary for the study of technical systems due to the separation of sciences into distinct disciplinary fields of knowledge. This is different from the knowledge that comes from action which is nearer to technological science, a research field calling upon physics and biology, technical sciences and social sciences. Technology gives particular scope to multidisciplinary methods.

Between the aspects of nature studied by the biological sciences and the aspects studied by traditional economics, the organization of social groups in their environment is a neglected field. Within this field we give high priority to the study of technical systems created for the development of these groups, and thus undergoing constant change. Different approaches are needed to exploit this field. Disciplines such as anthropology, sociology and management collaborate with biological disciplines (ecology, genetics) and technical disciplines (agronomy, livestock research) in the study of 'hybrid' or 'borderline' subjects also important to the farmers and technicians with whom we work. This new approach results in new questions being put to the traditional disciplines; for example product quality can no longer be limited to biotechnical aspects but also needs to be considered as a wider social construction. Water quality, for example, does not only depend on soil, subsoil and input characteristics, but also on the type of plant cover and the farming practices used.

The fragmentation of livestock research sciences which has arisen from detailed work on biological mechanisms, best synthesized as physiological functions, means that new questions are being asked about the herd and its performances, or an animal's lifetime performance. These are facets of management for the livestock farmer and must be formulated and handled as scientific topics. Similarly the study of pastureland, especially on heterogeneous territory, can result in a review of approaches to animal feeding. The concept of permanent adjustment between input and needs must be put behind us, along with traditional concepts of digestibility, nutritive value, individual animal behaviour and feeding efficiency, giving way to a management approach using concepts of new plants or new pasture areas, appetite activators, 'menu', land configuration compared with herd size and the variety and time span of micro-climatic conditions. It is an admission that pasture management creates the value of the pasture resource, as Meuret pointed out in 1996[15], it is just as sentence and context give words their meaning!

Similarly in farm production economics, SAD studies have highlighted the importance of cash-flow practices, leading to their being researched in their own right. Multidisciplinary activity was essential to stop economists working alone and pluridisciplinary activity stopped them working solely with traditional economic flows based on theoretical principles totally different from the everyday reality of actual farms. It became essential for management economists to work on identifiable cash and monetary flows as well[16].

Most economics researchers involved in this systems science adventure are not satisfied with certain principles of economic theory, particularly production theory. Production theory was the first effort to create a model representing producer behaviour, but research work of several systems teams have shown its limitations in understanding farmer behaviour and activity. For these research teams, 'the company' is not an abstract entity in micro-economics, but is the subject of, and the participant in, the research. These findings have resulted in the creation of new models within the theory of adaptive behaviour[17], or action models[18]. The research approach adopted by SAD has, in short, offered a new look at traditional disciplinary tenets.

6.1.8 Conclusion

The institution of this field of research within INRA has given greater visibility to the sciences of complexity. SAD is founded on the interdisciplinarity of natural, technical and social sciences, and their interface with training. Its scientific field has been built on the study of technical systems which incorporate biotechnical and social dimensions. SAD researchers feel that they have an essential role to play in 'innovation networks', where the different viewpoints of actor-groups converge, where mutual learning takes place and where know-how is elaborated and combined.

The interdisciplinary experience of SAD does not aim to create a new disciplinary field (though we are well aware that at least 40 years are necessary for the creation of a new scientific field!) but it has given itself the task of legitimizing and validating a research culture which gives priority to relevance rather than academic excellence in the strict sense of the term. Validation indicators come from the constituent disciplines working as a group, and from the field, culminating in an overall management approach. Thus the department is not caught up in discussions where aspirations to exclusive academic excellence override the research priorities imposed by the needs of French farming.

The fact that our work was initiated on the farm level, then widened to include technical innovation in the overall development process at the level of the landscape and social and economic organizations, means that it is easy for SAD to conduct a dual dialogue and engage in indispensable partnerships with disciplines ranging from biotechnical to social, and with all actor-groups involved in development.

The systems approach means, above all, an attitude and methods of working with colleagues and other partners. The example of multidisciplinary research carried out on water quality, agricultural practices and changes in farming and agrarian systems is a good illustration of collaboration between disciplines and can be examined in Chapter 11. It is vital to emphasize the importance of collaboration between research institutions. It is not sufficient to state that objective-oriented research cannot exist without the systems approach: the approach must be promoted and explained and its usefulness made plain. The fact remains that, within an institution such as INRA, collaboration between an interdisciplinary department and other departments organized around specific disciplines can never be taken for granted, and is by definition an on-going challenge.

REFERENCES

1. **Parain, C., 1956**. Un mot du vocabulaire de synthèse historique: agriculture. Revue de Synthèse, 77 (1), 43–54.
2. **Brossier, J., 1994**. Recherche-système in USA. Courrier de l'environnement de l'INRA, no. 19. INRA, Paris.
3. **Brossier, J., 1987**. Système et système de production. Cahiers ORSTOM, Série Sciences Humaines 23, 377–90.
4. **Dumont, R., 1956**. Voyages en France d'un Agronome. Librairie de Médicis, Paris.
5. **Henin, S. & J.P. Deffontaines, 1970**. Principe de l'étude des potentialités agricoles régionales. Comptes Rendus de l'Académie d'Agriculture de France, 8, 463–72.
6. Ibid.

Deffontaines, J.P., 1964. Essai de détermination des potentialités agricoles regionalés. Acta Geographica, 28–34.

Deffontaines, J.P., 1967. Une méthode de détermination des facteurs techniques limitant la production agricole en montagne. Fourrages, 31, 36–52.

Deffontaines, J.P., 1973. Analyse de situation dans différentes régions de France. Freins à l'adoption d'innovations techniques. Etudes Rurales, 53, 80–90.

7. **INRA-ENSSAA, 1973**. Conditions de Choix des Techniques de Production et Evolution des Exploitations Agricoles. Région de Rambervillers (Vosges). Etude no. 4. INRA-SEI, Versailles.

INRA-ENSSAA, 1977. Pays, Paysans, Paysages dans les Vosges du Sud. INRA, Versailles.

Petit, M., 1975. L'adoption des innovations techniques par les agriculteurs. Plaidoyer pour un renouvellement de la théorie économique de la décision. Pour, 40, 79–91.

Brossier, J. & M. Petit, 1977. Pout une typologie des exploitations agricoles fondée sur les projets et les situations des agriculteurs. Economie Rurale, 122, 31–40.

Osty, P.L., 1978. L' exploitation agricole vue comme un système. Diffusion de l'innovation et contribution au développment. Bulletin Technique d'information du Ministère de l'Agriculture, 326, 43–9.

8. **Cristofini, B., J.P. Deffontaines, C. Raichon & B. de Verneuil, 1978**. Pratiques d'élevage en Castagniccia: exploitation d'un milieu naturel et social en Corse. Etudes Rurales, 71–2, 89–109.

Deffontaines, J.P. & C. Raichon, 1981. Systèmes de pratiques, terroirs, moyens d'analyse d'une agriculture régionale. Economie Rurale, 142, 30.

9. **Le Moigne, J.L., 1990**. La Modélisation des Systèmes Complexes. Dunod, Paris.

10. **Petit, M., 1975**. (Op. cit.).

Petit, M., 1981. Théorie de la décision et comportement adaptatif des agriculteurs. Proceedings Colloque du 21 Janvier. ENSSAA, Dijon.

11. **Osty, P.L., 1978**. (Op. cit.).

12. **Teissier, J.H., 1979**. Relations entre techniques et pratiques. Bulletin de l'INRAP 38, ENESAD, Dijon.

13. **INRA-SAD, 1979**. Eléments pour une Problématique de Recherche sur les Systèmes Agraires et le Développement. INRA, Paris.

14. **Gras, R., M. Benoit, J.P. Deffontaines, M. Duru, M. Lafarge, A. Langlet & P.L. Osty, 1989**. Le fait Technique en Agronomie. INRA and L'Harmattan, Paris.

Osty, P.L., 1990. Le fait technique en agronomie. Points de vue et questions sur quelques concepts. In: Brossier J., Vissac B. & Le Moigne J.L. (Eds) Modélisation Systémique et Système Agraire. Décision et Organisation. INRA, Paris, pp. 19–28.

15. **Meuret, M., 1996**. A grazing route in Provence. In: Proceedings of an EC/EAAP/SCAWS Meeting on the Welfare of Farm Animals. Edinburgh, UK.

16. **Chia, E., 1987**. Les pratiques de trésorerie des agriculteurs. La gestion en quête d'une théorie. Thèse de Doctorat. Faculté de Sciences Economiques et de Gestion, Dijon.

Brossier, J., E. Chia & E. Marshall, 1988. Quelle gestion de trésorerie pour les agriculteurs? Gérer et comprendre Annales des Mines, 12, 41–55.

17. **Petit, M., 1981**. (Op. cit.).

Brossier, J., E. Chia, E. Marshall & M. Petit, 1991. Gestion de l'exploitation agricole familiale et pratiques des agriculteurs: vers une nouvelle théorie de la gestion. Revue Canadienne d'Economie Rurale, 39, 119–135.

18. **Attonaty, J.M. & L.G. Soler, 1995**. Renewing strategic decision-making aids. In: Brossier, J., Bonneval, L. & Landais, E. (Eds) (Op. cit.) pp. 291–309.

6.2 SENEGAL'S EXPERIMENTAL UNITS

Michel Benoit-Cattin

The move off-station was accelerated in response to three factors: the extremely slow pace of adoption of technologies coming from research, the criticism that research was 'ivory towered' and the perception that research was controlled by foreign capitalist interests concerned only with export crops.

6.2.1 The political and institutional background

In 1968 the National Centre for Agronomic Research (CNRA) at Bambey in Senegal broadened its research agenda by beginning work in two groups of villages located in the south of the groundnut growing basin in the Sine Saloum region. The Kumbija Unit was located near Kounghel in the Department of Kaffrine, and was, in 1970, made up of 13 villages with 1972 inhabitants and 229 farms on 5600 ha of land. The Ciise-Kaymor Sonkorong Unit comprised of the two large villages of Ciise-Kaymor and Sonkorong located in Medina-Sabax in the Department of Nioro du Rip, with 1465 inhabitants and 131 farms on 4500 ha. These became known as the experimental units (EU).

In the 1950s and 1960s the CNRA had already been involved in initiatives in peasant farming areas. There had been a soil regeneration operation in Thienaba stimulated by rural economists[1] and the groundnut experimental block at Boulel north-east of Sine Saloum, set up to link the colonization of new land in the Mouride area with agricultural mechanization. Boulel was widened into a Secteur Expérimental de Modernisation Agricole (SEMA), an experiment in modernization, with participating farmers offered a plan for intensification based on research technologies and the use of cattle traction.

Although this move away from the research station and its artificial context went back some time, it was still the exception rather than the rule in agronomic research circles. The move off-station was accelerated in response to three factors: the extremely slow pace of adoption of technologies coming from research[2], the criticism that research was 'ivory towered' and the perception that research was controlled by foreign capitalist interests concerned only with export crops.

Agronomy history

The groundnut experimental station set up in Bambey as early as 1914 had a responsibility for the whole of French West Africa. It evolved into the Centre for Agricultural Research in the Sudan zone in 1938, and into a federal centre for agricultural research in 1950. Agronomy research was managed by the French Institute for Tropical Agronomic Research (IRAT) until 1975, when it was taken over by the Senegalese Institute for Agricultural Research (ISRA).

With independence, Senegal inherited the facilities, the staff and most of the lessons learnt from this research. These included varietal development in groundnuts, but also animal-drawn agricultural equipment, improved food crops, the use of mineral manures and methods of land preparation. Agronomic research was redeployed within Senegal, using the agricultural services' regional stations and setting up a network of testing sites called Support Points for Pretesting and Multilocational Experimentation (PAPEM – Points d'Appui de Prévulgarisation et d'Expérimentation Multilocale). This network supported experimentation on combinations of techniques on large sized plots for particular farming systems. Farming models were studied first on paper and then life-size in the stations and in the PAPEMs.

These 'farming units' were run in conjunction with so-called disciplinary component researchers and resulted in research technologies that had been evaluated in a local context and were interrelated. The approach, initially restricted to organic production methods using cattle traction, was broadened in the 1970s to include animal-breeding, mechanization and even irrigation using bore holes. Researchers working both at the multilocational sites and at the farming units on station were aware of the importance of socioeconomics and worked with 'peasant farmer correspondents'. It was hoped the peasants chosen for this role would be open, receptive and so on. In fact, they were representative of a minority owning large areas of land, earning significant non-agricultural income. They were, however, more open-minded, usually as a result of experience of urban or military life. While this bias in collaborators accounted for some of the programme content there is no doubt that the development as a whole was fundamental in recognizing peasant farmers as valued collaborators in research.

A socialist-inspired context

The launch of the EU project in 1968 by agronomic researchers makes sense only with an understanding of the political and institutional context of the times. Independent Senegal inherited an economy entirely dependent on exported groundnuts, a crop developed by the colonial administration since 1850. The new political order, socialist in tendency, sought to free itself from this dependence and underdevelopment. In agriculture the aim was a wide network of cooperatives managed by peasant farmers themselves but operating within a planning system run by the State. Initially the State assumed direct responsibility for the agricultural sector, restructuring and strengthening its administrative apparatus. This included the establishment of rural development and cooperation services.

One of the most interesting structures in this new 'development administration' was the Centre for Comprehensive Rural Expansion (CERP) which brought together staff from all the technical public services; agriculture, livestock, cooperation and rural extension at a local level. This team's role was to 'raise awareness' among the farming population, train extension workers within it, help get cooperatives working properly and promote grass-roots development units embracing groups of villages. It was planned to give these units responsibility for land management and they were explicitly referred to in the law on land nationalization then being drafted.

Thus the colonial 'cash cow' economy was replaced with economic planning, public technical assistance and marketing agencies required to service the nascent cooperative organizations.

The State's assumption of responsibility for agricultural development failed to yield spectacular results. 1964 marks a turning point in Senegal with a change from grass-roots development and agricultural extension work, towards agricultural development by project. The Yaounde Convention, signed that year, caused a 25% fall in the price of Senegal groundnuts as a result of the loss of the preferential rate given by France. It was hoped to compensate for this loss in income by an increase of 25% in the volume exported to be achieved by disseminating technical recommendations from agronomic research. Responsibility for managing this vast operation, known as the 'Millet–Peanut Productivity Project', was given not to CERP but to the Société d'Assistance Technique at Coopération (SATEC), a French public company providing technical assistance and cooperation. This company was also charged with setting up a national agricultural extension structure, SODEVA, and the training of the Senegalese staff who would take over its management.

Agronomic research initiatives

Agronomic research, as we have seen, was well established, forceful and controversial, with its very controversy stimulating a strong dynamic. When the second development plan was being drawn up in 1964 with the goal of regionalization, research bodies proposed that it should work at the level of village cooperatives to fully integrate the socioeconomic dimension of farming problems. The suggestion was that integrated regional development activities (ARDI – Actions Régionales de Dévelopment Intégrées) be devoted entirely to cooperative development, first at the level of pilot cooperatives, then extending to the whole of a homogeneous area, the development unit. These initial proposals were rejected, but in 1968 IRAT was able to relaunch the proposal, barely modified, for EUs to replace the ARDIs.

1968 marked the first deadline of the Millet–Peanut Productivity Project. It was also the date of the proclamation of a new ambition for Senegal: to join the industrial era by the year 2000. The research institutions reaffirmed they had the technical knowledge to break the existing constraints in the Millet–Peanut Productivity Project and to increase peasant farmers' income beyond the target set by the political administration, provided that all concerned accepted that the traditional farming environment would change.

The mass extension programme had a significant impact, though the way had been prepared by the earlier efforts of agricultural services, provident societies and mutual societies for rural development. Traditional farming was no longer hand tool farming but rather light yoke farming using donkeys and horses. The uptake of inputs, such as equipment, seeds and fertilizer, could not meet the production targets. The shortfalls led some staff to popularize cattle traction and more intensive production techniques. In doing so they were in effect implementing a systematic, life-size test of the proposals for intensifying farming systems which had been drawn from agronomic research since the late 1950s.

Thus the EUs emerged in a particular political and institutional context, which was not, by its nature, replicable. While the project provides a very useful experience in research and development of production systems, its uniqueness means it cannot be adopted as a model for research and development.

6.2.2 Implementation of the EU project

Project objectives

The objectives formulated by R. Tourte in 1968[3] were essentially the same as those used by industrial product manufacturers to identify and meet the needs of consumers.

- The relevance of technical and economic data from experimentation should be tested by application in intensive production systems at full-scale under real life conditions. Natural potential, development objectives, economic conditions, existing assured income and the farm's opportunities should all be considered.
- Working under actual production conditions the project should identify which technical and socioeconomic innovations farmers can be advised to adopt.
- The project should understand the constraints to the spread of technical progress and the stimuli which might be used to overcome these.
- The project should better understand the real possibilities.
- The project should seek a progression of steps in innovation from traditional to intensive production systems.

The level of intervention chosen was the cooperative. This was in line with political thinking but it was also because the cooperatives had exclusive rights both in product marketing and in supplying inputs on credit. For the first two units the research institution opted for a region which was favourable from most points of view; rainfall, population density, agricultural potential and diversification opportunities. The same area had also been chosen as a test site for cattle traction within the development plan and was the first area for intervention by the nascent SODEVA. Two unit locations were selected to introduce a degree of variety for this first experiment. The original plan allowed for several pilot units, but these were curtailed, leaving one EU per main ecological zone.

What was extended

The recommendations made to peasant farmers within the EUs varied over time and were interrelated to a greater or lesser degree. They included intensification practices: manuring, land preparation, ploughing in organic material; and land improvement practices: removing tree stumps, hedge planting, restocking woods and the extension of new crops – first cotton and then maize. Intensification practices involved the use of seeds, fertilizers and agricultural equipment provided on credit through the cooperatives. For extension purposes practices were organized into 'The Ten Commandments for Land Improvement', relating to either fields or farms.

- Land improvement (LI)
  Preconditions
  1. Plots grouped, field cleared of stumps, field treated with phosphate.
  2. Field capable of being put under LI.
  3. Farm capable of being put under LI.
- For a farm within LI
  Equipment
  4. Heavy equipment.
  5. Bovine traction.

  Annual rules
  6. Seeds of improved varieties.
  7. Heavy use of manure.
  8. Ploughing at the end of the cultivation cycle.

  Applied over several years
  9. Rotation.
  10. Tree planting.

The Project Mechanisms

Mechanisms evolved over time in the light of changing content and the human and financial resources available, but always included an extension and a research programme. The extension programme role was to popularize techniques among the peasant farmers and to organize supplies. Weak management in the cooperative system resulted in an extension service preoccupied with ensuring inputs were available. In each unit, there was a senior agricultural (or livestock) technician, assisted by agricultural or livestock deputies. These were the field staff working locally alongside the peasant farmers. Initially these staff were recruited from the Millet–Peanut Productivity Project, the same cadres who had initiated and implemented the cattle traction project. This gave continuity in the approaches used with the farmers, including group demonstrations in the field, tests, home visits and training sessions. Originally, the units were headed by the professional in charge of technology transfer at the Bambey CNRA who had been responsible for the multilocational experiments. The Project rapidly became more individual and autonomous after management was transferred to Kaolack in 1973.

The research programme supported researchers working full time on the project with their own staff. Until 1978 this included a socioeconomic unit, responsible for analysing farm performance. In 1971–72 an anthropologist joined the team to work on the social structures of the Wolof population of the Salum. A land organization unit operated from 1974 to 1977 and in 1978 a land improvement and conservation research unit was created.

A range of agricultural researchers were involved in a complementary way on a more *ad hoc* basis, using resources provided by the project. These were researchers from both inside and outside Senegalese agricultural circles. Trainees also worked on defined aspects within the framework of the scientific problems identified by the permanent team. Various Senegalese and foreign organizations were called in to address particular technical issues in soils mapping, agrostology, parasitology, demography and nutrition. Outside support was also sought in activities such as cooperation, literacy and welfare but proved unsatisfactory in these more general areas. Any successes were due to the initiative of motivated individuals, but it was practically impossible to sustain the interest of the institutions involved and there were no significant long-term results.

Research carried out in the EUs

There were four main areas of research:

- Work on understanding the physical and human environment, including soils, nutrition and health, agrostology and Wolof ethnology.
- Measurement and analysis of the agronomic and socioeconomic efficacy and feasibility of the agricultural techniques extended to farmers.
- Studies of traditional livestock breeding and its improvement.
- A variety of socioeconomic studies, including how the agricultural farm operates, extension methods, the cereal deficit of the region and women in development.

Researchers also pursued other activities that had been included in their original proposals aimed at enabling the adoption and dissemination of innovations. These included work with the cooperatives on accounting records and on their more democratic organization. This meant setting up grass-roots groups, broadening the cooperatives' activities into marketing cereals, retail and spare parts stores, setting up a livestock breeding section, recruiting a permanent store-keeper responsible for weighing, and teaching farmers to read and write. Despite the goodwill and determination of the project extension workers these activities had few tangible results, though their efforts brought a better understanding of the problems of agricultural development.

Research also had to provide a wide range of services for the units, beyond those in their original proposals:

- The sale of spare parts, agricultural tools, seeds and chemicals.
- The sale on credit of particular pieces of agricultural equipment.
- The construction on credit of drying screens and silos to store cereals.
- Setting up and managing a mechanization unit for threshing cereals and separating maize grains on the farm.
- The supply of veterinary services.

With the breakdown of the national supply organizations, these activities absorbed a great deal of staff energy. Activities to promote women were also initiated, which included reducing their agricultural workload, diversifying their sources of revenue, improving child nutrition and home-economics training. The only real success was a scheme for off-season market gardening in the Kumbija Unit – thanks to the income it generated.

Two major extension initiatives were launched in 1977; a management committee which met several times a year, and the organization of peasant farmers into producer groups. Neither was able to mobilize the resources required to sustain them, perhaps as a result of an overall failure of the system supplying the rural areas in 1981. Further initiatives in land management including reorganizing the plot system, correcting boundaries, exchanging fields and partial reallocations were initially also seen as parallel activities. As a result of the difficulties encountered these all became action research activities with the scientist responsible simultaneously researching into the land organization system and carrying out reallocations. This played a fundamental part in advancing knowledge and understanding.

6.2.3 Evaluations of the EU project

The EU project was evaluated many times, the last one in June 1982. Summarizing the conclusions is difficult and, to improve coherence, the conclusions of evaluations considering the EUs as a development project are summarized first, then the conclusions from evaluations considering it as a research project, and finally evaluations from those treating it as a research-development project.

EUs as a development project

The impact of the EU project can be discussed in terms of the dissemination of its technical proposals. The dissemination of crop intensification under the rubric of LI, as defined above, can be evaluated by a 21% increase in the area under LI and a 40% increase in farms with LI. Cattle traction is not exclusively restricted to those farms under LI, but these are the majority of users. At the regional level, farms are also classified as using bovine traction, and those using bovine traction and high dose manuring. Here the rate of increase is probably similar to that within the units but with the EUs showing a higher level of adoption. Clearly the fact that the EUs have faster population growth and a greater extension of cultivated area than the region as a whole has contributed to this relatively rapid progress.

The surface area under cultivation has increased over the period and the range of crops has been revolutionized. The area under groundnuts and cotton as cash crops has remained steady with a significant increase in the cultivated area of cereals through the introduction of maize and the extension of the area of millet. These changes correspond to changes in rotation practice, including less use of annual fallows. The changes in crop rotation combined with increased outputs due to the LI strategy, among other factors, undoubtedly had economic repercussions. Regional revenues have increased though per capita cash incomes have remained stable. The cereal situation has changed radically – from being barely self-sufficient the area is in regular surplus. These aggregates of course mask wide disparities between farms, a factor of continuing interest to researchers.

EUs as a research project

Most of the research carried out in the EUs has been on socioeconomic issues and the project has certainly contributed to the advancement of methodologies and knowledge. This has improved our understanding of the peasant farmer's way of thinking and our ability to monitor and evaluate projects. Though (except in its final years) no agronomic research was carried out in the EUs, the project has made a major contribution to a scientific understanding of peasant-farmer management. It replaces the traditional 'impressionist' perceptions of small farmers' practices. In both agronomy and socioeconomics the improved understanding has contributed directly or indirectly to the better training of researchers and of developers.

EUs as a research-development project

According to its initial objectives the project could be seen as typical research and development, and these early objectives can be considered to have been achieved. The constraints and motivations of the peasant farmers have been understood and intensive agriculture has been promoted in the units. Advice given to farmers has been structured into 'management sequences' allowing a progression towards more intensive systems. Real potential has been measured by the outputs experienced over different types of year and the varying management capabilities of the producers.

The varied impact of the approach is illustrated by three cases:

- The non-dissemination of end-of-cycle ploughing to bury stubble.
- The absence of veterinary research results.
- The successful introduction of maize.

End-of-cycle ploughing

This was the central pillar of the proposed intensification scheme. It included the turning of the soil at the end of the crop cycle, returning minerals to it and adding organic manure. It was presented as a take it or leave it package. Attempts to popularize the technique failed mainly as a result of constraints on the way the work was organized on the farm. Experience showed that this 'best technique' was not the *only* technique and that, to achieve acceptance, diversity of practice and flexibility in implementation were required. The experience led to a reorientation of station experiments and to other work being transferred to farmers' fields.

Veterinary research results

The veterinary work was not successful. It suffered from the lack of a foundation of experimental information. For example there was no data on food rations made up from local sources. All rations tested earlier were made up from industrial by-products, none of which were available to the small farmers involved.

Maize cultivation

The introduction of maize varieties and husbandry was one of the main successes of the project. Maize was a new crop in the units and all aspects were covered, including cooking and marketing. Its success resulted in maize being adopted as a main theme in the regional development programme and the staff of the units project collaborated closely in project planning and implementation.

The links between research and development in the EU project can be said to be excellent at the field level but increasingly tenuous as one moves higher up the levels in the respective hierarchies of the institutions involved. This demonstrates the preoccupation of the promoters of the EUs who were seeking direct interactions with peasant farmers. But development is not solely agricultural and cannot be reduced to 'projects' alone. The EU project has involved a certain number of development agencies: the cooperative, credit, marketing, welfare and land improvement agencies, right up to the Ministries of Agriculture and Planning. Generally speaking, in developing countries such services, like many projects, are preoccupied by the sheer difficulty of operating, rather than being tightly focused on national development goals.

REFERENCES

1. **Brochier, J., 1968**. La diffusion du progrès technique en milieu rurale Sénégalais. Etudes tiersmonde. Presses Universitaires de France, Paris.
2. **Tourte, R., 1968**. Recherche et Vulgarisation. IRAT, Sénégal.
3. **Tourte, R., 1968**. (ibid.).

6.3 TWENTY YEARS OF SYSTEMS RESEARCH IN SOUTHERN MALI – THE SIKASSO FSR EXPERIENCE

W.A. Stoop, D. Kébé, O. Niangado and T. Defoer.

Problems connected with variations in conditions cannot be solved without the active participation of the farmers. Action research approaches and participative training processes will be required if farmer participation in research and extension activities is to become more effective.

6.3.1 Summary and introduction

The agricultural situation across southern Mali (Fig. 6.3.1) is often presented as uniform, justifying a generalized research agenda and giving the impression that there exists a specific 'sustainable system', which could be achieved through the application of a standardized technological package. This is a rather distorted view of rural reality. There is a great deal of local diversity and agriculture is very varied and complex throughout this southern area. Added to geographical diversity, production systems are changing fast in response to increasing population, and are in a continuous state of flux.

If a research programme is to assist rural development, it must take into account this diversity and these on-going changes. The research programme of the Sikasso Production Systems and Natural Resource Management (PSNRM) team of the Institut d'Economie Rurale (IER) has been taken as an example. The programme covers most of southern Mali, including the Guinean zone in the south, a transitional zone and the Sudanian zone in the north. The team works in research villages that are grouped into four subzones: Bougouni and Kadiolo, representing the northern Guinean zone; Sikasso, the transition zone; and Koutiala, the southern Sudanian zone.

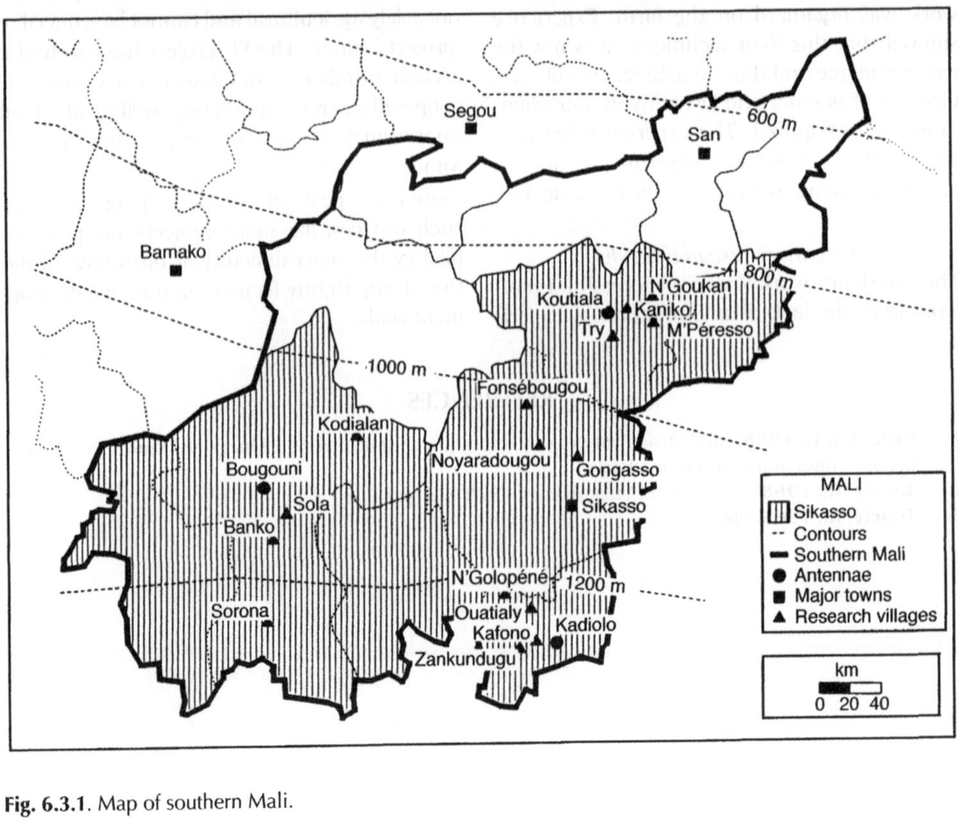

Fig. 6.3.1. Map of southern Mali.

The team works in various subzones and at various levels: regional, village (or group of villages), farm, plot and herd. Its research priorities were determined on the basis of the urgency of problems as expressed by end-users, relevance to the Institute's long-term strategic plan and complementarity between research topics. From these priorities a portfolio of research projects was formulated. The results from this programme in the form of technologies, methods and information have provided support to rural development in southern Mali and to the national programme, through a wide range of publications and other communications.

6.3.2 Agricultural dynamics in southern Mali

Agriculture and rural development have evolved rapidly in southern Mali since cotton was introduced as a cash crop in the 1950s. The high rate of population growth has also had a fundamental impact on land occupation. This growth, combined with increases in cattle herds and the development of the increasingly monetarized rural economy, has had far-reaching effects. With less farmland available per capita, agricultural systems will tend to become settled[1]. If these are to be sustainable, small farmers have to intensify their practices, for example through a closer integration of livestock and farming activities such as increased recycling of manure and the use of mineral fertilizer and pesticides. As long as there is a surplus of useful land available, farmers tend to be unwilling to intensify[2], preferring to occupy remaining virgin areas first and concentrate on mechanization. However, the maintenance of soil productivity requires new investments that are, in turn, dependent on the presence of markets. Although these initially lead to increased yields, this change to intensification can lead to serious degradation of natural resources.

The changes currently taking place can be divided into various phases, all of which are found in different parts of southern Mali:

- Phase 1: Extensification – increased production through expansion of the cultivated area (shifting system); the recommended quantities of mineral fertilizer are used, but only on cotton as in the Kadiolo and Bougouni areas of the northern Guinean zone.
- Phase 2: Transition – the main phase for land degradation. Farmers have not yet mastered the new land and water management techniques and no effective farmer organizations exist. In this phase, a first awareness of the need for soil fertility management on individual farms appears, as in the Sikasso area of the intermediate zone.
- Phase 3: Intensification – higher farm production through increased yields per hectare in a permanent system as in the Koutiala area of the southern Sudanian zone. This is the result of increased and improved production of organic manure, widespread use of mineral fertilizer not only on cotton, but also on cereals (in particular maize), and erosion control measures.

The adoption of these techniques by farmers is related to local population density, the availability of arable land, and the predominant soil types occurring along the toposequence[3]. Farms in phases 1 and 2 probably account for 80–90% of the total in southern Mali. The transition to commercial farming also means that production systems are more readily influenced by external economic conditions (for example devaluation of the CFA franc in 1995), that are beyond the control of small farmers. In response the rural population will start to realize that it needs to be organized if it is to cope with such outside factors. As population density increases the farming community will be confronted by complex problems such as conflicts between ethnic groups, agriculturalists and herders, neighbouring villages, and between urban and rural communities. A need for new management systems for communal natural resources (such as pastureland and woodland) will evolve as well as new rules, laws and/or organizations to ensure a sustainable use of these resources. All these developments must, however, take place within a public, legislative framework. The changes in agricultural systems as described above can be examined in two ways:

- In space, by comparing different subzones representing different systems.

- In time, by analysing changes in conditions and farming methods for a specific subzone.

6.3.3 Variability in the rural environment

Southern Mali has a wide range of environments. The Guinean zones of the south are relatively wet, while the Sudanian zones of the north are drier. Within these zones large soil-type variations occur along the toposequences. Socioeconomic differences between farm units are also considerable, due to different levels of management and also because of varying farm sizes[4]. The inhabitants of the northern Guinean zone, and to a lesser extent those of the intermediate zone, will have different views on the priorities for agricultural development and on the question of sustainability. Therefore, the needs will vary from one subzone to another and planned interventions must fit these varied needs to be acceptable to the local communities.

An appreciation of the diversity in the rural environment and the dynamics in the various farming systems is provided by the Monitoring and Evaluation Programme (MEP) data, an important information source for the formulation of the Sikasso team's research programme, contributing greatly to the definition of research priorities to meet the needs of farmers from different subzones.

6.3.4 Institutional framework for systems research: regional agricultural research institutes

In 1991 the IER was reorganized through the merger of two research institutes, and the regionalization of their activities through six regional agricultural research centres (RARCs). While regionalization is the only way to deal effectively with the many existing agricultural and rural development problems, the decision to regionalize (and hence decentralize) has fundamental implications for the management of the IER and its research programmes.

Effective agricultural research – research in support of local small farmers – must cover two domains: the natural resources of the zone or subzones; and the present production systems. These are interdependent and directly linked to the agroecological conditions of the zone in question. Any production systems research has to include, therefore, a component of natural resource management research[5] and research priorities must be specific to each (sub-) zone.

Regionalization also means that researchers can make more direct contact with the end-users, working with them to gain a better appreciation of the conditions, potentials, needs and constraints of the rural environments. IER took two major initiatives in this respect:

- The establishment of a PSNRM team within each regional agricultural research centre.
- The creation of regional users commissions.

As an interdisciplinary unit combining socio-economic and biotechnical disciplines, each PSNRM team has to play a key role in its regional agricultural research centre by liaising with the different partners of research. The teams intervene at all levels from the plot to the farm and village to the region. A degree of overlap between the natural resource management, production systems and thematic research activities of subject-specific programmes is vital to ensure effective communication between the various research groups.

6.3.5 The case of the Sikasso PSNRM team

The Sikasso PSNRM team is often seen as an example for teams in Mali and elsewhere. However, it should be remembered that its experience was built up over a period of nearly 20 years, and with considerable foreign technical and financial assistance. This allowed the development of a broad interdisciplinary team with solid financial and human resources.

Programme development and its contribution to agricultural development in southern Mali

In the early days between 1975 and 1985, FSR was seen as a quick and easy way of linking disciplinary research activities to the practical problems of small farmers. It became clear, however, that systems research is neither simple nor fast, because of the wide variations and the many types of interaction that occur in the 'small-farmer' environment. Systems research had to evolve in order to cope with this complexity and diversity, and to find out how its results could be exploited. This involved building up four key aspects:

- Interdisciplinary approach.
- Participation of small farmers.

Table 6.3.1. Organization of the Sikasso PSNRM team's research programme (1996).

Levels	Main research sector	Research projects
Region/(sub-) zone	Basic data	Ongoing monitoring and evaluation
	Rural economics	Rural financial systems
Village land	Village-based land management	• Village-based land management approach • Analysis & planning for agroforestry zone • Improved management of wood • Improved management of natural pasture land
Farm	Farm management	• Improved productivity of lowlands • Farm management advice • Improved management of cereal stocks • Labour associations • Role of women in cotton
	Intensification of animal husbandry systems	• Improved management of cattle herds • Improved goat breeding
Plot	Intensification of cropping systems	• Soils in the toposequence • Improved soil fertility management • Improved maize productivity • Integrated striga control • Karite and locust-bean plantings

- Systems perspective, so that constraints and their solution, as well as the various types of interactions, can be seen in a comprehensive context.
- A database (see MEP) for each main zone (or subzone) including variations in natural resources, technical levels of agriculture and stages of agricultural development.

Systems research activities at Sikasso began in 1978 when the Division de la Recherche sur les Systemes de Production Rurale (DRSPR) project initiated socioeconomic studies in the intermediate Sikasso zone[6]. The research programme and its organization were gradually expanded from case studies on various types of farm units, to the management of all the land belonging to a village territory, as well as to subregional problems. Although the team collaborated closely with the Compagnie Maliene de Developpement des Textiles (CMDT) from the outset, its structural relations with subject-specific research remained relatively weak. Only after the establishment of the Sikasso Regional Agricultural Research Centre in 1991 was the collaboration with commodity research programmes strengthened.

Over the years, the project developed a comprehensive framework for activities to clarify the complexities of the farming situation. Four interdependent intervention levels were recognized (Table 6.3.1): the region, with its agricultural and environmental zones; the village (or group of villages); the farm; and the plot or herd. This framework contributed much to the systematic identification of the complex problems and interactions that the production systems in southern Mali have to face.

The recent programme of the Sikasso PSNRM team

The recent programme is very comprehensive, covering rural development problems for the four intervention levels and for different types of systems, from shifting to permanently settled. It is organized into six main sectors and 18 long-term research projects (Table 6.3.1). The target group and the specific nature of problems and actions change, depending on intervention level (Table 6.3.2). It also follows that different work methods are required at each of the intervention levels.

The 'village-based land management' research sector: content and approaches

After the first years of the DRSPR project, scientists realized that a whole series of problems required solutions at community level[7], apart from the production constraints on individual farms. If soil degradation is to be halted, management systems have to be developed other than those traditionally practised such as bush fires, shifting cultivation, stray grazing of livestock and the uncontrolled gathering and cutting of firewood. This cannot happen without parallel changes in the local rules that guide the access to and control of resources, particularly land.

It became clear that technical interventions would fail unless they were introduced in consultation with the local population. Villagers therefore needed to organize themselves, and to participate in the formulation of local rules and regulations for farming operations, including a system of monitoring, protection and sanctions. This had to happen in the context of the national legal framework.

Through joint analysis of the agroforestry zone, the PSNRM team and field staff of development agencies described the present land use of a particular subzone and its major constraints[8]. Joint analysis of the Koutiala subzone revealed that the degradation problems were caused to a large extent by uncontrolled grazing – notably by the herds of outsiders – and by woodcutting to meet urban needs. Villagers had to deal with overexploitation of their village territory, without the institutional authority to enforce sustainable management. This led to a trial programme in the Koutiala zone, which drew up new local rules to regulate the use of village natural resources[9].

It follows that technical studies on the exploitation of various wood species and their regeneration will be useful only if these are carried out in a systems perspective. The population has to be involved in order to ensure that results are adapted to local conditions and needs. This implies that the cropping calendar and its labour demands must be taken into account as well as existing socioeconomic and legal conditions. The introduction of sustainable production systems will be a complex process that will evolve gradually as the local population assumes greater responsibility. Participative approaches are vital, together with the establishment of discussion groups or mixed groups of small farmers and scientists[10]. The most helpful instruments for mobilizing these groups will be village territory mapping, the transect and the cropping calendar[11], which can also provide links with other research activities, particularly cropping systems intensification, focusing on individual farms and plots.

The 'cropping systems intensification' research sector

This sector was largely reformulated in 1992[12]. Six long-term research projects (Table 6.3.3) were chosen on the basis of their relevance to producers, potential for short-term practical solutions and complementarity with thematic research and other research levels of the PSNRM team, including village-based land

Table 6.3.2. Interrelations between intervention level, main target group and type of problems encountered.

Intervention level	Main target group	Main problems
Region/(sub-) zone	General population	• Socioeconomic • Judicial • Institutional
Village area	Village community	• Socioeconomic • Local organizations • Biophysical and technical
Farm	Family	• Biophysical • Technical • Socioeconomic
Plot	Individuals/family	• Biophysical • Technical • Socioeconomic

Table 6.3.3. The 'cropping systems intensification' research sector: priority projects based on needs and conditions in three zones.

	Zone		
Research project	Southern	Sudanian intermediate	Northern Guinean
Soils in the toposequence	XX	XX	XX
Karite and locust–bean plantations	XX	X	X
Improved soil fertility management	XX	XX	X
Improved maize productivity	XX	X	XX
Fodder crops	XX	X	–
Integrated striga control	X	X	XX

XX, first priority for the subzone (based on actual farming system, the relative importance of a crop, or the urgency of a problem).
X, secondary priority.
–, not a priority.

management. While earlier fieldwork with replicated trials rarely gave significant results because of the large variability within farmers' fields, these projects use more participatory methods for farmer tests and surveys. Informal surveys, farmer maps and hierarchization of production criteria are seen by small farmers as complementing the various trials and farmer tests[13], creating a comprehensive research programme with the capacity to respond to the main questions of end-users. The programme for this sector, for example, included soils in the toposequence with a view to their potential and actual use. This relates directly to other subjects including agroforestry, fertility managements, striga control and maize systems, as well as interactions with village-based land management. Research projects on the dynamics of karite and locust-bean plantings and the use of swamp lowlands for rice were recently added, on the basis of exploratory analysis of the Bougouni subzone[14]. Actual farming conditions in the various (sub-) zones are taken as the starting point for action by the research projects, so that the results are adapted to farmers' needs and the development level of their individual farms. Table 6.3.3 gives a provisional indication of the research priorities as related to the different subzones.

6.3.6 A future orientation for systems research? 'action research for diverse and variable rural environments'

There is at present considerable disillusionment among donors about the impact of systems research in general (particularly FSR). With this in mind, PSNRM teams are advised to place greater emphasis on 'action research', rather than the general studies and large-scale surveys often conducted in the past.

Constraints include the fact that the strategy and roles of PSNRM teams may be clear, but the execution of their programmes is complicated. Moreover, most available scientists are trained as disciplinary specialists and often have limited experience with cropping/farming systems and the many internal and external factors that affect them. The diversity and variability in field conditions can make it hard for research to develop broadly valid technological packages and agricultural recommendations. Likewise, the many types of interactions at the farm level often make it difficult for PSNRM teams to formulate their feedback to thematic research, and their recommendations to development and political decision makers, in concrete terms.

The history of stagnant yields and small farmers' failure to adopt recommended farming techniques suggest that the techniques in Mali are often unsuitable, or have been offered at the wrong time for farmer's current needs and their level of development. Research and extension work must become more aware of differences between subzones and the specific needs of their farmers.

Problems connected with variations in conditions cannot be solved without the active participation of the farmers. Action research approaches and participative training processes will be required if farmer participation in research and extension activities is to become more effective[15]. There is still a great deal to be done, particularly in the training of officers from the various services, as well as the members of the regional users commissions. Nevertheless, an assessment of 18 years of research at Fonsebougou has highlighted the considerable progress made and its impact on agricultural development[16]. Even if the advances cannot be attributed solely to research activities, the latter have clearly had a fundamental influence on extension techniques, the expertise and information available, and technologies used:

- Knowledge about farming in southern Mali has improved greatly, as well as farmers' ability to manage their farms and plan their work.
- A wide range of development actions have been introduced, such as erosion control, fertility management, animal husbandry intensification, management counselling, mechanization and loans for the minimum farm equipment requirement.
- The approaches and methods have evolved:
  – the focus has shifted from studies of individual farm units to studies involving the entire village area, including village-based land management and management of natural resources;
  – the perspectives of individual farmers, farmers' groups and women have been introduced.

Problems of diversity can be tackled more realistically in the future if research proposes a range of solutions from which the small farmer can choose. This may mean an additional task for extension workers: as well as organizing the simple transfer of standardized techniques, they will have to advise farmers on the best choice given the specific conditions of individual farms. Extension should therefore encourage farmers to conduct their own informal trials, in addition to training on specific subjects. Training materials – for example on such priority subjects as striga control and fertility maintenance – will be one of the products of a PSNRM team.

REFERENCES

1. **Boserup, E., 1965.** The Conditions of Agricultural Growth: the Economics of Agrarian Change Under Population Pressure. Aldine, Chicago, IL.
2. **Binswanger, M.P. & P. Pingali, 1986.** Resource endowments, farming systems and technological priorities for sub-Saharan Africa. The World Bank, Agricultural and Rural Development Department, Research Unit Report no. ARU 60, Washington, DC.
3. **Hart, D., W. Brinkman & W.A. Stoop, 1995.** 'The toposequence concept'. Methods for linking partners in on-farm research for rural development. Background paper for IER/KIT workshop in Sikasso, Mali, KIT, Amsterdam.
4. **Brons, J., S. Diarra, I. Dembele, C. Bagayoko & Djouara, 1994.** Diversite de gestion de l'exploitation agricole. Etude sur les facteurs d'intensification agricole au Mali-Sud. Document ESPGRN 94/33. ESPGRN, Sikasso, Mali.

   Jansen, L. & S. Diarra, 1990. Mali-Sud vu Superficiellement. KIT, Amsterdam.
5. **Joldersma, R., A. Blokland, O. Niangado, S. Ouedraogo, B. Teme & W.A. Stoop, 1994.** L'integration de la recherche-systeme et la gestion de ressources naturelles: un mariage de bon sens. Proceedings International Symposium, Recherches-Systeme en Agriculture et Developpement, Montpellier.
6. **Kleene, P., B. Sanogo & G. Viestra, 1989.** A partir de Fonsebougou. Systemes de production au Mali, Volume 1. IER/KIT, Bamako-Amsterdam.
7. **Diarra, S., T. Hilhorst & N. Coulibaly, 1994.** Ver un gestion durable de la zone sylvo-pastorale: l'approche du DRSPR au Mali-Sud. Contribution ESPGRN-Sikasso au Symposium International Recherche-Systeme, Montpellier.
8. **Kone, Y., N. Coulibaly, I. Dembele & K. Bengaly, 1994.** Diagnostic exploratoire de la region CMDT-Bougouni. Document 94/15. IER-ESPGRN-Sikasso.
9. **Diarra, S., T. Hilhorst & N. Coulibaly, 1994.** (Op. cit.).
10. **Norman, D., 1995.** Mission d'appui en renforcement de l'implication des paysans dans la Recherche-Systeme. ESPGRN, Sikasso.
11. **Defoer, T., S. Kante, T. Hilhorst & H. De Groote, 1995.** Towards more sustainable soil fertility management. Paper presented in support of 'Nutrient Cycling Project', Farm Africa/IIED. IER-ESPGRN, Sikasso.

12. **Stoop, W.A., 1992**. Rapporte d'une mission d'appuiau projet DRSPR-Sikasso au Mali-Sud. KIT, Amsterdam.
13. **Kamara, A. & T. Defoer, 1995**. Experimentation paysanne: un defi professionel. Document ESPGRN 95/1. ESPGRN, Sikasso, Mali.
14. **Kone, Y., N. Coulibaly, I. Dembele & K. Bengaly, 1994**. (Op. cit.).
15. **Defoer, T., S. Kante, T. Hilhorst & H. De Groote, 1995**. (Op. cit.).
16. **Kone, A.F. & T. Vershelde, 1995**. Bilan des activities de recherche a Fonsebougou: 18 ans de recherche-systeme en Zone Mali-Sud. Document 95/06. IER-ESPGRN, Sikasso.

Blokland, A., M. Goita & R. Joldersma (Eds), 1994. La recherche-systeme au Mali: Une Bilan. IER/KIT, Bamako/Amsterdam.

6.4 THE INSTITUTIONALIZATION OF FSR IN EAST AND SOUTHERN AFRICA: AN OVERVIEW

S.A. Kean and M. Creasy Ndiyoi

The long-term sustainability of institutionalizing FSR is seriously threatened by the decline in donor funding which has placed additional burdens on already overstretched government coffers.

6.4.1 Introduction

The institutionalization of FSR has received much less attention than issues about FSR methods, farmer participation, indigenous knowledge, local organizations and technical interventions. It is these subjects, for example, that have dominated AFSRE international symposia. There have been important studies of the institutional implications of FSR, notably ISNAR's nine country comparative by ISNAR, which examined on-farm client-oriented research[1], and a number of individual country studies[2]. This contribution offers an overview of the institutionalization of FSR in east and southern African NARIs after more than 20 years of effort by a number of stakeholders with a diverse range of interests.

6.4.2 Institutionalization of FSR in NARIs in East and southern Africa

FSR is a problem-solving, systems-oriented, interdisciplinary process which promotes farmer participation and OFE. By gaining an understanding of small-scale farmers, FSR can play several roles: to help formulate research priorities, to design appropriate technology, to focus extension work on areas ripe for change, to design more relevant projects and to create better policies. To fulfil these roles it needs effective links with OSR, extension, development agencies, planners and policy makers[3].

Strengthening NARIs has been seen as an important achievement of FSR[4]. In East and southern Africa FSR has increased awareness of the circumstances and needs of resource-poor farmers among biological scientists. It has provided scientists with increased opportunities to interact with farmers, particularly through on-farm trials. It has provided social scientists with a role in the technology generation process. It has also improved research priority setting and targeting of research in some countries, although this has generally been less successful. In some, FSR has served as a bridge from research to extension organizations and has improved collaboration between them. However, it is in this area of institutionalization where serious difficulties remain, notably FSR's generally weak links with disciplinary and commodity research, extension, policy makers and other development services[5].

Perhaps the most important indication of some success in institutionalizing FSR is the fact that the debate about FSR's key functions is now centre-stage in most NARIs. In many, including those with separate FSR sections, there is now virtually unanimous agreement, at least in research policy rhetoric, that all scientists must conduct research aimed at solving farmers' problems. For example, Uganda's policy statement of 1991 declared that 'prioritisation of research activities will be guided first by client specifications'[6]. Some, such as Swaziland, state that priority must be given to adaptive OFR[7] and others, including Zambia, hold that all scientists should be working together in interdisciplinary problem-focused development teams[8].

6.4.3 Country histories

The following overview of experiences in institutionalizing FSR in East and southern Africa outlines progress in eight countries of the region which have used different structural options and have had varying levels of external support.

Botswana

The influence of donor stakeholders has been evident for many years in Botswana's FSR activities, beginning in 1969 with the donor-funded Dryland Farming Research Scheme. In the 1980s two more donor-funded FSR projects began in other areas but all three of these projects were located in different government departments. There was some informal coordination through the initiatives of individuals, but no formal channels of communication with the Department of Agricultural Research (DAR)[9]. Only in 1992 was DAR restructured in a way that formalized relationships across the FSR programmes, and brought them into a Production Systems Programme (PSP) under a single leader. While all had their representatives on the PSP committee after the restructuring, each of four teams operated independently in their regions, having their own administrative structures and linkages with various other agencies. Moving towards more standardized team composition and research approach still remains a challenge for the PSP. With the reduction in donor funding DAR also faces the challenge of ensuring the long-term financial sustainability of FSR.

Ethiopia

The institutionalization of FSR in Ethiopia has proceeded in a relatively organized manner as part of the Institute of Agricultural Research (IAR). IAR initiated an OFR programme in 1976 through its Department of Agricultural Economics (DEA) to demonstrate available technologies in farming communities. In the following years, multidisciplinary surveys were undertaken to identify farmers' circumstances, formulating a package of technologies for on-farm testing. From 1984 onwards, diagnostic surveys were used to identify farmers' production constraints and look for potential solutions[10]. IAR first implemented FSR by including different disciplines in the DEA, a decision reflected by the renaming of the department as the Department of Agricultural Economics and Farming Systems Research (DAEFSR). However, between 1984 and 1988 only two agronomists ever worked with up to 10 economists. Although FSR activities were successfully undertaken, this arrangement was found to be inefficient and created tensions between on-farm and on-station researchers and between DAEFSR and the Agronomy Department (AD)[11]. This alerted IAR management to the fact that once scientists from different departments had been exposed to the FSR approach it was possible to initiate interdepartmental research projects. It was no longer necessary to base the multidisciplinary FSR team within DAEFSR. The agronomists were transferred back to the AD and DAEFSR reverted to being the DAE. FSR is now a function of IAR as a whole rather than the preserve of a single department[12]. The critical issue in this institutional arrangement is one of coordination to ensure that scientists in the different sections work together. Several mechanisms for promoting interdisciplinary collaboration have been established including interdepartmental training courses in FSR, joint review of research proposals, annual planning meetings and a formal directive outlining the mode of collaboration. It has been pointed out that strong leadership in both departments was important for building effective FSR, as was the support of IAR's general manager. The combination of centralized directive management and IAR's relatively limited dependence on donor funding and expatriate staff has assisted the task of coordinating FSR activities. An important challenge for FSR institutionalization in IAR is how it responds to the opportunities and problems posed by the government's major programme of decentralization to the regions, which includes research.

Kenya

FSR has gone through several phases of institutionalization in Kenya[13], but two are of major importance. The first phase lasted from 1975 to 1990 and saw the introduction of FSR by the CIMMYT and the establishment of socioeconomics capacity in the Scientific Research Division (SRD) of the Ministry of Agriculture. Senior research managers remained largely unconvinced of the relevance of FSR, no decision was made as to how the process might be institutionalized, and FSR initiatives remained on the margins. CIMMYT was allowed to mentor newly recruited farm systems economists posted to the main research stations and to organize large in-country training programmes for research and extension staff. The economists undertook many diagnostic surveys, but attempts to develop collaborative programmes with research station agronomists were largely unsuccessful, not least because of scepticism amongst station directors. After 3 years, 10 of the 12 newly recruited economists had left SRD. For much of this time FSR was viewed as a role for economists[14]. Perceived as a separate discipline operating in isolation from mainstream research and extension, it did not fulfil expectations.

The second phase began in the late 1980s when some senior research and extension managers recommended steps for institutionalizing FSR in both research and extension systems. A workshop to harmonize FSR terminology and concepts took place in 1990, which produced national guidelines for a Farming Systems Approach to Research, Extension and Training (FSARET) for use by all Kenyan institutions undertaking FSR. Subsequent institutionalization has seen FSR as an approach, emphasizing the full range of FSR characteristics, and not as a separate discipline. This second phase has seen a proliferation of FSR activities in research, extension and training institutions and each of the many stakeholders has had considerable latitude in how it has been institutionalized.

Within Kenya's NARI, the Kenya Agricultural Research Institute (KARI), FSR activities are conducted within both the Regional Research Programmes (RRP) and the complementary commodity focused National Research Programmes (NRP). The interdisciplinary RRPs are planned as a comprehensive set of adaptive research activities on problems from each regional research centre's mandated area identified through the participatory diagnosis of farmer problems. RRP activities are being supported through the National Agricultural Research Project NARP II by ODA, the World Bank, The Netherlands, EU, USAID and the Rockefeller Foundation, each with different areas of emphasis. The NRPs aim to bring together all research activities related to a specific commodity or factor and their output is expected to feed into adaptive research. Some donors are also encouraging the NRPs they support to undertake their own adaptive research through NARP II, including pasture and fodder crops, maize, sorghum and crop protection. Beyond this a number of donor-funded projects are conducting FSR outside both the RRPs and NRPs, for example the ODA-funded Dryland Agricultural Research and Extension Project (DAREP). In addition, the Ministry of Agriculture and Livestock Development and Marketing is responsible for the District Farming Systems Teams that have been established in six pilot districts with support from the Swedish International Development Assistance (SIDA) and the Food and Agriculture Organization of the United Nations (FAO). To facilitate the process of institutionalizing the FSARET guidelines, three formal committees were established at national, regional and district levels: the National Farming Systems Coordinating Committee, Regional Research-Extension Advisory Committees and District Farming Systems Teams. Within KARI, adaptive research is guided by a Research Coordinating Committee and is the responsibility of an Assistant Director for RRCs supported by an Adaptive Research Coordinator. In spite of the various mechanisms for coordination, the potential for individual interpretation and emphases among the different stakeholders implementing FSR poses a considerable challenge for KARI to develop effective mechanisms for institutional learning.

Malawi

FSR was formally institutionalized in Malawi when the Adaptive Research Programme was established in 1983 to ensure that the DAR addressed smallholder farmers' priority problems. The decision to institutionalize FSR was suggested by a group of national scientists who had monitored CIMMYT's approaches.

Emphasis was given to adapting available technical research results, feeding unsolved technical problems back to commodity research teams and linking research closer to extension[15]. In 1985 DAR was reorganized into commodity and specialist groups which included the renamed Adaptive Research Teams (ART), each consisting of an agronomist and socioeconomist using the CIMMYT approach and methods. Although the ART National Coordinator was based in DAR and responsible for ART's technical programme, the teams became the administrative responsibility of the Agricultural Development Divisions in the NRDP, with a strong extension mandate. This split the responsibility, and NRDP's problems of coordination at ministerial and field levels[16] contributed to the marginalization of ART's activities. ART's unpopularity with commodity team scientists was apparent from the outset[17] and working relations had improved very little by the early 1990s[18]. Managers were reluctant to admit that they did not understand farmers' problems and the ARTs effectively became teams demonstrating conventional answers.

Since the late 1980s DAR's commitment to FSR has declined while its commitment to production-oriented commodity research has increased. The 1993 World Bank-funded Agricultural Services Project was designed to strengthen formal linkages between research, extension and farmers as well as conducting on-farm trials. It made no explicit mention, however, of ART. While the formal institutionalization of FSR in Malawi has declined, several commodity programmes, particularly those with donor support, have increasingly incorporated FSR characteristics into their research programmes, including on-farm trials, diagnostic and quantitative surveys and farmer participation[19].

Tanzania

The process of institutionalizing FSR in Tanzania had a long gestation period. CIMMYT first introduced the approach in 1976 and in 1982 a national policy commitment stated that research should be problem-oriented and adopt an FSR approach. Emphasis in the zonally based interdisciplinary FSR teams has been on the full range of FSR functions and characteristics. A national FSR coordinator was appointed in 1985 but it was only in 1989 that FSR was fully institutionalized with defined roles, its own structure, a budget, staff and a leader at assistant commissioner level in the Ministry of Agriculture's Department of Research and Training[20]. Since 1986 rhetoric has been matched by resource commitments as demonstrated by the growth in the number of FSR research staff from six in 1986/87 to 34 in 1994/95. Similarly, while donor financial support has waned, the government has continued to allocate funds for FSR, albeit in modest amounts. Interestingly, although research managers have clearly supported the institutionalization of FSR, most have not been proactive but have left many issues, including links with commodity scientists, up to the FSR section.

The present structure of FSR teams complementing commodity and disciplinary research is 'not felt to be enough, rather it is deemed important for all researchers to use a farming systems approach to their work, whether it is on-station or on-farm'[21]. The logic of this argument is that some research managers have tended to regard FSR as an approach and consider that the programme should be dissolved. The future institutionalization of FSR in Tanzania faces many challenges, particularly the programme's sustainability if funds remain limited, the coordination issue – especially if further integration into zonal research-at-large takes place – and low staff morale if poor staff conditions continue.

Uganda

FSR has never been formally institutionalized in Uganda. Interest in incorporating the characteristics of FSR into research programmes has, however, increased rapidly in recent years among a variety of stakeholders from the NARI and NGO sectors. Multilocational trials on farmers' fields dates as far back as the 1930s[22]. This research, however, focused principally on non-traditional cash crops and in 1985 Opio-Odongo described agricultural research in Uganda as having 'evolved with little or no consideration of the interests of local producers'[23]. From the late 1980s several commodity research programmes began to incorporate characteristics of FSR into their activities. The stakeholders with an interest in FSR included some of the first group of Ugandan scientists who had attended CIMMYT FSR training workshops in the early 1980s, and several donor agencies. The initiatives were regarded with some scepticism by officials within the Ministry of Agriculture, Animal Industry and Fisheries who criticized FSR's cost-effectiveness. In 1993 the NARI was restructured as a unified research organization and agreed to follow the 1991 Research Strategy which stressed that: 'prioritisation of research activities will be guided first be client specifications'. Its current commitment to FSR can be judged by its policy of on-farm testing. Farmer acceptance of new varieties and practices must take place before they are released or recommended. In addition, all NARI-funded research proposals are expected to contain an FSR component. To increase awareness and skills in FSR among NARI scientists, several workshops and training courses have been organized. NARI has followed the model of institutionalizing FSR whereby individual scientists combine both experiment station and OFR activities in their work programmes. The focal point for FSR activities are therefore numerous commodity, area and specialist programmes such as banana, root and tubers, beans, coffee and the Agricultural Development Program in eastern Uganda. Many of these activities are being undertaken as part of donor-funded projects and their approaches and methods vary widely. Despite management's professed commitment to FSR, there is concern about its sustainability. This and concern about institutional learning have prompted the government to establish a National OFR Coordination Committee.

Zambia

Many actors have played roles in the institutionalization of FSR in Zambia. In the mid 1970s some policy makers and research managers in Zambia's Ministry of Agriculture were concerned that resource-poor farmers were receiving only limited attention from the disciplinary focused research branch. FSR activities thus began in Zambia in 1978 when CIMMYT was invited to demonstrate its procedures for an interdisciplinary approach to agricultural research based on FSR methodology[24]. In 1980 the Adaptive Research Planning Team (ARPT) was established as a separate provincially based team in the research branch to focus on each province's farming systems. ARPT was designed to complement the work of the then recently formed multidisciplinary Commodity and Specialist Research Teams (CSRTs) which have commodity or factor foci and national mandates. ARPT was intended to emphasize the full range of FSR characteristics and over time even these have been broadened to include such activities as advocacy for small-scale farmers. By 1988, with considerable donor assistance, ARPT was working in eight of Zambia's nine provinces and the last provincial team became fully operational in Southern Province in 1995. Each provincial ARPT team comprises an economist, agronomist and research extension liaison officer (RELO). Originally, these were coordinated by a single ARPT national coordinator. In the early 1980s a study identified the need for sociologists to be included in the team as regionally based support staff. Similarly, adaptive livestock systems specialists have been included in some provincial ARPTs and a nutritionist at headquarters. In the early 1990s it was felt that the ARPT national coordinator should be assisted by a national support team to include an agronomist, economist, sociologist and information/RELO and nutritionist.

Although the team initially followed CIMMYT's concepts and methodology it has modified and expanded on these whenever appropriate, particularly in the areas of sociology, farmer participation, adaptive livestock research, sustainable farming methods, household food security, policy analysis, women in development, monitoring, evaluation and impact assessment. It has made considerable progress in developing effective collaboration with extension services, planners and policy makers. Institutionalizing FSR in a separate team has meant that much effort has been put into developing effective links with CSRT scientists and, although many mechanisms have been tried, it has been difficult to make sustained progress. The conflicts and jealousies which have beset this issue have posed a continual threat to ARPT's very existence. From the mid 1980s, as part of a World Bank initiated restructuring of the research branch, attempts were made to curtail the scope of ARPT's activities.

A critical review of ARPT in 1993 by SIDA, which has been one of ARPT's major funders, suggested a major restructuring of ARPT along regional lines and hence in 1994 ARPT became the Farming Systems Research Team (FSRT). These recent structural changes are also set against the background of a fundamental review of Zambian agricultural development as the Ministry of Agriculture has been reassessing its role in the light of the new government's liberalization and structural adjustment policies.

Zimbabwe

OFR was initiated in Zimbabwe's Department of Research and Specialist Services (DRSS) in 1980 when radical policy changes were made following independence. These required DRSS to expand its mandate from serving 4000 large-scale commercial farmers to address the needs of some 850,000 resource-poor peasant farm families. Within 7 years, nine of its 17 research institutes or stations had developed major OFR efforts and about 20% of research staff time was allocated to OFR. Institutionalization of FSR in DRSS has meant adding the function of OFR to research institutes and commodity teams. However, a small multidisciplinary FSR Unit was established in 1984, with donor support, to conduct crop and livestock systems research. Perhaps the most important change was the recognition among scientists that resource-poor farmers were an important and distinct client group requiring different technologies[25]. FSR activities were strengthened by the establishment of the Committee for On Farm Research and Extension which, among other things, was responsible for the coordination of joint priority setting, planning and review of on-farm operations in both research and extension[26]. However, in recent years the institutionalization of FSR has been threatened by the drastic reduction in operating funds forced on DRSS and the attitude that, as Shumba said in 1993, that: 'on-farm trials, often regarded as expensive "add-ons" to established station activities, were the first to be cut'[27]. This rather minimalist approach to institutionalizing FSR has involved only a relatively small number of stakeholders and donor agencies only to a limited extent.

6.4.4 A synthesis of regional experiences

The *ad hoc* institutionalization of FSR generally began in the region in the mid 1970s and became more formalized throughout the 1980s and 1990s. It remains difficult to say how successful FSR has been in any particular country and the regional experience shows the vagaries of the institutionalization process. In some cases there has been clear progress, particularly where key characteristics of FSR have been adopted by senior managers for use by all scientists working in a NARI as in Ethiopia, Tanzania and Zambia. Elsewhere the institutionalization process has regressed, as in Malawi, and, while commitment from management may have increased, declining resources may have meant stagnation, as in Botswana and Zambia. The review clearly indicates that the pace of institutionalization has been both variable and uncertain. This challenges the impression given by some that FSR has in some way been permanently institutionalized[28]. Support for FSR has risen and fallen as various stakeholders have come and gone and the history of institutionalization is littered with battles between fervent supporters, sceptics and critics as in Kenya, Uganda and Zambia.

Opposition to FSR has been greatest when it has been organized separately, and has been driven by various factors. FSR has been perceived as a criticism of past research efforts to meet the needs of small-scale farmers and there was jealousy at the level of donor resources it attracted in the 1980s. Some biological scientists were sceptical of the value of social scientists who were perceived to dominate FSR programmes, and it is true that FSR suffered from a lack of credibility because of the inexperience of many of its scientists, often junior recruits. The notion of successful institutionalization is itself highly subjective and the impression given in any evaluation or progress report can depend on whether the author supports or opposes FSR[29]. Critically, it is apparent that support for the FSR approach cannot be taken for granted, a point which becomes clearer the more one disaggregates the institutionalization process to understand the role played by different stakeholders as in Zambia.

It is clear that NARIs in the region have brought FSR into their organizational structures in different ways. At a general level it is possible to categorize the introductions into two broad models: specialized regional on-farm teams, and individual scientists combining both experiment station and OFR activities in their work programmes[30]. An attempt to categorize initiatives in Table 6.4.1 shows the NARIs split roughly equally between the two models. However, for most countries this dual categorization oversimplifies the situation. Most FSR organizational arrangements, like most NARIs, have undergone change and some have been in a permanent state of flux, such as Zambia. In some NARIs old FSR structures may exist in name but in practice are moribund, as in Malawi, or a new FSR structure may have been created while the old one continues to function under a new name, as in Zambia. A national structure may obscure the reality that donor projects with multiple objectives may determine what actually happens on the ground in, for example Botswana, Kenya and Zambia. In all of these cases the reality is far more diverse than a brief overview allows. Part of the problem is the tendency for many institutional analyses to emphasize theoretical structures and the rhetoric about what should be happening rather than relating the diversity of practice. The advantages and disadvantages of the two main models have been thoroughly discussed elsewhere, but in general this review finds that separate FSR sections tend to be marginalized, as happened in Malawi. It was also found that if the FSR approach is to be taken up by all scientists in a NARI, it is essential to have effective coordination actively supported by senior managers, as in Ethiopia, Kenya and Tanzania.

National coordination of FSR activities is now an important issue facing FSR managers in many of the NARIs reviewed. The purpose of coordination varies between different countries. In some NARIs, such as Botswana, it is seen as essential for introducing a more standardized team composition and approach, thereby reducing the diversity of approaches found within the previously donor-funded projects[31]. In Tanzania and Zambia, effective coordination between FSR scientists themselves is seen as important for promoting institutional learning about all aspects of their working practice. This type of coordination aids reflection, self-monitoring and evaluation for the purpose of programme improvement, and also helps to reduce duplication and repetition of mistakes. In some NARIs coordination is needed to facilitate the interdisciplinary function of FSR when scientists of different disciplines are not working in the same section, e.g. Ethiopia and Kenya. Coordination is also seen as important for ensuring that effective linkages are developed between FSR scientists and on-station scientists, extension workers and policy makers.

Experience from the region indicates that national coordination has been difficult to achieve, in spite of increasing awareness of its importance. Many NARIs have established official coordination mechanisms, with individuals and committees given responsibility for coordination, but these have not been very successful. It has been difficult to achieve intrasection coordination even where FSR is conducted in separate sections, as in Zambia. When the FSR team has straddled both research and extension as in Malawi, coordination of links with research has been problematic[32]. The tasks confronting countries where FSR functions are found within regional teams, commodity programmes and miscellaneous donor-funded programmes, as is the case in Kenya, Uganda and Zimbabwe, are quite enormous. The proposed integration of FSR activities into zonal research-at-large in Tanzania also poses serious challenges for national coordination. The difficulties of effectively coordinating the activities of numerous influential stakeholders are explored in more detail in the Zambian analysis in Chapter7.

Just as the organizational structure of FSR has varied in the different NARIs, so too has the emphasis given to the different FSR roles and characteristics. Most NARIs with separate FSR sections have emphasized a wider range of roles than those NARIs in which FSR functions and characteristics have been added to the mandates of scientists conducting on-station experiments (Table 6.4.1). Station-based scientists have tended to emphasize technology design and research priority setting and OFE in particular. Less emphasis has been given to farmer participation, use of a farming systems perspective and interdisciplinary research. Most of the NARIs with separate FSR sections have tried to follow the CIMMYT approach. Some such as those in Tanzania and Zambia, have developed it to include specific components for: livestock, nutrition, an interface with policy makers and even lobbying for small-scale farmers. Progress on institutionalizing these roles has not occurred in some theoretical or apolitical

Table 6.4.1. Progress with institutionalizing FSR in eight countries.

Country	FSR starts *ad hoc*	Full FSR institution starts	Current FSR structure; OF & OS: *same* or *separate* teams	Main FSR characteristics emphasized	Level of donor involvement: Past	Level of donor involvement: Current	Current institutional diversity
Botswana	1969	1992	Separate (*ad hoc*)	Variable: OFR feedback to OSR, linkages and farmer participation	High	Low	Moderate
Ethiopia	1976	1984	Was separate, now same	Standard CIMMYT methodology	Low	Low	High
Kenya	1975	1990	Same but mandate split between RRCs and NRCs	Variable from OFR to full range	Moderate	High	High
Malawi	NA	1983	Separate and *ad hoc* in commodity research	Standard CIMMYT methodology	High	Low	Moderate
Tanzania	1976	1982	Separate	Full range	Moderate	Low	Moderate/ high
Uganda	1930s	1991	Same	Variable: OFR to full range	Low	High	High
Zambia	1978	1980	Separate, now with regional mandate	Full range	High	Moderate	High
Zimbabwe	NA	1980	Both same and separate	OFR	Low	Low	Moderate

Notes:
1 'Full range' of FSR characteristics include: problem solving, system oriented, interdisciplinary, adaptive, on-farm, promoting farmer participation and with effective linkages to OSR, extension, development agencies, planners and policy makers[3].
2 OF, on-farm; OS, on-station; RRC, Regional Research Centre; NRC, National Research Centre (usually commodity mandated).

manner but has often reflected the interests of influential FSR stakeholders, sometimes evolving through processes of institutional learning.

This review concludes that external agencies have had a very significant impact on the institutionalization process. CIMMYT, in particular, played a key role in creating the initial interest in the concept and its training programmes have meant that the approach and methods continue to form the basis of much of the FSR in the region[33]. Many other agencies including IARCs, as well as numerous donor agencies, have influenced the approach, structure, speed, and methods by which FSR has been institutionalized. While donor assistance to FSR remains at a significant level in many of the countries reviewed, it has decreased in most of them in recent years (Table 6.4.1). This assistance has affected aspects of FSR in most countries including the approach taken, the characteristics and organization adopted, and the pace of institutionalization. Some external agencies have used almost missionary zeal to promote particular characteristics of the approach, for example farmer participation.

The long-term sustainability of institutionalizing FSR is seriously threatened by the decline in donor funding which has placed additional burdens on the already overstretched government coffers. In Tanzania, while the FSR programme has expanded from four FSR researchers in 1986 to over 30 today, funds per researcher have decreased and the sustainability of the programme in its present size is questionable[34], despite considerable government funding.

In Zimbabwe the drastic reduction in operating funds has forced DRSS to cut some programmes, especially those on-farm trials which were regarded by some as 'add-ons'. FSR in Botswana, which for many years was highly dependent on donors, has also come to depend on government contributions[35]. Since ARPT began in Zambia the government has rarely contributed more than 30% of the team's funds and its dependence on donors makes its survival, and indeed that of the entire research branch, quite precarious. In Kenya, Malawi and Uganda the funding of research activities, including FSR, is highly dependent on donor funding in the form of a multiplicity of small and large external research projects. In these countries, as in Zambia, scientists are being actively encouraged by research managers to look for funding, thereby decentralizing funding decisions[36]. Such an approach may well ensure the survival of FSR but it will, at the same time, increase the difficulties of national coordination.

REFERENCES

1. **Merrill-Sands, D., P. Ewell, S. Biggs & J. McAllister, 1989**. Issues in institutionalizing on-farm client-oriented research: a review of experiences from nine national agricultural research systems. Quarterly Journal of International Agriculture, 28 (3/4), 279–300.
2. **Stroud, A., 1995**. Institutionalization of FSR in Tanzania: Past, Present and Future. Paper presented at the Southern African Association FSR/E Conference, October, 1995.

   Franzel, S., 1992. Impact, institutionalization and methodology: research with farmers in Ethiopia. In: Franzel, S. & van Houten, H. (Eds) Research With Farmers: Lessons from Ethiopia. CAB International, Wallingford.

   Matata, J.B.W., Undated. Farming Systems Approach in Kenya: Historical Perspective and Current Status. Nairobi. (Mimeo).

   Heinrich, G. & E. Modiakgotla, 1992. Farming Systems Research in Botswana. Report no. 2 Farming Systems Programme, Gaborone.

   Baguma, S.D., 1993. The impact of on-farm research on government, donors, and other agencies in establishing agricultural development priorities: the Uganda experience. In: Heisey, P. & Waddington, S. (Eds) Impacts of On-Farm Research: Proceedings of a Networkshop on Impacts of On-Farm Research in East and Southern Africa, Harare. CIMMYT East and Southern Africa On-Farm Research Network Paper no. 24, Harare.

   Shumba, E., 1993. The future of on-farm research activities within national agricultural research systems. In: Heisey, P. & Waddington, S. (Eds) (op. cit.).
3. **Merrill-Sands, D., 1988**. Introduction to the ISNAR study on organization and management of on-farm client-oriented research. In: Kean, S.A. & Singogo (op. cit.).

   Tripp, R., P. Anandajayasekeram, D. Byerlee, & D. Harrington, 1990. Farming systems research revisited. In: Eicher, C.K. & Staatz, J.M. (Eds) Agricultural Development in the Third World. John Hopkins, Baltimore.
4. **Tripp, R., *et al.*** (Ibid.).
5. **Low, A.R.C., S.R. Waddington & E. Shumba, 1991**. On-farm research in southern Africa: the prospects for achieving greater impact. In: Tripp, R. (Ed.) Planned Change in Farming Systems: Progress in On-Farm Research. John Wiley & Sons, Chichester.
6. **Republic of Uganda, 1991**. National Agricultural Research Strategy and Plan: Volumes I and II. Uganda Working Group 9A, Agricultural Policy Committee. ISNAR, The Hague.
7. **Seubert, C., 1989**. On-Farm Research and Extension Linkages: Experience from Swaziland. CIMMYT Network Report no. 19. CIMMYT, Harare.
8. **NORAGRIC, 1993**. Evaluation of the SIDA Funded Adaptive Research Planning Teams, Zambia, 1993. Evaluation Report by Norwegian Centre for International Agricultural Development. Agricultural University of Norway, Norway.
9. **Heinrich, G. & E. Modiakgotla, 1992**. (Op. cit.).
10. **Seyoum, K., H. Beyene & W. Mwangi, 1993**. The impact of on-farm research in the Institute of Agricultural Research in Ethiopia. In: Heisey, P. & S. Waddington (Eds) (Op. cit.).

11. **Franzel, S., 1992.** (Op. cit.).
12. (Ibid.).
13. **Matata, J.B.W., undated** (Op. cit.).
14. (Ibid.).
15. **Nyirenda, F.M., 1985.** Farming Systems Research Approach: the Malawi Experience. Paper at a workshop on methodologies, practical approaches and potential contribution of farming systems research for rural development in Sub-Saharan Africa, held at Egerton College, Njoro, Kenya, August, 1985.
16. **Arnold, S. & C. Bryant, 1983.** Managing Coordination in Malawi's National Rural Development Program. Washington, DC, World Bank, Economic Development Institute, Case no. 035/014, May.
17. **Nyirenda, F.M., 1985.** (Op. cit.).
18. **Jones, R.B., 1993.** The development of technologies for farmers by traditional commodity research teams: the Malawi experience. In: Heisey, P. & S. Waddington (Eds) Impacts of On-Farm Research, Proceedings of a Networkshop on Impacts of On-Farm Research in Eastern and Southern Africa, Harare. CIMMYT, Eastern and Southern Africa, Network Paper no. 24, Harare.
19. (Ibid.).

    Kean, S.A. & M. Iles, 1995a. The Institutional Dimensions of Integrated Pest Management Research in Malawi. Unpublished Mimeo for IPM Working Group, Chatham.
20. **Semgalawe, Z.M. & J. Kaaya, 1992.** Farming Systems Research in Tanzania. Report no. 3 Farming Systems Programme, Gaborone.
21. **Stroud, A., 1995.** (Op. cit.).
22. **Baguma, S.D., 1993.** (Op. cit).
23. **Opio-Odongo, J.M.A., 1985.** Some Thoughts on Critical Entry Points for Farming Systems Research Perspective in Training of Agricultural Scientists. Paper presented at a workshop on methodologies, practical approaches and potential contribution of farming systems research for rural development in Sub-Saharan Africa, held at Egerton College, Njoro, Kenya, August, 1985.
24. **CIMMYT and Government of Zambia, 1978.** Demonstrations of an Interdisciplinary Approach to Planning Adaptive Agricultural Research Programmes; Report no. 3, Serenje District, Zambia. CIMMYT, Nairobi.
25. **Merrill-Sands, D.M., S.D. Biggs, R.J. Bingen, P.T. Ewell, J.L. McAllister & S.V Poats, 1991.** Institutional considerations in strengthening on-farm client-oriented research in national agricultural research systems: lessons from a nine-country study. Experimental Agriculture, 27, 343–74.

    Avila, M., E.E. Whingwiri & B.G. Mombeshora, 1989. Zimbabwe: Organization and Management of On-Farm Research in the Department of Research and Specialist Services, Ministry of Lands, Agriculture and Rural Resettlement. Study report no. 1 on the organization and management of on-farm client oriented research, International Service for National Agricultural Research, The Hague, Netherlands.
26. **Fenner, R.J. & E.M. Shumba, 1989.** Linking research and extension through on-farm research and demonstrations: the Zimbabwe experience. Paper presented at the ISNAR workshop on making the link between agricultural research and technology users, 19–25 November, The Hague.
27. **Shumba, E., 1993.** (Op. cit.).
28. **Low, A.R.C., S.R. Waddington & E. Shumba, 1991.** (Op. cit.).

    Semgalawe, Z.M. & J. Kaaya, 1992. (Op. cit.).
29. **Biggs, S.D. & G. Smith, 1995.** Contending Coalitions in Agricultural Research and Development: Challenges for Planning and Management. Paper prepared for the international evaluation conference: Evaluation for a New Century: a Global Perspective, held in Vancouver, November 1995.
30. **Merrill-Sands, D.M., *et al.*, 1991.** (Op. cit.).
31. **Heinrich, G. & E. Modiakgotla, 1992.** (Op. cit.).
32. **Nyirenda, F.M., 1985.** (Op. cit.).
33. **Byerlee, D., M. Collinson, *et al.*, 1980.** Planning Technologies Appropriate to Farmers – Concepts and Procedures. Economics Program, CIMMYT, Mexico.
34. **Stroud, A., 1995.** (Op. cit.).
35. **Heinrich, G. & E. Modiakgotla, 1992.** (Op. cit.).
36. **Kean, S.A. & M. Iles, 1995a.** (Op. cit.).

Chapter 7

Some Dimensions of the Organization of FSR

7.1 INSTITUTIONALIZING FSR IN TANZANIA: A CASE STUDY

Ann Stroud

In theory, the primary product of FSR is farmer-focused technology that will provide a greater number of options for small farmers to solve problems, to relieve constraints and to exploit new opportunities for development in a sustainable way. These should be maintained or supported locally without causing further stress to the environment.

7.1.1 Introduction

Tanzania is not unlike many other African countries in terms of potential and challenges. Bonte-Friedheim, writing in 1994[1], highlighted the challenges facing developing country researchers: location-specific production systems due to diverse natural conditions; crop combinations and specific human preferences; the short history of food crop research; the lack of resources and trained research staff, despite the constant expansion of research agendas; and the isolation of research institutions.

The general conditions in Tanzania are dynamic and complex. Structural adjustment is transforming the economy and fluctuating global markets mean that income from agricultural products is unreliable and frequently devalued in comparison to manufactured, imported necessities. A formidable list of other factors directly affect the functioning of government-supported institutions including democratization, corruption, inflation, reductions in the size and effectiveness of government revenues, the uncoordinated influences of many donors, growing populations, increased environmental degradation and diseases, particularly human immunodeficiency virus (HIV).

7.1.2 The agricultural research system

Agricultural research is presently handled by two major government organizations in Tanzania: the Department of Research and Training (DRT) of the Ministry of Agriculture (MOA) and Sokoine University of Agriculture, with the Commission for Science and Technology acting as a coordination body for these and other research entities in the country.

One of five departments in the MOA prior to reorganization, DRT was formed in 1989–90 from the merger of two parastatal organizations, the Tanzania Agricultural Research Organisation and the Tanzania Livestock Research Organisation. DRT now consists of five sections: livestock, crops, training, special programmes (soils, agroforestry, agricultural engineering) and farm systems research (FSR), each headed by an assistant commissioner (AC). Initially, the DRT received major support from the World Bank and the African Development Bank to rehabilitate a decayed physical infrastructure and to set up a new organizational structure. This led to the establishment of seven research zones, each covering between two and four administrative regions and headed by a Zonal Director of Research and Training (ZDRT). Each station focused on either livestock or crops, with the exception of the integrated Uyole Agricultural Centre (UAC) serving the southern highlands, which joined the DRT in 1993. This historic separation of crops and livestock made wider integration more difficult and left inadequate coverage of some important commodities in selected zones; such as rice in the Lake Zone and cereals and legumes in the southern zone. Each zone differs in its programming according to its history and the major commodities grown in the zone.

The new DRT maintained livestock and crop commodity and disciplinary programmes, each headed by a national coordinator, as well as major sections for planning and budgeting. National priorities, using a weighted scoring technique, were set under the National Agriculture and Livestock Research Master (NALRM) plan in 1989[2], to guide budget and manpower allocations. FSR was designated as one of the first priority programmes, with zonal priorities set more recently.

7.1.3 Evolution of the FSR programme

Although FSR began in Tanzania more than 20 years ago, when the approach was introduced by the International Maize and Wheat Improvement Center (CIMMYT) in eastern and southern highlands zones, it has taken a long time to be accepted. The Tanzanian National Agricultural Policy of 1982 stated that agricultural research should be problem-oriented, and should adopt the FSR approach. In 1985, a FSR national coordinator was appointed to implement the policy as part of a USAID-funded project. Various *in situ* projects have influenced the adoption of FSR (Table 7.1.1). Generally, the FSR team approach was adopted, with a small multidisciplinary group charged with conducting local diagnostic studies to guide research and on-farm testing and farmer evaluation. However, in the central zone, support from the International Development Research Center (IDRC) fostered a station-wide involvement and in southern highlands a small agroeconomics unit was charged with creating links with commodity sections.

A national FSR programme was institutionalized in 1989–90 with fully budgeted multidisciplinary teams of three to six researchers in each research zone, and headed by an AC/FSR who is both the technical leader and the head of a National Coordination Unit (NCU) which manages the programme. Each zonal team is headed by a zonal FSR coordinator, who is responsible for providing technical guidance, management and monitoring of FSR activities in the zone and the team is administratively answerable to the ZDRT at each centre[3].

Between 1986 and 1996 the number of FSR research staff grew from six to 36, accounting for almost 10% of DRT research staff. Of these, 17 were economists, 13 agronomists and nine livestock specialists. Some 25 had an MSc and eight BSc degrees. In 1992, two-thirds of the staff had less than 5 years experience in FSR[4], but donors, notably USAID, the Overseas Development Administration (ODA) and The Netherlands, have helped to upgrade the level of training. Donors have, at one time or another, given support to FSR activities in most zones and also at the national level. However, only two zones and the NCU presently receive outside assistance. The Government of Tanzania

Table 7.1.1. FSR team formation, number of professional staff and donor assistance.

Zone	Year established	Current number of professional staff	Donor assistance
S. Highlands	1981[1]	4	FINNIDA (1982–92)
Eastern	1983	3	USAID (1983–86)
Northern	1984	4	USAID (1983–86)
Central	1985	4	IDRC (1985–93)
Southern	1987	4	ODA-CRP (1991–96)[2]
Lake	1988	12	Netherlands (1988–97)
Western	1992	2	IFAD (1993–96)
NCU	1989	3	Netherlands (1991–96)

[1] The team did not become multidisciplinary until 1991, prior to this it was an agroeconomic unit.
[2] Partial support provided.

Table 7.1.2. Operational funding for FSR from the Government of Tanzania, 1986/87 to 1995/96.

Year	US$	Number of active zones[1]	Number of FSR researchers[2]	Av. no. FSR researchers per zone	US$ per zone[3]	US$ per researcher[4]
1986/87	26,522	2	4	2.0	13,261	6,631
1987/88	18,147	3	7	3.5	6,049	1,728
1988/89	21,595	4	10	2.5	5,399	2,160
1989/90	64,951	5	20	4.0	12,990	3,248
1990/91	75,000	5	19	3.8	15,000	3,947
1991/92	32,550	5	23	4.6	6,510	1,415
1992/93	28,250	6	28	4.7	4,708	1,002
1993/94	19,130	7	28	4.0	2,733	683
1994/95	15,400	7	30	4.3	2,200	512
1995/96	10,000[5]	7	34	4.9	1,429	292

[1] Although southern highlands was not included until 1992/93.
[2] This figure does not include those out for long-term training.
[3] Western and Lake zones are excluded. Although active they did not receive recurrent funds.
[4] Does not include researchers in Western and Lake zone due to reason in[3].
[5] Allocation estimated.

has supplied regular but dwindling financial support for operational costs to FSR as well as other programmes. Budget allocations were given according to priority, e.g. all first priority programmes received the same amount of money. However, no account was taken of the number of staff and type of research and associated costs when assigning the budget. This meant that FSR received less per scientist than some third priority programmes. Because of the programme's expansion, inflation, and falling government support, operational resources declined to Tsh530,000 per researcher (< US$1000) per year. As a consequence the number of activities for teams without external aid has diminished. The sustainability of the programme in its present size is questionable if resources remain so limited (Table 7.1.2).

7.1.4 Recognized and adopted FSR approach

The National FSR Strategy document of 1992 describes the FSR approach that is followed in Tanzania today:

> FSR in Tanzania is defined as an approach designed to generate relevant technologies for specific clients, namely resource-limited households. To accomplish this primary objective a research process developed by CIMMYT is followed: an interdisciplinary systems perspective is used involving farmers to diagnose problems and needs which are prioritized, then potential solutions are designed, developed and evaluated. This technology development process considers: existing aspects and relationships among components of the farming system (e.g. crops, livestock, trees, off-farm activities, etc.) and variations existing among farmers (e.g. socio-economic differences, resource levels, farming practices, etc.). The process emphasizes on-farm adaptive research and interactions among components in the farming system, and complements applied commodity and disciplinary research, which is conducted on research stations. The FSR process is iterative and involves five major stages: description/diagnosis, priority setting and planning, experimentation, evaluation, and recommendation and diffusion.

In theory, the primary product of FSR is farmer-focused technology that will provide a greater number of options for small farmers to solve problems, to relieve constraints, and to exploit new opportunities for development in a sustainable way. These should be maintained or supported locally without causing further stress to the environment. The technology should make better use of, or expand the use of, existing resources.

7.1.5 Indicators of institutionalization

As David Norman wrote in 1991: 'Institutionalization is the process by which new ideas or

Table 7.1.3. Evolutionary stages of FSR in Tanzania: major achievements and shortcomings.

Evolutionary Stage[1]	Main characteristics	Characteristics of FSR	Achievements and shortcomings
Early (pre 1988)	• Low interest by decision makers • Mainly donor driven • Organized at programme level • Linkages *ad hoc*	• Small number of researchers • Young and inexperienced • Lacked credibility • Viewed as 'extension' by other researchers • Isolated from other programmes • Economics new and unfocused	• Learning techniques • Many surveys but limited use of results • Poor linkages with other programmes • Mostly variety testing and package demonstrations • No user differentiation
Past	• Donor support moderate • Decision makers gaining interest • Increasing interest by NGOs • Encouraged by East and southern African commodity networks • Linkages *ad hoc*	• Exists as separate programme • Staff gaining experience • More economists • FSR teams in every zone • 'Extension' oriented but more proactive • Systems thinking in infancy	• Collecting feedback on technology • Start using FPR techniques • Start collaborative work with commodity scientists • Gender awareness starting • Still low-level farmer adoption
Present	• Donor support limited • Donors focused on farm-level impact and development • Commodity researchers increase farmer interaction • Decision makers start to embrace FSR approach • Effort to improve linkages	• FSR considered experts for village-level work • Appreciation for economic/social aspects • Increased collaborative research • FSR still separate programme but has more collaboration with others	• Use farmer research groups to give research direction • Use of PRAs and farmer assessments • Feedback incorporated into research programmes • Linkages with commodity programmes receiving attention • Adoption and therefore impact limited due to poor services, costly inputs

[1] It is difficult to put dates on this continuum.

practices are accepted as valuable and become incorporated into normal routines and ongoing activities of the society'[5]. Institutionalization therefore embodies not only organizational change but also operational changes, where ways of conducting research change due to a different 'mind-set' or paradigm. The following responsibilities were accepted in Tanzania as the role of FSR in the DRT[6]. Until 1995, these were seen as the role of the FSR teams. They are now accepted as the role of all research scientists[7]:

- Characterize farming systems and client groups, diagnosing priority problems and identifying research opportunities, thereby improving planning and priority setting.
- Provide a systems orientation to research, having an interdisciplinary systems perspective, and provide integration across programmes.
- Technology testing with farmers with the aim of developing usable recommendations:
  - adapt existing technologies to farmer conditions with inputs from farmers;
  - focus on specific research themes that involve system interactions, e.g. soil fertility management, integrated pest management, crop–livestock interactions.
- Promote farmer participation in all stages of research.
- Methodology development to bring various types of farmers into the research process.
- Perform linkage functions: from farmers to commodity and disciplinary programmes on-station; from research to technology transfer and development agencies; from farmers to planners and policy makers.
- Monitor, analyse and document dissemination, adoption and impact of technologies.

Table 7.1.4. Three stages of institutional evolution of FSR compared over time using indicators to analyse progress.

Indicators	Evolutionary stages		
	Early	Recent past	Present
Organizational			
Budget	–	+	+
Structure conducive	–	+	+
Priority programme	–	++	++
Considered in planning	+	+	++
Considered in M + E	–	–	–
Multidisciplinary (including economics)	+	++	++
Field work facilitated	+	+	+
Linkages functional	+	+	+
Manpower development	+	++	+
Operational			
Participative surveys	–	+	++
Farmer input into setting research agenda	+	+	+
Farmer input into trial design	–	–	+
Farmer input into trial management (collaborative)	+	+	++
Farmer assessment of technology	–	+	++
Research targeted to beneficiaries	+	+	+
Farmer input to extension messages	–	–	+
Use of farmer research groups	–	–	+
Joint agenda with linkage partners	–	+	+
Multidisciplinary team/group	+	++	++
Use of diagnostics in planning and siting research	+	+	+
User differentiation	–	+	+
Widening technology options	–	+	+

–, not present; +, partially present; ++, present.

NB: The scores are given to the research system as a whole, not only FSR teams. An intermediate score may indicate that across the zones only a few researchers are practising a given technique or that in only a few zones something is happening. The rating is subjective and based on personal knowledge of the situation and history.

In an attempt to track the institutionalization of FSR in Tanzania, key points from this list are identified as indicators. Other indicators are added that show organizational acceptance of the approach. These two types of indicators, for operation and organization, are compared across the three evolutionary stages that have been outlined for Tanzania.

Organizational indicators are:

- Allocation of budget for FSR and linkage activities.
- Organizational change to ensure client-orientation such as creation of FSR teams, units or committees to address farmers' needs.
- Multidisciplinarity such as team or collaborative research.
- Planning mechanisms to foster client-orientation, involving farmers in some stages, and ensuring that research is based on farmers' needs in a systems context.
- FSR is given a high priority.
- Indications in the monitoring and evaluation system that researchers are using a systems approach and participatory methods (beneficiaries are acknowledged, farmer participation is measured).
- Feedback mechanisms in-built (adoption, farmer assessment, impact studies).
- Field research is facilitated (researchers are motivated, transport provided).
- Manpower development is facilitated to build appropriate skills.
- Integrated, multidisciplinary teamwork with a systems perspective.

- Linkage agreements and working modes are developed.

Operational indicators in the Tanzanian context are:

- Research activities at various stages use farmer participatory techniques such as farmer assessment sessions, participatory surveys conducted and information used, farmer-managed trials, farmer involvement in trial design, farmer involvement in extension message development.
- Research targeted to farmers needs (on and off station): researchers are aware, for example, of farmer goals, preferences and limitations.
- Use of farmer differentiation techniques (including gender analysis).
- Consider several options in relation to different types of users and situations, rather than only an optimum level.
- In-the-field linkages with farmers and technology transfer agents using Farmer Research Groups (FRGs), attend fora and have joint agenda, NGO and extension involvement in meetings and field work.
- Use of diagnostic information in planning and siting research.

In 1995, a self-assessment survey in selected research stations determined to what degree researchers felt they were using aspects of the farming systems approach (Table 7.1.5). Several conclusions were drawn. Over time, definite progress has been made in institutionalization of FSR in both operational and organizational aspects. In terms of operationalizing FSR, participatory research had risen in terms of the number of techniques and the number or researchers practising them. Many researchers had been exposed, but few are using many of the techniques, economic evaluation is a widely underutilized example. However, there was little change in some indicators on organizational aspects. Monitoring and evaluation in particular was weak. Structural features fostered nationally oriented planning and budgeting, but these did not necessarily take regional (zonal) needs into account. Budgeting and general facilitation was limited to research necessitating donor input, and manpower development was also directly related to donor interest. Zones that had donor assistance have gone further in adopting, testing and practising many of the techniques indicating adoption of the process.

Table 7.1.5. The use of the farming systems approach and specific FSR techniques in the Department of Research and Training, MOA, Tanzania.

Research zone	Number of responding scientists	% Scientists exposed to farming systems approach	% Scientists with on-farm trials involving farmers	% Scientists using survey information to plan OFT	% Scientists using economic analysis	% Scientists using farmers assessment
Central	25	100	92	52	12	52
Eastern	34[1]	82	44	44	15	29
Lake	31	58	77	55	39	61
Northern	48	19	27	27	10	27
Southern	19	47	63	95	21	63
S. Highlands	37	65	49	49	30	32
Western	9	78	67	78	56	78
Total	203[2]	60	55	50	22	37

Source: Strengthening Client-oriented Research Workshop, DRT – Zonal self-assessments, June 1995.

[1] In the eastern zone, the total number of scientists represents those programmes located at Ilonga. Other stations not represented include: Tanga livestock, Mlingano, Kibaha and NCDP. Programmes represented are: maize, grain legumes, sorghum and millets, FSR-E, sunflower, crop protection, rice and sugarcane.

[2] The total number of scientists in DRT is 350. Some who are on training are not included here as they could not be interviewed.

7.1.6 The future

Research institutions are constantly under pressure to change, with economic forces, government policy, donor influence, research management and scientist exposure to new ideas all adding to these pressures. In the near future it seems that donor support will continue to be contingent on incorporation of a client-oriented research (COR) approach. DRT management accepts this and is planning for capacity building in this area and for further institutionalization. Acceptance of the COR approach will probably result in eventual integration of FSR team members into other sections, perhaps with the exception of economists, bringing the FSR mandate to all programmes. The economists will assume a more professional role rather than acting as generalists. Station committees will be created to coordinate on-farm research (OFR) and review systems thinking.

Outside forces also influence effectiveness. If linkages with farmers and extension continue to improve, one can expect greater impact. If privatization takes hold, one may see eventual improvement in enabling services and a better trained extension service would be able to target recommendations and assist farmers to form action groups. If research is increasingly client-driven then orientation will improve. Finally, the fruits of COR will be contingent on the 'killer' assumption: the availability of funds and motivation to do the job.

7.1.7 Conclusion

Tanzania's story offers many lessons: with hindsight, the use of a team approach rather than trying to achieve general exposure was a good way to start. This allowed for small 'modules' of expertise to develop locally and a critical mass nationally, and as others became convinced of the positive aspects of the FSR approach, they were able to draw on this expertise. It also served as a starting place for economists who slowly gained the acceptance of biological scientists so that collaboration became the norm rather than the exception. It was necessary to have donor support throughout to help generate interest and also to absorb the risk when adopting a new way of working. Consultancy opportunities for farmer-based research from development NGOs also catalysed changes in attitude and practice in some zones. Over time, the increased exposure of researchers involved in commodity networks and in regional and international workshops helped legitimize the FSR approach.

Various factors slowed the process of acceptance and institutionalization. The original source of technology for on-farm testing was generated on-station without the use of FSR. Many of these technologies were rejected by farmers as station researchers had a poor understanding of farmer constraints and recommended technologies were, perhaps, too labour intensive or too costly. It was often wrongly assumed that inputs would be available to farmers. Feedback from FSR teams was weak, poorly communicated, and therefore only slowly accepted by on-station researchers who were still a long way from their clients.

OFR was perceived by managers to be expensive – vehicles, petrol and per diem payments were all needed to sustain early efforts. In the early years, researchers rather than farmers managed most of the trials, hence the need for frequent follow-ups, increasing operational costs. Many trials were started but were abandoned due to logistical difficulties or budget problems and few gave reasonable results. Difficulties arose in timing of treatment application, resulting in failures when researchers were not able to get to the site. Farmers were targeted as individuals, not as groups, so that the risk of data loss was high. Considerable effort was made with little return for this effort. The lack of useful results again slowed acceptance of the methodology by non-practitioners. OFR techniques were either still under development or the practitioners had limited experience in using the more sophisticated design and analysis techniques required to adjust for losses or interpret variability.

The skills and management required to successfully implement an OFR programme were underestimated by donors, managers and researchers alike. Although individual attitudes and work habits did change slowly, it was more difficult to adjust the institutional processes. Planning mechanisms, farmer-based priority setting, monitoring and evaluation and government financial support, all lagged behind. In more recent times, once the DRT accepted the approach as part of its operational and organizational framework, most of these processes have become more supportive. Tanzania provides strong evidence that institutionalization is not an easy or straightforward process, but takes time and considerable effort.

REFERENCES

1. **Bonte-Friedheim, C., 1994**. Editorial: the future of agricultural research. Tropicultura, 11 (3).
2. **Department of Research and Training (DRT), 1991**. National Agricultural and Livestock Research Masterplan (NALRM). MALDC, Dar es Salaam.
3. **Semgalwe, Z.M. & J. Kaaya, 1992**. Farming Systems Research in Tanzania, FAO/SIDA Farming Systems Programme. Harare, Zimbabwe.
4. **National Coordination Unit (NCU), 1993**. Summary of Planning and Priority Setting Workshop. DRT and MOA, Dar es Salaam, Tanzania.
5. **Norman, D., 1991**. FSR in a declining donor environment. A workshop for research and extension administrators for East and Southern Africa, CIMMYT, Nairobi.
6. **National Coordination Unit (NCU), 1990**. A summary of participants' recommendations during the FSR orientation workshop, ALCC, Arusha, 17–19 September, 1990.
7. **National Coordination Unit (NCU), 1995**. Proceedings from Two National Strengthening Client-oriented Research Workshops for DRT. DRT and MOA, Dar es Salaam, Tanzania.

7.2 INSTITUTIONALIZING FSR IN ZAMBIA: A STAKEHOLDER PERSPECTIVE

Stuart A. Kean and M. Creasy Ndiyoi

Strategies for ensuring the long-term sustainability of FSR must be tailored to the institutional power structures in specific NARIs. In those where institutional power continues to be relatively centralized, FSR managers can generally manage through directives. However, in NARIs where decision making is dispersed, FSR managers must become strongly proactive stakeholders.

7.2.1 Introduction

For well over 20 years in Zambia, ever since the notion of an FSR team was first suggested, the subject has created passionate debate both in support and opposition. To its supporters the Adaptive Research Planning Team (ARPT) represents the first major attempt since Zambia's independence to focus research branch attention on the needs of small-scale farmers[1]. It has received international acclaim, including the Food and Agriculture Organization (FAO) World Food Day Silver Medal in 1992, and it has been seen internationally as an example of how farming systems research might be institutionalized. Its supporters claim that it has helped to bridge the many gaps in the technology continuum between farmers, extension workers and scientists and it has increased the relevance of research for small-scale farmers[2].

Some, however, view ARPT as a donor-funded 'white elephant', acting autonomously as if it were a separate organization within the research branch[3]. They feel that it has acted as a barrier in the technology continuum by limiting Commodity Research Team (CRT) scientists to on-station trials and they do not believe that it has had any impact on the relevance of technologies for small-scale farmers. What is beyond dispute is the fact that from 1980 ARPT was a significant section of the research branch, growing until, in 1987, 20% of research branch scientists worked within the team[4].

This contribution analyses the process of institutionalizing FSR in Zambia. It first examines the concept of institutional sustainability by considering the role played by different stakeholders in establishing the team and support for ARPT varied over time. An analysis of national coordination shows how this was undermined by the influence of various stakeholders.

7.2.2 Establishing and sustaining the ARPT

The decision to establish ARPT was not the result of any major policy statement or directive issued by senior government officials. In fact there was no mention of ARPT in any research branch policy document until it was well established in 1984. Several policy statements written in 1990 and 1993 failed to mention ARPT, although it continued to be a significant section in the branch. This apparent absence had no bearing on ARPT's establishment. Indeed, the first ARPT national coordinator was able to use the vague policy rhetoric to justify the establishment of ARPT, citing the statement in the Third National Development Plan that: 'Regional research stations will continue to pay more attention to local problems'[5]. This suggests that policy rhetoric does not always deserve the significance it is sometimes accorded. It also shows that existing rhetoric can be utilized by stakeholders to suit their interests.

The team's establishment was a gradual process over the course of 3–4 years. It was the product of a series of independent decisions (Table 7.2.1), none of which would have created ARPT on their own. There was a general agreement about the need to reform the research organization among some individuals at senior management levels in the MOA. They showed their support for reform in their response to the demonstration of FSR methodology by the East African Economics Programme of CIMMYT in 1977. The support of some decision makers was only important for a short time but it was an absolutely critical time. The Director of Planning, for example, who invited CIMMYT to undertake its demonstrations, set in motion a chain of events with the potential to produce significant change long after he had left the MOA. The support of other stakeholders was more continuous, such as that of the planning unit economist who acted as liaison officer between MOA and CIMMYT, and eventually became the first ARPT national coordinator.

7.2.3 Sources of support for ARPT 1977–96

From its inception ARPT needed supporters to break the mould of conventional research. Once established, its sustainability depended on the support of stakeholders who could influence decisions on resource acquisition and allocation. Most of this support was serendipitous, resulting from informal contacts. Some, however, was carefully cultivated in response to the opposition to ARPT which grew as its activities expanded.

The very high level of support from the incumbent Director of the Planning Unit in the Ministry of Agriculture between 1977 and 1979 was critical in initiating developments which resulted in ARPT's establishment. His goal was a research programme that would be more relevant to the needs of small-scale farmers, and his decision to welcome CIMMYT's demonstration of adaptive research planning procedures[7] bestowed ministry approval on the initiative. However, his support was much less important once the Steering Committee had been established, and he had left the post before ARPT was in existence. Subsequent planning unit directors, while not as interested in the work of the research branch, were broadly supportive.

Some senior managers within both the research and extension branches of the Department of Agriculture had come to the same conclusion as the Director of Planning about the need for change. Four senior managers in particular played key roles: the Director

Table 7.2.1. Critical decisions in the establishment of ARPT.

Date	Decision	Key stakeholder(s)
June 1977	Response to CIMMYT's request	Director of Planning
July 1977	Set up Steering Committee	Director of Agriculture
December 1978	Agreement to appoint two economists	Steering Committee
July 1979	Temporary home for two economists in weed research team	Weed team leader
December 1979	Agree to institutionalize CIMMYT procedures	Steering Committee
February 1980	Agree to set up ARPT	Steering Committee

Source: Kean, 1994[6].

of Agriculture, the two Assistant Directors of Agriculture for Research and Extension – ADA(R) and ADA(E) – and the Chief Agricultural Research Officer (CARO).

- The Director of Agriculture was generally supportive throughout the period, particularly in 1979, when the incumbent Director recruited two economists to continue the work demonstrated by CIMMYT, prior to ARPT's formation.
- Vital support was given to ARPT in its first 4 years by the incumbent ADA(R). His wholehearted commitment was instrumental in its rapid establishment and in the selection of the more radical institutionalization option of a separate FSR section. His aim was to have 60% of the activities of commodity station research teams (CSRT) scientists originating from farmers' problems identified by ARPT. He delegated authority to the Coordinator and urged the allocation of resources to the team.
- The support of the CARO for ARPT was variable. The incumbent CARO during ARPT's establishment was highly supportive. Later incumbents were, however, less enthusiastic and some became focal points for criticism.
- The role of the ADA(E) was important prior to ARPT's formation when the incumbent supported the concept demonstrated by CIMMYT. Subsequent ADA(E)s have also been generally supportive, particularly in recent years.

Scientists in the Weed Control Research Team, having undertaken a small farms weeds project, backed the concepts behind ARPT, providing a temporary institutional base for the first economists recruited to conduct FSR[8]. By the time provincial ARPTs were functioning, some of the provincial agricultural officers and subject matter specialists in the extension branch regarded ARPT as their local research team and lobbied the ADA(E) and provincial authorities on its behalf.

Most of the nine donors who supported ARPT did so because it had a strong mandate to work for small-scale farmers, and they provided scientists, as well as funding. The support of the World Bank and USAID was generally short term, ending after 3–5 years. Other donors, including Swedish International Development Assistance (SIDA), The Netherlands, Norwegian Aid (NORAD) and the International Fund for Agricultural Development (IFAD) provided longer term support. Like the research branch as a whole, ARPT needed donor resources to function.

CIMMYT helped to establish ARPT by organizing the early demonstration of procedures, which ARPT was created to operationalize. CIMMYT initially provided ARPT with concepts and methodology, supported through training, later supplemented by financial support for the coordinator's office and ARPT-Lusaka Province. Its support remained strong even after its direct financial support ceased.

The National Coordinator increased support for ARPT by approaching potential donors for support to short-circuit the bureaucracy for formal donor relations. His previous experience in the MOA Planning Unit provided contacts in several ministries and donor organizations. He contacted stakeholders in the MOA Planning Unit to speed decisions on resource allocation. More recently, the coordinator's office has lobbied for the interests of small-scale farmers and, by implication, for ARPT's survival.

7.2.4 Sources of opposition to ARPT

There was no serious opposition to ARPT during its initiation or in the early stages of implementation. However, subsequently some CSRT scientists expressed scepticism about the role of ARPT, its methodology for conducting on-farm trials and the limited experience of ARPT staff. The most serious and sustained criticism was that ARPT was being favoured in the allocation of resources. Similarly, when the Research Management Team (RMT) was formed in 1990, some members were critical of ARPT's favourable resource position, perceiving it to be acting autonomously, and believed it should be dissolved[9]. ARPT national meetings were cancelled and the National Coordinator was downgraded by exclusion from the RMT. The idea of an ARPT national support unit was opposed by the RMT. A former member of the RMT acknowledged the opposition to ARPT. He said that many in the management team lacked an understanding of ARPT's role and felt threatened by the changes it represented.

By 1993 the opposition was such that no one was able to say definitively whether ARPT existed or not! A mid-term review recommended that it should be dissolved, a view backed by some RMT members. SIDA's Agricultural Sector Support Programme (ASSP) review, and another specific review of SIDA's support to ARPT, recommended its continuation in a modified and renamed form. In 1994 ARPT became the Farming Systems Research Team (FSRT), although its structure and functions remained the same, despite strong SIDA recommendations for reform. ARPT staff continued to function in most provinces, producing work considered by many to be both worthwhile and innovative[10].

7.2.5 Lessons for sustaining the institutionalization of FSR

The ARPT case highlights important lessons for sustaining the institutionalization of FSR in the region.

- Policy rhetoric may not be an accurate guide to the actual priority given to FSR and may bear little relation to the level of support available for its institutionalization. Stakeholders promoting FSR should not devote too much time to getting FSR included, but they should make use of whatever relevant rhetoric exists to support the FSR case.
- The level of support for FSR from any particular stakeholder can fluctuate over time and cannot be taken for granted. In situations where there is rapid turnover of staff the support base can change very rapidly. The best strategy is to continue to develop as wide a base of supportive stakeholders as possible.
- It is essential that FSR managers act proactively and strategically, making the most of every opportunity to garner support from other stakeholders. They can do this through advocacy activities which increase the size of their constituency and by linking formally and informally with other like-minded groups and individuals.
- These last two points indicate that neither research nor institutionalization are apolitical processes but are ones in which institutional power and politics play crucial roles and in which no stakeholder is neutral. It is therefore helpful for FSR managers to understand the interests and levels of influence of all stakeholders in key decision-making areas, including their own power base, so that they can assess their room for manoeuvre.
- Whilst the support of managers was important it was possible for relatively junior stakeholders to play crucial roles in the institutionalization process even if it was only for a relatively short time, in the case of ARPT the weed scientists.

The experience of ARPT supports the notion of contending coalitions 'which look for points of leverage and spaces for manoeuvre to further their causes'. It is necessary for strategic managers to 'take stock of the political environments in which ... work is conducted, and take decisions about which coalitions ... to be part of at a specific time'[11]. The case of ARPT indicated the importance, particularly for the coordinator, of having informal contacts in key positions who were able to help resolve difficulties. In recent years the coordinator's office has also undertaken a range of advocacy activities for small-scale farmers, such as presentations during agricultural shows, which have provided opportunities to publicize the work of ARPT.

7.2.6 Experiences with national coordination in the ARPT

The national coordination of FSR has been given varying degrees of emphasis within the region. National coordination was needed to:

- Provide technical and methodological guidance to create a coherent programme.
- Provide a mechanism for the supervision of the decentralized provincial teams to create an environment for institutional learning and reflection and reduce duplication and the repetition of mistakes.
- Take on substantial administrative responsibilities, especially during the establishment of ARPT.
- Coordinate linkages between ARPT staff with those of other sections of the research branch and in other organizations.
- Maximize the support of ARPT's nine donors.

The National Coordinator made efforts to influence the activities of all stakeholders working with ARPT, especially team members. Many formal and informal management mechanisms were developed including team reviews, provincial exchange visits, a variety of committees and in-service training sessions[12]. These aimed to encourage institutional learning, team building, information exchange and facilitate interaction within the team and between team members and other organizations. The coordinator increasingly focused on participative mechanisms as typified by the reflective experiential workshops. In recent years less use has been made of formal mechanisms apart from 6-monthly internal team meetings which encourage reflection and self-criticism.

The national coordination of ARPT's activities was rather disappointing. Several reviewers of ARPT in the late 1980s noted areas where methodology had not been standardized. The provincial teams were inconsistent in their levels of interaction with farmers, with CSRT scientists and extension workers, in their use of CIMMYT methodology, their interdisciplinary cooperation and their identification with the ARPT as a national team.

The main reason for the lack of cohesion was the coordinator's limited authority in key decision-making areas easily influenced by other stakeholders. While research managers, including the coordinator, had influence in structural matters, donors had the most direct influence over the allocation of funds and staff, giving donor-funded staff greater autonomy. Donor-funded technical assistance scientists often worked to donor project job descriptions and project objectives, identifying more closely with their projects than with ARPT. Even if they were responsible to the ARPT coordinator their loyalty remained with their employer or recruiting agency. ARPT scientists in general, but particularly those funded by donors, had great influence over the content of their activities, their links with other stakeholders, and the methods they used. CIMMYT and the coordinator, for example, advocated the use of a particular research process but scientists had the freedom to decide the extent to which they would follow it. Staff could be allocated to a province by managers or a donor, but were often able to decide where they worked within the province. The wide range of stakeholders meant that research managers, including the coordinator, did not have sufficient power to be able to control important aspects of institutionalizing FSR.

The coordinator was influential in decisions on the team's structure and provided continuity and momentum during the team's establishment. Although he had little power to control where the staff and funds were allocated, the first coordinator was heavily involved in recruitment and in discussing resource allocation. In spite of having a clear sense of direction for the team, he had to compromise with donors, usually on what assistance was provided. The coordinator was less successful in coordinating the activities and methods of ARPT scientists. Some of the donor-funded technical assistance scientists had strong opinions and little interest in working as part of a newly formed national team. Even if their project documents specified that they were responsible to the ARPT coordinator their loyalty remainined to their employer or recruiting agency. A reviewer commented: '... it is open to question to what extent he (the coordinator) is in a position to effectively execute his tasks. ... Since the coordinator does not control the funds of the different ARPT projects, his authority will have to come from his professional capacity and his position as coordinator. This may not bring him far in directing expatriates with strong opinions and with the leverage which their position and funding arrangement provide them'[13].

7.2.7 Lessons for the national coordination of FSR

The Zambia case highlights several lessons about national coordination useful in the region. First, that those managers charged with coordination may not have the level of influence over resource allocation decisions to be able to coordinate effectively. Donor authority may convey sufficient influence to relatively junior stakeholders to have them ignore coordination efforts; the tail can wag the dog under certain circumstances. Second, although FSR managers can use a variety of coordination mechanisms these are unlikely to be effective unless they take stakeholder's interests and influence, including their own, into account.

FSR managers should analyse the interests and influence of stakeholders in different areas of decision making to find the most effective points of leverage on them, e.g. job descriptions, personnel assessments, incentive systems, etc. Third, although a range of management styles may be appropriate, the Zambian experience generally indicates that less directive, more participatory and experiential mechanisms have been more effective in situations where decision-making power is highly dispersed across many stakeholders. That is voluntary approaches involving reflective team learning are more likely to be effective in coordination than heavy-handed dictates. Some useful mechanisms included: provincial team exchange visits, bi-annual team meetings on specific themes organized on experiential lines, and study tours.

National coordination is increasingly a key issue in the process of institutionalizing FSR for countries in the region. The indications are that it may well be as difficult as in Zambia. Until recently Botswana experienced similar difficulties with national coordination due to the dominance of donor activities[14]. In countries where there are many different FSR projects being undertaken within a national agricultural research system (NARS), irrespective of the level of donor funding, there is already a sufficiently high dispersion of decision-making power to exacerbate coordination difficulties, e.g. Kenya, Uganda, Zimbabwe, Zambia and increasingly Malawi. In these countries many scientists, research centre directors and project leaders are implementing FSR each with different interests, organizational settings, methodologies, training and work experience. In Tanzania the success of national coordination to date may be jeopardized if the NARO moves too fast towards integrating FSR into zonal research-at-large[15].

7.2.8 Operational implications of a stakeholder perspective for institutionalizing FSR

Interest in the characteristics and functions of FSR has never been greater in East and southern Africa NARIs. A growing number of diverse FSR organizational arrangements and structures are being created within these National Agricultural Research Institutes (NARIs), and this diversity had created a double-edged challenge for the institutionalization of FSR. It is decentralizing responsibility for the long-term sustainability of FSR into the hands of many more stakeholders. It is also, however, increasing the need for effective national coordination. The inherent contradiction in these challenges can best be understood through an analysis of the interests and influence of all the stakeholders involved.

Across the region over the least 20 years, power structures have broken down and centralizing forces over resource acquisition and allocation have become extremely weak[16]. Systems are, to varying degrees, not under the control of any particular stakeholder[17]. Many managers and some donors have tried, without success, to recentralize the power structure. An alternative, more pragmatic, approach is for stakeholders to accept that institutional power is dispersed, and they must work proactively within these limitations, creating coalitions with like-minded stakeholders to achieve their goals.

Strategies for ensuring the long-term sustainability of FSR must be tailored to the institutional power structures in specific NARIs. In those where institutional power continues to be relatively centralized, FSR managers can generally manage through directives. However, in NARIs where decision making is dispersed, FSR managers must become strongly proactive stakeholders. This can be achieved by:

- Creating interest in FSR among all stakeholders, as individuals and groups, including relatively junior stakeholders as well as ministers and senior managers.
- Not being content with stated policy commitments from senior stakeholders.
- Continuing to campaign to increase the number of stakeholders interested in FSR even in systems where considerable FSR activities are being undertaken.
- Undertaking a wide range of FSR advocacy activities including the development of a strong national network, promotional materials, meetings, in-country training and study tours.
- Campaigning for resources.

This final, vital activity could include analyses to understand how formal and informal resource decision-making processes work and stakeholders' relative levels of influence, building skills in resource acquisition such as the writing of project proposals, and building of resource networks including databases of individuals and organizations providing resources for FSR.

Incorporating a stakeholder perspective into national coordination for FSR requires a similar understanding of the interests and influence of all stakeholders. FSR managers should:

- Use all formal coordination mechanisms that may exist, such as job descriptions, programme meetings, supervision, monitoring and evaluation. They must understand the power structures respected by different stakeholders and endeavour to influence these.
- Provide opportunities for regular self-monitoring on a participatory and voluntary basis. This would enable stakeholders to reflect on their experiences in relation to those of their co-workers and their colleagues from other organizations.

REFERENCES

1. **World Bank, 1983**. Zambia: Agricultural Research and Extension Review. Regional Mission in East Africa. World Bank, Nairobi.
2. **Kean, S.A. & L.P. Singogo, 1988**. Zambia: a case study of organisation and management of the Adaptive Research Planning Team, Research Branch, Ministry of Agriculture and Water Development. Study Report no. 1. On-farm client oriented research. ISNAR, The Hague, The Netherlands.
3. **Government of Zambia, 1993**. Mid-Term Review of Zambia Agricultural Research and Extension Project (ZAREP). Ministry of Agriculture Food and Fisheries, Lusaka.
4. **Kean, S.A. & L.P. Singogo, 1988**. (Op. cit.).
5. **Government of Zambia, 1979**. Third National Development Plan 1979–83. National Commission for Development Planning, Lusaka.
6. **Kean, S.A., 1994**. The Institutional Politics of Agricultural Research Policy in Zambia: a Model of Contingent Innovation. PhD thesis, University of East Anglia.
7. **Kean, S.A. & L.P. Singogo, 1988**. (Op. cit.).
8. **Parker, J.M.H., 1981**. End of Contract Report for the Period 1977/78 to 1980/81. Mimeo, Chilanga.
9. **Government of Zambia, 1993**. (Op. cit.).
10. **Dobel, R., 1996**. Participation and Sustainability – the Case of the Adaptive Research Planning Team. Mimeo, Lusaka.
11. **Biggs, S.D. & G. Smith, 1995**. Contending Coalitions in Agricultural Research and Development. Challenges for Planning and Management. A paper prepared for the international evaluation conference: Evaluation for a New Century: a Global Perspective. Vancouver, November 1995.
12. **Kean, S.A. & L.P. Singogo, 1988**. (Op. cit.).
13. **Birgegard, L.E., 1988**. The Zambian National Agricultural Research System – Issues and Problems. Issue Paper no. 5. Uppsala: International Rural Development Centre, Swedish University of Agricultural Sciences.
14. **Heinrich, G. & E. Modiakgotla, 1992**. Farming Systems Research in Botswana. Report no. 2 Farming Systems Programme, Gaborone.
15. **Stroud, A.S., 1995**. Institutionalisation of FSR in Tanzania: Past, Present and Future. Paper presented at the Southern African Association FSR/E Conference, October 1995.
16. **Kean, S.A., 1994**. (Op. cit.).
17. **Kean, S.A. & M. Iles, 1995b**. Promoting IPM Implementation in National Agricultural Research Systems: Reviewing Dispersed Power Structures in Malawi and Uganda. Unpublished Mimeo for International IPM Working Group, Chatham.

7.3 COSTS OF ON-FARM RESEARCH: A COMPARISON OF EXPERIENCES IN SIX COUNTRIES[1]

Elon H. Gilbert

On-farm research suffers from the common perception that it is more expensive. These case studies provide strong evidence that the perception is incorrect.

7.3.1 Introduction

Few aspects of FSR have received as little attention as the costs of its application. In large part this dearth of cost analysis reflects a general lack of accessible data on institutional operating costs. The costs of the earliest projects, widely supported by donor funds, were generally high, relative to comparable services provided through national agricultural research systems (NARS). The fact that the transition to locally staffed and supported programmes was painful (or simply failed) in many countries is attributed by some to the costs of maintaining FSR.

FSR itself was developed to complement, rather than substitute for, conventional agricultural research. Its supporters saw it as a means of improving the linkage between research systems and their clients, to better identify relevant priorities and to determine which innovations would best address the needs of specific farming systems. Had FSR been widely perceived in this way it might have become a valued step in the activities of NARIs almost immediately. In reality, many FSR programmes were viewed as competition, or even as an alternative to the conventional process, and this raised strong opposition.

One result was the failure to develop a functional symbiosis envisaged between FSR units and existing commodity and subject matter research programmes, linkages between FSR and other NAR programmes were notably weak in most countries[2]. The relatively heavy dependence of most FSR programmes on donor funds placed them in a particularly vulnerable position. The support of NARI managers was required to extend such funding, and donors often insisted that the NARI assume an increasing share of FSR costs. NARI managers found it hard to do so when it was already difficult to cover existing NARI expenses[3]. With the conclusion of donor funding, many FSR programmes were forced to curtail their operations, some drastically.

This contribution addresses the following questions:

- How much do different types of OFR cost and how are these costs structured?[4]
- What is the relationship between costs on the one hand and the scope, scale and programme type on the other?
- How do the costs and cost structures of OFR programmes compare with those of the NARIs as a whole in the same countries?

This contribution examines cost information from seven programmes in six countries; Ecuador, Guatemala, Panama, Nepal, Zambia and Zimbabwe[5]. It is important to remember that this data is well over 10 years old. It remains, however, the best data available. This very fact highlights the urgent need for more work in this area, as decisions are being made about the future of FSR programmes without any recent costings.

The data were gathered during a study of on-farm client-oriented research experiences by the International Service for National Agricultural Research (ISNAR) in the mid 1980s. The data has not been published earlier, partly because of the real difficulties in reconciling data sets within NARIs and across countries. These seven cases were selected from the 21 programmes studied because of availability of adequate data sets illustrating the range of costs and cost structures of OFR. The coverage in terms of numbers and types of OFR programmes is less than would be desirable, but even this limited information does provide useful insights into the relationships between costs, programme type, scale and scope. OFR suffers from the common perception that it is more expensive. These case studies provide strong evidence that this perception is incorrect.

7.3.2 Three types of OFR programmes

Ewell, in 1988[6], divided the 21 programmes covered in the ISNAR On-Farm Client-Oriented

Research (OFCOR) study into three broad types – individual technician, minimal pair and multidisciplinary team – according to the size and disciplinary breadth of the teams. At least two examples of each are found in each of the six countries examined in this contribution.

Solo generalist

The solo generalist programme type consists of an individual with training in the biological/extension sciences and familiarity with survey work and simple economic analysis. The generalist is resident in the area of research activity, and is often assisted by one or two technicians, backed up by specialists in specific fields based at central stations. The geographic scale tends to be small with a focus on villages near the base. The field 'team' is capable of implementing elementary FSR activities, including informal surveys, and both researcher- and farmer-managed trials. The purpose is to determine the acceptability of specific technologies to farmers in the area, often as a precursor to their widespread dissemination. Examples of the solo generalist among the OFR programmes studied include the Instituto de Ciencia y Tecnologia (ICTA) in Guatemala and the Programa de Investigacion en Production (PIP) in Ecuador.

Minimal pair

The minimal pair team consists of at least one agronomist and a socioeconomist with the capacity to carry out surveys and conduct a range of on-farm trials in specific areas. The coverage may still be limited, but it is broader than that of the solo generalist. The disciplinary scope is significantly greater with the inclusion of a full-time socioeconomist. The pair is normally resident in the area where the research is carried out, assisted by one or more field supervisors. The geographic scale may often include an entire province. Staff from research programmes and/or a national FSR team may provide back up. The pair is distinct from the solo generalist in its broader scope, and the increased research content made possible by the addition of a socioeconomist. Examples of the minimal pair in the case studies include the Lusaka ARPT in Zambia and the Caisan Project in Panama.

Multidisciplinary Team

The multidisciplinary team consists of three or more disciplines such as animal science, agronomy, anthropology, agricultural economics, forestry and nutrition. An extension agronomist or research/extension liaison officer might also be involved and the team may have a leader focusing on management and administrative tasks. Multidisciplinary teams are capable of carrying out OFR on a wider range of commodities and issues, utilizing different types of interactions with farmers, other than OFR configurations. The research content tends to be high, with a comprehensive orientation towards the farming systems of a region. Consequently, there is the potential for greater feedback of a general strategic nature to NARIs and policy makers. They may also operate in separate administrative units with the NARI. Examples from the case studies include the Farming Systems Research Unit (FSRU) in Zimbabwe, the Farming Systems Research and Development Division (FSRDD) in Nepal and the Luapula Province ARPT in Zambia.

The most important distinctions concern the number of individuals or team members who regularly interact with one another as well as the number and diversity of disciplines represented.

7.3.3 Costs and cost structure

There is no single answer to the question of how much OFR costs since there are different types of OFR, and individual programmes each have their own unique histories. This analysis of costs seeks to identify patterns and relationships between OFR costs on the one hand and scope, scale and programme type on the other.

The data

For comparative purposes programme expenses are expressed in terms of costs per scientist. All costs have been converted to 1980 US$ using the deflators and exchange rates for the purchasing power parity (PPP) expenditure series in the ISNAR database[7]. The use of PPP facilitates comparisons between the six countries given the wide variations in their costs of living. Adjusting for PPP dramatically increases costs in Guatemala, Nepal and, to a lesser extent, Zambia, compared to the three other countries.

Costs of OFR

Tables 7.3.1(a) and 7.3.1(b) summarize the available information from the seven OFR programmes according to three cost categories: salaries and wages, other operating costs and capital costs. The accounting systems of OFR programmes do not, as a rule, make provision for the general management and administrative services provided by the NARI. Accordingly, as a rough estimate of overheads, a flat 15% of total NARI costs per researcher have been added to the OFR costs per researcher for each programme. The overhead charge has been integrated into each of the three cost categories for the seven OFR programmes in accordance with the cost structure for the corresponding NARI.

Table 7.3.1(a) gives total costs for the seven programmes, though a comparison of totals is not very meaningful, since some programmes are national in scope (e.g. Guatemala, Nepal and Zimbabwe) while others cover a single province or area (Panama and Zambia). Accordingly, costs per researcher were calculated for each programme. Costs per researcher/year, in 1980 PPP-adjusted US$, ranged from just under $19,000 for the Lusaka ARPT to just over $60,000 for FSRU. The most striking feature of Table 7.3.1 is the relationship between OFR programme type and costs. Not surprisingly, the two programmes with the highest total costs, FSRU in Zimbabwe and FSRDD in Nepal, both have national responsibilities and are multidisciplinary teams. The five programmes with lower costs are either solo generalists (PIP in Ecuador, ICTA in Guatemala) or minimal pairs (Lusaka and Luapula ARPTs in Zambia and Caisan in Panama) with local or regional responsibilities. The geographic coverage is a major factor in delineating scale while programme type is a composite of several elements including scope and functions/nature of research as well as scale.

The two ends of the cost spectrum deserve special comment. FSRU (Zimbabwe) is at the top end, partially a function of the high salary levels. At the bottom end, the Lusaka ARPT was specifically designed as a low-cost operation[8]. Overall the figures suggest that programmes which are heavily dependent on donor funding tend to spend more. The two most expensive programmes, FSRU (Zimbabwe) and Luapula (Zambia) were primarily supported by donors. Two of the low-cost programmes, Lusaka ARPT and the Ecuador PIP, received most of their funds from the government.

A major disadvantage of donor funding is that it usually comes to an end. The incorporation of previously donor-funded programmes into the government financial system can be a traumatic process. The end of donor funding is often marked by the simultaneous withdrawal of funding and technical assistance personnel, special benefits for local staff and the special accounts which buffered research activities from irregularities in the release of funds. Maintaining the momentum of research activities under such conditions is a formidable challenge.

In the case of PIP in Ecuador, the transition from donor funding in 1982/83 was far from successful and resulted in fundamental changes in the character of activities. The incorporation of PIP staff and activities into the experiment stations resulted in a reduction in levels of compensation and travel budgets. One-third of the staff had left by 1986 and were replaced by younger people with significantly less experience and training in OFR. Most important were the changes in the planning and supervision of the field activities as described by Ewell, writing in 1987:

> The PIP programmes responded formally to the Technical Committees named by the directors of the stations. These changes had the favourable effect of improving potential

Table 7.3.1(a). Total programme costs and numbers of researchers (annual cost per programme in 1980 US$ adjusted for PPP).

	Lusaka, Zambia	I'bura, Ecuador	FSRDD, Nepal	Caisan, Panama	ICTA, G'tmala	L'pula, Zambia	FSRU, Z'bwe	Average
US$								
costs	56,730	45,171	265,980	33,960	157,750	136,800	272,655	138,455.00
No. staff	3.0	2.1	13.0	1.2	5.0	4.0	4.5	4.70

Table 7.3.1(b). OFR programme costs and cost structures (annual cost per researcher in 1980 US$ adjusted for PPP).

Country	Lusaka, Zambia	I'bura, Ecuador	FSRDD, Nepal	Caisan, Panama	ICTA, G'tmala	L'pula, Zambia	FSRU, Z'babwe	Average
Salaries	6,960	10,510	8,820	16,940	18,940	11,450	42,650	16,800.00
Operation	4,030	6,640	8,900	4,690	6,650	13,630	14,780	8,530.00
Travel	(2,610)	(1,960)	(4,450)	(2,960)	(4,060)	(11,530)	(7,470)	(5,220.00)
Misc.	(1,420)	(4,680)	(4,450)	(1,730)	(2,590)	(2,100)	(7,310)	(3,310.00)
Capital	7,920	2,960	2,740	6,670	5,960	9,120	3,160	5,450.00
Total	18,910	21,510	20,460	28,300	31,550	34,220	60,590	30,800.00

> coordination with on-station research, but corresponded to a weakening of the client-oriented, interdisciplinary character of the program. ... As that pressure (for a broad, interdisciplinary farming systems approach) has been relaxed, entropy has set in. The PIP's methodology has stagnated and the program is in danger of becoming simply a technology screening arm of the experiment stations.[9]

Weaning a programme away from donor funding can be difficult. This is especially true with the more demanding types such as the multidisciplinary team. Solo generalist and minimal pair programme types that are integrated into research or development institutions may find it easier to reduce or eliminate dependence, simply because their costs are lower.

Cost structure

Costs are broken down into three major categories: personnel costs (salaries and wages), operating expenses and capital costs. Of the total OFR costs per researcher-year, salaries and wages account for 55% of the total on average, with 28% for operating costs and 17% for capital.

SALARIES. These include all regular support staff in addition to researchers. Some programmes use permanent labourers (included under salaries) and others use seasonal labour, so the wages of casual labour have been grouped with salaries, rather than under operating costs, to facilitate comparisons between the programmes. Technical assistance staff are costed in local salary equivalents. Salary costs may be understated to the extent that programmes use seconded staff from other departments, as with research and extension field staff working with the ARPTs in Zambia, and are generally provided with services by the NARI. These costs are included in the NARI budgets, but not in the OFR field-team accounts. An overhead charge is included to cover the support services provided by the NARI.

The two Zambian ARPTs are at the low end of the spectrum, with less than 50% of costs going on salaries, while Zimbabwe is the opposite, with salaries accounting for nearly 70% of OFR costs. While this is chiefly the result of higher salary levels, FSRU Zimbabwe also has more support staff than programmes in other countries.

With the exception of Luapula ARPT in Zambia, the salary and wages component is larger, both in absolute and relative terms, for multidisciplinary team type programmes. This is partially a function of skill levels, since the low-cost programmes, particularly the three Latin American cases, rely primarily on researchers with bachelors degrees, while the majority of research staff in the other programmes have advanced degrees. The difference is also the result of the more extensive use of both permanent and casual hired labour, particularly by the multidisciplinary teams, to assist with trials and other tasks. In contrast, solo generalist and minimal pair programmes rely more on farmers for the range of operations associated with carrying out on-farm trials. The use of hired labour is related to the importance placed on researcher-managed, as opposed to farmer-managed, trials.

OPERATING COSTS. These account for roughly one-quarter of OFR costs on average, grouped into two categories: travel-related expenses and supplies and miscellaneous expenditures. Travel-related expenses depend on the scale of operations, including area covered and the number of research sites, as well as the location of the staff

Table 7.3.2. NARI and OFR: total costs per researcher (1980 US$'000 expressed in PPP).

Country	Nepal	Zambia	Ecuador	Panama	Guatemala	Zimbabwe	Average
NARI	22.7	46.7	30.5	46.0	40.0	106.7	50.5
OFR	20.5	27.7	21.5	28.3	31.5	60.6	31.7

Note: Average costs per OFR researcher here are US$31,700 compared to US$30,800 in Table 7.3.1. The Zambia average is increased, weighted by the larger and more expensive Luapula programme.

base. Then there are costs related to the frequency of visits to sites; and the importance of extension and liaison duties that may not be associated with specific sites. Programmes with frequent site visits have better supervision of field sites – important for the quality of the research performed. Not surprisingly, the programmes with larger networks of sites (FSRU in Zimbabwe, FSRDD in Nepal and Luapula ARPT in Zambia) have higher travel-related costs, while all three Latin American programmes, with staff posted in the field research areas, have reduced travel requirements.

The Luapula and Lusaka ARPTs in Zambia are an interesting study in contrasting operational styles and costs in the same country. Both teams covered an entire province, but Lusaka was specifically designed as a low-budget operation with one vehicle for three researchers, who often used local transport[10]. Luapula, in contrast, had four researchers, each with a vehicle. This, combined with the large geographic area of responsibility, produced the highest travel-related costs per researcher of all the programmes.

CAPITAL COSTS. These are the smallest major category of OFR costs, accounting for just 17%, on average, of the total costs. Vehicles are the largest component by far. The available data on capital costs is, however, the least satisfactory. Expenditures are recorded in the year in which the capital items are purchased, but are then 'consumed' over a number of years. Housing construction can be an important component of capital costs when accommodation is required for an entire team in a remote area. Construction is a major portion of the capital costs of the Zambia projects, but has been largely excluded from the calculations on the grounds that the laboratories and houses were not for the exclusive use of the ARPTs.

7.3.4 A comparison of NARI and OFR costs

Comparisons of the costs of NARI and OFR programmes have serious limitations. OFR and OSR should be complementary parts of the same research process. NARI data inevitably includes the costs of OFR and OSR as well as of general administration. However, NARI managers must make resource allocation decisions. A sense of the relative costs of different phases of the research process will help them in this task.

Table 7.3.2 compares NARI and OFR programme total costs per researcher in the same countries. NARI costs include at least portions of OFR programmes in most instances, but these are not a major component of total NARI costs.

The NARIs have higher costs per researcher than the OFR programmes, though in Nepal the two come very close. On average, OFR costs are 62% of the NARI costs per researcher. Salaries are the major factor, with lower salary scales for OFR researchers and fewer support staff per researcher. This is clearly the case for the three Latin American programmes, but is less clear for the other three countries[11]. A comparison of the structure of costs across the seven OFR programmes and the six NARIs is presented in Table 7.3.3.

The OFR cost structure is almost identical to

Table 7.3.3. Cost structures for NARI and OFR programmes (% of total expenditures).

Category	NARI	OFR
Salaries	60	55
Operations	25	28
Capital	15	17

the average for the NARI. This is particularly surprising in view of the common impression that OFR is more dependent upon operating costs than NARIs in general. In absolute terms, average operational expenses for OFR programmes are less than two-thirds of those for the NARIs. Although OFR travel costs per researcher may be equal or greater to those of the NARIs, OSR can be expected to be more expensive in virtually every other category, including equipment, supplies, structures and administrative overheads. This, with the lower salaries for the more junior professionals usually involved in OFR, and a larger support staff for on-station researchers, leads to a general conclusion that the NARIs as a whole have higher costs per researcher than OFR.

7.3.5 Costs in relation to scope, scale and programme type

Scale and Scope

Scale (area and number of sites) is perhaps the least controversial and most conceptually simple of the factors affecting costs. Similarly, the scope of investigations can be expected to have a positive relationship to costs. The elements of costs which should be most affected by scope are salaries and wages. Although it is possible for a single individual to have responsibility for several commodities and subjects, as may be the case with the solo generalist type of programme, research expertise in any specific area can be expected to dilute as scope widens. A critical consideration is the relationship between economies of scope and scale in OFR activities. Larger teams can cover wider geographic areas and more research sites. The range of expertise that can be accommodated in a big team can be spread across a region or an entire country, whereas resource constraints may make it impossible to support the same team to work exclusively on one area within the country.

Two programmes, the PIP (Ecuador) and Caisan (Panama), have a narrow scope which is reflected in the low expenditures for salaries and wages. PIP and Caisan are essentially local programmes of trials on specific commodities. Of the other five programmes, the two that are national in scale also have the highest scope rankings (FSRDD in Nepal and FSRU in Zimbabwe).

Activities and Methods

The activities and methods of an OFR programme indicate the intensity of the application of resources. Some programmes include formal surveys and a range of experiments normally associated with station-based research and social science field research. The more sophisticated the activities and methods used are, the greater the intensity of research involvement is likely to be. One would expect greater researcher involvement to be reflected in higher operating and capital costs per researcher. On the experimental side, since most of the programmes contain a mixture of different types of trials and degrees of research and farmer involvement, it is difficult to relate functional spread to costs in any systematic way. Programmes that use formal surveys might be expected to have higher operational costs than those that do not.

Table 7.3.4 summarizes the relationship between activities and methods on the one

Table 7.3.4. OFR programmes: research functions and costs.

Programme	Type	Surveys	Trials	Cost per researcher (000 US$, 1980 PPP) Total	Operating
Zambia, Luapula	MP/MDT	Formal	OSR, RM/RI, RM/FI	32.2	23.3
Zimbabwe, FSRU	MDT	Formal	RM/FI	55.2	17.1
Nepal, FSRDD	MDT	Informal	RM/FI	34.9	12.7
Zambia, Lusaka	MP	Informal	RM/FI	16.9	11.6
Panama, Caisan	MP	Informal	RM/FI	26.0	12.4
Guatemala, ICTA	SG	Informal	RM/FI	29.6	14.8
Ecuador, PIP	SG	Informal	RM/FI	19.2	10.2

hand, such as OSR and the OFR programme costs (with and without salaries), on the other. Costs are shown with and without salaries because these vary considerably between countries and tend to overshadow other relationships. Formal surveys and research managed/research implemented (RM/RI) trials are examples of research functions of higher research intensity compared to informal surveys and farmer managed/farmer implemented (FI) trials.

Table 7.3.4 suggests at least two relationships between costs and programme-type/research functions. First, programmes with a broader range of research functions have higher non-salary costs per scientist than those with narrow sets of functions. Formal surveys appear to be a particularly critical feature since the two most costly programmes, Luapula ARPT in Zambia and FSRU in Zimbabwe, are the only ones who report this activity.

Second, the non-salary costs per scientist for all solo generalist and minimal pair programmes are surprisingly similar, a fact that may be related to the common range of research functions (i.e. largely RM/FI trials and informal surveys). The essential difference of the minimal pair programme type is the inclusion of a social scientist. Although this does increase total programme costs and change programme activities, costs per researcher are not affected. If minimal pair programmes have a substantially higher productivity per researcher and/or important qualitative differences, this programme type must be preferable to the solo generalist.

7.3.6 Conclusions and principal findings

Cost structure

Wages and salaries are the most important component of costs, accounting for 55% on average for the seven programmes. Operating costs, including travelling expenses, consume 28% of total expenditures. The structure of costs is, on average, virtually identical to the cost structure of the NARI as a whole in the same country.

Costs, scale, scope and research functions

Costs per researcher of OFR programmes appear to be a function of scale; the number of and distance to sites; scope; the diversity of subject matter, including commodity coverage; and the range and complexity of research functions. With reference to types of programme teams multidisciplinary teams have the broadest scope and functional capacity and cost more per researcher than the other types, particularly in terms of non-salary expenses. Costs per researcher are very close for minimal pair and solo generalist programmes, which suggests economies associated with the larger team size of minimal pairs. Further, the qualitative differences in capacity with the addition of a social scientist suggests that the minimal pair will be preferable to a solo generalist in most situations.

NARI and OFR costs per researcher

OFR is less expensive per researcher than NARI research as a whole. This evidence is clear on this point, even including a 15% overhead charge for NARI services to OFR which makes up 23% of OFR expenditures: OFR costs average roughly 60% of the NARI expenses per researcher.

Operational costs

OFR programmes are not significantly more dependent on operational expenses than NARIs. In absolute terms, OFR operational expenses per researcher are slightly more than half of those for the NARI as a whole. As a percentage of total costs per researcher, NARI and OFR operational expenses are virtually identical, at 25% and 28%, respectively.

These findings contain several implications for FSR programmes:

- Cost is less of a barrier to initiating or maintaining an FSR programme than commonly perceived. There are certainly costly forms of FSR, notably the multidisciplinary teams, but there are also modestly priced, readily affordable programmes.
- Scope, scale, activities and methods all offer flexibility in managing the reconciliation of OFR needs with available resources. Variation in geographic scale, in subject matter (scope) and in the methods used, all influence both the quantity and quality of services. If, as the evidence suggests, minimal pair programmes are no more expensive

per researcher than the solo type, teams of two (or possibly more) researchers representing different disciplines appear preferable from both cost and quality perspectives.

- The type of OFR programme and its costs can be changed over time in a given country or region in response to changing needs and resource constraints. An area about which little is known may benefit from a multidisciplinary team of three or more during the initial years, but could be adequately served by a minimal pair (possibly backstopped by a national team and research programme staff) thereafter.
- Various forms of FPR which have received increased attention in recent years offer additional options and possibilities for achieving cost reductions in OFR.
- Some form of OFR may the only option for maintaining significant geographic coverage in many developing countries. This is especially important for countries and regions with high ecological diversity. Small countries, as well as regions within larger countries, that face serious resource constraints and whose research needs are largely adaptive in character should seriously consider strategies which emphasize OFR[12].

From a strict perspective of cost, the analysis clearly suggests that OFR, particularly minimal pairs or solo generalist programme types, is cheaper than almost any form of station-based research. The minimal pair seems desirable, especially where geographic coverage can approach that of two solo generalists. However, both these types of OFR rely on backstopping from researchers elsewhere, putting a premium on collaboration throughout the research process. Presumably, at least some support could (and does) come through regional networks, institutes and IARCs.

Farmer participation was a feature in all but one of the eight OFR programmes included in this study, but researchers played the leading roles in all instances. As well as ensuring that OFR activities are relevant to local needs and conditions, greater farmer involvement can significantly expand the geographic reach of a limited budget. It is neither feasible nor essential for OFR research staff to be intimately involved in activities at a large number of locations, especially where farmers can define local priorities and test possible solutions. This has been the case in many countries in the years since these data were collected. As with the NARIs as a whole, the deployment of OFR resources is defined in large part by national and regional priorities. For those commodities, subjects and areas that are not among the priority concerns, some form of farmer participatory research offers a low-cost mechanism for linking local communities to sources of innovations that address their specific needs.

The main considerations are not so much cost as performance. Much of the station-based research, especially in small, resource-starved NARIs, contributes little to national development and research failure rates are very high because of equipment, funding and human malfunctions. There is a better chance of obtaining usable results with some form of OFR as a stage in the research and development process, especially using the more participatory forms. As one moves along the spectrum of conditions from poverty stricken and poor infrastructure to civil disorder and war, as in Somalia and Liberia, some form of farmer-operated OFR may be the only option that can serve as a complement to relief activities.

A final note

The costs and cost structures of OFR programmes in different countries reflect their origins which, in most cases, involved donor-funded projects and outreach programmes of international agricultural research centres. With the notable exception of ICTA in Guatemala, the programmes included in this analysis did not emerge in response to reforms or initiatives within the NARIs and were, in varying degrees, separate from the organizations that they were formally part of. This 'distance' enabled OFR programmes to develop in different directions and illustrate a fair degree of variation in scope, scale and range of functions. The distance also produced tensions, so that as special funding was withdrawn it became difficult for the parties most directly involved to objectively assess the strengths and weaknesses of different approaches for communicating with clients and adapting research results to the requirements of specific farming systems.

Today, several years after the peak period of FSR projects, it is perhaps possible to focus more on functions that are needed and less on the jurisdictional disputes between research and development and between public and private or voluntary sectors. Although the importance of OFR activities may have declined in several countries in response to the withdrawal of external funding and general public sector financial problems, the need for the functions which FSR programmes sought to offer remains as strong as ever. As more local initiatives emerge to address these needs, it is hoped that the lessons offered by the experiences recorded here can be weighed in considering OFR options.

REFERENCES

1. **Gilbert, E., 1990**. Costs and Financial Management of On-Farm Client Oriented Research: a Review of Experiences in Nine Countries. Unpublished report, ISNAR, The Hague.
2. **Merill-Sands, D., & J. McAllister, 1988**. Strengthening the integration of on-farm client-oriented research and experiment station research in national agricultural research systems (NARI): management lessons from nine country cases. OFCOR Comparative Study Paper no.1. ISNAR, The Hague.
3. **Kean. S.A. & L.P. Singogo, 1988**. Zambia: Organization and Management of the Adaptive Research Planning Team (ARPT), Research Branch, Ministry of Agriculture and Water Development. OFCOR Case Study no. 1. ISNAR, The Hague, The Netherlands.

   Bingen, R.J. & S.V. Poats, 1989. Staff management issues in on-farm client-oriented research. OFCOR Comparative Study no. 5. ISNAR, The Hague, The Netherlands.
4. **Merill-Sands, D. & J. McAllister, 1988**. (Op. cit.).

   Ewell, P.T., 1988. Organization and management of field activities in on-farm research: a review of experience in nine countries. OFCOR Comparative Study Paper no. 2. ISNAR, The Hague, The Netherlands.

   Ewell, P.T., 1989. Linkages between on-farm research and extension in nine countries. OFCOR Comparative Study Paper no. 4. ISNAR, The Hague, The Netherlands.

   Biggs, S.D., 1989. Resource-Poor Farmer Participation in Research: a Synthesis of Experiences from Nine National Agricultural Research Systems. OFCOR Comparative Study no. 3. ISNAR, The Hague, The Netherlands.
5. **Gilbert, E., 1990**. (Op. cit.).
6. **Ewell, P.T., 1988**. (Op. cit.).
7. **Pardey, P.G. & J. Roseboom, 1989**. ISNAR Agricultural Research Indicator Series: a Global Data Base on National Agricultural Research Systems. Cambridge University Press, Cambridge.
8. **Kean. S.A. & L.P. Singogo, 1988**. (Op. cit.).
9. **Ewell, P.T., 1987**. Summary of Ecuador Case Study. Memorandum dated 1 July, 1987. ISNAR, The Hague, The Netherlands.
10. **Kean, S.A. & L.P. Singogo, 1988**. (Op. cit.).
11. **Bingen, R.J. & S.V. Poats, 1989**. (Op. cit.).
12. **Gilbert, E., P. Mation & P. Eyzaquire, 1994**. New Perspectives for Vulnerable Institutions: Agricultural Research Systems in the Small Countries of West Africa. Discussion Paper no. 94–17. ISNAR, The Hague.

Chapter 8
Training for FSR

8.1 THE HISTORY OF FSR TRAINING IN EAST, CENTRAL AND SOUTHERN AFRICA

Ponniah Anandajayasekeram

Training in FSR methods stresses practical work, team work, learning by doing and personal development. During training, the multidisciplinary nature of the procedures and the team effort required should be clearly demonstrated to introduce a much needed change in the attitudes of researchers, both to small farmers and to other specializations.

8.1.1 An overview

The institutionalization of the FSR process in eastern, central and southern Africa (ECSA)[1] has led to changes in the organization of research, in research/extension linkages, in planning and priority setting processes, in resource allocations and, not least, in the training needs of countries in the region.

For most of 1970s and 1980s, the majority of the FSR training activities, both national and regional, were organized, sponsored and conducted by the International Maize and Wheat Improvement Center (CIMMYT) and other International Agricultural Research Centres (IARCs), particularly the International Center for Tropical Agriculture (CIAT), the International Potato Council (CIP), the International Centre for Research on Agroforestry (ICRAF) and the International Livestock Center for Africa (ILCA), as well as the technical support teams of bilateral donors. Since the late 1980s much of the effort from donors, the IARCs and National Agricultural Research and Extension Services (NARES) has been in incorporating FSR concepts and methods into the curricula of national academic institutions and building national training capacity.

This contribution reviews the evolution of farm systems research (FSR) training activities in the region, with an emphasis on CIMMYT activities in the early days, the current status, and the practical difficulties involved in developing and sustaining the FSR related activities at the national and regional level.

Special skills in FSR training

Several terms have been used to describe the set of procedures involved in FSR[2]. In this contribution, FSR is interchanged with other terminologies commonly used in the region. Despite this variety, all the procedures used emphasize farmer-oriented and problem-focused, multidisciplinary research.

Training in FSR methods stresses practical work, team work, learning by doing, and personal development. During training, the multidisciplinary nature of the procedures and the team effort required should be clearly demonstrated to introduce a much needed change in attitudes of researchers, both to small farmers and to other specializations. Providing training to scientists and extension staff together as a multidisciplinary research team anticipates that these participants will learn to value each other's contribution as essential to successful research and extension programmes. The training also requires a better understanding of the small-farmers' production system and decision-making process which entails a substantial amount of field work and interaction with farmers. The systems perspective of the procedures is highlighted by actually working within the system concerned. Training procedures cut across disciplinary boundaries and the participation of several disciplines promotes interdepartmental and interdisciplinary collaboration. Finally FSR training is very field based, and trainers need adequate practical exposure to the methods used. However, such experience is no substitute for the training skills that allow trainers to communicate with trainees. FSR training therefore requires two sets of skills, both of which have been scarce in many developing countries.

8.1.2 The evolution of FSR training in the region

Training in FSR procedures to scientists from developing countries began in the early 1970s at the headquarters of some IARCs, including CIMMYT, the International Institute for Tropical Agriculture (IITA) and the International Rice Research Institute (IRRI). CIMMYT and other centres have also organized in-country courses during critical periods in the crop season so that centre staff could work in the field with participants at these key times. In-country trainers were also brought to the IARC headquarters for in-service courses. These training programmes have three distinguishing characteristics: learning by doing, an emphasis on finding short-term solutions to production problems and a commodity-based research focus[3]. Trainees learn to conduct farm-level surveys aimed at identifying the most important production problems of farmers, and to design, execute and analyse a set of on-farm trials. The basic purpose has been to help participants develop the ability to identify the most relevant research opportunities as well as the appropriate solutions to production problems.

CIMMYT'S Regional Economics Programme

In the mid 1970s, the Consultative Group on International Agricultural Research (CGIAR) promoted the establishment of regional programmes by the IARCs. One of the first, established in 1976, was a CIMMYT Regional Economics Programme for ECSA. In its initial phase, diagnostic and experimental planning techniques were implemented in six countries of the region in collaboration with National Agricultural Research Service (NARS) scientists. These demonstrated the value of the systems approach in addressing the production problems of the small farmers[4]. In the early 1980s, there was a rapid increase in donor interest in the use of FSR to improve the adoption rates of new technologies by small farmers. There was a massive expansion of FSR projects in the region, many of them funded by USAID under Title XII grants, with technical support coming from American universities. This surge of activity created a demand for trained FSR researchers, but there was no training capacity in the region to meet it.

The most immediate and effective way for training practitioners to meet demand was through short courses followed by on-the-job training in the region. From 1979 the balance of the CIMMYT programme shifted from demonstration and promotion to training, an emphasis which continued until the close of the programme in 1992. A two-level strategy was developed; exposure and capacity building for all interested NARS in the region, with a concentration of effort on six countries: Ethiopia, Kenya, Malawi, Tanzania, Zambia and Zimbabwe, where commitment seemed strongest and the circumstances most suitable.

Regional training in FSR methodology

The first regional FSR training activity was organized by CIMMYT in Nairobi in 1979. The primary objectives of the workshop were:

- To create an awareness for the need for, and potential of, a systems-based, farmer-oriented, problem-focused research approach among researchers and research administrators in the region.
- To develop a nucleus of FSR-oriented scientists for the NARS of the region.

After 2 years of workshops in Nairobi, it became clear that they should be affiliated with a regional institution to ensure their continuity. The University of Zimbabwe took on this responsibility and the regional workshops moved to Harare. From 1982 to 1992 its Department of Agricultural Economics in the Faculty of Agriculture ran FSR training workshops in collaboration with CIMMYT[5]. Two workshops were run annually and were open to researchers, mostly first degree holders working in, or about to join, on-farm research (OFR) projects. The first workshop, normally held for 3 weeks in February/March, covered the diagnostic and experimental planning stages of FSR. The second, for 2 weeks in September, covered the implementation and evaluation of on-farm trials. A case study approach was used in these workshops based on an ongoing FSR programme in Mangwende communal area 80 km north of Harare. The workshops were attended by 25–30 researchers from NARS of up to 10 countries each year.

When developing short training programmes, it is difficult to balance conceptual and practical aspects. It is essential to include a minimum amount of concepts, while practical aspects absorb most of the training time. The regional training programme has undergone consolidation and refinement in this area over the years. Reviews of the training programme identified the need for new course segments, for more locally specific training materials and the expansion of participants to include extension staff, university teachers and some professionals from NGOs as well as NARS researchers. Reviews also identified disciplinary topics that required reinforcement, and CIMMYT began special training workshops to meet this need.

The regional training programme at the University of Zimbabwe provided initial exposure to FSR procedures, but could not meet the demand for trained FSR researchers in countries with an increasing commitment to the process, including those with large donor-assisted FSR projects. The young national scientists assigned to these, often as counterpart staff, needed more intensive training. Such staff would be responsible for a sustained effort in FSR once bilateral donor projects, with technical support from expatriate scientists, were completed. In several such projects even the expatriate technical support had no hands-on experience of FSR and benefited from participation in the regional training courses.

As interest in FSR increased among countries and donors, in-country training began. Introductory workshops were held for research and extension managers, and an in-country 'call system' of training was developed. The annual regional training workshops in association with the University of Zimbabwe continued to offer an introduction to researchers from newly interested countries, and also helped countries that already had in-country training programmes to initiate new staff joining the FSR cadre. Finally, these regional workshops became a vehicle for the training of trainers when countries decided to develop their own training capacity in FSR. National trainers were among those running the workshop.

In-country national orientation workshops

The in-country orientation workshop lasted from 3–5 days, providing an overview of FSR procedures and the implications of adopting them for the research and extension process. This longer workshop programme included a field visit to bring out, on the ground, the sort of insights FSR offered for the development of appropriate technologies. It was initially attended by project managers, top and middle level national research and extension administrators, and representatives of academic institutions. Later, as national experience accumulated, scientists from other countries in the region participated as resource persons and shared the experiences from their own countries.

In-country call system training

The most intensive FSR training offered through the CIMMYT programme was the 'in-country call system'. This began in the region in 1983 and, as its name suggests, it consists of a sequence of 'calls' at the main stages of the sequence from farm system diagnosis to experimental evaluation. Three or four senior scientists from CIMMYT would teach, often drawing on specialized help from other IARCs and bilateral programmes. A country had to commit a minimum of 10 scientists to an 'OFR' cadre to qualify. Trainees were mainly from the national research and extension services, but also included university and NGO staff, as well as technical support professionals from donor-funded projects. Six such courses had been completed by 1987 in five countries[6].

In the 'call' system trainees are convened at a particular location near an OFR site several times during a crop cycle, where they follow the steps of an OFR programme. The sequence involves four to six calls over a 12–15-month period to capture one full cropping season. During that time, experiments are planned, implemented and evaluated by participants. The duration and the number of calls varies with the local situation but 'calls' were normally timed to allow:

- Diagnostic survey work.
- Survey interpretation and experimental planning.
- Statistical design and the laying out of experiments in the field.
- Experimental observation and recording keeping.
- Harvest of experiments, data compilation, evaluation and interpretation.

The formats used included classroom teaching with an emphasis on group work, field work, field demonstrations, and on-the-job training. Ideally, participants would complete the same step in their own 'home' field locations before the next call. Such training at the national level reduced travel costs, reached more participants, allowed training courses that were more relevant to national conditions, and accelerated the development of national training capabilities.

Topical training

At the end of each in-country training cycle participants were asked to assess their ability to conduct independent FSR and identify the areas in which they needed additional training. These evaluations highlighted gaps in capabilities and, in consultation with managers, short courses were organized around these topics. These included diagnosis, planning, report writing, farmer participation in OFR, crop–livestock interaction, gender analysis in FSR and trial evaluation techniques.

To improve the balance of the regional training workshop at the University of Zimbabwe, two supplementary training workshops were initiated. The first responded to a demand for more time to be devoted to methods of survey data collection, management, analysis and interpretation for social scientists. An annual workshop was started in 1987 to cover these aspects using computer software and meet a demand for computer training. Initially conducted jointly by CIMMYT and ILCA in Addis Ababa, it was subsequently run in collaboration with Egerton University in Njoro, Kenya. The second workshop was initiated to bolster the capacity of agronomists to move from a set of on-farm trial data sheets through to the documentation of results and conclusions. It taught trial analysis, interpretation and report writing, and, again in response to the demand for computer training, used the Michigan State University Microcomputer Statistical (MSTAT) programme to carry them through the process. This workshop was conducted in collaboration with the University of Zimbabwe. In both these supplementary workshops participants were required to apply the analytical tools to their own data sets. Since the participants came from diverse backgrounds with different levels of experience, great flexibility was needed in managing the programme.

Reviews of the OFR programmes in several countries in ECSA showed that while trial planning was done by the research team, the monitoring of the trials was left in the hands of technical assistants, a cadre with little experience in on-farm trial management and data collection. In many instances the quality of management was poor and data were lost; trials were either abandoned, or the information collected was too inadequate to draw meaningful conclusions. Monitoring of the farming system was often completely ignored by trial assistants. It was not a part of traditional experimental recording and, not understanding its purpose, they undervalued the improved understanding of local farming that it provided. Responding to the need to train trial assistants, a programme was created to:

- Improve the understanding of technical assistants (TA) and field assistants (FA) of management in field experimentation and data collection procedures.
- Expose TAs and FAs to techniques that would improve the quality and efficiency of field experiments.
- Allow research officers an opportunity to gain experience in training research assistants.

8.1.3 FSR training in academic institutions

The regional and in-country training activities were organized and conducted by CIMMYT, by other IARCs, and by bilaterally supported donor programmes for a considerable time. All of these channels depended on donor funding and the training was jointly administered by the IARCs and NARS. Although the in-country training programmes and short courses organized and conducted by the centres and outside agencies were clearly of value, they lacked the continuity to ensure a flow of suitably trained researchers.

In the mid 1980s it became apparent that training should become an in-country responsibility, and that the national higher learning institutions must develop their own training capacity. Capacity building efforts to help these institutions develop national higher learning institutions were given high priority. Initially, both regional training workshops and in-country call system training were used in the training of local trainers, but it was clear that institution building should be based on strengthening national agricultural universities to ensure long-term support and continuity in manpower development.

Some interesting distinctions developed. First, between the 'special' and 'part' courses. While 'special' is self explanatory, 'part' describes the modification of existing courses to offer some FSR content. Second, the distinctions between FSR teaching in a single department, often agricultural economics, and its teaching across a range of disciplines. Tanzania, with the support of a Netherlands bilateral project, has extended FSR teaching to its institutes offering

Table 8.1.1. History and current status of FSR training in universities in the region.

National universities	Year of start	Institutions involved	Current mode	Students involved	Teaching department
Zimbabwe	1983–90	CIMMYT, IDRC, Ford	Part courses	All taking course	Crop science & agric. econ.
Alemaya, Ethiopia	1983	ORSTOM	Special course	All in animal & plant science, agric., eng. & econ.	Agric. econ., FSR unit
Egerton, Kenya: • Diploma • Degree	 1976 1987	CIMMYT, FAO/UNDP SIDA/FAO	Special course	All degree in agric. & econ. All dipl. in farm management	Agric. econ.
Nairobi			Part courses	All agric. under- and postgrad.	Agric. econ., animal prodn.
Moi, Kenya			Part courses	All in faculties agric. & forest	Soil science & agric. econ.
Kenyatta, Kenya			Part courses	All	Food sci. & P.H. technology. Ag. ext. & Rur. soc.
Makerere, Uganda	1989	USAID, CIMMYT	Part courses	All agric., forestry, food science & technology	Rural econ. Crop science. Agric. ext & education
Malawi	1988	USAID	Part courses	All opting for rural development	Rural dev.
Zambia	1986	CIMMYT	Special course	Selected 5th & final year in crop & anim. science, agric., eng. & econ.	Agric. econ.
Sokoine, Tanzania	1978 formally in 1988	IDRC, CIMMYT, SIDA/FAO	Special course	All students	Rural econ. Crop science, agric. ext. & education
Diploma Inst., Tanzania	1991	Netherlands, CIMMYT, SIDA/FAO			
College Agriculture Botswana	*c.* 1988		Part courses	All agriculture	Agric. econ. & ext. education

agricultural diplomas, and all students receive the training. Three universities in the region – Sokoine in Tanzania, Zimbabwe and Nairobi – offer postgraduate degrees in FSR, the first two through field research.

There are plenty of examples of the kind of issues confronting the popularization of FSR training in the universities:

- In Zambia it was found difficult to persuade disciplinary departments to accept interdisciplinary reports from their students. Chinene, in 1991, stated that 'there was pressure to demonstrate a disciplinary emphasis in the work'[7].
- Institutions looked at a variety of options[8] and found that the best way to introduce OFR methods into academia is by adding systems-oriented sections in various disciplinary courses. A wider range of students is thus exposed to the perspective.
- The recurrent expenditure needed for fieldwork is beyond the budget of all regional universities. Practical training is best provided on an in-service or vocational basis. This calls for a close working relationship between the universities and the NARS and the attachment of students to NARS programmes in FSR.
- Despite much effort from staff at the University of Zambia, the early studies did not merit publication and staff were reluctant to participate in FSR programmes because of the poor publication opportunities[9].
- University staff are often overloaded with projects, many well paid. Unless parallel incentives can be offered there is not much to attract staff to FSR.

The progress noted in Table 8.1.1, however, demonstrates that these drawbacks can be overcome.

8.1.4 Crop management research training (CMRT)

Although the in-country call system type of training had been effective in imparting FSR skills, it required a very intensive input from IARC staff and, to be cost-effective, required a minimum of 10 national scientists working full time in FSR. Many countries in the region did not have this number but still needed training in FSR procedures, especially in diagnostic techniques and experimental management, both on-farm as well as on-station. In keeping with the emphasis on building up local training capacity, some IARCs began to move their headquarters-based training activities into the regions in collaboration with the larger, more mature, NARS. CIMMYT began exploring the possibility of building up production agronomy capability across the ECSA region.

Towards the end of the 1980s the reorganization of the NARS in Kenya placed a heavy emphasis on adaptive research. The Kenya Agriculture Research Institute (KARI) felt that a national training capacity should be developed within Kenya to increase the training opportunities and to provide regular in-service training for its staff. The Egerton University of Agriculture, with long experience in training diploma holders to assist the farming community, and a new mandate to create a research and extension division, was very eager to undertake a farming systems related research programme. In addition, Egerton had a history of catering for students from the region as a whole. The idea of a regional training facility for CMRT was proposed by a group of research and extension administrators, donors, and representatives of academic institutions at a meeting in Addis Ababa in 1989. The NARS representatives welcomed the idea, as did the regional representatives of CGIAR, and both CIMMYT and CIAT were willing to commit personnel to support this training. Donors were approached and USAID and Canadian International Development Assistance (CIDA) came forward to fund the new venture. As a result, a regional joint KARI-Egerton-CIMMYT Crop Management Training Programme was initiated at Egerton University in 1991, with a strong systems perspective and farm-level production orientation. The training objectives were:

- To develop a sustainable national and regional crop management training capacity.
- To improve the quality of crop management research through training and collaborative research.
- To provide research supervision for postgraduate student training.

The first group of trainees began their studies in March 1991. Initially the technical input for the course was provided mainly by CIMMYT, KARI,

Egerton University and CIAT and the project was administered by CIMMYT. It was anticipated that the project administration and coordination responsibilities would be handed over to the national institutions within 5 years and they would also support the teaching. A national coordinator now administers the project and national trainers carry out most of the teaching.

8.1.5 FSR training; impacts and lessons learned

Impacts of FSR Training

The introduction and institutionalization of training in OFR concepts and methods have brought many positive changes. National academic training has become much more sensitive and responsive to the needs of the NARES and small resource-poor farmers. It has helped to draw the NARES and agricultural higher learning institutions closer together, and established better working relationships, coordination and linkage mechanisms, including joint planning and execution of training activities and joint participation in curriculum development. Most of the regional academic institutions have successfully incorporated FSR concepts and procedures into their curricula.

Joint training of research and extension staff has given them an appreciation of the complementarities of their roles in day-to-day activities. This has helped harmonize the relationship between the research and extension services by emphasizing technology generation and dissemination as a continuous process. It has demonstrated the need for feedback mechanisms between research and extension. Beyond this the NARES have begun to play a leading role in coordinating training activities at the national level, resulting in better coordination of donor-supported training activities in relation to FSR and creating a pool of regional trainers who can undertake effective training at the national level. Overall, system concepts, farmer orientation and problem-focused research have been widely accepted by the research and extension community.

Lessons learned

The form in which the FSR process has been institutionalized within NARES has, to a large extent, determined the type of training needed. It has become clear that if institutional affiliations and linkages are not clearly defined, it is difficult to know where training activities should be located, what kind of training to provide, who should do the training or the resource implications. Emerging models have demonstrated that it is possible to introduce FSR methods into academic institutions and this complements the in-service training offered by NARES. It has also proved possible to develop national capacity and to offer FSR training on a continuous and sustainable basis. However, training activities that are totally supported by external agencies are not sustainable. Even regional training, though it can play a crucial role in the early stages of manpower development, is not sustainable in the long run unless:

- The cost of administrative and logistical support is provided by the countries themselves or by some external agencies.
- Training fellowships are provided on a regular basis either by the NARES or external agencies such as IARCs and donor projects.

Unless there are tangible benefits and incentives for national institutions, regional initiatives will wither. The sustainability of national FSR capacity and FSR-related training requires long-term commitment by the policy makers in the appropriate ministries. As Eicher pointed out in 1988[10], investment in research, training and extension should be viewed as a part of an interactive investment package in a 20–30-year framework. To provide such a framework a clear manpower development strategy including training needs (number, type, duration, frequency, etc.) should be jointly developed by the NARES and academic institutions at the national level. Donors and international agencies should operate within this framework. Institution building should be given a higher priority and a significant part of the efforts of IARC and donor projects should go to the building of national training capacity. Even though most of the resources to support the training facilities and activities can be provided by the IARCs and donors during the early stages, a careful plan must be pursued to increase the national budgeting support so that the external input can be phased out. In parallel, incentive and reward systems should be changed to hold skilled staff as trainers, and opportunities should be provided to upgrade their knowledge base in FSR.

The issues that should be considered when embracing FSR in the curriculum at an academic institution include a needs assessment, the approaches to be followed in introducing the concepts and procedures, developing interdepartmental collaboration, course contents, the balance between formal training at academic institutions and in-service training at NARES, sequencing of courses, training of trainers and the development of training material.

8.1.6 The status of FSR training in the region

The current status

The donor-funded, CIMMYT-implemented FSR project came to an end in 1992 and most of the FSR training supported by CIMMYT stopped at the same time. The SIDA-funded, FAO-executed regional FSR-E project (covering Botswana, Zambia, Kenya and Tanzania) is currently supporting the regional FSR training workshop at the University of Zimbabwe, in-service in-country training on participatory rural appraisal (PRA) methods and the institutionalization of FSR training at academic institutions[11]. This project is expected to continue. The CIDA-supported Regional Crop Management research training is also expected to continue. In southern Africa, the Southern African Centre for Co-operation in Agricultural and Natural Resources Research (SACCAR) is currently coordinating 13 regional research projects and networks, most of them implemented by IARCs and offering training on aspects of FSR, usually within a commodity framework.

As shown in Table 8.1.1, most of the academic institutions in the region are currently offering FSR training. In addition, the Faculty of Agronomy and Forestry Engineering, Eduardo Mondlane University of Mozambique, has incorporated FSR concepts and procedures. At a 1994 National Workshop on FSR-E strategy development for the Republic of South Africa (RSA) the urgent need for training in FSR at all levels was emphasized and it was agreed that top management must be briefed on the concept, implementational issues and costs and benefits of adopting FSR-E practices. Middle management require short training courses on the scope and nature of FSR-E methodologies including operationalization and implementation, while research and extension practitioners need in-depth training. It was recommended that FSR-E concepts should be incorporated into the curricula to provide the continuous training needed. Training of trainers was identified as a priority activity, and it was proposed that SACCAR, with assistance from national and international experts and donors, should coordinate this initiative. Given the nine provincial departments of agriculture and millions of small-scale farmers in RSA, there is a great demand for FSR training. The University of Pretoria is beginning revision of its curriculum and the University of Natal and Zululand is also in the process of initiating FSR training. Such developments require continuous efforts in training in FSR procedures.

The future of FSR training in the region

Despite the efforts made in the past, the demand for FSR training will continue in the region. While regional training activities entirely supported by donors are not likely to be sustainable once donor support is withdrawn, short-term regional initiatives, as 'starter' courses, may be desirable in selected areas. The requirements of the individual countries will vary with their stage of FSR development. Observing the current trend, it is becoming clear that the FSR methods will become an integral part of the research and extension services. The implication is that all research and extension staff involved in technology development and transfer will require some training in FSR procedures.

Academic institutions are now much more sensitive to the needs of the NARES, and they are continually assessing their role in FSR training. In the past, the departments of agricultural economics have taken the lead role in organizing and conducting FSR training in most academic institutions. Given the wider awareness and commitment of the institutions this trend will change and other related departments will play a greater role in FSR training. A recent review of the institutionalization of FSR at the academic institutions showed a growing demand for training of trainers and locally specific training materials, as well as greater budget allocations to these functions. As the knowledge system concept develops, new technologies are evolving to refine the FSR procedures, and the research and extension staff will need to employ new techniques in performing their functions. Course content must change to reflect these developments. The efforts of the academic institutions should be complemented by other initiatives. Thus the need for topical training activities and for on-the-job training will expand.

REFERENCES

1. **Anandajayasekeram, P. & M. Rukuni, 1995**. Institutionalisation of FSR Training at Academic Institutions in Eastern and Southern Africa: Experiences and Issues. SAA FSR-E Bulletin, 2, 4–20.
2. **Harrington, L.W., M.D. Read, D.P. Garrity, J. Woolley & R. Tripp, 1989**. Approaches to On-Farm Client-oriented Research: Similarities, Differences and Future Directions. Paper presented at the International Workshop on Developments in Procedures for FSR/OFR, March, Bogor, Indonesia.
   Simmonds, N.W., 1985. Farming Systems Research. World Bank Technical Paper no. 43. World Bank, Washington, DC.
   Merrill-Sands, D., 1986. Farming systems research: classification of terms and concepts. Experimental Agriculture, 22, 87–104.
3. **Dowswell, C.R., 1986**. Strengthening National Research Programs Through Training: a Twenty-Year Progress Report. CIMMYT, Mexico.
4. **Collinson, M.P., 1982**. Farming Systems Research in Eastern Africa: the Experience of CIMMYT and some National Agricultural Research Services 1976–81. MSU International Development Paper no. 3.
5. **Anandajayasekeram, P., 1987**. CIMMYT's Mode of Collaboration in OFR Training for National Research Programs. A paper presented at the Senior Research/Extension Administrators Workshop, May, Lilongwe, Malawi.
6. **Tripp, R., P. Anandajayasekeram & G. Sain, 1990**. The Design and Management of Call System Training in On-farm Research. CIMMYT Economics Working Paper 90/01, Mexico, D.F., Mexico.
7. **Chinene, V.N.N., 1991**. Current Status of the institutionalisation of on-farm research training in the University of Zambia. Paper presented at the Research and Extension Administrators Workshop, April, Nairobi, Kenya.
8. **Mattee, A.Z., 1989**. Introducing the OFR/FSP approach into the SUA Curriculum. In: Minde, I.J., P. Anandajayasekeram and A.Z. Mattee (Eds) Proceeding of the Workshop on the Institutionalisation of OFR/FSP training in National Training Institutions, May, Sokoine University of Agriculture, Tanzania.
   Minde, I.J., 1991. Current Status of the Institutionalisation of On-Farm Research Training in Tanzania. Paper presented at the Regional Research and Extension Administrators Workshop, April, Nairobi, Kenya.
   Opio-Odongo, J.M.A. & S. Zziwa (Eds), 1989. Critical Issues in On-Farm Research in Uganda. Proceedings of a Workshop held at the Faculty of Agriculture and Forestry, Makerere University, March, Kampala, Uganda.
9. **Chinene, V.N.N., 1991**. (Op. cit.).
10. **Eicher C.K., 1988**. Sustainable Institutions for African Agricultural Development. Paper presented at a Seminar on The Changing Dynamics of Global Agriculture: Research Policy Implications for National Agricultural Research Systems. September, sponsored by ISNAR, CTA and DSW at Feldafing, Federal Republic of Germany.
11. **Anandajayasekeram, P., 1992**. Curriculum Development for FSR, a Consultancy Report Submitted to the Regional Farming Systems Programme (SIDA/FAO), UNDP, Gaborone, Botswana.

8.2 ORIENTING RESEARCH TO AGRICULTURAL DEVELOPMENT: THE ICRA TRAINING PROGRAMME

Richard Hawkins

A real job to do provides a relevant and real experience, which can be evaluated by the impact and follow-up of the product. Participants are more motivated when there is pressure to produce a genuinely useful output. And the practical application of participatory appraisal or planning techniques is impossible – as well as immoral – when a study remains just an academic training exercise, or stops at just an analysis of farming systems.

8.2.1 The origins and goals of ICRA

The International Centre for development oriented Research in Agriculture (ICRA) in The Netherlands was established as an initiative of the European donor members of CGIAR in the late 1970s. Donors were concerned that the pool of European scientists who had tropical training and experience from colonial days was drying up, and that it was increasingly difficult to give young scientists overseas experience as recipient governments became less willing to accept people short of work experience. In those European countries without a colonial past, the problem was seen as even more acute. An additional concern was the need for a new generation of agricultural scientists who were well trained in their own discipline, sensitive to the broader environmental, social and economic implications of their work, and able to work in a multidisciplinary context[1].

The donors saw the need therefore, for a training programme that could provide a cadre of agricultural scientists able to apply their specialized knowledge to the development needs of small-scale agriculture. As Ken Anthony has said to me: 'We had no difficulty in selling the idea: the climate was just right at that time, with a more general awareness of the importance of economic and social constraints to change. A few years earlier, it might not have been'. The donors set up a working group to examine training needs in more detail, and how these could be met. In addition, this group recognized the need to provide training for scientists from developing countries, whose academic formation was also considered too disciplinary. They foresaw that equal participation from donor and developing countries would lead to a valuable interaction between those from different environments and backgrounds. It was envisaged that the training programme would result in an international corps of agriculturalists in overseas development and also help contribute to international cooperation[2].

The aims of ICRA, summarized in Box 8.2.1, have not changed significantly since its foundation. Neither has the Centre's perception of the basic type of training needed, or the overall shape of the training programme. However, ICRA's strategy, as well as the detailed contents of the training programme, have continued to evolve to meet these objectives more effectively.

8.2.2 ICRA's evolving operational strategy

During the first decade of ICRA's existence, half the participants were from European donor countries and half from developing countries. Selection of developing country participants was largely on an individual basis: the brightest and the most motivated, from national agricultural research institutes, universities and NGOs. Selection of European participants in the 1980s was according to the interests of donor countries. However, the trend towards more autonomous development agencies and use of private consultancy firms has loosened the

Box 8.2.1. The purpose of ICRA is:
- To assist in strengthening the international cadre of agricultural research workers
- To provide young scientists with the necessary background knowledge and awareness of research needs that will enable them to contribute effectively to national, regional and international programmes designed to produce results
- To produce results which are useful, relevant and acceptable to low income farmers in developing countries

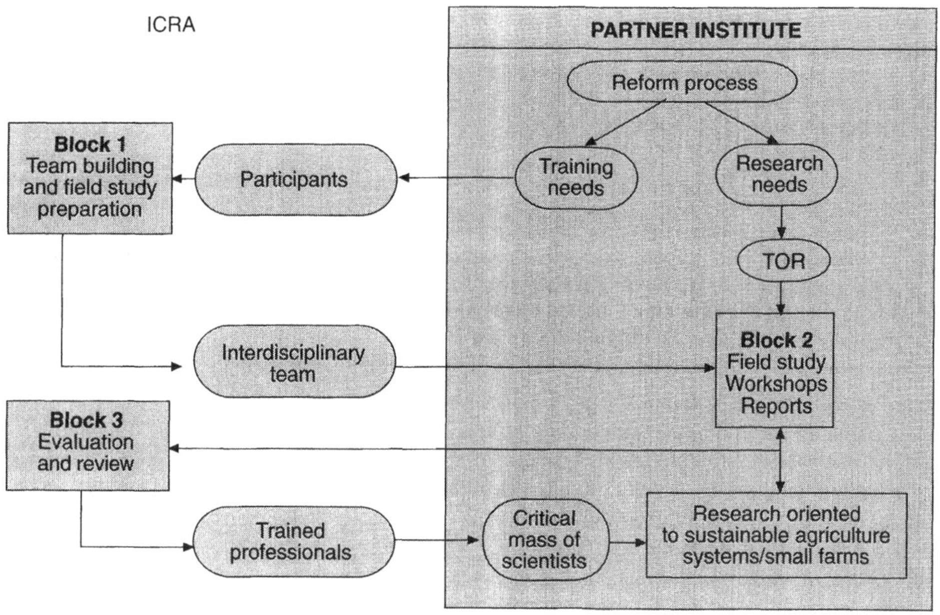

Fig. 8.2.1. ICRA training as a capacity building service.

contact between individuals and the overseas development cooperation 'systems'. Eventually, the priority given by many of the donors to training of their own nationals decreased, and some ceased funding for this altogether.

The addition of the Francophone Programme in 1991, and the increasing percentage of funding through bilateral programmes or projects has also increased the proportion of developing country participants, which stood at over 85% in 1996. This changing clientele, the need to create groups of trainees who then have the critical mass to effect institutional change, and the related need to demonstrate the impact of ICRA's programmes, all led to an increasing focus on NARS in developing countries. ICRA has also strengthened its relationships with specific institutions in these systems. Partner institutes are taking an increasing role in the selection of participants, within the scope of collaborative agreements.

The working group that designed the original ICRA programme recognized the need for a period of in-service training, consisting of survey work useful to the development programmes of host countries. This 3-month assignment, which is carried out by interdisciplinary groups of five to six participants, has been a crucial aspect of the training programme, and one that continues to distinguish it from many others. Increasingly, this field assignment has been seen as a means of integrating service with training, and as part of a more useful package for collaborating NARS than simply training individuals.

This integration of fieldwork service and training is shown in Fig. 8.2.1. ICRA responds to requests from – and looks for – institutions that are undergoing a reform process. Where there is not such a favourable institutional attitude or a shift towards a more client-oriented and interdisciplinary approach, training individuals has proved to be largely a waste of time and resources.

ICRA therefore seeks to establish collaborative agreements with partner institutes, which include the two components of research and training. The research needs are developed into terms of reference (TOR) for the field assignment to be carried out by ICRA groups. The training needs lead to identification of one or two key individuals to participate in the ICRA programme, who will return to their mandate area to carry out the assignment, together with three or four participants on the ICRA programme from other institutes or countries.

Box 8.2.2. TOR for ICRA-EMBRAPA field study in Brazil, 1996.

General objective:

- To assess the competitiveness and sustainability of milk production on family farms in the Serra do Sudoeste of southern Rio Grande do Sul State, given the projected inclusion of milk products under the regional trade agreement MERCOSUR

Specific objectives:

- To identify problems that affect milk production on family farms in the region, taking Cangucu as a representative municipality
- To identify development options for family-farm milk producers
- To recommend modifications to ongoing research programmes at CPACT-EMBRAPA and identify possible future research projects or subprojects
- To recommend modifications in the organization of research programming at CPACT-EMBRAPA, to permit a more holistic and interdisciplinary view

These agreements often last for 3 or 4 years, partly because of the staff time and costs involved in establishment, but mostly because of the objectives of training a critical mass of scientists and achieving a measurable impact.

The field studies normally aim to identify and prioritize research needs, and develop research proposals that can be integrated into the programmes of the partner institute and its collaborators. This output represents an important advance on early ICRA studies, which often ended at a 'farming systems analysis', and draws the institutional dimension in more strongly. The TOR typically focus on a specific geographical mandate area, or on production systems within such an area, and encompass not only natural resource aspects, production and marketing, but also information flow and organizational aspects (Box 8.2.2).

Within the TOR agreed with the partner institute, groups have to carefully plan the study: define the clients and other interested parties, outputs, data needs and efficient methods for gathering and analysing the required data. The identification of research needs and development of research proposals requires close interaction with farmers, input supply and marketing services, extension, researchers from the partner institute and other agencies involved in research and development in the area. Teams also need to develop appropriate decision-making and internal management procedures (Table 8.2.1).

In effect, therefore, the ICRA programme consists of a double cycle of research planning. During block 1 of the programme, participants plan the field study. During the field study, participants plan a longer-term programme of research for the partner institute. Passing through the planning cycle twice reinforces the critical elements of the planning process. Field studies thus provide an important output – a detailed identification of research needs and a

Table 8.2.1. Field study plan; Brazil, 1996.

Week	Objectives	Outputs	Activities	Actors
1–2	Introduce team, review TOR, identify main actors	Refine TOR, establish contacts	Visits, interviews; review secondary information	EMBRAPA, farmers, cooperatives, extension, etc.
3–4	Definition of major farming systems	Preliminary typology, qualitative model of farming systems, select case study farmers	Workshop, group interviews	Farmers, cooperative
5–6	Definition of constraints	Problem-causal tree, interim report	Individual interviews	Selected farmers
7–8	Prioritize constraints, identify R & D interventions	Prioritized list of R & D interventions	Group and individual interviews	Cooperatives, EMBRAPA, large-scale farmers
9–11	Formulate research projects	Draft report	Workshop	EMBRAPA, cooperative
12–13	Finalize report	Final report		

research plan – as well as a concentrated learning event for the ICRA participants and for collaborating staff of the partner institute.

There is, to some extent, an inherent conflict here: learning requires space and opportunity to experiment and make mistakes, a professional assignment requires a high quality product in the shortest time. Nevertheless, ICRA has found that these duel objectives are generally mutually reinforcing. A real job to do provides a relevant and real experience, which can be evaluated by the impact and follow-up of the product (see below). Participants are more motivated when there is pressure to produce a genuinely useful output. And the practical application of participatory appraisal or planning techniques is impossible – as well as immoral – when a study remains just an academic training exercise, or stops at just an analysis of farming systems.

In recent years, ICRA has made a number of organizational changes – and compromises – to improve the integration of the field studies with the programmes of partner institutes. Originally, the participants that made up the field teams were drawn from countries other than the country where the study was to be carried out, allowing participants to gain experience of agricultural systems and research institutes in other parts of the world. However, the absence of full participation of staff from the host institute limited its 'ownership' of the study and results, and hence reduced follow-up and implementation of the findings. ICRA experimented with the idea of accepting complete teams from partner institutes for participation on the programme, but we concluded that this was likely to lead to less innovation. Current practice is, therefore, to include between one and three team members from the partner institute in the total team of five or six, depending on the circumstances.

Another change to field teams made by ICRA is to drop the use of tutors. These were staff specifically hired for the duration of the field assignments (and part of block 1 of the programme when planning of the assignment took place). However, the inclusion of 'tutors' inevitably led to the perception of team members as 'students', and not as professionals doing a worthwhile task. The tutor's role was also difficult: as well as being ICRA's representative with the host institution, they were expected to provide guidance without stifling the development of participants' own leadership and group moderation skills. As an alternative to tutors, ICRA staff, or hired consultants familiar with ICRA, now make two review visits during the field assignment period. It is interesting to note that since ICRA dropped the use of tutors, the sense of responsibility and achievement by the team members has grown. Finally, ICRA has increasingly required partner institutes to assume the local costs of the field studies, or part of the training fee. Apart from the need to diversify funding, this is based on the principle that free goods are usually undervalued.

8.2.3 The elements of capacity building in development-oriented research

As well as the 3-month field assignment, the first ICRA programme designed by the working group included 18 weeks of course work; 13 before the field assignment (block 1) and 5 weeks afterwards (block 3). These sessions covered a comprehensive range of topics including physical, biological, ecological, social and economic constraints to agricultural development, research methods (with emphasis on rural survey techniques), planning and organizational aspects of research. However, participants were critical of the number of subjects covered, and of the lack of integration of the many lecturers with the fieldwork (in total, some 40 outside lecturers were used from a wide range of European universities and research institutions).

After the first course, therefore, the formula was modified to shorten the lecturing component, make greater use of case studies and group discussions, and direct the reduced component of lectures more towards the field studies. This trend towards integrating the 'course work' with the field studies has continued to the present. Our experience has shown that concepts and methods which are not consolidated or operationalized through their practical inclusion in the diagnostic work and research planning that forms the fieldwork are neither appreciated nor remembered.

Currently, block 1 still consists of 12 weeks, but block 3 has been reduced to 2 weeks of debriefing, exchange of experiences and evaluation of the programme by participants. With the advent of portable computers and printers,

Table 8.2.2. Preparatory classroom sessions – block 1.

	Week	Module	Outputs
Knowledge acquisition phase	1	Agricultural development computer skills	Conceptual frameworks and increased understanding of:
	2	Knowledge systems	• Role of research in rural development
	3	Agricultural systems	• Linkages with other development 'actors'
	4	Team management and group planning	• Interrelationships within and between systems • Sustainability of agricultural systems and development
	5	Research evaluation	• Professional and disciplinary roles
	6	Surveys & participatory appraisal methods	• Biological, economic, social, environmental criteria for evaluating research impact Improved teamwork skills
	7	RAAKS exercise	
Field study planning phase	8	Secondary data analysis	Reviews of secondary data
	9	Systems modelling	Initial farm typology; qualitative systems models; definition of performance criteria
	10	Problem analysis, priority setting	Problem-causal models (trees); proposed solutions with *ex ante* cost–benefit analysis, first pass
	11	Proposal formulation guidelines, definition of field study objectives	Defined research themes, hypotheses, study objectives
	12	Fieldwork activity design	Defined data needs, survey methods
	13	Plan finalization	Detailed research plan, report outline, redefined TOR

participants are expected to finalize and present the field reports to the partner institution before returning from the field assignment. Therefore, the overall programme is now 27 weeks. About 15 invited lecturers now complement the four full-time ICRA technical staff, whose workload has inevitably increased with these changes and as a result of the more intensified field team review visits after dropping the use of tutors. The annual Anglophone and Francophone programmes each accept 20–25 participants, and involve a total of eight or nine field assignments spread across different countries each year.

Over the years, ICRA has experimented with the detailed structure of block 1, in an attempt to integrate this initial block with the fieldwork. One approach was to progressively include all subject matter in the context of detailed planning for the field studies. However, this meant that in practice, participants had to be assigned to field groups before ICRA staff and participants themselves had an appreciation of individual qualities and hence the optimal composition of the field teams. The current structure of block 1 compromises by including an initial 6-week phase of knowledge acquisition, during which key concepts are introduced using 'external' case studies and participants' own experiences, before reviewing these concepts in a process of detailed planning for the field studies (Table 8.2.2). Block 1 is therefore very intensive, including a heavy individual reading load as well as lectures and group discussions. As a change of scene, participants are taken on an exercise in a rural area of Europe during week seven. During this 'rapid appraisal of agricultural knowledge systems' (RAAKS), they can begin to appreciate the multifaceted stimuli to agricultural change, as well as methods to investigate the linkages between the different actors involved.

Although ICRA attempts to integrate the topics covered with research planning, these are still wide ranging. Topics were consolidated into an ICRA textbook by Mettrick in 1993[3]. It illustrates ICRA's emphasis on eclecticism. We try to expose participants to different research approaches and methods, as well as different intellectual traditions and points of view, in spite of the confusion that this causes in some participants. In the 1980s, much emphasis was given to comparing different FSR approaches, especially the 'Anglophone' vs 'Francophone' schools and the different models promoted by the various IARCs. A comprehensive review is included in the textbook. However, the proliferation of approaches and terms used to describe FSR and participatory research has led us to rather emphasize the fundamental concepts and the similarities, rather than the (often quite subtle) differences. Do we therefore now promote an 'ICRA methodology'? We do not like to think so, although many participants refer to it.

The type of research that ICRA promotes has much in common with characteristics usually ascribed to FSR (interdisciplinarity, holistic, problem orientation, farmer participation), and ICRA has often been seen as a 'FSR course' – although it has never specifically attempted to be one or be advertised as such. The programme does, however, contain the key elements of FSR training identified by Tripp and Woolley in 1990: a farming systems perspective, FSR methods and research organization. It also places strong emphasis on interdisciplinary communication, team management and priority setting which these authors considered particularly important for the interests of FSR.

FSR programmes and projects have been widespread in the last two decades. Nevertheless, their impact has often been disappointing[4]. There are a number of reasons for this poor performance, chiefly the problems of achieving an interdisciplinary approach[5], and linkages with more strategic/on-station research (OSR) and with extension/other development actors[6]. These problems are in addition to those of diagnostic or OFR methods *per se*, which have perhaps been the focus of most FSR training programmes.

ICRA strongly believes that concepts, methods and skills cannot be divorced from a consideration of professional attitudes, roles and responsibilities. Problems of communication and interdisciplinary interaction require some reflection about professional and disciplinary paradigms, as well as intensive coaching in group tasks. ICRA is heavily oriented towards experiential learning. Practising, making mistakes, reflecting on these mistakes, are all necessary steps of learning. This needs time: there are no shortcuts to behaviour change. We see the relatively prolonged nature of the ICRA programme (7 months, compared to the 2–6 weeks of many FSR courses) as vital in giving 'space' to reflect upon these issues, and the intensive 3-month fieldwork as crucial in learning how to work within intercultural and interdisciplinary groups and with rural people.

8.2.4 Impact

The ICRA programme is not cheap: the total costs of the Centre in 1994 (excluding special projects, not directly related to training), divided by the number of trainees, gives a figure of about US$35,000 per participant. Donors or institutions providing fellowships naturally want to know what they are getting for their money. There are two obvious products of ICRA: trained researchers and field study recommendations and proposals.

During its first 15 years, 1981–96, ICRA trained 409 scientists from 64 countries. Demand for places on the programme(s) has remained strong, with about 10 applicants for every fellowship available – although this may not be a reliable indicator of demand as it also reflects ICRA's changing advertising and recruiting strategy. Many of the alumni have risen into positions of prominence – although again this could be due to astute selection rather than the benefits of the programme *per se*.

An external review panel in 1992 mailed questionnaires to all the 214 ex-participants and their employers to canvas them on the usefulness of the programme; 74 participants replied; about 40% of the total. About 90% of the respondents reported that the ICRA programme had been useful to them (almost all the respondents from developing countries), and over 90% (including all 49 replies from developing countries) had recommended the programme to their employers and colleagues. Most negative responses came from the few Europeans who were no longer working in developing countries. Of the 13 respondents from developing countries who had been promoted, 11 related this to their participation in the programme.

When asked if and how the programme had improved participants' effectiveness, typical replies included comments such as: 'how to organize, plan and evaluate', 'see things as a whole', 'broader outlook', 'how to work in a team' and 'more flexible'. These replies indicate that the programme is achieving its objectives in changing the perspectives, attitudes and general ability of participants to plan and implement development-oriented research. But quantifying such benefits is, of course, problematical.

The other product of the programme is the field study reports and findings. Up to 1996, 82 field studies have been implemented in 28 different countries as an integral part of the programme. In 1994, ICRA conducted an assessment of 12 of these studies, representing 25% of the studies conducted until 1992. Four countries were selected, in each of which ICRA had conducted several studies with the same partner institution: Burundi, Mexico, Tanzania and Thailand. Fifteen replies were received to questionnaires mailed to 38 scientists from the partner institutes and development workers most closely associated with the studies. This survey indicated that the analyses and recommendations published in the study reports had been useful – about 70% of the answers were positive – and that action had been taken on about half of the recommendations in the reports. As with many consultancies, this does not prove cause and effect: teams may have recommended actions that would have been developed anyway. The survey also showed awareness of the studies and reports to be generally disappointing; no reports had been translated into local languages[7].

The results of this assessment, and the need to more effectively monitor ICRA's impact, were among the reasons for the measures taken to more closely integrate with partner institutes noted in the previous section. Preliminary results are encouraging: many of the findings and research proposals developed in the 1995 field studies were incorporated into subsequent work plans of the partner institutes.

8.2.5 Future directions

More than 20 years ago, Dillon, writing in 1976, predicted a training pattern for agricultural researchers in which an initial 1-year introduction to the systems approach would be followed by a 2-year period of disciplinary specialization and finally a period of 1 or 2 years of interdisciplinary work in a relevant agricultural system. However, with notable exceptions, such as the University of Western Sidney Hawkesbury described by Bawden in 1992[8], and the graduate team research project at Cornell and CIMMYT described by Contreras *et al.* in 1977[9], the basic pattern preparation of agricultural research through individual disciplinary thesis has remained much the same. The continued stress on individual output shows that even when institutes base their teaching on systems (such as the emphasis on *systèmes agraires* by CNEARC and INRA in France), there remains the assumption that interdisciplinarity is more a matter of individual thinking than team work. Scientists continue to be trained as individuals and are then expected to work in teams.

Most universities are therefore not organized to respond to the challenges of 'soft' approaches, and exclude from core agricultural curricula the skills and competence in communication, negotiation, team work and participation[10]. According to Knickel, in 1994[11], 90% of teaching at universities is carried out in the form of one-way lectures – hardly likely to develop the sort of skills promoted by ICRA. As argued by Ison in 1990[12]: 'teaching threatens sustainable agriculture'. And in 1994, Gibbon and Bell wrote: 'In short, the people who are expected to develop a sympathetic understanding of farmers and farming systems, and to seek ways of improving their situations are singularly ill-equipped to do this, because they have never learned how to learn ... until we radically rethink the ways in which we approach our education programmes, we will continue to be obliged to mount reorientation programmes for inadequately trained scientists'[13].

We believe that the need for an interdisciplinary team approach continues to grow. The recent emphasis on sustainability implies nothing if not a call for multiple criteria and a 'soft systems' approach when promoting rural change and developing agricultural technology. The continuing fixation of research scientist on yield per hectare and biological production is just not sufficient as we enter the 21st century – however important this aspect may be in a world of finite resources and rapidly increasing population. In the words of one applicant for the 1996 ICRA programme: 'although I have received formal instruction in scientific methodology in agricultural applied sciences, I was not trained to deal with the multitude of complex problems I have to face as the head of a medium sized experiment station in a developing country; that is to say research and transfer methodology, social and economical restraints and opportunities, interaction with decision makers and the political sector, growing environmental issues and change as a constant in all aspects of technological development'.

After the 1992 external review, the ICRA Board stated its belief that 'the philosophy and the original vision that led to the foundation of ICRA are still valid ... that the course is still unique, has no direct competitors and that the need for it is greater than ever'. That said, there is an undeniable tendency for donors to be more hard-headed about the use of public funds and perhaps to be less idealistic. Like many institutions, ICRA has had to diversify its funding and consider its partner institutions and participants more as clients than simply as beneficiaries. Although there have been sequences of participants from individual institutes in previous years (in an attempt to form the 'critical mass' that is often needed for change), the trend of focusing training on specific NARS – where a demonstrable impact can be achieved – is likely to continue.

Currently, the Centre is looking into requests to establish similar programmes at a regional level in North Africa and Latin America, in addition to the European-based programmes at Wageningen and Montpellier. These new initiatives offer scope to tailor the programme to the concerns, policies and institutional arrangements of individual countries. One example is the programme I am now heading in Mexico, which began in 1998. This aims to train groups of participants from priority states and the preparation of proposals for submission to a federal competitive research fund. The intention is that these groups will represent 'interinstitutional consortia', comprising universities, research institutes, NGOs and farmer organizations, which will then collaborate to carry out the research and associated development programmes funded.

The institutional changes underway in Mexico and other Latin American countries also offer a pointer – and a warning – to researchers in other regions. The change in funding mechanisms of publicly supported agricultural research from an institutional model to a one of competitive bidding and channelling through client groups will force researchers to be much closer to clients and their needs than has often been the case in the past. This client-orientation will require a shift in the planning procedures and research methods, as well as in abilities in communication and professional attitudes. ICRA sees its training and research planning programmes as playing an important role in the changes that are taking place to achieve the unchanging goal of more effectively applying research to agricultural development.

REFERENCES

1. **Anthony, K.R.M., H.K. Jain & G. Vallaeys, 1985**. Report of the Review Team of the International Course for Development-oriented Research in Agriculture (ICRA). International Centre for Development-oriented Research in Agriculture, Wageningen, The Netherlands.
2. (Ibid.).
3. **Mettrick, H., 1993**. Development-oriented Research in Agriculture: An ICRA Textbook. International Centre for Development-oriented Research in Agriculture, Wageningen, The Netherlands (French edition also available).
4. **Tripp, R. & J. Woolley, 1990**. Training in farming systems research: review and prospects. Experimental Agriculture, 26, 247–62.
   Merrill-Sands, D. & M.H. Collion, 1993. (Op. cit.).
5. **Byrnes, K.J., 1989**. A review of AID experience: farming systems research and extension projects 1975–1987. AID Evaluation Highlights, No. 4. USAID, Washington, DC.
6. **Merrill-Sands, D., *et al.*, 1991**. (Op. cit.).
7. **Torquebiau, E. & U. Knoche, 1995**. ICRA field study impact assessment: a summary of findings. International Centre for Development-oriented Research in Agriculture, Wageningen, The Netherlands.
8. **Bawden, R., 1992**. Systems Approaches to Agricultural Development: the Hawkesbury Experience. CIAT, Cali, Colombia.
9. **Contreras, M.R., D.L. Galt, S.C. Muchena, K.M. Nor, F.B. Peairs & P. Rodriguez, 1977**. An Interdisciplinary Approach to International Agricultural Training: the Cornell-CIMMYT Graduate Student Team Report. Program in International Agriculture, Cornell University, Ithaca, New York.

10. **Jiggens, J., 1994.** Systems approaches: the challenge to university teaching. In: Systems-Oriented Research in Agriculture and Rural Development (Op. cit.) pp. 791–5.
11. **Knickel, K., 1994.** Application of systems thinking to improvement of higher education in agricultural sciences. In: Systems-Oriented Research in Agriculture and Rural Development (Op. cit.) pp. 796–802.
12. **Ison, R.L., 1990.** Teaching threatens sustainable agriculture. IIED Gatekeeper Series no. 21. International Institute for Environment and Development, London.
13. **Gibbon, D. & S. Bell, 1994.** Farming systems and learning systems. In: Systems-oriented Research in Agriculture and Rural Development. Proceedings of an International Symposium, 21–25 November, 1994. CIRAD-SAR, Montpellier, France, pp. 807–12.

8.3 A NOTE: THE STORY BEHIND THE 'GUIDELINES'[1]

William W. Shaner

Several issues confronted us during the course of the study. Perhaps the most pernicious was the belief among traditional researchers that FSR was nothing new.

8.3.1 Introduction

By the second half of the 1970s a small group of researchers around the world had begun attracting attention to their experimental research aimed at improving production of small farmers in the less developed countries (LDCs). A curious thing about these individuals was that they were addressing a common problem, often in a similar fashion, but without much interchange of ideas. Some of the IARCs had begun testing various aspects of FSR as an interdisciplinary complement to their normal focus along disciplinary lines of research.

Recognizing the potential benefit of these new approaches for the majority of developing-world farmers, USAID called for 'a set of integrated, multidisciplinary farming systems R & D methodologies adapted to the personal and financial constraints of the LDCs, packaged for easy delivery in the form of a comprehensive handbook or handbooks to LDC institutions'[2]. The Consortium for International Development (CID), based in Logan, Utah, won the bidding and, in collaboration with the University of Hawaii, initiated work on the study at the close of 1978.

8.3.2 The early stages

CID encountered an early problem when its choice as principal investigator, Donald Plucknett from the University of Hawaii, declined the position. This meant that another person had to be found quickly, as USAID had accepted CID's proposal and expected the project to get started without delay. CID headquarters called me in Nairobi, where I was working for them on a short-term assignment. It took me about 5 minutes to accept the position of principal investigator. While I knew little about the subject, I accepted the challenge because of my long-standing involvement with agricultural projects in the LDCs. Returning to CID headquarters in Logan, a small group of us set about defining the project's scope and preparing a plan of work to be submitted to USAID's Ken McDermott. We started with the intention of preparing separate guidelines for Asia, Africa, Latin America and the Middle East. However, setting up four teams to simultaneously collect and analyse the data and prepare a coordinated set of guidelines was simply too big a task. So, we settled on a single set of guidelines that would combine information from around the world, while still maintaining the integrity of data from the several regions.

A key factor favouring the study was the support we received from McDermott. He, along with Plucknett, had tracked FSR's evolution for some time. He was up-to-date on the studies by CGIAR on FSR's effectiveness, and was convinced that it was time for USAID to support this study. He passed information on to us as the study evolved, introduced us to those with knowledge about FSR, such as the International Agricultural Development Service (IADS), and provided us with the names of those researchers around the world who were generating the growing interest in FSR. Another key factor was the willingness of these researchers to talk to us about their work.

8.3.3 Our approach

The complexity and diversity of our study dictated that we draw on a range of disciplines and, because this was to be a set of guidelines for use by practising researchers in the LDCs, those participating in the study needed to have extensive practical, overseas experience. To direct this effort we decided to have three senior authors: one to cover the biological sciences, another to cover the social sciences, and myself to look across the disciplines and coordinate the work. We fulfilled our commitment to the University of Hawaii by choosing Perry Philipp as one of the authors; an agricultural economist with experience in extension who showed a keen interest in the project. We chose Bill Schmehl, an agronomist and soil scientist from Colorado State University, to fill the agronomic position from within the CID organization.

So that we might get to know each other better, and to create a common base of understanding, the three of us visited those involved in FSR in Mexico and Central America. Stops included CIMMYT in Mexico, the University of Florida team in El Salvador, the Instituto de Ciencia y Tecnologia (ICTA) in Guatemala and the Tropical Agricultural Research and Education Centre (CATIE) in Costa Rica. The advantages of this trip included its proximity to the USA, the presence of national, regional and international agricultural research programmes, and the opportunity to talk with some of the most active researchers in FSR. Shortly after this trip we reviewed our findings, planned the next phase of data gathering, and began a worldwide search for more information. We wanted to talk to those actively engaged in FSR and gather examples of their work, and each of us focused on one geographical part of the world. We visited national and university programmes in Brazil, Colombia, India, Indonesia, the Philippines, Senegal, Taiwan, Tanzania and Thailand; the regional centre of the Asian Vegetable Research and Development Center (AVRDC) in Taiwan; and the headquarters of:

- CIAT.
- CIP.
- The International Center for Agricultural Research in the Dry Areas (ICARDA).
- The International Crops Research Institute for the Semi-Arid Tropics (ICRISAT).
- IITA.
- ILCA.
- The International Laboratory for Research on Animal Diseases (ILRAD).
- IRRI.
- The West Africa Rice Development Association (WARDA).

Later, some members of our expanded team visited other countries, such as Mali and Jamaica, and paid repeat visits to Mexico and Central America. With strong backing of USAID and IADS we were able to assemble a dozen or so of the most active and well known of these researchers for a 3-day workshop in 1980. Many had heard of each other's work but had not met face to face. We used this occasion to give them an opportunity to discuss their individual approaches to FSR, debate points of interest to them, and advise us on how our study might best proceed.

From this point, our path was relatively clear. We brought in additional expertise in animal sciences, anthropology, biometrics, agroclimatology, communications, cropping systems, extension, interdisciplinary teamwork, organizational and management theory, sociology and training and backed this up with the requisite support staff. Helping us were two outstanding names among FSR researchers: Peter Hildebrand and David Norman. A key decision resulted from a suggestion by our editor, Don Zimmerman, that we contact Westview Press for possible publication of our study as a paperback and Ken McDermott agreed to this use of project funds. As a result we were able to produce copies for worldwide distribution – quite a contrast with our contractual obligation to provide USAID with only 200 copies.

8.3.4 Guiding principles

We were guided by a few key principles. Reference has already been made to our work being a synthesis of applications of FSR. To our way of thinking, it was particularly important that the resulting Guidelines be a collection of successful practices, suitably tested in the field, and appropriate for the level of expertise commonly found among national agricultural research organizations. This was not to be a theoretical treatise or an opportunity for 'university professors' to introduce their ideas about what might be suitable procedures. Our most important contributions, if we were to be successful, would be:

- Gathering information on diverse approaches to agricultural research targeted for small farmers operating under a variety of conditions.
- Synthesizing this information.
- Presenting the results in a logical, comprehensible manner.

Our task was *not* to generate guidelines out of our own experiences and ideas.

Another principle guiding us was a focus on activities in support of research trials, such as preliminary data gathering, planning, conducting the trials, their evaluation and follow-up. Much work had been carried out to conduct on-farm trials suitable to small-farmers' conditions and to link these trials to supportive on-station trials. But little of this work had received the attention it deserved at the time of our study. We gave only passing attention to descriptions of physical, biological and economic systems. While such work could be useful in the long run, especially by staff in the international and regional centres, we did not pursue approaches that relied on mathematical modelling of farming systems or the farm environment. In this sense, we downplayed the 'systems' part of FSR.

Ken McDermott warned us not to be 'captured' by research station managers to avoid the risk of producing a set of guidelines not much different from traditional agronomic research practices. Our study was funded because these very methods had failed small farmers in the LDCs. The distinction between FSR and traditional research can be illustrated by an example from Southern Africa. There, agronomists in varietal improvement were concerned with how the materials in off-station trials performed across different agroclimatic zones. Some varieties were quite unsuited to some local conditions but the trials provided them with data for their across-site analyses of varieties. The researchers had no interest in how a variety might meet the needs of local farmers, even where these farmers were hosting a trial.

8.3.5 Some issues

Several issues confronted us during the course of study. Perhaps the most pernicious of these was the belief among traditional researchers that FSR was nothing new. Applied researchers they said, had been conducting experiments on-farm for years; consequently, they did not need to learn about FSR, even in a LDC setting. Or again, from the economists' side, FSR was no more than farm management. Some professionals insisted that agricultural economists had always met farmers on their farms to suggest and help plan improvements.

Despite our interest in the whole-farm setting, we focused heavily on cropping systems simply because that was the state of the art as we found it. Little work had been done on livestock systems, partly because of the difficulties and risks associated with research on large animals. Sociological research suitable for small farmers in the LDCs was in its formative stages. And we did little with farmer groups because the bulk of current techniques on participatory data gathering, experimentation and decision making were still evolving.

The Women in Development group was critical of one of our early drafts because we had not said enough about women's issues. At that time we thought that we had covered this issue by studying the farm household, followed by suitably identified technologies. Their argument, however, was that women's issues really needed promotion because, historically, they had been overlooked. As a result of these criticisms we included a section on women and switched to gender-neutral wording.

8.3.6 Other insights

Our own study team bought into FSR, for a whole variety of reasons. First were the unbridled enthusiasm, competence, imagination and dedication of those in the forefront of this work. One only needs to track their careers since to validate this early appraisal of ours. Second was the evidence we found that recommendations coming from research stations and promoted by extension were not being adopted by small farmers. There were gaps between what was being promoted and what farmers needed.

Typically, on-station trials tested for genetic potential, while on-farm trials in FSR took specific account of farmers' constraints. Third was our conviction that what works in countries like the USA does not necessarily work in the LDCs. In the USA, for example, many farmers have an educational background similar to agriculturalists in extension and research; information on suitable farming practices is widely disseminated; and the private sector serves farmers' needs effectively. Most LDCs lack these advantages, so that the approach to research must be different. Also, researchers typically favour working within their own disciplines and are rewarded according to their list of publications. Finally, FSR seemed, above all, to be appropriate – identifying small farmers as the target group and introducing procedures that addressed their needs, such as use of recommendation domains, on-farm trials, and post-trial investigations of farmer acceptance of introduced technologies.

We bought into Hubert Zandstra's idea that rather than study a subject to death, one must sometimes 'kick the dog'. We interpreted this to mean that the best way to see if a modification to a farmer's way of doing things is good or not is to try it out under the farmer's conditions. And we bought into Donald Winkelmann's admonition to seek 'better not the best'. We agreed with the idea that farmers accepted changes to their farming practices cautiously and incrementally. While whole farm analyses might point the direction in which the farmer might be headed, they did not offer a solution to the farmers' immediate pressing problems.

The guidelines stress the distinction between multidisciplinary teamwork and interdisciplinary teamwork. A disadvantage of multidisciplinary teamwork can be the tunnel vision that occurs when one addresses a problem purely from the perspective of one's discipline. Interdisciplinary teamwork, on the other hand, brings disciplines together synergistically to address a commonly identified problem. We found that the FSR specialists we contacted were so adept at interdisciplinary teamwork that one simply could not discern their disciplines from the way they talked.

Coming up with acceptable definitions and diagrams can be difficult. While the three of us spent considerable time on them, some could undoubtedly be improved upon. If we were to do it all again, I would like to come up with a more engaging definition of FSR. The figure on the five basic activities of FSR in the chapter on the conceptual framework seems, however, to have stood the test of time, at least in our eyes.

Aside from the normal introductory material, the contents of the guidelines really flowed out of the conceptual framework and these activities to address the key topics:

- Gathering data about farmers and their conditions.
- Planning and conducting on-farm trials, backed by suitable on-station trials.
- Evaluating the results.

We did not discuss typical experiment-station research because that was the convention and already well known. We did include a chapter on extension, because of its importance in diffusing the results of successful research. We rounded out the guidelines by addressing organizational approaches and training needs and opportunities. The body of the guidelines contained the central arguments, while the appendices contained many of the details and examples. Because we had so many people working simultaneously on so many different topics, we catalogued our data by subject, author, country or region, date and possible application. Then, through cross-referencing we could search our database as needed. Strange as it may now seem, it was only after a lengthy search that Tom Sheng, my research assistant, found a programme developed by the US Forest Service that could meet this need. In the end, we were so pressed for time that we failed to make adequate use of this capability. For the observant and the curious, the photos in the guidelines are without captions simply because in the rush to get to press, the photos and their descriptions got separated.

A conclusion we drew towards the end of our project was that FSR is not all that special. Sure, it has certain requirements and tricks of the trade. But, in the end, it is simply good research methodology. What is surprising is that other approaches to research aimed at the small farmer make so many mistakes: they do not recognize the farmers' particular interests, conditions or constraints.

As a personal footnote, after completing the

guidelines, I actively sought overseas assignments in FSR. This effort resulted in one long-term and a few short-term assignments that totalled about 3 years. From this experience, I found the logic of FSR approaches and procedures to be as valid as ever, but that their implementation was far more difficult than I had anticipated. It simply is not an easy thing to change long-standing patterns of work and organization.

REFERENCES

1. **Shaner, W.W., P.F. Philipp & W.D. Schmehl, 1982**. Farming Systems Research and Development: Guidelines for Developing Countries. Westview Press, Boulder, CO.
2. **USAID, 1978**. Farming Systems Research and Development Methodology Contract. USAID, Washington, DC.

Part IV
FSR: the Professional Dimension

EDITORIAL INTRODUCTION

Mike Collinson

Part IV of this book examines the history of the professional organizations in farming systems research (FSR), the evolution of the Association for Farming Systems Research and Extension (AFSRE), the emergence of regional associations and, more recently, the reformulation of the relationship between the international and regional associations. The first four contributions on the histories of the international and regional associations make up Chapter 9, while the three contributions of Chapter 10 explore how FSR relates to the disciplines of many of its practitioners.

THE CONTRIBUTIONS

The first contribution in Chapter 9 is a history of the AFSRE written by Hal MacArthur, a founder member, who highlights key events in its evolution. The contribution on African professional organizations in FSR is by Professor James Olukosi who describes continuing efforts to build networks and associations in western, eastern and southern Africa, highlighting the growing strength of SAAFSRE, the South African Association. While Latin America has no formal professional association. Julio Berdegué, the President of the International Farming Systems Research Methodology Network (RIMISP), one of five FSR-based networks in Latin America, describes their evolution. Nimal Ranweera, a past president of AFSRE, describes the history of the Asian Farming System Association.

In Chapter 10, four eminent professionals have bravely analysed how their respective disciplines have impacted FSR and, conversely, how FSR has impacted on their disciplines. David Norman examines farm management, Peter Hildebrand and Dennis Keeney agronomy, and Constance McCorkle anthropology and sociology.

A COMMENTARY

The International Maize and Wheat Improvement Center (CIMMYT) training exercise used in East and Southern Africa in the 1970s and 80s, mentioned in the editorial introduction to Part III of this book, illustrates the dilemma of current research organizations with their disciplinary and commodity loyalties. During their training, research scientists would visit small farmers and would, invariably, identify priority problems linked to their own disciplines. Similarly, it is difficult for example, to see an International Centre for Research on Agroforestry (ICRAF) scientist offering anything but agroforestry as a solution to soil fertility or animal feed problems, or to see a CIMMYT scientist pointing to improvements for anything other than maize or wheat. The same is true of National Agricultural Research Institutes (NARIs) organized on commodity and disciplinary lines producing recommendations for extension programmes.

Much of the blame for this must fall on the reward systems for scientists. Where they are rewarded by journal publications, the easy route for fieldwork, writing up and getting published lies in simple 'reductionist' monocropping trials. Even intercropping trials are harder. Trying to conduct multienterprise systems trials using a hotly contested participatory (constructivist) paradigm and getting the results published is much more difficult. Such narrow allegiances create inappropriate recommendations for farmers who operate multienterprise systems and highlight the importance of allegiance to the beneficiaries as the appropriate driver of research programming and organization.

The multidisciplinary nature of FSR offers a perspective or an approach, not a discipline. It acknowledges and seeks to articulate the interactions across the subject matter of several disciplines. There has been success; more social scientists can now, courtesy of FSR, communicate with biophysical scientists, as Constance McCorkle points out. This synergy goes several ways. One impact of FSR on agronomists is that they have learned enough economics and sociology to be able to work effectively in interdisciplinary teams. Since many people believe that such teams are necessary for dealing with many of today's problems this is a major contribution. Exactly the same can be said about farmer participation. Farmers need to participate in problem-solving research. Involved disciplines have learned new ways of interacting with farmers. Some, of course, had more to learn than others.

Similarly, the potential strengths of the FSR Associations are their allegiance to people, to the systems perspective, and their cross-disciplinary membership. These characteristics form the perfect foil to the commodity focused, technically dominated research institutions and the many disciplinary based professional organizations. They underpin a strong case for publicly funded, regionally deployed FSR teams, with loyalty to the communities they serve, as an appropriate interface for both public and private R & D institutions.

THE CHALLENGE FOR THE FSR ASSOCIATIONS

Although only Nimal Ranweera is explicit about the crisis in the Asian association it is clear that all the FSR associations, with the exception of SAAFSRE, are in crisis created by low membership, tired leadership and low funding. In 1993, an article by Lightfoot and Noble in the Journal of Farming Systems Research and Extension[1] offered the view that FSR had failed to remain innovative: '(the) imbalance of single-enterprise, researcher managed, on farm experiments versus whole farm, farmer managed experiments has contributed to disenchantment with FSR. Moreover disenchantment has spread into our own ranks. Participatory Rural Appraisal, Participatory Technology Development, Low-Input Agriculture and Agroecosystems Analysis all distance themselves from FSR'.

Van Eijk, in 1998[2] and in Chapter 11.1 of this book, poses an alternative view; that the constant innovation in methodology and recurring claims of new paradigms has destabilized practice and confounded training. I side with van Eijk. First, I see weak implementation, mainly due to poor education and training, as miles apart from current best practice. Second, infighting has been an all too common occurrence, personalities with new flags have sought to take over or displace the movement. Underlying the new flags, new paradigms and new methods is an evolution in process that has been less than coherent, and that cadres on the ground have lacked the information and training to exploit. Take Lightfoot and Nobles' own field of agroecosystems analysis, a diagnostic method that responds well to the widening awareness of environmental needs is one of its innovative features. A second is that it seeks multiple innovations for a whole farm transformation. Yet each new nutrient flow identified as a possible improvement in natural resource management (NRM), will be a new component for an existing system. As such they will be subject to small-farmers' adoption and diffusion processes. Where several are readily compatible with current resource use, just as in the 'old' FSR or indeed in the old 'Transfer of Technology' (TOT), these will be absorbed as a package. At the other extreme however, even single components which threaten farmers' priorities because of their resource requirements will be rejected. Efforts to introduce a multiple component package which implies a wide reallocation of resources will require very heavy supervision, this must raise doubts about its replicability and diffusion.

In my view Berdegué's article, in the Journal of Farming Systems Research and Extension in 1993[3], hits the nail on the head: 'all disciplines need to be able to sort out the useful from the nonuseful, the sound from the unsound, the true improvements from the background noise. It is not clear how this is done in the FSRE movement, with its unrivalled disposition to welcome all aboard ... new developments need to be integrated effectively into the nucleus of current thought. Otherwise, (these) ... become simply appendices that are not internalized within the conceptual and methodological framework of FSRE.'

Berdegué goes on to say: 'The efforts of the Board, symposia, journal and newsletter must reflect the position of the members of the Association. ... I wonder how many of us think of the Association as an active leadership forum in the development of new ideas.' He believed that AFSRE had reached a turning point and saw four ideas as useful in charting a future course:

- The AFSRE must come out strongly and convincingly as a forum for change and renewal.
- The AFSRE must strive to integrate different currents of thought that apply systems concepts to the problems of agricultural development.
- The AFSRE must become a truly international body.
- The AFSRE must be able to obtain adequate resources.

It is a challenge that, 7 years later at the turn of the millennium, still deserves thought and action. It must be remembered that Berdegué was talking from a Latin American, and particularly a Chilean, perspective. Much has changed there in the last 15 years, more so than in the rural areas of Africa and south Asia. The movement itself has to follow FSR principles and acknowledge that national and local circumstances provide a context for the choice of the appropriate FSR approach.

Circumstances have created a challenge for the Association: it needs to keep its eye on the prize: the promotion of the FSR function in the R & D process. It seemed to me that the 1994 Montpellier Symposium showed how anything with the word system affiliated was readily embraced. It exposed a more fundamental diversity issue. There is a clear need to separate out diversity in approach due to the need to accommodate different national circumstances (for example, local empowerment remains politically unacceptable in some countries), and diversity due to confusion in roles and goals. I see three issues underpinning the challenge, each perhaps partly stemming from the huge amount of donor funding associated with FSR in the 1980s:

- The AFSRE has provided a home for pioneers and campaigners in both human and environmental dimensions of development: empowerment, experiential learning, indigenous technical knowledge (ITK), gender differentiation, agricultural knowledge and NRM.
- The AFSRE has sheltered a wide array of 'systems people'. Others in the systems business saw it as a vehicle for promoting their own roles.
- The AFSRE has been used as a haven by professionals who have found Western-dominated professional organizations, which lead the field in their disciplines, too sophisticated, but who continue to make essentially disciplinary contributions in AFSRE fora.

Embracing these pioneers and campaigners has been highly stimulating: we certainly needed to learn from the empowerment school the tools for effective participation and partnership. We needed to take in the merits of experiential learning when training both FSR professionals and farmers. We are a part of the agricultural knowledge system. Equally clearly we are a part of the 'systems invasion' of process in agricultural R & D. All these invasions are important to increase the relevance of R & D and of policy to small farmers. However, each of these campaigns also has its own goals, and embracing each added goal has confounded the mandate of AFRSE, diluted its focus, and rendered it less effective in moving FSR forward.

Perhaps the classic example is participation and empowerment which now has such a grip on local process, particularly in the NGO community. Participators say they work on what farmers and communities think are important and some would argue that a professional understanding of the farming system is redundant. This leads me to resurrect the argument used in earlier days against FSR by the traditional research establishment: if farmers know so much why are they so poor? The FSR process, in bringing relevant *outside* wisdom to bear on local situations, requires an understanding of both the local system and of research, policy and their products, and is greatly enhanced by a 'multiple source' strategy.

I don't believe the Association can continue trying to be everything to everyone, fitting anything with a human angle or a systems perspective under the acronym AFSRE. There seem to me to be three options:

- FSR was born to solve the problem of lack of adoption by small farmers. It has not done so yet. The Association may be better to narrow itself and to refocus on this problem.
- There is a shift to 'systems approaches' at many levels of the systems hierarchy in R & D. It may be appropriate to widen the associations into an umbrella body and draw in agricultural systems professionals promoting the whole range of systems approaches.
- Smallholder agriculture has become a cause of frustration in both national and international communities, its development needs a strong professional constituency

Financial backing for the narrow option may be thin, FSR has perhaps fallen too far down the donor's agenda. However, there may be other ways of attracting funding. One might be to mandate the Association to take up consulting projects within its competence that are attractive to funding agencies. The challenge of organizing regional databases to pursue the multiple source strategy of technology supply is one possibility. If such databases adopted the agroecological framework painstakingly defined by FAO, and sought to build on this, it might capture an important partner for the Association and improve the funding outlook.

The second alternative is an expanded role for the associations. Other systems based initiatives are also facing severe financial constraints. A good example is the creation of the International Consortium for Agricultural Systems Applications (ICASA), an attempt by crop modellers to close ranks in the aftermath of USAID winding up the International Benchmark Systems Network for Agricultural Technology Transfer (IBSNAT) it had supported for many years. A wider mandate for the association might attract up-and-coming systems-based constituencies including crop modelling, economic modelling, geographical information systems (GIS) and the geographers, some of which, like FSR, have been struggling to find an entry into the R & D process in agriculture. Extension is also looking for new paradigms, and FSR, with allegiance to the community as a watchword, might sit well there, as an alternative to its traditional home in research.

A third alternative looks even wider, beyond systems applications to agricultural development strategy. There is some frustration with smallholder development in some countries and among some members of the international community. The idea of sector transformation, a takeover of agriculture by large-scale commercial or public companies is being resurrected. Many countries have tried the public route in commercial agriculture and failed. In the current free-market atmosphere, with many major multinationals increasingly dependent on developing world markets for expansion, and finding greater favour in the developing world, the climate for private sector initiatives in commercial agriculture is improving. A professional association dedicated to the development of smallholder agriculture would embrace all related functions, including, importantly, policy formulation and implementation, and would help balance judgements on when large-scale agriculture is and is not appropriate for sector development.

Whichever option is pursued, the professional associations need to provide leadership. An immediate requirement is the codification of systems applications across the systems hierarchy and the assessment of alternative FSR processes for countries of varying sizes with different political and institutional circumstances.

REFERENCES

1. **Lightfoot, C. & R. Noble, 1993**. Participatory experiment in sustainable agriculture. JFSRE, 4(1), 11–34.
2. **Van Eijk, T., 1998**. Farming systems research and spirituality: an analysis of the foundations of professionalism in developing sustainable farming systems. Thesis, Landbouwuniversiteit, Wageningen.
3. **Berdegué, J., 1993**. Challenges of farming systems research and extension. JFSRE, 4(1), 1–10.

Chapter 9
The Regional and International Associations

9.1 TEN YEARS IN THE MAKING: THE ASSOCIATION FOR FARMING SYSTEMS RESEARCH AND EXTENSION

Hal MacArthur

Although located deep in the American heartland, these meetings quickly became international events, with the participation of researchers and farming systems practitioners from Africa, Asia, Europe, Latin America, the Middle East, North America and Oceania.

9.1.1 Early background: the Kansas State years (1981–86)

It began with the ring of a bell. On the morning of 11 November 1981, Dr Vernon Larson, Director of International Agriculture Programs at Kansas State University, stepped to the podium on the stage of the Little Theatre in the K-State Union and rang his ubiquitous cow bell to call to order the KSU Farming Systems Research and Extension Symposium. Little did he or the organizers, Cornelia Butler Flora and David Norman, realize that this 2-day meeting would launch a movement that would culminate in the formation of an international group: the AFSRE.

The enthusiasm generated at this first symposium prompted Kansas State University to host a total of six annual FSR-E symposia as part of its Title X11 Strengthening Grant programme. Although located deep in the American heartland, these meetings quickly became international events with the participation of researchers and farming systems practitioners from Africa, Asia, Europe, Latin America, the Middle East, North America and Oceania. While the average attendance was around 250 at each of the meetings, it is significant to note that between 1981 and 1986, the last year that Kansas State University hosted the programme, the proportion of non-American participants rose from just 2% to 51%.

The contribution of USAID assistance

The initiation of the symposia and the growing international attendance can be attributed to three factors:

- Farming systems approaches were given increased attention by the International Agriculture Research Centres (IARCs).
- The US Agency for International Development (USAID) was funding FSR projects overseas.
- A global Farming Systems Support Project (FSSP) was created by the USAID Bureau for Science and Technology.

The FSSP and individual projects provided funding support for increasing numbers of scientists from developing countries to attend the annual symposia. With regular participation by key African, Asian and Latin American scientists, and the emergence of a widely circulated FSSP newsletter, the symposia evolved from an annual gathering of American professionals interested in learning about FSR into a forum for networking and sharing of information, methods and experiences across the different regions and cultures of the world.

Box 9.1.1. The Kansas State period (1981–86).

1st Symposium, 11–13 November, 1981

Focus:	Small Farms in a Changing World: Prospects for the Eighties
Attendance:	249 registered: 243 North America, five Asia, one unknown
Highlights:	The role of small farmers, the role of technology development and transfer, bridging the gap between researcher and farmer

2nd Symposium, 21–23 November, 1982

Focus:	Farming Systems in the Field
Attendance:	207 registered: 174 North America, 10 Latin America, nine Asia, two Africa, one Europe, one Oceania, 10 unknown
Highlights:	FSR-E methodologies and case studies of their application in the field

3rd Symposium, 31 October to 2 November, 1983

Focus:	Animals in the Farming System
Attendance:	262 registered: 215 North America, 19 Latin America, 11 Africa, 10 Asia, four Middle East, two Oceania, one unknown
Highlights:	Formation of the FSSP, first FSSP Newsletter. Overview of the status of farming systems research, links between crops and livestock in the research process

4th Symposium, 7–10 October, 1984

Focus:	FSR-E Implementation and Monitoring
Attendance:	272 registered: 226 North America, 19 Latin America, 14 Africa, six Asia, four Middle East, two Oceania, one Europe
Highlights:	American applications of FAR, relationship between FSR and extension, institutionalization and evaluation of FSR

5th Symposium, 13–16 October, 1985

Focus:	FSR-E Management and Methodology
Attendance:	122 registered: 88 North America, 12 Africa, 10 Asia, 10 Latin America, one Europe, one Middle East
Highlights:	Management of FSR-E projects, methods for problem diagnosis

6th Symposium, 5–8 October, 1986

Focus:	FSR-Extension; Food and Feed
Attendance:	241 registered: 124 North America, 35 Asia, 34 Africa, 25 Latin America, 22 Europe, one Middle East
Highlights:	Consumption, post-harvest storage and processing, impact of FSR-E projects on human nutritional status

Symposium content, a progression of themes and issues

During the early Kansas State years, the symposia focused on such key topics as the role of small farmers, technology development and the development and transfer of field methodologies. Early debates focused on the application of statistical methods and traditional research controls to on-farm research (OFR) and the comparison of data from researcher and farmer managed field trials. By the third year of the series, attention was given to special issues within FSR. One symposium focused, for example, on the problems of integrating a livestock component into FSR. Another highlighted the domestic (American) applications of FSR, while a third meeting targeted problems related to food and feed (including the relationship between production and human and animal nutrition and health). Over the years, there was a general evolution in the content of symposium presentations, from discussions of methods to case studies and preliminary reports on actual field experiences from overseas projects.

In addition to hosting the symposium, Kansas State University launched a Farming Systems Research Paper Series in 1981 with support from its USAID Title XII Strengthening Grant. The series includes a background paper prepared by the Kansas State FSR Program Associates, published symposia abstracts and proceedings (1981–1992) and the FSR-E

Bibliography series which contains over 3900 published and unpublished items.

Although USAID funding terminated in September 1993, the KSU Farrell Library Information Support Services for Agriculture (ISSA) maintains and updates this collection as a service to AFSRE and interested researchers and FSR-E practitioners. The unique collection currently contains nearly 6000 items, most of which are analysed and indexed in either a bibliography or microcomputer database. Many of the publications in the FSR Paper Series may still be purchased for minimal cost to help defray reproduction and foreign postage expenses and out-of-print papers can be borrowed from KSU through Interlibrary Loan.

The influence of the FSSP

A major source of encouragement and support for the symposia came from the FSSP, a USAID-funded consortium of universities and private development groups with headquarters at the University of Florida. In 1983, the FSSP bolstered the symposium by furnishing travel grants to support participation of developing-world researchers and practitioners. Beginning in 1984, the FSSP decided to hold its annual meeting in conjunction with the symposium. It helped to strengthen the growing FSR movement by providing technical assistance, developing training materials and producing a newsletter and other publications to promote a consensus of 'theory and practice' in FSR-E. With the FSSP annual meeting, and Newsletter, the FSR Research Papers Series, and the continued availability of USAID project and Strengthening Grant funds to support the attendance of African, Asian, Latin American, and American researchers, the symposium facilitated the emergence of a core body of FSR-E practitioners who established strong personal and professional relationships.

By the autumn of 1986 it was clear that something unique was happening when there was a ground swell of support for continuation of the annual meetings and the newsletter in

Box 9.1.2. The Arkansas/Winrock period (1987–89).

7th Symposium, 18–21 October, 1987

Focus:	How Systems Work
Attendance:	356 registered: 232 North America, 61 Asia, 33 Africa, 16 Latin America, 14 Europe
Highlights:	Five subthemes: 1. Agroforestry Systems 2. Crop Systems 3. Crop/Livestock Systems 4. Information and Communication Systems 5. Macro Systems An *ad hoc* Association Task Force was set up to consider the future of the Association

8th Symposium, 9–12 October, 1988

Focus:	Contributions of FSR-E Towards Sustainable Agricultural Systems
Attendance:	340 registered: 220 North America, 60 Asia, 30 Africa, 15 Europe, 14 Latin America, one Oceania
Highlights:	Task Force recommendations; the development of FSR-E in different regions of the world Symposium papers were organized into regional contributions from Africa, Asia, Latin America and so on

9th Symposium, 8–11 October, 1989

Focus:	Impacts of Farming Systems Research/Extension on Sustainable Agriculture
Attendance:	321 registered: 215 North America, 59 Asia, 19 Africa, 15 Latin America, 13 Europe
Highlights:	FSR-E role in sustaining: 1. Productivity 2. Farmer Participation 3. Institutional Development 4. Environmental Quality The AFSRE was created at the first Association meeting, FSRE Newsletters nos 1–4 were planned and the Journal of the Association of Farming Systems Research and Extension was launched

spite of USAID's decision to reduce support for the FSSP and for FSR in general. At the same time, Kansas State University regrettably announced that it could no longer continue to host the symposium and that it was time to pass the leadership to another institution. The individual determination and institutional commitment to sustain the movement was reinforced on the last day of the 1986 Symposium, when the University of Arkansas at Fayetteville presented a proposal, in partnership with the Winrock International Institute for Agricultural Development, to host the annual symposium for the next 3 years.

9.1.2 From annual meeting to global association; the Arkansas years (1987–89)

The symposium faces a transition in support and content

The first Arkansas meeting, convened in Fayetteville (18–21 October, 1987) ushered in a period of transition and soul searching among the 'family' of core symposium participants. USAID funding was diminishing just as regional FSR-E groups were beginning to emerge in Africa, Asia and Latin America. This changing reality was reflected in the Arkansas decision to organize the symposium around a series of regional presentations. This new format was a critical step towards discussion and organization at the regional level and towards cross-regional comparisons in the application of different FSR techniques.

The decline of USAID support meant that funding for the symposium became a critical issue. With the help of Winrock, the University of Arkansas could meet the immediate challenge and raised sufficient funds from different private foundations and government agencies to keep the symposium alive. While travel funds were generally available from international donors, administrative support for each meeting became increasingly difficult to secure. By the end of the first Arkansas meeting, it was clear that some type of independent institutional base was needed to keep the movement going. As a strategy to broaden the base of support for FSR, Arkansas sought the active involvement of a growing USA farming systems network in the symposium. This was the first attempt to build domestic programme links and laid the foundation for future liaison between AFSRE and the Sustainable Agriculture Research and Extension (SARE) programme of the US Department of Agriculture.

At the last FSSP meeting in 1987, an *ad hoc* committee was formed to explore ways of continuing the farming systems network and newsletter that had been created by the FSSP. At the eighth Annual Symposium, in 1988, after a year of reviewing various options, this group recommended that a task force be empowered to develop a plan for the creation of an 'Association for Sustainable Farming Systems Research and Extension'. This group of 15 individuals representing institutions in Africa, Asia, Latin America and six land-grant universities in the USA, was organized into five subcommittees:

- Future Symposia Committee, chaired by Harold J. MacArthur.
- Editorial Committee, chaired by Cornelia B. Flora.
- Finance Committee, chaired by Robert E. Hudgens.
- Constitution Drafting Committee, chaired by Timothy J. Finan.
- Nominating Committee, chaired by Donald E. Voth.

Under the leadership of Dr George Axinn of Michigan State University, a draft plan for the formation of an international association was prepared and presented at the 1989 symposium.

The AFSRE is Born

The task force report was accepted in principal by a voice vote in the plenary session of the Ninth Annual Symposium. The Association for Farming Systems Research and Extension (AFSRE) was officially launched at 10:00 a.m. on 11 October, 1989 by a unanimous endorsement of the draft constitution. AFSRE was to be a non-profit association with an international membership comprised of individual researchers and practitioners, and various regional associations involved in farming systems-related work around the world. The Association's primary function would be to foster and support the growing number of regional farming systems associations and networks. Its key services and functions would be to regularly convene international symposia and to publish an international journal and newsletter.

The Nominations Subcommittee of the Task Force presented a slate of candidates for the positions of President, President-elect, Treasurer, Editor and members of the Board of Directors. Following additional nominations from the floor, a written ballot elected a President, President-elect and Secretary-Treasurer, along with a 10-member Board of Directors. In addition, an Association Editor was appointed to assume production of the FSSP Newsletter and to launch the AFSRE journal.

When these officers were formally assigned their respective duties following the election the Association had no membership or operating budget. By 6 p.m. on the same day, in excess of $600 had been raised from the membership dues of individuals who became charter members of AFSRE. These funds were received by the University of Arkansas on behalf of the Association and were delivered to the newly elected Secretary-Treasurer for deposit to an account that would be opened on behalf of the Association. At 4 p.m. that same afternoon, newly elected President Peter Hildebrand chaired the first open meeting of the Board of the AFSRE.

Major items of business covered in this meeting included the drafting of a Mission Statement and a policy concerning membership and dues; dates and location for the next symposium and issues related to the language, style and format of the AFSRE newsletter and journal. The following general mission statement was presented and provisionally accepted:

> The association is an international society organized to promote the development and dissemination of methods and results of participatory on-farm systems research and extension. The objective of such research is the development and adoption, through the participation by both male and female farm household members, of improved and appropriate technologies. Such technologies will meet the socio-economic needs of farm families; adequately supply global food and fibre requirements; and utilise resources in a sustainable and efficient manner.

Dues were set at a rate of $40 per annum for members resident in the USA, Canada or Europe and $10 for members residing in developing countries. A proposal from Michigan State University to host the 1990, 1991 and 1992 symposia was approved by unanimous vote and dates for the 1990 meeting were set for 14–16 October. It was agreed that the hosting institution would be responsible for all logistical and administrative arrangements, including raising the required funds to cover all conference expenses, including international travel. On programme development, the host institution would identify a symposium Chair who would work closely with the President-elect of the Association to develop an appropriate theme and format for the meeting.

It was agreed that the FSRE newsletter, previously published by the FSSP at the University of Florida, would be continued by the Association as the AFSRE Newsletter and that a peer-reviewed Journal for Farming Systems Research Extension would be launched by the Association under the guidance of editor, Tim Frankenberger. The first volume of the new journal would contain a selection of the best papers presented at the 1988 symposium.

9.1.3 The Association in transition: the Michigan State period (1990–92)

Between October 1989 and the first Michigan State symposium in October 1990, the AFSRE Officers and Board members worked hard to publicize the new Association, increase membership, secure a funding base and to develop an exciting programme for the 10th Annual Symposium. On the opening day of the 10th Annual Symposium held at Michigan State University, Peter Hildebrand announced in his 'state of the association' report that during its first year of existence:

- Charter membership (as of 31 August, 1990) had grown to include 379 individuals from 62 different countries.
- Two issues of the AFSRE Newsletter had been published and distributed to members.
- Volume 1 (1990) no. 1 of the Journal of Farming Systems Research-Extension had been published and distributed to members.

Evolution of the annual symposia

During the 3 years that the symposium was held at Michigan State there was a marked evolution in its content, from an open symposium for exchange of information on FSR-E methods and results, to a more focused meeting with specific outputs expected. This change resulted from new developments in AFSRE and the results of independent evaluations of the 1990 and 1991 symposia, leading to increased involvement of the Association in the planning of the symposium.

As far back as the first symposium at Kansas State the programme and format had largely been the creation of the hosting institution. During the FSSP period, meetings of the Project were held on the morning after the final day of the symposium and were independently organized. This pattern continued through the period of the formation of AFSRE at Arkansas. Every year, the host institution established a planning committee, and chose a theme for the following symposium based on its own sense of the needs for information exchange among FSRE practitioners. For example, the 1990 symposium focused on sustainability. By 1991, however, it became clear that the symposium could, and should, do more than facilitate an exchange of information on FSR-E methods and results.

During the Michigan State period, the Kellogg Foundation provided crucial funding for the necessary administrative support, based on the assumption that the AFSRE meeting was not simply a professional conference, but a critical element in a communication process that served a growing clientele around the world. At each of the MSU meetings (1990, 1991 and

Box 9.1.3. The Michigan period (1990–92).

10th Symposium, 14–17 October, 1990

Focus:	The Role of Farmers in FSR-E and Sustainable Agriculture
Attendance:	250 registered: 105 North America, 40 Europe/Australia, 38 Asia, 37 Latin America, 30 Africa
Highlights:	1. Farmer organizations in agricultural research 2. Issues in Cropping Systems Research 3. Theoretical considerations and Methodological Approaches to FSR 4. The Political and Institutional context of FSR. 5. Farmer Participation in Management and Evaluation 6. Sustainable Agriculture 7. Gender Issues in FSR

11th Symposium, 5–10 October, 1991

Focus:	Critical Issues and Future Directions for FSR-E in the 1990s
Attendance:	241 registered: 124 North America, 35 Asia, 34 Africa, 25 Latin America, 22 Europe, one Middle East
Highlights:	1. Design and assessment of sustainable systems 2. Institutionalization of FSR-E within national agricultural research systems 3. Farmer participation in diagnosis and on-farm trials 4. Measuring technological change 5. FSR/E contributions to policy, development and trade 6. Gender analysis, making it into the mainstream 7. Methodological issues in impact studies 8. Mathematical and other formal methods

12th Symposium, 12–18 October, 1992

Focus:	Toward a New Paradigm for Farming Systems Research/Extension
Attendance:	206 registered: 124 North America, 31 Asia, 30 Africa, 11 Latin America, 10 Europe/Australia
Highlights:	The content of the symposium was indicative of changes taking place in the: 1. Historical foundations by regions 2. Strategic initiatives for FSR-E in on-farm methods and institutional linkages 3. Information exchange and networking The Association designated 1993 for regional meetings and decided to hold the next global AFSRE meeting in France in 1994

1992) a team of evaluators observed the different sessions of the symposia and interviewed a cross section of attendees from different geographical regions to determine how well the symposia and AFSRE were serving the needs of different client groups. Among the recommendations was the need for AFSRE to take more ownership over the symposium content and to develop programmes that would meet the needs of diverse groups of participants with different resources and levels of experience — from first time attendees to old-hand regulars and from new practitioners to senior farming systems researchers.

Regional initiatives

Two developments at the 1991 Symposium highlighted the need for new roles. First was the emergence and presence of regional farming systems associations or networks. Building upon the regional interactions that began at the Arkansas meetings, regional groupings began to take on a more formal structure and presence. Asian FSRE leaders organized a highly successful farming systems symposium in Bangkok, Thailand in November 1990, leading to the formation of the Asian Farming Systems Association (AFSA). This association, as well as regional groupings in Latin America, Africa and Europe all held spontaneous meetings during the 1991 symposium and each reported on plans for regional meetings during the coming years. The die had been cast and the international symposium would no longer be the only opportunity for FSR-E practitioners to exchange information on farming systems methods and results.

The European region offered to host the next international symposium – a reflection of increased interest in the potential of FSR-E for a role in the reorientation of agriculture in both Western and Eastern Europe. Structural transformation was imminent in agriculture in Western Europe as the European Community moved to a single unified market and an even greater transformation was beginning in Eastern Europe with the end of socialist management and the change to market economics.

The offer also reflected a general sense among the AFSRE global membership that it was time for the annual symposium to leave North America. By hosting the next international symposium, the Europeans would enrich their own region's understanding of FSR-E through the input of the global experience of the membership, while mobilizing the concerns of their region into concrete support for the symposium and its global exchange of information.

Second-stage difficulties of AFSRE

The second development was a recognition within the AFSRE leadership that the input of the membership was vital to establish a long-term plan for the future of AFSRE. Following its formation in 1989, with financial assistance from USAID, AFSRE had successfully launched the Journal for Farming Systems Research-Extension and the newsletter begun by the FSSP had been taken up and continued. The AFSRE had thus successfully passed through a first stage of initiating services for its membership.

Despite the success in establishing a quality publication, the Journal was not financially self-supporting. Developing country FSR-E practitioners made up a majority of AFSRE members and, as they could not afford the journal at a membership rate that would fully cover its publication costs, it was made available to them at less than cost. The Association experienced a major problem with membership renewals since most developing country members paid their dues during the year they were able to attend the symposium. Because of the cost and the difficulty in obtaining foreign exchange in some countries, very few renewals were received by mail from outside Europe and North America.

As the initial assistance for the establishment of the journal came to an end, the AFSRE leadership was faced with a second-stage difficulty – the gap that had developed between its mission to provide the journal to the membership, and the means available to achieve that mission given the nature of the membership that it served. The leadership came to the conclusion that the only way to resolve this difficulty was to present directly to the members for a decision that reflected their collective will.

A new role for the symposium: strategic planning for AFSRE

These developments stimulated the AFSRE leadership to change its relationship with the symposium. With a membership spread across the world, often in locations where communications are not easy or rapid, the annual symposium represented the only venue where a significant portion of the membership could come together to reflect and express its collective will. Hence, the AFSRE leadership decided to structure the 1992 symposium so that the membership could indeed express its will for the future of the Association. The leadership saw this expression as a step towards a strategic plan for AFSRE.

Based on this decision, the AFSRE leadership entered into a truly collaborative partnership with the 1992 symposium hosts, Michigan State University. The AFSRE President and the Michigan State Chair served as Co-chairs for the planning of the symposium. An outline for a strategic plan was developed, and a survey was sent to 77 leaders in AFSRE to get their views on several options for the 1992 symposium. The responses were overwhelmingly in favour of a symposium that focused on the development of a strategic plan, and which combined the traditional call for papers with invited papers that would address key issues and concerns in FSR-E.

Following the survey, the symposium was divided into two parts. The initial 3 days were focused on development of strategic directions for FSR-E. Two days of invited and submitted papers in seven theme areas (diagnosis and farmer participation, gender analysis, on-farm experimentation (OFE), systems perspectives in sustainable development, station research, public and private extension, and policy) were followed by discussion by the participants in small groups on these topics. The remaining 2 days took these strategic directions and considered how best to facilitate information exchange, networking and global linkages among FSR-E practitioners so that the state-of-the-arts of FSR-E advanced in those directions. Presentations by each region on its activities plans, and needs were followed by discussion by the participants on how best to meet those needs. Those discussions resulted in the general directions for the future of AFSRE that are expressed in this strategic plan.

Support for a stronger international base for AFSRE

The AFSRE leadership also sought to identify potential sources of support for its activities in serving its global membership. First, at the 1990 symposium, the AFSRE Board met with representatives of the Ford Foundation, Winrock International and the Kellogg Foundation. The Kellogg Foundation continued its support to the symposium during the 3 years that Michigan State hosted the meetings.

As plans for the immediate and long-term future of AFSRE were developing in 1991 and 1992, the leadership came to recognize that it faced a transition period. On the one hand, the leadership was encouraged by the growth of the regional associations and the initiative of the European region. It saw these developments as potentially strengthening the international base of support for a global association for FSR-E and possibly ensuring the long-term viability of AFSRE itself.

On the other hand, this internationalization meant a change in the relationships between AFSRE as an organization and the regions, and between the international symposium and the various regional meetings being planned. One conclusion of the 1991 symposium was to move the international symposium to an alternate year format following the 1992 and last symposium at Michigan State. This would allow regional meetings to take place independently in 1993 with an international symposium to follow in 1994. The international symposium had traditionally been the major source of new and continuing membership dues collection for AFSRE. The leadership was thus faced with a funding gap from 1992–94. Ultimately, the international symposium in Europe in 1994 would strengthen AFSRE's viability, but in the interim its support base would be weakened.

To assure the continuity of momentum that was built up through the 1992 symposium, the leadership decided to seek foundation support for the transition period. While the European group undertook responsibility for raising funding and developing a programme for the 13th annual meeting to be held in Montpellier, France, the Association leadership was still faced with a critical need to raise funds to keep the newsletter and journal alive. A proposal was prepared and submitted to the Ford Foundation for 3 years of support to AFSRE and, after a period of discussion, the Foundation generously agreed to assist. The African regional group, through the offices of the West African Farming

Systems Research Network (WAFSRN) offered to print the 1992 issues of the journal.

The challenge that faces AFSRE is to find a way in which regional farming systems organizations and networks can share in generating the necessary funds to keep AFSRE alive without compromising their own regional and national agendas. Clearly, AFSRE must be seen as performing a critical support and networking service for all farming systems professionals and their respective regional and national affiliations.

Under the original draft constitution the Board consisted of the AFSRE Officers, Committee Chairs and six members at large, with half the members at large elected to 1-year terms, resulting in a staggered process with half the Board elected at each annual meeting. In 1990 a change was made to include the outgoing President as a member of the Board, reduce the number of members at large and move to elected representatives from each of five regions. At the 1992 symposium it decided that only the officers would be elected by the AFSRE membership and that the regional representatives would be elected or appointed by their regional organizations.

9.1.4 The international years (1992 to date)

Because of the decision to move to a biannual global symposium, hosted in turn by the different regional associations, major changes had to be made in the way in which the AFSRE was organized and operated. It was decided that the President of the AFSRE should, along with the other representatives, be elected for 2-year periods, and should be located in the region which would host the next AFSRE global symposium. David Norman, the President at the end of the symposium at Michigan State University, therefore handed over to Janice Jiggens, at that time working in The Netherlands, who, along with European colleagues, took primary responsibility for the 1994 symposium in Montpellier, France, co-sponsored by a number of European organizations, mainly French (Box 9.1.4). The Montpellier symposium was perceived by the scientific and organizing committees as an opportunity to expose systems ideas more widely in Europe. Since Montpellier the European group has met in Granada in 1996, in Hohenheim in 1998 and plan to meet in Greece in 2000.

The advantages of a strong institutional support system in Montpellier and French institutions as co-sponsors were reflected in an all-time high attendance with some 700 participants, and very timely publication of the lectures and debates. The 1994 agenda reached beyond farming system research *per se* to wider applications of systems methods, particularly at the levels of community and landscape levels beyond the farm, with the issue of environmental sustainability a strong preoccupation.

It was unreasonable to expect the level of support and attendance in France to be maintained as the global symposium moved to the developing country regions. The December 1996 Colombo Symposium in Sri Lanka gained an attendance of 200 and can be viewed as a success in the face of limited funding, and support by volunteers with full-time jobs. It was also clear that civil unrest deterred some from attending. The Presidency of AFSRE then moved from Nimal Ranweera of Sri Lanka to Ponniah Anandajayasekeram, coincidentally a Sri Lankan, working in southern Africa on the preparation for the 15th global symposium in partnership with the Southern African Farming Systems Association and held in Pretoria, South Africa, in November/December 1998.

The revolving AFSRE Presidency, and the revolving biannual global symposium was one of a number of decisions which have shaped the Association. Another was the drafting and approval of a new constitution acknowledging the growing influence of the regional associations. At the end of 1993 the AFSRE was incorporated in the Philippines. Timothy Finan of the University of Arkansas, who had played a key role as secretary-treasurer of AFSRE since its inception, handed over to Gigi Cardenas of the University of the Philippines, Los Baños, who continues in this capacity. At the same time, Timothy Frankenburger, also originally at the University of Arkansas, stepped down as Editor of the Journal of Farming Systems Research-Extension and the AFSRE Newsletter, a role he had played so well since its inception. The Ford

Foundation supported the journal during the early 1990s, giving visibility to AFSRE and providing an important outlet for individuals working in farmer-focused farming systems work, particularly those from developing countries. Over the period 1990–94, 177 different authors had published in the journal, 55% of them from low income countries. Sadly, publication of the journal lapsed after the resignation of Timothy Frankenburger. It was revived by George Axinn stepping in as Editor, a post subsequently taken over by John Caldwell.

A 1997 Task Force reported to the AFSRE meeting at the 1998 Pretoria symposium. That meeting approved constitutional

Box 9.1.4. The international period (1994 onwards).

13th Symposium, Montpellier, France, 21–25 November, 1994

Focus: Systems-oriented Research in Agriculture and Rural Development

Attendance: About 700 professionals. The proceedings were sent to 583 addresses: 376 Europe (including 230 from France), 103 Africa, 38 Latin America, 32 Asia, 26 USA, eight Australasia

Highlights: Sponsored by AFSRE in association with AGRINET, a forum for European-based networks concerned with agricultural research and development in the South, and several French institutions: CIHEAM/IAMM, CIRAD, CNEARC, GRET and ORSTROM, supported by the French Ministries of Cooperation, Foreign Affairs and Agriculture, Fish and Food. The content widened from FSR to systems approaches in agricultural research and development, while maintaining a focus on the biophysical/human interface. Keynote lectures were supplemented by seven workshops:

1. Methods and scale of intervention
2. Environment
3. High input agriculture
4. Indigenous knowledge
5. Local organizations
6. Training
7. Agricultural policy

14th Symposium, Colombo, Sri Lanka, 11–16 November, 1996

Focus: Changing Agricultural Opportunities: the Role of Farming Systems Research

Attendance: 200 professionals, with no regional breakdown available

Highlights: Sponsored by AFSRE in association with AFSA and the Ministry of Agriculture, Lands and Forestry, Sri Lanka. The Symposium had five key themes:

1. Household food security.
2. The environment and agricultural resource management.
3. Innovations and social change: who is empowered?
4. Methodological issues: systemic questions to basic disciplines.
5. Policy and macro-economic issues

15th Symposium, Pretoria, South Africa, 29 November to 4 December, 1998

Focus: Rural livelihoods, empowerment and the environment: going beyond the farm boundary

Attendance: 511 professionals; 340 Africa, 79 Europe, 43 Asia, 18 Australasia, 17 Latin America, 14 North America

Highlights: Organized with the Southern African Association for Farming Systems Research and Extension, the symposium was supported by SACCAR, the University of Pretoria, FAO/SIDA, ODI, the South African Department of Agriculture, Land and Agricultural Policy Centre, AAFSRET and RIMISP

The programme was organized around five themes:

1. Ecologically sustainable development and farming systems
2. Short-term farmer survival vs long-term sustainability
3. Empowerment through capacity building
4. The institutional environment and farming systems
5. Methodological issues and challenges

changes to make the regional associations the sole members of the international association. In parallel the name of the AFSRE was changed to the International Farming Systems Association (IFSA), a change also applied to the journal. This mitigates the issue of dual membership, often beyond the pockets of individual professionals. Individuals will be members of the international association through their membership of a regional association. Funding of both the international association and the journal were discussed at the association meeting at the 1998 symposium in Pretoria. The goal is to put the international association and the journal on a firmer financial footing in a way which complements the role and activities of the regional associations.

9.2 FARMING SYSTEMS RESEARCH AND EXTENSION IN LATIN AMERICA

Julio A. Berdegué

FSR-E in Latin America has a long and fruitful history. Moreover, it is possible to identify a Latin American school of systems-oriented research under the general FSR-E umbrella ...

9.2.1 Introduction

This chapter presents a history of FSR-E in Latin America, from 1967 to 1996, focusing on projects which can be considered milestones in the development of the systems approach and methodology in Latin America. I have concentrated on earlier projects for one reason in particular: reading through several dozen original papers and documents, I found that those old projects broke moulds and blazed new trails, providing answers to current questions and a lot of common sense. They were also truly revolutionary.

The development of FSR-E in Latin America was shaped to a very large extent by the early influence of three key research and development projects: the Plan Puebla, in Mexico; the Caqueza Project, in Colombia; and the Cropping and Farming Systems Projects of CATIE (Tropical Agricultural Center for Research and Training), in Central America:

- They brought into the region the influence of events and ideas in other areas of the world.
- They trained and gave experience to groups of Latin American, North American and European scientists that would lead FSR-E projects in many of our countries.
- They developed the concepts and methods that remain the theoretical and operational foundations of Latin American FSR-E.

9.2.2 The Plan Puebla (1967–73)

The Plan Puebla, started in Mexico in 1967, was the first major systems-oriented experience in Latin America. Although it never adopted the label of FSR-E, it included, applied and developed many of the basic concepts and practices, and was perhaps the earliest significant attempt to take agricultural research to the farms and fields of small and poor peasants. Between 1967 and 1973, the project was a cooperative effort of the CIMMYT, the Postgraduate College of the National School of Agriculture at Chapingo and the National Institute of Agricultural Research, with the Rockefeller Foundation providing financial support. Since 1974 the Plan Puebla has been a programme of the Postgraduate College, funded by the Mexican Government.

In 1967 the creators of the project faced several basic questions[1]:

- Why did scientific and technical knowledge bypass traditional farmers?
- What operational strategies could increase their production of food?
- How could effective interaction among small farmers, technicians and national institutions be ensured?

The Plan Puebla was designed to increase the yield and production of maize, the staple crop of subsistence farmers in the Mexican state of Puebla. The project also had a strong training component. One of the first official reports of the Plan Puebla defined its 'basic philosophy' in the following terms: 'the generation and dissemination of information are part of a continuum, and it is not convenient to separate both components. To generate and disseminate information, the program requires a holistic approach, with constant interaction and feedback along this continuum'[2].

The strategy was defined as 'essentially a simultaneous and integrated plan of attack against the multiple problems that limit the use of technologies which are appropriate to farmers'. The project methodology included:

- A complete diagnostic study of the project area.
- OFR, stressing fertilizer use, plant densities and breeding of high yielding maize varieties adapted to the local conditions.
- Validation and demonstration plots.
- A strong 'technical assistance' component.
- A strong emphasis on building interinstitutional linkages.
- A project monitoring and evaluation component.

The truly innovative idea was that improved technologies need to be field tested before being recommended to farmers[3]. The notion of a continuum between the generation and dissemination of technologies was operationalized through the experimenting farmers. The Plan also broke with the idea that the dissemination of information would be sufficient to promote technical change.

The designers of the project recognized the complexity of the local farming systems. Turrent, in 1974[4], described several mixed cropping systems – maize and beans, maize and fruit trees – and concluded that the different combinations accounted for almost all of the maize acreage in the project area. However, under the influence of CIMMYT, the Plan concentrated on improving the performance of the maize component until the mid 1970s, a strategy later reproduced in several CIMMYT-led projects in Mexico and in Central and South America. Dr L. Jiménez, the person most closely associated with the Plan Puebla since its inception, recognizes that the maize improvement component had not reached its goals by the time it closed in 1972. Writing in 1988[5] he said that the most important contribution of the maize component was to show that the local germplasm could not be out-yielded by CIMMYT's improved hybrids and other varieties, thus 'saving farmers the unnecessary expense of investing in untested seed unlikely to produce benefits'.

From 1974, under the leadership of the Postgraduate College, a national institution, the focus gradually shifted from the maize crop to the cropping system and then to the farm system[6].

The Plan Puebla had a direct influence on several Latin American projects implemented in the mid and late 1970s, including the Caqueza and RioNegro Projects in Colombia; the Cajamarca-La Libertad Project in Perú; and the Francisco Morazán Project in Honduras. Its influence was also channelled through the CIMMYT staff that worked in the Puebla in the 1970s and in the 1980s in other systems-oriented projects, particularly in Central America.

9.2.3 The Caqueza Project (1971–75)

The second milestone in the development of FSR-E in Latin America, was the Eastern Cundinamarca Province Project (better known as Caqueza Project) in Colombia.

A Co-operative Program was established in 1970 by the Colombian Agricultural Institute (ICA), CIMMYT and Plan Puebla, to utilize the lessons learned in the Puebla Project in Mexico[7]. Four pilot projects were established: Caqueza, Rio Negro, Norte del Cauca and García Rovira[8]. CIMMYT, CIAT, the Ford Foundation, the Rockefeller Foundation and the Puebla Project were all involved and, in the specific case of Caqueza, Canada's newly formed International Development Research Center (IDRC) initiated its outstanding role in the promotion of the farming systems approach in Latin America.

The pilot projects were conceptualized as 'Rural Development Projects' emphasizing not only agricultural production, but also social well being[9]. Conceptual framework and research processes were gradually developed, later becoming Caqueza's main contributions to FSR-E development in Latin America. Caqueza began to shape what can be called a Latin American current within FSR-E[10] based on the contribution of Plan Puebla but developing two additional key ideas[11].

- Any technology has socioeconomic requirements, these requirements can become constraints and small farmers are unlikely to adopt a technology if they cannot overcome those constraints.
- It is difficult to design technologies that are within the 'degrees of freedom' allowed by the farmer's limitations. Projects should also attempt to remove or alleviate the limitations while improving the available technologies.

Because of these issues, socioeconomic research received a much stronger emphasis in Caqueza than in the first period of Puebla.

By 1972, the Caqueza project began to formalize the research process framework that would later become standard FSR-E: diagnosis, design, OFR, validation and dissemination. However, more important was the introduction and testing of a large number of specific research techniques and methods, many of which are still part of the basic toolbox of Latin American FSR-E practitioners. These methods were largely quantitative and thus the Caqueza project started a tradition for quantification that is still predominant among Latin American FSR-E initiatives.

The output of publications and field methods of the Caqueza project was such that it became a prime (if not the prime) source of information and analysis. Colombia launched a $350 million Integrated Rural Development (DRI) programme with World Bank support in 1975. The process and methods developed by the Caqueza project were the starting point for the technological development component of the DRI programmes in Colombia and other countries in Latin America[12].

At this point it is possible to identify a characteristic of FSR-E in Latin America: its 'anonymous ubiquitousness'. All the DRI programmes included components with names like 'technology adjustment' or 'production research' programme, which were nothing other than FSR-E.

9.2.4 The CATIE projects (1973–85)

In 1973, CATIE, with headquarters in Turrialba, Costa Rica, and with a regional mandate covering Central America and part of the Caribbean, started the first in a series of three notable projects, the last two of which in particular gave form to a distinct Latin American school of thought within the FSR-E approach.

The first CATIE project (formally known as the 'Central Experiment') began in 1973 in one of CATIE's research plots in Turrialba. It was based on a simple premise: food security. Basic foodstuffs were produced by small farmers who practised multiple cropping, so CATIE needed to understand multiple cropping systems[13]. This project was heavily influenced by the work of Richard Bradfield in Asia. Similar multiple cropping projects had started to appear a few years earlier in tropical American countries such as Bolivia[14], Colombia[15], El Salvador[16] and Mexico[17]. The CATIE Central Experiment caught the attention of the Latin American Bureau of USAID. According to Dr Jorge Soria, the first CATIE-USAID project originated from a chance encounter in a bar at the Eastern terminal in Miami's airport between himself and Dr Donald Fiester of USAID, who, after a couple of beers, committed himself to find funding for CATIE's fledgling cropping system programme. Those beers were eventually worth US$3.5 million!

In 1975 a 3-year cropping systems project was started by CATIE covering all the Central American countries using a new research team with young Latin American researchers who had recently finished their PhD studies.

Strong interaction was established between the CATIE group and the team from the International Rice Research Institute (IRRI) headed by Richard Harwood and later by Hubert Zandstra (formerly the main IDRC advisor at Caqueza), both of whom visited CATIE. Hans Ruthenberg had a strong influence, in particular over CATIE's new animal production systems team, and he was instrumental in obtaining a GTZ (German government aid) grant which allowed an expansion of the research programme in selected areas of Central America. David Norman's work was also influential, in particular on the economists in CATIE.

Unlike the Puebla and Caqueza 'rural development' projects, the CATIE initiatives explicitly assumed they were part of the FSR 'movement' that was also taking place in Asia and, to a lesser extent, in Africa. In strict and formal terms, the 1975–78 CATIE project was the first Latin American FSR-E project and was implemented in six Central American countries[18]. A second CATIE-USAID project (1979–85) moved beyond the cropping systems concept to adopt a whole-farm approach[19]. By then, a strong animal production systems groups had been established in CATIE, after a long conceptual and methodological 'fight' with the livestock specialists who, imbued with a US ranching perspective, could not visualize themselves working owning just a handful of cows, pigs and chickens under the best of circumstances.

Soon, a large number of FSR-E projects were being implemented throughout Latin America, some of them directly by the CATIE group with funding from such institutions as the European Economic Community[20], Canada's IDRC[21], the International Fund for Agricultural Development[22] or GTZ[23].

CATIE's first contribution was conceptual. For the first time there was a systematic conceptual framework available to explain small-scale agriculture in the American tropics. It was here that the methodology was formalized in a way that would remain the standard until the 1990s[24].

The methodological framework of CATIE follows essentially the well-known FSR-E scheme, but with a stronger emphasis on quantitative methods (dynamic farm analysis, bioeconomic studies, technology extrapolation, and so on). Quantitative research was important for both the agronomic and socioeconomic aspects and economic and bioeconomic models were frequently used. The method also gave prominence, however, to strong interactions with farmers; according to Raul Moreno activities similar to the *Sondeo* were always conducted at the start of any activity in a new area, and frequent meetings were held with collaborating farmers[25]. As should be expected, the second project was much stronger than the first in methodological terms; in particular, there were considerable advances in treating the issues of site-specificity[26], the research-extension linkage, and the relationship between the technical components and the whole-farm as the unit of study[27].

The strong and effective interdisciplinarity of CATIE's projects was also important. Animal production systems research, for example, was solidly placed in a farm system context, as is logical in a region where the interaction of animals and crops is the norm and not the exception[28].

Environmental issues were part of CATIE's projects from 1977, if not earlier, when such issues as reducing the use of imported inputs, minimum tillage, nutrient recycling and green manures, were recognized as key elements of a new technological identity[29]. The first alley cropping experiments, for example, started in the mid 1970s[30] and by the mid 1980s, agroforestry systems became a dominant concept[31].

Finally, the CATIE projects trained several dozen students at the MSc level. Almost all the members of the Crop and Animal Production Departments who left CATIE in the mid 1980s became national leaders of FSR projects and other initiatives in their countries. In Central America, the CATIE concepts and methodological approach were internalized at least to some extent by all the national institutions, thanks to the fact that the CATIE projects were carried out in close cooperation with local research and extension organizations in such countries as Costa Rica[32], El Salvador[33], Honduras[34], Nicaragua[35] or Panamá[36]. It is a paradox that, as CATIE's influence reached its peak in Central America and elsewhere, a new administration implemented a series of measures that effectively dismembered the interdisciplinary team and ended their work. CATIE entered into a period of decline that would not be reversed for several years.

9.2.5 Other initiatives

While recognizing the pre-eminence of Puebla, Caqueza and CATIE, other initiatives made important contributions to the development of FSR-E in Latin America.

The ICTA Project (1975–79)

Between 1975 and 1979 Guatemala's Institute of Agricultural Science and Technology (ICTA), with the support of the Rockefeller Foundation and USAID, implemented a project that differed from the CATIE model; it used the *Sondeo*, a method that would be a starting point for the rapid rural appraisal group of techniques[37]. Another important technique refined by this project was the use of farm economic records.

Most important perhaps, is the project's institutional experience. From 1973, ICTA was organized as a systems-oriented institution focusing on OFR, farmers' participation and regional multidisciplinary teams[38].

The PISA-PRODASA projects (1985–95)

In 1978–79, Canada's IDRC funded a cropping systems project implemented by three Peruvian universities, Ayacucho, Cuzco and Puno, with Peruvian scientists receiving training in CATIE. After several years, the project became a farming systems initiative, with a strong emphasis on indigenous knowledge and technologies and became the basis for the PISA-PRODASA series, as well as for several Andean cropping systems projects that IDRC supported until the early 1990s.

In 1985, Canadian International Development Assistance (CIDA) and IDRC approved a project to be implemented by Perú's National Institute for Agricultural and Livestock Research (INIA): the Andean Farming Systems Research Project, or PISA. In 1993, a three-year second phase was approved by IDRC, although the project's name was changed to Sustainable Highlands Agricultural Development Project, or PRODASA. The new CIP-led Consortium for the Sustainable Development of the Andean Ecoregion (CONDESAN) and the Natural Resources and Environment Center (CIRNMA), an NGO, played prominent roles.

The projects benefited from CATIE's decision to disband its farming systems team, attracting several of the displaced scientists to form a very strong technical team. The two PISA projects began with CATIE methodology, but developed it[39] to improve the sociological and economic studies[40], the dynamic analyses[41] and the systems modelling activities[42], as well as embracing micro–macro relationships and the analysis of the impact of policy on small farmers in the Andean highlands[43]. The issues of gender and intra-household dynamics[44] and environmental sustainability[45] were highlighted, and several methods tested to deal with them in an effective manner. The overall approach was more participatory than in the CATIE projects, thanks in part to the formation of groups of experimenting farmers. However, the projects made a serious attempt to combine greater farmer participation with a general methodology that emphasized quantitative methods or used participatory methods to define and refine the objective functions and the simulation scenarios used in mathematical models.

The projects also mobilized or directly conducted important applied research for the highlands of Perú and Bolivia[46], in Andean crops such as quinua[47], animal production systems[48], cereals[49], potato[50], pastures[51] and indigenous genetic resources[52]. IDRC supported at least two additional projects which replicated the project methodology, one in Bolivia (the IBTA-CIID Project), and one in Ecuador (REEPAN-CIID). Finally, in 1992, the project formed the basis for the formation of CONDESAN. The technical team that worked or supported the projects now forms the backbone of the CONDESAN activities in Bolivia, Colombia, Ecuador and Perú[53].

Projects in Brazil (1980–95)

FSR in Brazil had two simultaneous centres of origin. The first was in the Research Center for the Semi-arid Tropics (CPATSA), and the second in the Paraná Agronomic Institute (IAPAR).

In 1980 the Brazilian Institute for Agricultural and Livestock Research (EMBRAPA), launched a National Research Program for the Semi-arid Tropics, coordinated by CPATSA with the participation of over 20 national and international agencies[54]. The programme included three main components: natural and socioeconomic resource evaluation, natural and socioeconomic resource utilization and production systems research.

The resource evaluation component took the lead in developing an approach for agroecosystem and FSR, influenced by French agronomy and operational ecology[55] and the methods became the basis for most systems-oriented research in Brazil. By 1984 a number of the OFR and agroecosystem research projects supported by this programme were beginning to yield results.

In 1986 CPATSA, with advice from the French agencies Centre International de Recherches Agronomiques pour le Developpement (CIRAD) and the Institut Français de Recherche Scientifique pour le Developpement en Cooperation (ORSTOM), developed a new approach to focus more on rural development processes than on FSR. By

1989, its limitations had become clear and CPATSA refocused on agricultural development, using the Massaroca project as a laboratory with a 'recherche-developpement' perspective. In 1992 the concept of agricultural development at the local scale was expanded to include the notion of regional development, and underpins the current CPATSA programme[56].

In 1981, in the southern part of the country, IAPAR participated in a large integrated rural development programme[57]. 1982 and 1983 were dedicated to a complete characterization of the predominant farming systems in the state of Paraná and, in 1983, IAPAR contracted CPATSA to test its systems research methodology, providing the opportunity for the cross-fertilization of the IAPAR and CPATSA groups[58]. IAPAR developed its methodologies and its fields of expertise and is today well known for its diagnostic and farm typology studies[59] as well as its approach to managing soil erosion at the level of the micro-watershed, making extensive use of minimum tillage technologies adapted to the circumstances of small farmers[60].

Another important EMBRAPA group – with several of its leading members coming from the CPATSA experience – started work in the Sao Paulo region in 1985, establishing what is now the Environmental Monitoring Group of EMBRAPA, which works hand in hand with the NGO ECOFORCA. They specialize in the combined use of GIS, remote sensing and fieldwork, to study the environmental relationships and impacts of agriculture[61]. Of particular interest is their published work on the Amazon by Miranda *et al.* in 1995. For 10 years this team of researchers has made periodic surveys of a representative sample of about 500 small farms in an area of recent agricultural colonization within the Amazonian forest including about 250 agronomic, socioeconomic and environmental variables. The analysis of the data for the 10-year period has identified 36 relatively sustainable farm operations (defined as a group in which the values for the agronomic, socioeconomic and environmental indicators tend towards the optimum). The farms all have agroforestry systems with perennial and annual crops and cattle and, from an economic perspective, all make about three times the Brazilian minimum wage. Deforestation rates in these farms is decreasing rapidly and has reached zero in several cases. The International Farming Systems Research Network (RIMISP) and IDRC have both actively supported this important project.

An important development for FSR-E in Brazil was the formation in 1993 of the Brazilian Production Systems Society, launched with the strong support of the research agencies of the states of Paraná (IAPAR) and Santa Catarina (EPAGRI). It has organized meetings (1993 and 1995), which have become the most important fora for the discussion of FSR-E in Brazil.

In 1995 the national research agency, EMBRAPA, reorganized itself into 12 national programmes, one of which is the National Research Program on Production Systems of Small-Scale Agriculture. Fittingly, the programme is coordinated at the national level by CPATSA. The programme is attempting to promote improvements in three areas: interinstitutional cooperation for research and extension with small farmers, frequency of contact between researchers, their projects and small farmers, and the qualifications of researchers to work with small farmers. The programme is supporting the formation of regional networks. One innovative idea is the priority given to 'process technologies', an area highlighted in several countries as a key element of the search for environmental sustainability in agriculture. In 1996, the programme financed 20 research projects, with 82 subprojects implemented by 22 research centres. It is expected that this programme, which is entirely systems-oriented, will stimulate new FSR-E work in Brazil.

Projects following Francophone methodologies

CIRAD and ORSTOM have supported FSR in several Latin American countries.

In Brazil and Venezuela, CIRAD together with the respective national institutes EMBRAPA and the National Fund for Agricultural Research (FONAIAP) has promoted the formation of networks of 'reference farms'. In Venezuela[62] a network of 53 small farms was established to develop new technological alternatives for double purpose animal production systems. In the Cerrados of Brazil, 28 small farms form a similar network. These networks combine different types of activities: careful monitoring of technical, biophysical and socioeconomic variables; OFEs; validations and demonstrations; regular analysis of incoming data for a variety of users: policy makers, researchers and the farmers themselves. The aim is to offer a 'menu', as broad as possible, of tested technologies, from which the farmers can make a selection and adapt them to their particular circumstances.

A somewhat similar approach is followed in the semi-arid tropics of Brazil by a combined EMBRAPA-CPATSA-CIRAD project[63]. Here, an 'observatory of agrarian dynamics' has been developed – an interinstitutional scheme that allows the processing of information gathered and analysed at three different levels: regional, municipal and local. Using the concept of 'development itinerary', the main tendencies at each level are outlined and presented to the different institutional actors.

In Perú and Bolivia ORSTOM, in cooperation with the national agricultural research and extension institutions, has focused on the different systems levels from crop and animal production systems, through farming to agrarian systems. In the Andean ecoregion the work has included the project in the Cañete watershed of Perú documented in the book edited by Malpartida and Poupon in 1987[64], while research in Bolivia generated three books, by Hervé and Rojas in 1994, Hervé *et al.* in 1994 and Genin *et al.* in 1995[65].

The ORSTOM work has accumulated a wealth of knowledge about the relationship between crop rotations and soil fertility, soil moisture, soil pathogens, soil microbiology and natural vegetation growth. An interesting aspect is the interaction between anthropology and agronomy. The research has identified that, for the Quechua and Aymara indigenous communities, the land/soil is not only a physical resource but the centre of their world view and religion. The ORSTOM land use systems are guided as much by the culture of these peoples as by technical parameters. Another very fruitful line of work has been the study of the South American camelids (llama, alpaca, vicuña) and sheep pastoral systems. A third line of research concerns bovine animal production systems under the conditions of the Bolivian Andean highlands, with an emphasis on milk production. ORSTOM conducted several research projects in Perú to study the relationship between cropping systems and hillside soils, and organic soils of the Amazonian region and implications for erosion, utilization and productivity.

9.2.6 The Latin American FSR networks

There are three main regional FSR networks in Latin America: CONDESAN, RIMISP and the Latin American Animal Production Systems Research Network (RISPAL). A biannual Latin American FSR-E symposium has been organized since 1993, as a product of the collaboration of the existing regional networks and other important Latin American and international organizations.

International FSR methodology network (RIMISP)

With the support of Canada's IDRC, a group of institutions, projects and technical teams got together in 1986 to form a research network dedicated to developing innovative concepts, methods and instruments to support FSR-E. RIMISP was a response to the concerns about the efficiency and effectiveness of the FSR approach, and about the widespread use of the 'farming systems' label in badly designed research and development projects.

By June 1996, the RIMISP membership included 39 research and development organizations in 14 Latin American countries. Of these, 36% were NGOs and 64% were governmental organizations. In terms of institutional activity, 38% were in agricultural research, 21% were universities, 18% conducted both research and extension and 15% were dedicated to socioeconomic and policy research, 5% were in the field of agricultural development and 3% operated information systems. The network had a portfolio of projects with a total multi-year budget of about US$3 million and major donors included the International Fund for Agricultural Development (IFAD), the European Commission and several Latin American governmental institutions that have contracted specific projects with the network (67% of the total portfolio is financed by this type of contract with national institutions). Until 1995, IDRC was the major single RIMISP donor.

The main activities of the network are

research, training, information management, publications and international meetings with some occasional consulting work. The network was able to mobilize IDRC funding for methodological development components of applied FSR projects in many countries for many years. Typically, a single methodological problem would be researched simultaneously in several projects in different countries. Results would be compared after 2 or 3 years and the network would move on to a new research question. Today, RIMISP works to build *ad hoc* interinstitutional consortia with its member institutions, to jointly prepare, negotiate and implement multinational research projects. One example is a project to test different methodologies to evaluate agricultural sustainability in Argentina, Brazil and Chile, involving RIMISP and five other research organizations from Europe and Latin America.

The RIMISP approach has been to support methodology development components within applied farming systems research projects. In this way, the following topics have been discussed and tested by one or more of the RIMISP members within the context of network activities:

- Farm typologies and classification[66].
- Dynamic analysis based on farms' records[67].
- Reconceptualization of the design phase of FSR-E[68].
- Multiple objective programming to improve design and *ex ante* evaluation[69].
- analysis of farmers' decision making criteria with ethnographic decision trees[70].
- Approaches to extension and technology transfer[71].
- Evaluation of technology adoption[72].
- Gender analysis within FSR-E[73].
- Indicators of sustainability[74].
- Evaluation of farming systems according to productivity–profitability–sustainability trade-offs[75].
- Analysis of sustainability at the microregional scale using GIS[76].
- Assessments of sustainable land use systems research in South America[77].
- Assessments of farming systems research in Latin America[78].

Based on its research programmes, RIMISP has developed seven training modules covering methodological topics. Training workshops are designed on demand, and must be fully funded by the institutions requesting this service. RIMISP is also involved in a major 4-year training project contracted by Mexico's Ministry of Agriculture, Livestock and Rural Development. Information systems are a new area of work for RIMISP and its first project in this field has been funded by IFAD. This will design and coordinate a system to promote the exchange of information and knowledge between 28 IFAD-funded projects in 22 Latin American countries, making use of electronic communication tools. The network publishes the RIMISP Newsletter every 3 months for distribution to about 1700 subscribers and has published several books and monographic documents. RIMISP organizes its International Meetings every 2 years of so and the network has played a leading role in the organization of the Latin American Farming Systems Research/Extension Symposia, and in stimulating Latin American participation in the international AFSRE symposia.

In its fifth meeting in Colombia in 1995 the RIMISP board defined a new mission statement and a new set of strategic objectives, to equip the network for today's challenges. And there are several. Although it has been highly successful in evolving from dependence on a single donor, this means that RIMISP now lacks core funding and must derive 100% of its budget from specific projects – by definition an 'unsteady state'. However, its main challenge is to meet the very different needs of today's members, related to the interlinked issues of competitiveness, sustainability and social equity, in a context of very scarce resources. RIMISP must evolve in its role and in its organization and management in order to continue as a relevant player in systems-oriented research in Latin American agriculture and rural development.

RISPAL

RISPAL was formalized in 1989 by a group of animal production systems projects, again with the support of IDRC. An agreement between IDRC and the Interamerican Institute for Cooperation on Agriculture (IICA) provided the basic framework for its development. RISPAL was created for three main reasons:

- The cropping and farming systems methodologies and techniques required important

adjustments to deal with animal and mixed systems.
- There was a lack of well-trained staff that could apply the systems approach to the research and development of animal production systems.
- Small-scale farmers were receiving little institutional attention and support, particularly on small-scale animal production systems.

Its objectives are:

- To promote the development of the systems methodology applied to animal production research.
- To generate, validate and disseminate animal production technologies.
- To strengthen projects and institutions through technical support and training.

IDRC and RISPAL supported projects on dual-purpose cattle in the Dominican Republic, Guatemala, Panamá and Venezuela[79]. Milk production systems were developed in Chile and in Guyana[80]. A project on pig production was implemented in El Salvador[81] and on agropastoral systems in Costa Rica[82]. RISPAL also supported a project on goat production systems in Mexico[83]. In Perú, RISPAL-IDRC developed projects on South American camelids and cuyes[84]. Finally, RISPAL supported the animal production component of farming systems projects in Colombia and in the PISA-PRODASA projects.

Over 400 scientists from 11 countries have been trained by RISPAL. An extensive bibliographical database with about 17,000 entries has been compiled and is available in CD-ROM and magnetic media; the largest of its kind in Latin America. RISPAL has published five books on methodology in Spanish, and four in English, 11 bibliographical bulletins, 34 issues of its newsletter 'Cartas de RISPAL' and nine volumes of the Proceedings of its General Meetings. RISPAL is facing the challenge of adjusting its structure and orientation in order to respond to the changing institutional conditions, including the treatment of new research topics and objectives, the new demands of the membership and declining donor interest.

CONDESAN

The third major regional association is CONDESAN, organized in 1992 under the leadership of CIP to overcome the deterioration of land and water resources, prevent the loss of genetic diversity and to alleviate extreme poverty. Its main characteristics are:

- Full involvement of NARS, NGO, universities and other partners.
- Collaboration with national, regional and international institutions, to make a more effective use of scarce funds and existing scientific talents.
- Participatory programme planning by objectives (PPPO) as a mechanism to guarantee full participation of stakeholders for joint planning and shared monitoring of projects.
- Scientific rigour.
- A participatory decision-making structure.
- An information system (INFOANDINA) to improve the communication and information sharing links between researchers.
- Shared monitoring mechanisms for sustainable land use and the maintenance of biodiversity.

Since support for the Consortium is based on sharing both the costs and the benefits, stakeholders are expected to fund the activities that are undertaken. During 1996, CONDESAN has received funding from: Swiss Development Cooperation, IDRC, GTZ, the Governments of The Netherlands, Denmark and Spain.

CONDESAN's proposed activities in the Andean ecoregion cover thematic research areas: biodiversity of Andean crops, pastures and animals; land and water management; agricultural policy and rural development; and commodity systems. Much of the fieldwork is conducted in benchmark sites, representative of the Andean ecoregion, in La Paz (Bolivia), La Miel (Colombia), Carchi (Ecuador) and Cajamarca and Puno (Perú). Research focuses on measurements of the dynamic characteristics of systems, modelling to establish priorities for monitoring, evaluation of the impact of land use systems, biodiversity maintenance, and the design and implementation of policies to further these ends. The benchmark sites would also help in extrapolation of component technologies to similar agroecologies.

One example of the work of the consortium is with Andean root and tuber crops within the biodiversity theme, carried out with the financial support of the Swiss Development Cooperation and Germany's GTZ. A total of 52 projects are being developed by 24 institutions in Bolivia, Colombia, Ecuador and Perú. As a result of CIP's designation as convenor for the global implementation of research into Sustainable Mountain Development within the context of Agenda 21, CONDESAN is collaborating with institutions in the highlands of eastern Africa and the Himalayas to promote appropriate technologies and human welfare in mountain communities.

The work of these regional networks and consortia is complemented by the Latin American FSR-E symposia. The first meeting was held in Quito, Ecuador, in March 1993. The idea of forming a Latin American FSR-E Association was debated at this meeting, but it was agreed that it would be more efficient to rely on informal cooperation and contacts between of the existing regional networks.

9.2.7 Conclusion

FSR-E in Latin America has a long and fruitful history. Moreover, it is possible to identify a Latin American school of systems-oriented research under the general FSR-E umbrella, distinguished by the attempt to balance quantitative methods and farmer participation; research at the cropping and farm system level with micro-macro and policy analysis; and applied research which also carries conceptual and methodological contributions.

However, systems-oriented research in Latin America does not all fall under this one umbrella. The three CGIAR centres with headquarters located in the region (CIMMYT, CIAT and CIP), have conducted systems research within their own conceptual frameworks. The French organizations have had an important presence. Brazil is a continent by itself, and FSR there has developed in ways that do not just differ from those of other countries, but differ from region to region within the country itself.

No evaluation has been conducted to date to estimate the impact of the approach in stimulating technological change and technology adoption in Latin America. It is the author's biased impression that the overall effect has been positive, in particular if measured on the basis of the benefits for the population of peasant farmers. It can be argued that on-farm research, farmer participation and better research-extension linkages, have improved the relevance of the process of technology generation and extension.

What is clear is that participatory and systems-oriented approaches are the only viable options for Latin America. Privatized extension systems, farmer-to-farmer methods and NGO-led programmes are struggling to fill the void left by the virtual disappearance of the traditional extension systems.

A more recent phenomenon deserves documentation: the impact of a systems-based, bottom-up perspective on policy formulation. In all Latin American countries, and in most programmes, policies are still formulated at the national level, with little direct input from the rural populations. In recent years there have been important shifts in direction, including:

- Implementation of the principle of 'social control' in Argentina's Agricultural Research Institute.
- Municipalization of technical assistance in Colombia.
- Decentralization of most public expenditures in the field of agricultural technology development in Mexico (research, extension and training).
- Privatization of Chile's extension system with strong participation by NGOs and farmers' organizations in the provision of services.
- A strongly participatory process for the formulation of local rural development programmes and projects in Bolivia.

There is little doubt that the tide is starting to move in the direction of greater participation and decentralization in Latin America and this is rapidly affecting the institutions involved in policy formulation, research, extension, training and rural development. These shifts should improve the institutional and policy environments for effective FSR-E in future years. The most important conclusion is that today, FSR-E in Latin America is everywhere, thanks to its characteristic 'anonymous ubiquitousness' first recognized at the time of the Caqueza project. As Moscardi in 1992 pointed out: 'It can be said without any doubt, that most agricultural research institutes in Latin America have developed some form of on-farm or FSR capabilities or programs'. All the major governmental research and extension organizations formally declare themselves to be 'systems-oriented' today. If this is the situation in governmental institutes, it is more so in the world of the NGOs, where 'being systemic', or 'being holistic' is, strictly speaking, a dogma. A few years ago RIMISP calculated that there were no less than 5000 researchers and extensionists in Latin America working under the explicit label of the systems approach. Perhaps part of this success in institutionalization is due precisely to the anonymity of the approach. You did not have to declare yourself a believer of a new church, you just went ahead and took what you liked and worshipped it. Was this good or bad? It is 'bad' if we are purists and would like to be the Pope, or even just a priest in this church. It is very good if we want more on-farm work, closer to the small farmers, with a greater understanding of them and their agriculture.

Institutionalization has taken place gradually but persistently, because the approach captured the minds of several hundred bright and dedicated people. The effective institutionalization of FSR-E has taken place indirectly, through the gradual and persistent dissemination of the approach by means of training, meetings, publications, networking and, most importantly, through the practical activities of many projects that involved many hundreds if not thousands of researchers and extensionists. Direct efforts to institutionalize the systems approach by administrative fiat, have not only largely failed, but have often led to strong negative reactions by the people supposed to implement the changes. While many of these attempts failed because they were badly designed, their shortcomings can often be explained by the fact that the systems approach is a body of theory and methods that cannot be imposed.

Problems certainly remain, both in the substance of what has been done and in what we are still doing[85]. There is still too much emphasis on problems within the farm-gate, forgetting the lesson of Caqueza: 'generating technologies within the narrow confines of the existing constraints is under most circumstances a difficult undertaking'. Too little attention is given to the wider context in which technology operates, forgetting the lessons of Puebla: 'research and dissemination are a continuum' and 'for technology to work, there have to be markets, inputs, services and farmers' organizations'. Too much 'holistic paralysis', forgetting the lessons of CATIE: 'if theory and practice clash, practice wins'. Too much time spent in trying to simplify the complex lives that small farmers experience every day. We should look back more often, so that we can move forward faster.

REFERENCES

1. **Jiménez, L., 1988**. Mexico learns from the Plan Puebla. In: L. Jiménez, R. Schachter, Mongenthau & B. Peña (Eds) Fighting Rural Hunger in a World Full of Grain. CILCA-Colegio de Postgraduados, México, pp. 87–116.
2. **Plan Puebla, 1970**. El Projecto Puebla 1967–69. CIMMYT, México.
3. **Zandstra, H.G., K. Swanberg, C. Zulberti & B. Nestel, 1979**. Caqueza. Living Rural Development. IDRC, Ottawa, Canada.
4. **Turrent, A., 1974**. Tecnología de la producción. Colegio de Postgraduados, Chapingo, Mexico.
5. **Jiménez, L., 1988**. (Op. cit.).
6. **Turrent, A., 1974**. (Op. cit.).
7. **Tobón, J.H., 1985**. El Proyecto Rionegro. Sus experiencias en sistemas de producción y desarrollo rural. In: Seminario Sobre Investigación en Sistemas de Producción y su Contribución al Desarrollo Rural en América Latina, Turrialba, Costa Rica, 22–26 April, 1985. CATIE-FSSP, Turrialba, Costa Rica.
   ICA (Colombian Agricultural Institute), 1970. Proyecto piloto para la incorporación de la tecnología a nivel de campesino tradicional. Desarrollo de áreas Tipo 'PUEBLA' para acelerar la producción e incrementar los rendimientos de cultivos alimenticios. Planeación Documento no. 1. ICA, Bogotá, Colombia.
8. **Zandstra, *et al.*, 1979**. (Op. cit.).
   Chaves, L.E. (Ed.), 1991, Veinte años de tecnología agropecuaria al servicio del minifundio. Instituto Colombiano Agropecuario, Bogotá, Colombia.
9. **Chaves, L.E. (Ed.), 1991**. (ibid.).
10. **Berdegué, J.A. & G. Escobar, 1995**. New directions of the systems approach for the modernization of Latin American peasant agriculture. Journal for Farming Systems Research-Extension, 5, 1–30.

Escobar, G., 1995. Desarrollo metodológico para la aplicación del enfoque de sistemas en América Latina. In: J.A. Berdegué & E. Ramírez (Eds) Investigación con enfoque de sistemas en la agricultura y el desarrollo rural. RIMISP, Santiago, Chile, pp. 99–109.

11. **Zandstra *et al.*, 1979**. (Op. cit.).
12. **Chaves, L.E. (Ed.), 1991**. (Op. cit.).

Llumiquinga, I., H. Ortíz, F. Silva, C. Carrasco & C. Varela, 1989. Sistematización y difusión de tecnología generada y experimentada en proyectos de desarrollo rural integral. Ministerio de Bienester Social, Quito, Ecuador.

13. **Pinchinat A.M., J. Soria & R. Bazan, 1976**. Multiple cropping in Tropical America. In: R.I. Papendick, P.A. Sánchez & G.B. Triplett (Eds) Multiple Cropping. Proceedings of a Symposium Sponsored by the American Society of Agronomy, Crop Science Society of America and Soil Science Society of America. Knoxville, Tennessee, August 1975. ASA-CSSA-SSSA: Madison, Wisconsin, USA, pp. 51–61.

PISA, 1993. Informe final 1985–1992. PISA, Puno, Perú.

14. **Sánchez, P.A., 1973**. Manejo de suelos bajo el sistema de roza. In: P.A. Sánchez (Ed.) Un Resumen de las Investigaciones Edafológicas en América Latina Tropical. North Carolina State University Technical Bulletin, 219, 51–74.
15. **Gómez, J.A., 1968**. Rotación y rendimiento en maíz. Agricultura Tropical, 24, 204–20.
16. **Hildebrand, P.E. & E.C. French, 1974**. Un sistema salvadoreño de multicultivos. In: Conferencia sobre sistemas de producción agrícola para el trópico. CATIE, Turrialba, Costa Rica.
17. **Lepiz, R., 1971**. Asociación de cultivos maíz-frijol. Agricultura Técnica, 3, 98–101.

Moreno, R., A. Turrent & R. Nuñez, 1973. Las asociaciones de maíz-frijol. Una alternativa en el uso de os recursos de los agricultores del Plan Puebla. Agrotecnia, 14, 103–17.

18. **Moreno, R. & J. Saunders, 1978**. A farming systems research approach for small farms in Central America. CATIE, Turrialba, Costa Rica.
19. **Burgos, C. & L. Navarro, 1985**. Investigación para el desarrollo de tecnologías mejoradas para agricultores de recursos limitados. In: Seminario sobre Investigación en Sistemas de Producción y su Contribución al Desarrollo Rural en América Latina, Turrialba, 22–26 April, 1985. CATIE-FSSP, Turrialba, Costa Rica.
20. **CATIE, 1986a**. Caracterización ambiental y de los principales sistemas de cultivo, villa San Antonio, Honduras. CATIE Serie Técnica no. 69. CATIE, Turrialba, Costa Rica.
21. **CATIE, 1985a**. Diseño y evaluación de alternativas. Estelí, Nicaragua, 1978–1984. CATIE Serie Técnica no. 58. CATIE, Turrialba, Costa Rica.
22. **CATIE, 1987**. Sistemas de cultivo. Logros y avances de las investigaciones de CATIE en Nicaragua, 1980–1985. CATIE Serie Técnica no. 121. CATIE, Turrialba, Costa Rica.

Laguna, I.G., L.G. Salazar & J.F. López, 1983. Fungal and bacterial diseases of aroids (*Xanthosoma* sp. and *Colocasia esculenta*) in Costa Rica. CATIE Serie Técnica no. 10. CATIE, Turrialba, Costa Rica.

23. **van Tienhoven, N. & J. Lagemann (Eds), 1981**. La producción agrícola en Jinotega, Nicaragua. CATIE Serie Técnica no. 12. CATIE, Turrialba, Costa Rica.

van Tienhoven, N., J. Icaza & J. Lagemann, 1982. Farming systems in Jinotega Nicaragua. CATIE, Turrialba, Costa Rica.

24. **Burgos, C., 1977**. Bosquejo sobre la metodología seguida por CATIE en la ejecución del Proyecto de Sistemas de Cultivo para Pequeños Agricultores. CATIE, Turrialba, Costa Rica.

Escobar, G. & R. Moreno, 1984. Desarrollo de tecnología para sistemas de producción agrícola. Enfoque metodológico y aplicación empírica. CATIE, Turrialba, Costa Rica.

CATIE, 1986b. Reunión interna de discusión sobre validación/transferencia en la etodología de desarrollo de la tecnología apropiada. CATIE Serie Técnica no. 78. CATIE, Turrialba, Costa Rica.

CATIE, 1986c. Una metodología para definir áreas aptas para tecnologías agronómicas. CATIE Serie Técnica no. 65. CATIE, Turrialba, Costa Rica.

25. **CATIE, 1977**. Informe de la primera reunión de agricultores colaboradores en el Proyecto de Sistemas de Cultivo para Pequeños Agricultores, celebrada en el CATIE el 8 y 9 de diciembre de 1977. CATIE, Turrialba, Costa Rica.
26. **Hawkins, R., J. Henao, R. Alegria & E. Celada, 1986**. Empirical model of maize–sorghum intercrop production and agrotechnology transfer in Central America. Agricultural and Forestry Meteorology, 36, 269–80.
27. **Navarro, L., 1982**, Developing technology for small farms. A case study in Central America. Paper presented at the Symposium on 'Farming Systems in the Field', Kansas State University, Manhattan, Kansas, 21–23 November, 1982.

28. **Avila, M., D. Pezo, M.E. Ruíz & A. Ruíz, 1979**. Sistemas de producción en pequeñas fincas de Costa Rica. Paper presented at the VII Meeting of the Latin American Association of Animal Production, 23–29 September 1979, Panamá.

Avila, M., L.A. Navarro & J. Lagemann, 1982. Improving the small farm production systems in Central America. Paper presented at the XVIII International Conference of Agricultural Economists, Jakarta, Indonesia, August 24 to September 2, 1982.

29. **Moreno, R., 1977**. Fudamentos del Programa de Sistemas de Producción de Cultivos para Pequeños Productores. Paper presented at the International Meeting on Technical Cooperation CATIE-CIAT-CIMMYT-IICA, Turrialba, Costa Rica, 2–3 June, 1977.

Shenk, M.D., J. Saunders & G. Escobar, 1983. Labranza mínima y no labranza en sistemas de producción de maíz para áreas tropicales húmedas de Costa Rica. CATIE Serie Técnica no. 8. CATIE, Turrialba, Costa Rica.

30. **Kass, D.L., A. Barrantes, W. Bermúdez, W. Campos, M. Jiménez & J. Sánchez, 1989**. Resultado de seis años de investigación de cultivo en callejones (alley cropping) en 'La Montaña', Turrialba, Costa Rica. El Chasqui, 19, 5–24.

31. **Fassbender, H.W., 1989**. Cíclos de los elementos nutritivos en sistemas agroforestales de café con árboles de sombra en el 'Experimento Central' del CATIE. In: J.W. Beer, H.W. Fassbender & J. Heuveldop (Eds) Seminario Avances en la Investigación Agroforestal. CATIE, Turrialba, Costa Rica, pp. 176–86.

32. **Mateo, N. & R. Moreno, 1976**. Estudio de siete sistemas de producción agrícola en platanares de Pérez Zeledón, Costa Rica. CATIE, Turrialba, Costa Rica.

Holle, M. & J. Saunders, 1977. Descripción de los sistemas de cultivo y algunas características de los agricultores de San Isidro de El General, Costa Rica. Paper, the International Meeting on Technical Cooperation CATIE-CIAT-CIMMYT-IICA, Turrialba, Costa Rica, 2–3 June, 1977.

CATIE, 1986d. Alternativa de manejo para el sistema maíz-yuca. Pococi-Guacimo, Costa Rica. CATIE Serie Técnica no. 64. CATIE, Turrialba, Costa Rica.

33. **CATIE, 1986c**. (Op. cit.).

34. **CATIE, 1984**. Alternativa de manejo para el sistema maíz–firjol. El Rosario, Honduras. CATIE Serie Técnica no. 48. CATIE, Turrialba, Costa Rica.

35. **CATIE, 1985c**. Alternativa de manejo para el sistema maíz–frijol y tomate–frijol. Matagalpa, Nicaragua. CATIE Serie Técnica no. 59. CATIE, Turrialba, Costa Rica.

36. **CATIE, 1985b**. Caracterización ambiental y de los principales sistemas de cultivo en fincas pequeñas. Guarumal, Panamá. CATIE Serie Técnica no. 61. CATIE, Turrialba, Costa Rica.

37. **Hildebrand, P.E., 1979a**. Generating technology for traditional farmers. The Guatemalan experience. Paper presented at the Symposium on Socio-economic Constraints to Crop Protection, IX International Congress of Crop Protection, Washington, DC, 5–11 August, 1979.

Ruano, S., 1989. El Sondeo: Actualización de su metodología para caracterizar sistemas agropecuarios de producción. RISPAL, San José, Costa Rica.

38. **Waugh, R.K., 1977**. El Instituto de Ciencia y Tecnología Agríolas. Cuatro años de historia. ICTA, Guatemala.

Ruano, S., 1982. Investigación agrosocioeconómica en sistemas de cultivo. El ICTA, una experiencia en Guatemala. ICTA, Guatemala.

Fumagalli, A., R. Ortíz & M. Castillo, 1985. A new model for technology transfer within FSR/E. Paper presented at the Farming Systems Research and Extension Symposium, Kansas State University, 13–16 October, 1985, Manhattan, Kansas, USA.

Ruano, S. & A. Fumagalli, 1988. Organización y manejo de la investigación en finca en el Instituto de Ciencia y Tecnología Agrícolas. ISNAR OFCOR Case Study no. 2. ISNAR, The Hague, The Netherlands.

39. **Holle, M., 1992**. Articulación de la caracterización con el diseño de la investigación con enfoque de sistemas, para definir y validar alternativas tecnológicas. In: V International Meeting of RIMISP. RIMISP, Santiago, Chile.

40. **Estrada, R.D., 1992**. Análisis de información secundaria en el diseño de sistemas de producción. In: V International Meeting of RIMISP. RIMISP, Santiago, Chile.

41. **Quijandría, B., V. Agreda, J. Escobal & W. Twanama, 1990**. Análisis dinámico de pequeñas fincas en cuatro regiones del Perú: aspectos metodológicos. RIMISP, Santiago, Chile.

Arze, J. & J.L. Russo, 1991. Hoja de trabajo para la información dinámica de seguimiento en sistemas de producción. In: IV International Meeting of RIMISP. RIMISP, Santiago, Chile.

42. **Quiroz, R., B. Arce & M. Holle, 1991a**. Métodos de investigación con enfoque y análisis de datos de sistemas agropecuarios. Turrialba, 41, 1–14.

León-Velarde, C. & R. Quiroz, 1994. Análisis de sistemas agropecuarios: uso de modelos bio-matemáticos. CIRNMA-Editorial Edigraph, La Paz, Bolivia.

Arce, B., R. Quiroz & G. Mamani, 1990. Utilización de modelos de simulación. In: Taller sobre modelos de simulación. PISA, Puno, Perú.

43. **Estrada, R.D., 1987**. Potencial de adopción tecnológica en las comunidades de Puno. Análisis acro-económico. PISA, Puno, Perú.

Estrada, R.D., 1995. Incidencia de las políticas económicas en la conservación de los recursos naturales de la zona andina. REPAAN-CONDESAN, Quito, Ecuador.

44. **Claverías, R., H. Muñoz & J. Salas, 1991**. Tipología de las familias en comunidades campesinas en la región Sur. Instituto Pastoral Andino, Cusco, Perú.

Gladwin, C., 1991. Toma de decisiones en familias campesinas. Invited paper presented at the IV International Meeting of RIMISP, Salcedo, Ecuador.

45. **León-Velarde, C. & R. Quiroz, 1993**. Sostenibilidad de un sistema de producción de alpacas en el altiplano peruano. Abs. ALPA.

Estrada, R.D., 1996. Las limitaciones y potencialidades del medio natural para el desarrollo de sistemas de producción campesinos. In: Seminario Regional para la Promoción de Sistemas de Producción Agrícolas Sostenibles para el Sector Campesino de los Andes Centrales. CONDESAN-FAO-CIP-UNEP, Quito, Ecuador, January 1996.

46. **Argüelles, L. & R.D. Estrada (Eds), 1991**. Perspectivas de la investigación agropecuaria para el Altiplano. PISA, Lima, Perú.

47. **Risi, J., 1991**. La investigación de la quinua en Puno. In: L. Argüelles & R.D. Estrada (Eds) Perspectivas de la Investigación Agropecuaria Para el Altiplano. PISA, Lima, Perú, pp. 209–58.

48. **Li Pun, H. & O. Paladines, 1993**. El rol de las pasturas y la ganadería en la sostenibilidad de los sistemas de producción andina. In: CIP, El Agroecosistema Andino. CIP, Lima, Perú, pp. 187–212.

Quijandría, B., 1991. Perspectivas de la investigación en producción animal. In: L. Argüelles & R.D. Estrada (Eds) Perspectivas de la Investigación Agropecuaria Para el Altiplano. PISA, Lima, Perú, pp. 151–92.

Quiroz, R., G. Mamani, R. Revilla, C. Guerra, J. Sánchez, M. González & G. Pari, 1991b. Perspectivas de investigación pecuaria para el desarrollo de comunidades de Puno. In: L. Argüelles & R.D. Estrada (Eds) Perspectivas de la Investigación Agropecuaria Para el Altiplano. PISA, Lima, Perú, pp 357–406.

49. **Tola, J., 1992**. Comentarios y acciones de investigación de cereales de Puno. In: L. Argüelles & R.D. Estrada (Eds) Perspectivas de la Investigación Agropecuaria Para el Altiplano. PISA, Lima, Perú, pp. 259–72.

50. **Ezeta, F., 1991**. Perspectivas de investigación de la papa en Puno durante la década de 1990. In: L. Argüelles & R.D. Estrada (Eds) Perspectivas de la Investigación Agropecuaria Para el Altiplano. PISA, Lima, Perú, pp 273–84.

51. **Paladines, O., 1991**. Recomendaciones sobre investigación de pastizales en el PISA, Puno. In: L. Argüelles & R.D. Estrada (Eds) Perspectivas de la Investigación Agropecuaria Para el Altiplano. PISA, Lima, Perú, pp 285–304.

León-Velarde, C. & F. Izquierdo (Eds), 1993. Producción y utilización de los pastizales de la zona altoandina. REPAAN, Quito, Ecuador.

52. **Holle, M., 1991**. Los recursos genéticos para el altiplano. In: L. Argüelles & R.D. Estrada (Eds) Perspectivas de la Investigación Agropecuaria Para el Altiplano. PISA, Lima, Perú, pp 305–18.

53. **CIP, 1993**. El Agroecosistema Andino. CIP, Lima, Perú.

54. **EMBRAPA (Empresa Brasileira de Pesquisa Agropecuária), 1981**. Programas nacionais de pesquisa para a regiao de trópico semi-árido. EMBRAPA, Brasilia.

55. **de Miranda, E.E., 1981**. Métodos de pesquisa agronômica sobre as limitacões dos rendimentos culturais a nível de pequenos e médios agricultores de trópico semi-arido do Brasil. CPATSA-EMBRAPA. Documentos no. 10. CPATSA-EMBRAPA, Petrolina, Brazil.

de Miranda, E.E. & M.B.A. Silva, 1982. Tecnologia agrícola para o trópico semi-árido; análose do processo de geracão de tecnologia. R.Economia Rural, 20, 269–99.

56. **Caron, P., F. Prevosi & P.C. Gama da Silva, 1993**. A evolucão de um programa de pesquisa em sistema de producão no Nordeste Brasileiro. CPATSA-EMBRAPA, Petrolina, Brazil.

Guimarães, F.C., P. Caron & P.C. Gama da Silva, 1995. Enfoque sistemico no desinvolviemento rural: a experienca do CPATSA/EMBRAPA. CPATSA-EMBRAPA, Brazil.

57. **Neumaier, M.C. & S. Shiki, 1991**. Esaios metodológicos de pesquisa em sistemas de producão no Paraná 1981–88. IAPAR Boletim Técnico no. 38. IAPAR, Londrina (PR), Brazil.
58. **de Miranda, E.E., 1983**. Como melhorar o processo de generacão de tecnologia agrícola na área de PRORURAL. IAPAR, Londrina (PA), Brazil.
59. **Munguía-Payes, M.A., 1993**. Sistemas de producão predominantes na região de Irati, Paraná. Un estudio de tipologia e diferenciacão de produtores rurais. IAPAR, Londrina (PR), Brazil.
Yu, C.M. & V.J. Sereia, 1993. Tipificacão e caracterizacão dos produtores rurais do Estado do Paraná. IAPAR, Londrina (P), Brazil.
60. **Ribeiro, M.F.S., 1995**. A experiência do IAPAR en validacão de tecnologias. In: Anais II Encontro da Sociedade Brasileira de Sistemas de Producão. SBA-IAPAR, Londrina (PR), Brazil, pp. 53–63.
Darolt, M.R. & M.F.S. Ribeiro, 1995. Pesquisa e desemvolvimento sustentável: experiências com o plantio direto para pequena propiedade. In: Anais II Encontro da Sociedade Brasileira de Sistemas de Producão. SBS-IAPAR, Londrina (PR), Brazil, pp. 291–301.
61. **de Miranda, E.E., A.J. Dorado, M. Guimarães, J.A. Mangabeira & J.R. Miranda, 1995b**. Impacto ambiental sostenibilidad agrícola. La contribución de los Sistemas de Informaciones Geográficas. RIMISP, Santiago, Chile.
62. **Bonnal, P., J. Castillo & V. Dollé, 1992**. Una red de fincas de referencia como instrumento de observación y gestión en el medio rural. El proyecto Aroa-Bajo Tocuyo. Revista Investigación Desarrollo para América Latina, 1, 40–58.
63. **Guinmaraes Filho, C., P. Caron & P.C. Gama da Silva, 1995**. Enfoque sistêmico no desinvolvimento rural: a experiência do CPATSA/EMBRAPA, Brazil.
64. **Malpartida, E. & H. Poupon (Eds), 1987**. Sistemas agrarios en el Perú. UNALM/ORSTOM, Lima, Perú.
65. **Hervé, D. & A. Rojas (Eds), 1994**. Vías de intensificación de la ganadería bovina en el altiplano boliviano. ORSTOM, La Paz, Bolivia.
Hervé, D., D. Genin & G. Riviere (Eds), 1994. Dinámicas del descanso de la tierra en los Andes. ORSTOM/IBTA, La Paz, Bolivia.
Genin, D., H.J. Picht, R. Lizarazu & T. Rodríguez (Eds), 1995. Waira Pampa. Un sistema pastorial camélidos-ovinos del altiplano árido boliviano. ORSTOM-IBTA, La Paz, Bolivia.
66. **Escobar, G. (Ed.), 1988**. Clasificación de sistemas de finca para generación y transferencia de tecnología apropiada. IDRC, Ottawa, Canada.
Escobar, G. & J.A. Berdegué (Eds), 1990. Tipificación de sistemas de producción agrícola. RIMISP, Santiago, Chile.
67. **CE&DAP (Centro de Estudios y de Desarrollo Agrario del Perú), 1989**. Análisis Dinámico de Datos de Finca (Perú). CE&DAP, Lima, Perú.
68. **Holle, 1992**. (Op. cit.).
Berdegué, J.A., 1992. Objetivos de la etapa de diseño en la metodología de sistemas de producción. In: V International Meeting of RIMISP. RIMISP, Santiago, Chile.
Arze, J., 1992. Sistemas expertos y fase de diseño de la metodología para la investigación y desarrollo de alternativas tecnológicas. In: V International Meeting of RIMISP. RIMISP, Santiago, Chile.
Estrada, 1992. (Op. cit.).
Escobal, J. & M. Castillo, 1992. Algunos ejemplos de la vinculación económica en la etapa de diseño. In: V International Meeting of RIMISP. RIMISP, Santiago, Chile.
Maino, M., 1992. Programación matemática y diseño. In: V International Meeting of RIMISP. RIMISP, Santiago, Chile.
de Miranda, E.E., 1992. Posibles aportes de los Sistemas de Información Geográfica al diseño de sistemas e producción. In: V International Meeting of RIMISP. RIMISP, Santiago, Chile.
Monardes, A., 1992. Algunos comentarios sobre el análisis económico en el diseño de sistemas de producción. In: V International Meeting of RIMISP. RIMISP, Santiago, Chile.
Ramírez, E., J. Durán & L. Mora, 1992. Estudios de caso en profundidad y diseño. In: V International Meeting of RIMISP. RIMISP, Santiago, Chile.
69. **Maino, M., J. Pittet & C. Köbrich, 1993**. Programación multicriterio: un instrumento para el diseño de sistemas de producción. RIMISP, Santiago, Chile.
70. **Gladwin, C., 1991**. (Op. cit.).
71. **Monardes, A., G. Escobar & G. González (Eds), 1994**. Transferencia de tecnología agropecuaria: de la generación de recomendaciones a la adopción. RIMISP, Santiago.
72. **Monardes, A., T. Cox, D. Narea, E. Laval & C. Revoredo, 1993**. Evaluación de adopción de tecnología. CEDRA, Santiago, Chile.

73. **Campaña, P., 1992**. El contenido de género en la investigación en sistemas de producción. RIMISP, Santiago, Chile.
74. **Berdegué, J.A. & E. Ramírez (Eds), 1995a**. Operacionalización del concepto de sistemas de producción sostenibles. RIMISP, Santiago, Chile.
75. **de Miranda, E.E., C.O. Mattos & J.A.C. Mangabeira, 1995a**. Na forca das idéias. Indicadores de sustentabilidade agrícola na Amazônia. NMA-EMBRAPA/RIMISP, Campinas, Sao Paulo, Brazil.
76. **de Miranda, E.E., *et al.*, 1995b**. (Op. cit.).
77. **Berdegué, J.A. & E.E. de Miranda, 1991**. Assessment of sustainable land-use systems research in South America. In: Sustainable Land Use Systems Research. Proceeding of an International Workshop. New Delhi, India, 12–16 February 1990. Rodale Institute – USDA – Indian Council of Agricultural Research, pp. 201–18.
78. **Berdegué, J.A. & E. Ramírez (Eds), 1995b**. Investigación con enfoque de sistemas en la agricultura y el desarrollo rural. RIMISP, Santiago, Chile.
79. **Isidor, M.E., M. Hernández, M. Mueses, J. Richardson, J. Mercado, M. Taveras & J. Bodden, 1989**. Proyecto de investigaciones sobre sistemas pecuarios de doble propósito en la región Este de la República Dominicana. In: M.E. Ruíz & A. Vargas (Eds) Informe VIII Reunión General de RISPAL. RISPAL, San José, Costa Rica, pp. 41–54.
Cubillos, G., H. Vargas, M.A. Gutierrez & C. Gándara, 1989. Mejoramiento de sistemas de producción bovina de doble propósito. In: M.E. Ruíz & A. Vargas (Eds) Informe VIII Reunión General de RISPAL. RISPAL, San José, Costa Rica, pp. 15–39.
Quiel, J. & R.A. Quiroz, 1989. Sistemas de producción de bovinos de doble propósito en Panamá. In: M.E. Ruíz & A. Vargas (Eds) Informe VIII Reunión General de RISPAL. RISPAL, San José, Costa Rica, pp. 55–64.
Vaccaro, L., R. Vaccaro & O. Verde, 1991. Evaluación y mejoramiento genético de rebaños de doble propósito. In: A. Ruíz & M.E. Ruíz (Eds) Informe IX Reunión General de RISPAL. RISPAL, San José, Costa Rica, pp. 43–46.
80. **Pichard, G., J. Ortega & J.A. Alcalde, 1991**. Proyecto de investigación en sistemas de producción de leche para pequeños productores en La Unión. In: A. Ruíz & M.E. Ruíz (Eds) Informe IX Reunión General de RISPAL. RISPAL, San José, Costa Rica, pp. 47–59.
Smith, J., P. Osuji & H. Muñoz, 1991. Milk production systems. In: A. Ruíz & M.E. Ruíz (Eds) Informe IX Reunión General de RISPAL. RISPAL, San José, Costa Rica, pp. 61–79.
81. **Ventura, J.A., 1989**. Proyecto de mejoramiento de la productividad del cerdo criollo. In: M.E. Ruíz & A. Vargas (Eds) Informe VIII Reunión General de RISPAL. RISPAL, San José, Costa Rica.
82. **Romero, F., L.A. Camero & L.A. Sánchez, 1989**. Proyecto sistemas silvopastoriles para el trópico húmedo bajo de Costa Rica. In: M.E. Ruíz & A. Vargas (Eds) Informe VIII Reunión General de RISPAL. RISPAL, San José, Costa Rica, pp. 223–42.
83. **Salinas, H., Quiroga, P. Sáenz, M. Martínez, A. Guerreo, J. Espinosa, J.F. Cano & J.L. Avila, 1991**. Sistemas de producción caprina en México. In: A. Ruíz & M.E. Ruíz (Eds) Informe IX Reunión General de RISPAL. RISPAL, San José, Costa Rica, pp. 81–98.
84. **Leyva, V., A. Nuñez & J. Choquehuanca, 1991**. Sistemas de producción de camélidos sudamericanos. In: A. Ruíz & M.E. Ruíz (Eds) Informe IX Reunión General de RISPAL. RISPAL, San José, Costa Rica, pp. 117–34.
Zaldivar, M. & L. Chauca, 1991. Proyecto sistemas de producción de cuyes. In: A. Ruíz & M.E. Ruíz (Eds) Informe IX Reunión General de RISPAL. RISPAL, San José, Costa Rica, pp. 99–116.
85. **Berdegué, J.A. & Escobar, G., 1995**. (Op. cit.).

9.3 AN OVERVIEW OF FSR-E, AND FSR-E NETWORKS IN AFRICA

James O. Olukosi

The strongest aspirations of FSR-E associations in Africa are towards ensuring FSR's sustainability and Association efforts are geared towards gaining long-term support for this innovative step in the research and development process.

9.3.1 The historical perspective

The research and extension services for technology transfer established in most African countries during the colonial era and soon after independence largely promoted the production of export commodities, such as cash crops, and forest products. From the mid 1960s the research efforts of many African countries aimed to improve the yields of the principal food crops. The emergence of a food crisis of alarming proportions since the 1980s has re-emphasized the need for higher productivity in African farming systems, showing that research results to increase yield are not enough to achieve technological change in general, or sustainable rates of growth in food production in particular. The commodity research approach characterized by the Green Revolution did not cope with either the diversity or complexity of the farming systems in Africa; as a result, except in a few countries and for selected commodities, an African 'Green Revolution' has yet to be realized.

Great efforts have been made over the last 30 years in the development of FSR methods, but relatively little effort has been put into integrating the FSR approach in the whole research process and into its institutionalization in NARIs. The challenges on institutionalization were grossly underestimated and many NARIs encountered confusion and considerable problems in attempting to integrate FSR into their research and extension activities. Some lessons learnt include:

- The institutionalization of FSR within NARIs has been slow and this has delayed progress in improving the relevance of research results to farmers' needs.
- Over the last 30 years in many African countries the scope of FSR has been limited to cropping systems, failing to integrate complementary production subsystems, particularly livestock and agroforestry.
- The effectiveness of external technical support in FSR was greatly influenced by the research context of particular NARIs; their institutional setting, their linkages with extension and development agencies as well as their mandate and research priorities, and their perception of FSR – often seen as a threat to the traditional establishment.

9.3.2 Emerging national FSR capacity

Much progress in FSR has already been swept away, caught in the tide of on-going socioeconomic, political and institutional change taking place in Africa. Rwanda and Ethiopia are two particularly tragic examples. Nevertheless national FSR capacity has grown, has begun to penetrate both the NARIs and, more recently, NGOs active in rural development. This national capacity has probably been a prerequisite to the emergence of networking and associations important to professionals staking their reputations on work in this new field.

The institutionalization of FSR into African national systems took place mostly in the 1980s. Several summary country histories and three case studies are to be found in Chapters 6 and 7. West Africa is represented by the case of Senegal, East and Central Africa by the cases of Tanzania and Zambia.

West Africa is unique in the sense that both Anglophone and Francophone schools were well represented in the region. The pattern of FSR development in the different countries demonstrates the dichotomy between the two. In 1985 Fresco noted that in Francophone countries FSR had been closely associated with development agencies and that national or regional policy issues received more emphasis than in Anglophone countries[1]. FSR activities formed an integral part of the long-term countrywide rural development effort. In Anglophone countries, however, FSR activities were concerned with the adaptation of existing agricultural research findings to provide technologies relevant to resource-poor farmers. Links with on-station and component technology research were generally stronger in Anglophone countries while in the Francophone countries linkages with extension/development were a strength.

Thumbnail sketches for both Mali and Nigeria are included here. The history of FSR in Nigeria, a country with an Anglophone tradition, offers a comparison with Senegal (Chapter 6.2), a country with a Francophone tradition. Nigeria is of particular interest as its quest for FSR capacity stimulated an in-country network that brought a degree of cohesion to efforts across the country and synergized the capacity building process. Mali is perhaps something of a hybrid. Despite a French colonial history its FSR initiative was planned by a task force led by David Norman following his earlier experience in Nigeria.

9.3.3 Nigeria

As early as 1958, researchers at the research division in Samaru were already talking FSR. The pioneering work of David Norman, who used a multidisciplinary approach and established the need to work in farmers fields, started in 1965, when the Rural Economy Research Unit (RERU) was established at the Institute for Agricultural Research at Samaru. In 1981 the Federal Ministry of Science and Technology (FMST) directed that all institutes concerned with food production should evolve a FSR programme. In 1982, the ministry organized a training workshop on FSR for the researchers from the national institutes. In 1983 it appointed a national coordinator for FSR.

Before the Ministry's call some institutes had no FSR activities at all, at those that did, their organization varied widely from institute to institute. Some FSR programmes worked side by side with commodity programmes, others had a separate division within a research department with responsibility for socioeconomic research and/or extension and technology delivery. Each of the major NARIs has responded to the Ministry's call, albeit in different ways.

FMST created nine nationally coordinated research projects (NCRP) in 1983, and although eight were commodity based, the ninth was in FSR. Each received token start-up funds and those for FSR were allocated to the four research institutes which were then operating FSR programmes; the Institute of Agricultural Research and Training (IART) at Ibadan, the National Cereals Research Institute (NCRI) at Badegi, the Institute for Agricultural Research (IAR) at Samaru and the National Root Crops Research Institute (NRCRI) at Umudike. Those institutes with relatively new FSR programmes received assistance from IITA and the Federal Agricultural Co-ordinating Unit (FACU) in implementing diagnostic surveys in selected Agricultural Development Project (ADP) areas followed by on-farm trials. Newly acquired FSR capacity in these institutes, coupled with the existing FSR capacity at IAR, marked the emergence of a peer group of scientists committed to institutionalizing an FSR process within their respective national research institutes.

The Nigerian FSR Network

The creation of a National FSR Network for Nigeria was prompted by the desire to improve coordination of the emerging FSR activities of the groups of research scientists, both in socio-economic and technical disciplines. The Network was created in 1983 around the five research institutes mandated for research on food crops. It soon became obvious that the severe funding constraints faced by FMST would jeopardize the effective institutionalization of the Network within the National Agricultural Research Systems (NARS). A proposal from FMST to the Ford Foundation, which had already started funding network activities in a limited way through a grant to IITA, was approved in 1985 and provided support for 3 years of network operation. The mandate of the NFSRN has four main goals:

- To improve the flow of research results by regular meetings among researchers through workshops and the publication of newsletters.
- To demarcate the country into farming systems zones and allocate responsibility for these to groupings of national research institutes, universities and ADPs.
- To coordinate FSR activities in the various institutions and organizations and stimulate collaboration.

- To identify sources of financial, logistic and technical assistance for FSR in Nigeria.

The network is run by a steering committee made up of representatives from FMST, the Federal Ministry of Agriculture Rural Development and Water Resources, IITA, the Federal Agricultural Research Unit, the universities and the five zonal coordinators, with a secretariat implementing its decisions. Its achievements include:

- The government's adoption of the zonation delineated by the network.
- The annual review of Zonal On-Farm Adaptive Research (OFAR) programmes, bringing together participants from universities, ADPs and national and international research institutes.
- The many training sessions conducted both at zonal and national levels on FSR topics, including participatory rapid appraisal and indigenous knowledge systems.
- The 11 issues of a newsletter that have been published and distributed free to members. Many publications on FSR have been financially and logistically supported by the network.
- The network's development and maintenance of linkages with regional and international FSR networks.

Network activities have helped the institutionalization of FSR in Nigeria. All the agriculture-related research institutes now have FSR programmes as mandated by FMST, and conduct research into the farming systems in their respective zones.

9.3.4 Mali

The FSR programme in Mali was a response to the need to find more effective means of transmitting research findings to peasant farmers. There was a need to understand why some historical research findings were rejected, and how to reorient programmes to be more relevant to the priorities of Malian farmers. The Institute of Rural Economics (IER) was set up in 1960 under the Ministry of Agriculture to carry out research, evaluation and conceptualization. IER also acted as coordinator and permanent liaison office between the various service organizations and bodies responsible for research, publications and the organization and evaluation of agricultural development programmes in Mali. But the approach still failed to take into account the peasant-farming strategy. It neglected the constraints on farmers in the socioeconomic environment such as price, marketing, land and tenure problems, did not distinguish between different types of farmers, and could not understand the whole farm as a 'system'.

A systems perspective was gradually introduced in both research and development. The earlier approach, which centred around cash crops, gave way to a more integrated approach embracing all the crops of a given zone, and taking into account factors such as health care, education, hydraulics and road and track construction.

Two conferences were organized in Bamako, in November 1976 and in February 1978. These brought together scientists from neighbouring countries and from national and international institutions to help the IER build up a methodology. As a result of the conferences a 5-year programme was drawn up for Mali in July 1977 with the aid of a task force led by Professor David Norman. This proposed southern Mali (Mali-South) as the start-up location for a 'systems' team. The training of Malian senior staff in the locale is perhaps its most noteworthy achievement and it has also contributed a number of new ideas on the research-development linkage. Researchers meet together in technical coordinator committees, bringing together boards of development and research that examine, independently of the regular technical committees, the problems of a particular zone. This has been a major step towards the regionalization of research programmes[2].

The structure of the project meant a radical change in the dialogue between researchers and peasant farmers as research was no longer required to go through administration in order to reach the farmer. A better understanding by researchers of the needs of and constraints on the peasants made it possible for programmes to be more closely aligned with realities. Until this time, in planning interventions, all farmers were looked upon as identical, in terms of both constraints and possibilities. Those who did not follow along were consequently written off as recalcitrant or rebellious. Through the work of the 'systems team' the assumed physical and human homogeneity of the Mali-South zone was exposed as quite false, and peasant farmers were recognized as highly diverse. There is a full account of the activities of a systems team in Mali in Chapter 6.3.

9.3.5 East, central and southern africa

Countries in the region with the strongest FSR programmes include Botswana, Ethiopia, Kenya, Malawi, Tanzania, Zambia and Zimbabwe. Outline histories are to be found in Chapter 6 with detailed case studies of facets of institutional organization in Tanzania and Zambia in Chapter 7. The East and Southern African Regional Programme set up by CIMMYT in 1975, which ran until 1992, pioneered the introduction of FSR to national agricultural research services in the region under the rubric of OFR with a farming systems perspective (OFR/FSP). The CIMMYT programme was essentially a networking activity which operated at two levels; all countries in the region were exposed to FSR through regional workshops and a newsletter, and six countries; Ethiopia, Kenya, Malawi, Tanzania, Zambia and Zimbabwe were targeted for an intensive effort in capacity building[3]. A similar pattern was followed in each country:

- The demonstration of a farming systems approach to understand small farmers, and the design of experiments based on that understanding.
- Promotion of the need for social scientists in the agricultural research establishment.
- In-country training of human resources in OFR using a systems perspective.
- The implementation of cycles of OFEs with results evaluated using farmers' criteria.
- Guidance on institutional structuring to use OFR/FSP as an adaptive research process and to link it with the traditional research structure.

The years of the CIMMYT programme split into two parts. From 1976–81 efforts focused on raising awareness in the region. Demonstrations of a farming systems perspective for planning OFEs were carried out in collaboration with local farmers and, by 1980, eight studies had been completed in six countries. A shift from demonstration to capacity building began. By 1982 training, linking OFR into the research process, advice on institutional arrangements for its management, and networking, were the main programme emphases. As well as one-on-one collaborative research with national scientists, several training modes were used, described in more detail by Ponniah Anandajayasekeram in Chapter 8.1:

- Introductory workshops allowed national research managers to explore the potentials and organizational implications of FSR.
- Annual regional training workshops, in collaboration with the University of Zimbabwe, helped to build a cadre of practising FSR scientists in the region. Later, as more in-country training was completed, they served to train new members for the established FSR teams.
- In-Country Training Courses on a 15–18 month 'call' system covering diagnosis and a full experimental cycle. Between five and seven visits by a minimum of two CIMMYT scientists covered each stage in the OFR cycle, teams then returned to their own areas and implemented the taught stage locally.
- Specialised training workshops were used to boost necessary skills, such as computer analysis of survey records, experimental design and statistical analysis.

By the time the programme concluded in 1992, most countries in the region had some capacity in FSR. USAID invested heavily in FSR in the region during the 1980s with some 12 bilateral Title XII projects, staffed by teams from American universities, with Botswana, Rwanda, Sudan and Swaziland in particular all benefiting from its bilateral support. This heavy but short-lived investment raised criticism that technical assistance staff were learning on the job, rather than building local capacity[4]. Both The Netherlands and IDRC have had a more sustained policy of investment in FSR in the region.

9.3.6 WAFSRN

WAFSRN (West African Farming Systems Research Network) was created in 1982 as a professional association of West African scientists interested in involving farmers in research to improve agricultural productivity. Its general objective is to promote and facilitate Cupertino among the national, international and external scientists, programmes and institutions working in FSR across the region. The Network supports FSR scientists and strengthens national programmes through training, exchange of methodological experiences, comparison of results and better access to information. USAID provided early technical backstopping through the Farming Systems Support Project (FSSP) managed by the University of Florida, Gainesville. This project produced massive documentation for training in both diagnosis and OFE[5].

After a symposium in Dakar in 1986, an agreement was signed with the Scientific Technical and Research Commission of the Organization of Africa Unity (OAU/STRC) to give WAFSRN a legal status and to help to access financial resources[6]. A secretariat was later established within the office of the Organization for African Unity and SAFGRAD in Ouagadougou, Burkina Faso, with a full-time coordinator. The network is directed by a Steering Committee made up of a Chairman, a Coordinator, representatives of the international research centres operating in West Africa and an OAU/SAFGRAD representative. The 2-year work-plans and requests for funding prepared by the Steering Committee were welcomed by several donors. Early funding was provided by the French and Dutch Ministries of Cupertino, the Ford Foundation, GTZ, the International Center for Research in the Semi-Arid Tropics (ICRISAT) and the International Institute of Tropical Agriculture (IITA). IDRC provided funding in two phases – the first to support various WAFSRN bodies, the second to continue support for the network and to strengthen national programmes through training, promoting dialogue among researchers, and improving access to information. It had five specific objectives:

- To stimulate collaborative planning and evaluation of FSR throughout West Africa.
- To improve methods used by organizing meetings, monitoring tours and other activities to facilitate exchange of relevant experiences.
- To help organize and institutionalize training programmes to meet the needs of a farming systems approach to research and development.
- To encourage national FSR programmes to publish and disseminate the results of their research to all parties in a way that presents clear policy alternatives to government.
- To assist national FSR teams in obtaining financial assistance for their programmes.

At the 1989 Accra Symposium, participants asked that WAFSRN seek to stimulate collaborative research activities, bringing together country teams that would work on common themes of interest across the various member countries. These teams were incorporated into what became known as the Phase II Research Project and the following teams became operational:

- The Collaborative Group on Maize-Based Farming Systems (COMBS), with the help of IITA.
- The Collaborative Group on Farming Systems of Sudano-Sahelian Zone (GREF-MASS) with the help of ICRISAT and IDRC.
- The Collaborative Group on Root and Tuber Improvement Systems (CORTIS) with the help of IITA.
- The West African Animal Traction Network (WAATN) with the help of ILCA (now ILRI).
- The Women in Agriculture Group (WIAG).

One of the major areas of achievement of WAFSRN has been information dissemination. The peer-reviewed journal entitled 'Agricultural Systems in Africa' is published twice a year, with articles that emphasize FSR concepts, methods and results in the participating countries. It aims to bring FSR results to researchers, development agents, policy makers and others in the subregion and elsewhere. To better equip young scientists in the art of scientific writing WAFSRN has co-sponsored two scientific writing courses. The WAFSRN Bulletin was published twice a year to provide regular information for all members about the activities of the network and the collaborative research groups. Other publications include the proceedings of the 1986 Dakar and the 1989 Accra symposia and two volumes of 'Abstracts of Literature on Maize-Based Farming Systems in West Africa'. A Directory of Members was produced in 1990 and updated in 1991. The activities of the network continued until 1994 when the IDRC and Ford grants came to an end. As of 1996 the membership of WAFSRN was about 600 professional from 17 countries in West Africa, but from 1995 to date WAFSRN activities have been limited to the publication of the journal. Attempts to find further funding proved impossible as donor interest in FSR-E fell. None of the collaborative research groups is functioning at present and the network secretariat is currently housed at Ahmadu Bello University, Zaria, Nigeria to minimize operating expenses.

9.3.7 Southern African Association of Farming Systems Research & Extension (SAAFSR-E)

At the FSR-E conference in Johannesburg in February 1992, delegates decided that a Southern African Association for FSR-E should be formed, and the new association was officially launched in 1993 during the next conference, held in Swaziland. The secretariat, which is located and incorporated in Swaziland, supports about 500 registered members and receives financial support from FAO and SIDA to publish and distribute its newsletter. The major aim of the SAAFR-E is to promote FSR-E and to strive to improve networking between individuals, institutions and professional societies active in southern Africa. In South Africa itself, FSR-E had not received much attention in the past. With the post-apartheid political emphasis on African farming, however, keen interest is developing.

SAAFSR-E has made tremendous progress since its establishment. It has held annual conferences and its newsletter appears on a regular basis. Membership has been increasing by about 100 every year for the past 6 years and the association is planning to launch a southern African journal of FSR-E in the near future. A Members Directory has been finalized to enhance linkages between FSR professionals and with other professionals in the various fields closely related to FSR-E.

At the 1995 Ordinary General Meeting in Harare a decision was taken to appoint area or country coordinators to provide a linkage between members, the permanent secretariat and the Council and President, in particular supporting the work of the secretariat. Their task is to:

- Update the mailing list and membership directory.
- Identify libraries and institutions to receive the SAAFSR-E newsletter and bulletin.
- Provide a mailing link between the members and the secretariat.
- Support the collection of articles, publications and information for the newsletter and bulletin.
- Collect fees from country members.
- Collect, administer and manage membership fees from country members.
- Maintain and manage the local chapter account according to regulations and guidance provided by the secretariat.
- Solicit potential donors in their countries for various activities, and link them to the association president where necessary.
- Organize in-country networking activities.
- Organize activities for country members to facilitate information sharing in FSR/OFR methodologies for technology development and transfer. Such activities may include symposia, workshops and field visits.
- Recruit members.

At the same meeting in Harare it was resolved that SAAFSR-E would engage in policy dialogue in the region. Responding to the apparently weak understanding in South Africa of FSR-E and its institutionalization, SAAFSR-E arranged a national workshop in Pretoria in 1994 on institutionalizing FSR-E. It was attended by top and middle management of the National and Provincial Departments of Agriculture and the Agricultural Research Council. Members of SAAFSR-E gave their time freely to present case studies on the institutionalization of FSR-E in Kenya, Tanzania, Zambia and Zimbabwe and a report on the findings of the workshop was sent to decision makers as well as those present, meeting with positive acclaim. At a further regional meeting in Arusha in 1996 a number of excellent papers on livestock systems were presented, bringing livestock colleagues further into the FSR-E fold. SAAFSR-E has commissioned case studies on the institutionalization of

FSR-E in 10 East and southern African countries from Ethiopia southwards. These were presented at a 1997 workshop and a book synthesizing the lessons learned was published in October 1998[7]. The Association successfully hosted the 1998 AFSRE global symposium at the University of Pretoria in South Africa in December 1998 which over 500 hundred FSR professionals attended.

Members are proud of their individual professional status, non-aligned with either governments or the private sector. This has allowed the Association to engage in dialogue with national governments as has been the case in South Africa. No government funds have been received in support of core activities, though governments have supported the annual conferences in their countries. SAAFSR-E's future plans include the streamlining of FSR-E methodologies and procedures and the integration of the livestock component in the small-scale farming systems. FSR-E case studies from the region will continue to be documented and disseminated by the Association, which will try to bring coherence to the training programmes mounted across the region, record successful institutional arrangements, and document results achieved and lessons learned.

9.3.8 African Association of Farming Systems Research Extension and Training (AAFSRET)

At the 1991 international AFSRE symposium held at Michigan State University, the FSR-E professionals working in Africa met for the first time to take stock of current state of the arts. An interim steering committee was set up at that point to persuade members in Africa to come together under one umbrella. At the next symposium in 1992 the decision was taken to form an association. An interim Chair and Secretary were elected along with some other members to organize the first all African FSR-E symposium which took place in October 1993 in Nairobi, Kenya. The association links organizations and individuals involved in the farming systems approach in Africa and is open to all individuals and organizations interested in AAFSRET's objectives. Membership is currently estimated at about 1500 professionals.

The objectives of AASFRET are:

- To promote a farmer-based approach to agricultural development in Africa.
- To promote the formation and maintenance of subregional and national FSR/extension/training networks in Africa.
- To strengthen and promote links between individuals, organizations and farming systems (FS) networks in Africa and elsewhere.
- To stimulate and promote the dissemination and exchange of information, research findings and experiences in farming systems.
- To facilitate coordination and collaboration in agricultural research and development activities.
- To enhance the performance and professional development of practitioners in farming systems at all levels through appropriate training initiatives.
- To assist national and subregional FSR networks on technical and organizational matters relating to the planning, preparation, funding, implementation and evaluation of their FSR programmes.
- To organize farming systems activities, including biennial meetings, study tours and professional visits.

A second AAFSRET symposium was held in August 1996 at Ouagadougou, Burkina Faso and here a further attempt was made to continue FSR-E activities in West and Central Africa with the birth of the West and Central Africa Association of Farming Systems Research and Training.

9.3.9 Future Aspiration and Constraints

The future directions for the African based FSR associations; WAFSRN, SAAFSR-E and AAFSRET will address the following areas.

FSR for policy use

Most early FSR programmes in Africa focused on technology development and ignored agricultural policy questions. Many were criticized for failing to use their understanding of African farming to influence policy and release technological opportunities. There are still no formal mechanisms for adaptive research scientists to interact with planners, policy makers and development agencies. Some linkages have been made, largely through informal initiatives of the adaptive research programmes, but these are no substitute for a national agricultural policy framework that considers such interaction essential.

Working links and information flows between planners and FSR programmes are needed throughout Africa to encourage the use of micro-level FSR data for policy analysis and design. Mechanisms would include exchanging reports, carrying out joint micro-level policy related work and FSR seminars for policy planners.

Staff development

Most FSR-E staff have been new graduates with very little or no experience in research, and their ability to design, conduct and analyse surveys and experiments has been limited. In the early days these youngsters had to be protected from senior colleagues in the traditional disciplines of the research services until they learned their skills. The AFSRE associations have long-term training plans, which aim to have all FSR-E researchers skilled in methodologies and FSR practice by the year 2000. Training will also seek a better balance between disciplines in FSR teams, as poor disciplinary complementarity is still a problem.

Cost-saving methods

Most national FSR programmes, in common with most research programmes, have faced problems of inadequate resources. To ensure quality in FSR work with limited resources at hand, the associations are seeking cost-saving methods in conducting OFR such as the use of farmer groups and clustering techniques.

Farmer involvement

According to Biggs in 1989, farmer involvement in FSR has been classified into contractual mode, consultative mode, collaborative and collegial mode[8]. The collegial mode is the ideal situation in which the farmer is regarded as an equal partner with the researcher and in which the knowledge of the farmer counts equally with the knowledge of the researcher. Most African countries still operate in a consultative or collaborative mode. Increased farmer participation is being given more emphasis to help ensure positive impact from FSR activities. Various farmer participatory methods are being introduced to FSR and it is planned to expose staff to these through short courses and workshops on a continuous basis.

Livestock and livestock–crop integration

Little research has been done on livestock and livestock/crop interactions[9]. This was attributed to the donor influence, and the historical separation of crop and livestock research at national level. There was also limited knowledge on the feasibility of OFR with livestock, though this improved at the end of the 1980s[10] and more emphasis will be placed on integrated crop–livestock systems in the future.

Information exchange

Documentation of case studies, research results and methodological advances in FSR are being disseminated through the publication of bulletins, journals, books, monographs, conference proceedings and other documents. The exchange of information will continue to be encouraged by the associations between individual members, networks and associations.

Sustainability of FSR activities

The problems of inadequate funding, lack of reliable transport and shortages of research facilities have been common problems in FSR and have reduced the quality and impact of FSR activities in most African countries. In the last two decades resources for FSR have been provided in large part by foreign aid agencies, with African governments contributing only a small proportion of budget requirements. Donor support is often hedged with conditions seeking short-term impact and this dependency poses a threat to the sustainability of an FSR approach.

There is a significant imbalance in resource endowments between the nascent FSR programmes and the more traditional research and development components of commodity research and extension. However, making inroads into these traditional government budget allocations, particularly when this requires the addition of new disciplines to the establishment, is a slow process at best. It is made particularly difficult when government revenues are in decline. The strongest aspirations of FSR-E associations in Africa are towards ensuring FSR's sustainability and association efforts are geared towards gaining long-term support for this innovative step in the research and development process. As with FSR-E programmes, associations also suffer from a funding problem, particularly with the fall off of donor interest. Attempts to charge membership fees in AAFSRET following the decline of WAFSRN highlighted the dilemma. Professional salaries are so low that an acceptable level of fee cannot support the activities of an association. The difficulty of communication both within and between countries in Africa is a further inhibition to the transfer of information and to the organization of an effective, attractive, regional programme.

REFERENCES

1. **Fresco, L., 1985**. Comparing Anglophone and Francophone Approaches to Farming Systems Research and Extension. Networking Paper no. 1. International Programs, University of Florida, Gainesville.
2. **Berte, A.T., Diarra & M. Tangara, 1988**. Farming Systems Research in Mali. A Case Study of the Mali South Zone. In: Proceedings of the West African Farming Systems Research Workshop, Dakar, Senegal 10–14 March, pp. 47–61.
3. **Collinson, M.P., 1982**. Farming Systems Research in Eastern Africa: the Experience of CIMMYT and some National Agricultural Services, 1976–81. MSU International Development Papers no. 3. Department of Agricultural Economics, Michigan State University, East Lansing, Michigan.
4. **Byrne, K., 1989**. A Review of AID Experience: Farming Systems Research And Extension Projects, 1975–87, Center for Development Information. AID Evaluation Highlights no. 4. USAID, Washington DC, USA.
5. **FSSP, 1987**. Diagnosis, Design and Analysis in Farming Systems Research and Extension, vols I-III. Farming Systems Support Project, University of Florida, Gainesville, USAID Cooperative Agreement (no: DAN-4099-A-00–2083–000).
6. **Baker, D.C. & D.W. Norman, 1989**. A Framework for Assessing Farming Systems Activities in National Setting in West Africa: with Special Reference to Senegal, Nigeria and Mali. In: Proceeding of the West African Farming Systems Research Workshop, Dakar, Senegal 10–14 March, pp. 62–87.
7. **P. Anandajayasekeram and T. Stilwell (Eds), 1998**. Institutionalization of Farming Systems Approach in Eastern and Southern Africa. SAAFSR-E and FARMESA, Swaziland and Zimbabwe.
8. **Biggs, S., 1989**. Resource-Poor Farmers Participation in Research: a Synthesis of Experience from Nine National Agricultural Research Systems. OFCOR Comparative Study no. 3. ISNAR, The Hague, The Netherlands.
9. **Olukosi, J.O. & A.O. Ogungbile, 1992**. Impact of FSR-E in Africa. West African Farming Systems Research Network Bulletin, 13, 10–13.
10. **Amir, P. & H.C. Knipscheer, 1989**. Conducting On-farm Animal Research: Procedures and Economic Analysis, p. 244. Winrock Institute for International Development, USA and the International Development Research Center, Canada.

    ILCA, 1990. Livestock Systems Research Manual, Vols 1 & 2. ILCA Working Paper no. 1. ILCA, Addis Ababa, Ethiopia.

9.4 THE ASIAN FARMING SYSTEMS ASSOCIATION

Nimal Ranweera

It was agreed that in the future, the wider application of the FSR-E paradigm should become the core agenda for sustainable agriculture in the 1990s. The Asian Farming Systems Association would strengthen its efforts to address equity and economic efficiency issues by promoting convergence in the efforts of government and NGOs, and the universities, in improving productivity and profitability through sustainable farming systems.

9.4.1 Introduction

For the last two and a half decades Asia has been at the forefront of the development of FSR in the agricultural sector. Historically, it was in the early 1970s that IRRI first formulated a 'systems' approach to agricultural research, with its cropping systems programme. The objective was to increase productivity per unit of land, through crop intensification and diversification with the ultimate aim of increasing income levels of the farming community. This programme, initially funded by the IDRC, played an important role in sensitizing researchers to a 'systems thinking' process.

9.4.2 Pathfinders in Asian farming systems programmes

This seminal work at IRRI provided the 'first cut' of a methodology for systems work. The associated networking brought together a number of researchers from around Asia and their combined efforts led to coordinated programmes of research and development within the framework of the Cropping Systems Network and later the Asian Rice Farming Systems Network (ARFSN). By 1994 this covered 16 participating countries, as well as other regional programmes. Based on a novel methodological sequence – baseline surveys, component technology testing, pre-production technology testing and large-scale testing on farmers' fields – new cropping systems were developed. The most significant of these were:

- Mungbean before rice in Thailand.
- Triple cropping of rice in Bangladesh.
- An early rice crop followed by rice or other field crops in Sri Lanka.
- The Gogo ranch technology in Indonesia.
- The Iloilo-Pangasinan technology which underpinned the Kabsaka programme in the Philippines.

A number of interest groups gradually evolved to concentrate on specific activities and issues. One was the group on agroecosystems in FSR, which later became a network of South-East Asian Universities (SUAN). Similarly the Participatory Rapid Appraisal group, initially from Khon Kaen University, and the research activities funded by Ford Foundation and USAID in Nepal, VISCA in the Philippines and Bangladesh at the Bangladesh Agricultural University are all important examples. More recent examples include the work at Cantho University in Viet Nam, and the GTZ-sponsored activities in northern Thailand. In the late 1980s and early 1990s, the farming systems development (FSD), concept was introduced into Asia by FAO, bringing an added dimension to FSR.

As a result of the activities of ARFSN, and the growing number of practitioners, many countries began organizing their own FSR programmes and workshops. At the same time universities began providing short-term courses of up to 6 weeks, medium-term courses up to 6 months and degree programmes at undergraduate level, and later at postgraduate level, on farming systems. The programmes at the Asian Institute of Technology in Thailand and the Universities of Khon Khen and Kaesetsart, at the Farming Systems and Soil Research Institute (FSSRI) at the University of the Philippines, at Los Baños (UPLB), the Masters and PhD programmes in collaboration with IRRI and the programmes at the Bangladesh Agricultural University (BAU) in Bangladesh and the Bogor Campus in Indonesia, all reflected this enthusiasm. Formal institutionalization of FSR increased and began to involve the NGO sector and a few private sector institutions. In addition, a large number of universities in many Asian countries undertook FSR programmes.

9.4.3 Some background to the Asian Farming Systems Association (AFSA)

The increase in the number of practitioners led to demands for the publication of their results in refereed journals and for participation in workshops, symposia and seminars. The AFSA was the culmination of a number of efforts along these lines. This increased interest coincided with the Kansas State University (KSU) farming systems conferences being organized in the USA. With the expansion of the Kansas State and later the University of Arkansas symposia there was increased opportunity and increased interest on the part of southern practitioners to present their work, showing the need for some sort of association, preferably formal, to provide a means to bring FSR practitioners in Asia closer together. In 1988 at the 8th Annual Farming Systems Symposium at the University of Arkansas, a group of Asian participants formed a group that decided to organize the first Asian symposium in 1990.

The first Asian symposium

An informal International Organizing Committee (IOC), in collaboration with the Agricultural Food and Engineering Division of the Asian Institute of Agriculture in Bangkok, Thailand, organized the first Asian farming systems symposium in November 1990. The overall theme was Sustainable Farming Systems in 21st Century Asia, and 220 participants from 18 Asian and five other countries attended. Four tasks were identified for the programme:

- Appraisal of past FSR-E experience in Asia, and assessment of various FSR-E procedures for addressing future needs.
- Examination of alternative methods for assessing the impact of FSR-E on households, the environment, and on research and development institutions.
- Discussion of future directions for FSR-E in Asia, with a focus on increasing research and extension efficiency, contributing to policy debates and addressing environmental issues.
- Stimulation of linkages among FSR-E practitioners through associations and networks.

The Symposium was opened by His Excellency Khun Chuan Leekpai, Thailand's Minister of Agriculture and Cooperatives and each of the three major themes were introduced by a distinguished speaker. For the keynote addresses, Dr H.G. Zandstra, Deputy Director General of IRRI, noted problems faced by FSR-E practitioners and the improvements that had been made. Dr Jock Anderson, Principal Economist of the World Bank, challenged FSR-E practitioners to consider the views of investors and the Hon. Lalith Athulathmudali, Sri Lanka's Minister of Education and Higher Education, reminded participants that research should be for farmers and not for scientists. Over 100 papers and 26 posters were presented during the 4-day programme.

At each session panellists and participants developed future directions for FSR-E work in Asia. It was felt that rapid population growth and environmental degradation posed grave threats to 21st century Asia, endangering agricultural productivity, farm incomes and food prices. New sustainable farming systems are needed, as well as farming systems that will increase the income of farm families and reduce the real cost of food for low-income consumers, while preserving or improving the quality of the environment. Some practitioners felt that the actual impact of FSR-E had fallen short of expectations and called for a greater focus on efficiency in FSR-E, especially the streamlining of tools to hasten and increase its impact. Only modest efforts had so far been made to assess the effect of FSR-E activities on farmers, consumers, research institutions and the environment, and these efforts should be strengthened. Farmers had been, and should continue to be, the primary users of information generated by FSR-E, but this information should also be packaged for use by policy makers, research managers and other audiences. Participants agreed that the future directions for FSR-E should be guided by the following:

- **Broadening scope**: FSR-E practitioners should pay more attention to non-technical factors that affect technology generation and adoption, such as institutional constraints or agricultural policy.
- **Affecting policy**: Policy makers should be seen as clients for research results.
- **Promoting equity**: Research focusing on resource conservation can generate recommendations that endanger the economic

interests of low-income farmers and researchers must weigh the likely distribution of benefits arising from resource-conserving innovations.

- **Incorporating external factors**: FSR-E practitioners should be prepared to deal with such external factors as transaction costs, common property resources, pricing policies, land degradation, agroecosystems diversity, and links between agriculture and global pollution.
- **Setting priorities**: Researchers must set priorities among technical research themes and among directions for expansion, requiring careful matching between specific FSR-E approaches, institutional restrictions and environmental conditions.
- **Involving farmers**: The role of farmers in FSR-E should be expanded. Sustainable practices often have their roots in farmers' experiments and the recommendations of innovative farmers. Farmers can also be used to train other farmers and to collect and report data when appropriate.
- **Working for sustainability**: Researchers and Research Managers must understand and examine agroecosystems diversity, links between agricultural enterprises and the vulnerability of agricultural systems to global pollution and climate change. FSR-E should be especially useful in recommending strategies for sustainability.
- **Integrating systems**: Restructuring emphasizes nutrient and energy cycling, and requires integration of new enterprises. An integrated system maximizes the recycling of waste and by-products between different enterprises within the farm.
- **Building models**: Conceptual and mathematical models must be developed and improved. Researchers should not, however, become overly reliant on complex and expensive formal models, and modellers should be aware of the need for substantial farmer input in model construction.
- **Assessing impact**: More attention must be paid to assessing the impact of FSR-E activities. Although methods of monitoring and evaluation are well developed, few FSR-E practitioners are familiar with them.
- **Emphasizing gender**: FSR-E practitioners should understand and emphasize the role of women and reflect this in research design and evaluation. Research managers must target women as beneficiaries of FSR-E activities, and must work to increase female participation as FSR-E professionals.
- **Strengthening research links**: Links between FSRE researchers and disciplinary and commodity scientists must be strengthened to convince disciplinary scientists of the value of FSR-E concepts and tools.
- **Working with extension**: FSR must link more directly with extension services. Farmers themselves should be called on to perform more extension functions through farmer-to-farmer training activities.
- **Teaming up with NGOs**: Governmental FSR-E programmes should strengthen their links with NGOs. When research resources are limited, NGOs can be of immense help in directing technology generation activities towards small farmers. Moreover, NGO participation can foster creativity in the development of FSR-E procedures.
- **Improving training**: The new directions suggested for Asian FSR-E will require new skills from professionals. Universities and agricultural education systems must be prepared to meet this challenge.

The Symposium concluded with the resolution by the participants that a symposium of this nature for Asian FSR-E practitioners would be appropriate every 2 years.

9.4.4 AFSA

The same symposium unanimously endorsed the idea of an organization to institutionalize the aspirations of systems practitioners in Asia, providing a forum for the exchange of professional experiences as well as field practices. The International Organizing Committee for the 1992 Symposium was given the mandate to form an Association for the AFSA. AFSA was formally incorporated with the Securities and Exchange Commission of the Philippines as a non-profit organization in 1992 based at the South-East Asian Centre for Graduate Study and Research in Agriculture (SEARCA) with its registered office at the Asian Institute of Technology (AIT) in Bangkok, Thailand. The broad objectives of the Association are to:

- Provide a forum for the exchange of latest activities in the field of FSR and sustainable agriculture in Asia.
- Facilitate through linkages of existing networks and new interest groups, easy access to other practitioners.
- Facilitate the obtaining of project funds to research institutes and individuals to undertake systems-oriented research.
- Sensitize policy makers to the importance of addressing resources and policy initiatives to the small-farmer groups.

Initial membership was confined to the founder members and a voluntary membership fee of US$5.00 for Asians and US$10.00 for non-Asians was agreed. By the end of the symposium the Assocation had 98 members, which had risen to 274 from 32 countries by the end of 1992. The membership fee covers the 2-year period between symposia. The fee structure was later adjusted to reflect the costs of operations and to widen the services for members. The present fees range up to US$300 for members representing Foundations.

Board selection considered subregional representation and a diversity of organizations from government, universities, NGOs and the private sector.

In pursuing its objectives the AFSA was to undertake three types of activity:

- Organize a symposium in an Asian country every 2 years to provide a forum for FSR-E practitioners to share their experiences in FSR.
- Publish and circulate a newsletter for members to keep them updated on recent events.
- Publish a journal (four issues every 2 years) to contain the papers from the symposia as well as contributed papers; an opportunity for researchers to publish in refereed publications.

Second symposium

The first Board of Governors organized the second AFSA symposium in Sri Lanka in November 1992, with the overall theme 'Sustainable Agriculture: Meeting the Challenges Today'. The Mission of the symposium was to:

- Discuss ways to improve the linkages between FSR-E and other agricultural research and development programmes and institutions.
- Propose ways to improve the cost-effectiveness of FSR-E.
- Integrate and apply knowledge from FSR-E to such issues as sustainable agriculture, furthering efficiency of research and improving productivity of agriculture.
- Stimulate and nurture partnerships among FSR-E practitioners in Asia and other parts of the world.
- Identify proposals for future thrusts in FSR-E in the Asian context.

The symposium emphasized the promotion of agricultural development patterns that meet the needs of the present without compromising the ability and aspirations of the future generations to meet their own needs. It focused on three themes:

- Farming systems and sustainability.
- Institutional design and linkages.
- The effectiveness and impact of FSR-E.

Some 90 papers and 80 posters were presented to a total of 234 participants representing the whole gamut of institutional membership from a wide range of countries, from Australia to Nepal to Viet Nam. Regional and International Agricultural Research Centres were also well represented. The symposium was opened by Hon. R.M. Dharmadasa Banda, Minister of Agriculture Development and Research in Sri Lanka, who called for a greater recognition of the efforts of the farmers in their role of ensuring that Asian countries are well supplied with food. Recommendations emerging from the discussions of issues surrounding the three themes were presented to the participants and approved in a plenary session chaired by Dr Clive Lightfoot.

The resolutions adopted represented the consensus of the participants from diverse national and scientific backgrounds and interests. Participants affirmed that equity concerns form the core of the concept of sustainable agriculture. Emphasis was given to enhancing resource-neutral and scale-neutral agricultural technology generation and dissemination to assist both economically and ecologically disadvantaged farm families. Attention was called for sustainable advances in total farm productivity, while preserving or improving the quality of the environmental capital stock. Suitable integration of research strategies and public policies was recognized as a prerequisite to achieve this goal. The participants affirmed the need for continued and vigorous advancement of the FSR-E systems approach with a conscious bias towards resource-poor farming households.

Institutional mechanisms and linkages were proposed to furnish community structures for implementing their specific functions. Greater farmer participation in the research process and effective integration of the options offered to farmers with the programmes of existing local, non-governmental and national institutions were encouraged. The need for greater participation of women in FSR-E decision-making process was emphasized. The importance of impact assessment was recognized and a set of tasks was recommended for implementation in the future.

It was agreed that in the future, the wider application of the FSR-E paradigm should become the core agenda for sustainable agriculture in the 1990s. The AFSA would strengthen its efforts to address equity and economic efficiency issues by promoting convergence in the efforts of government and NGOs, and the universities, in improving productivity and profitability through sustainable farming systems. AFSA also conducted business in the course of the symposium, including the election of new board members.

Third symposium

The third symposium organized by AFSA was held in Manila, Philippines in 1994, with the overall theme 'Conservation and Equitable Growth: the Challenge for Farming Systems'. Again, defining the purposes of the symposia showed the dominance of sustainability as an issue;

- Provide a forum for discussion and debate among farming systems practitioners on the current thematic issues of sustainability and environment with special reference to Asian agriculture.
- Undertake a critical examination of long-term issues in rainfed upland agriculture particularly in relation to settled forms of agriculture.
- Stimulate and strengthen linkages among farming systems practitioners throughout Asia.

The third symposium, declared open by His Excellency President Fidel V. Ramos, President of the Republic of the Philippines, was attended by 235 participants from 19 countries. The keynote address was made by Mr A.Z. Obaidullah Khan, the Assistant Director General of the FAO for Asia and the Pacific.

Fourth symposium

The fourth symposium was organized collaboratively with the International Association for Farming Systems Research and Extension following changes in its constitution to reflect the emerging maturity of the regional farming systems associations. It was held in Colombo, Sri Lanka, in November 1996.

The background to the symposium was the dawn of a new century with farming in a rapid state of change brought on by globalization and market liberalization on the one hand and resource degradation on the other. These dynamics are intensifying the pressure on the land and water resources available and altering the future for both large and small farmers. New realities must be effectively reflected in agricultural research and development programmes. The farming systems profession can help with the development of the agricultural sector in this increasingly complex context. The purposes of the symposium were to:

- Encourage debate on the challenges facing farming systems practitioners as they attempt to consider issues of long-term sustainability, the linkages of FSR-E with other segments of agricultural research and the cost-effectiveness of the FSR-E process.
- Strengthen the farming systems profession and stimulate linkages among practitioners in order to develop the agricultural sectors with particular reference to resource-poor and food-insecure households.

The Symposium was opened by Hon. D.M. Jayaratne, the Minister of Agriculture, Lands and Forestry of Sri Lanka, with the keynote address given by Dr N.F.C. Ranaweera, then President of the Association of Farming Systems Research and Extension. The sessions included 80 selected papers and 48 posters and the thematic workshops, which formed the major part of oral presentations, were developed around five themes, namely:

- Household food security.
- Environment and agricultural resource management.
- Innovations and social change: who is empowered?
- Methodical issues: systematic questions to basic disciplines.
- Policy and macro economic issues.

Important cross-cutting issues, particularly the role of agricultural and trade policy, and impacts on the environment, were explored in each workshop. Outside the workshops, smaller fora addressed a wide variety of issues including: sustainability indicators, human resource development for farming system approaches, indigenous knowledge, modelling and participatory methods. Panels on 'Differing Institutional Experiences and Perspectives' and 'Information Technology and Data Management' occupied important plenary sessions.

9.4.5 The Journal of the Asian Farming Systems Association (JAFSA)

JAFSA was created at the time of AFSA's formation. The journal filled a gap that existed for a number of years and provided the forum for Asian practitioners to have their work published in an accepted publication for professionals in farming systems research and extension in Asia. The journal has presented well over 100 major articles, and has, in the past, received funding from IDRC and logistical support from the International Center for Living Aquatic Resource Management (ICLARM), where Clive Lightfoot, the editor was on the staff. The journal is now on hold due to a lack of funding, the high costs of a publisher and, recently, the lack of an editor.

9.4.6. The impact of AFSA activities in Asia

The symposia have been extremely popular since they are one way to bring together a large number of Asian participants within Asia itself to share their experiences. They have also brought together the diverse views of researchers practising in different agroecological zones on issues of sustainability of upland agriculture, rainfed agriculture and the general concerns and issues of poverty, nutrition and hunger. The average attendance of over 200 professionals is in itself a manifestation of the interest in the farming systems approach. Clearly the symposia have been an important binding force, bringing like minds together across the region every 2 years.

AFSA has been fortunate in having a large number of individuals who are committed to fostering the philosophy of farming systems. With the increased emphasis on sustainable agriculture, the environment and gender, the relevance of farming systems concepts has increased this commitment and the adoption of FSR is evident in the change in priorities in national agricultural research systems. The interactions fostered by the AFSA have resulted in the spread of some key technologies beyond the countries of their introduction. The mungbean before rice technology developed in Changmai Province in 1985–86 is now being practised on almost 100,000 ha that had previously carried a single crop of rice. The mungbean crop is harvested early, providing the cash necessary for investment on to the rice crop. In Thailand the rice–fish technology has helped improve farm incomes. In Indonesia the groundnut before rice system has been accepted by farmers on millions of hectares around the country. Vast strides have been made on rice–fish culture and the needs of the whole of Jakarta and its suburbs are being met by the performance of the rice crop after the fish programme. In Bangladesh the three rice crops technology has been developed by intensifying the Aman and Boro rice seasons, largely due to the efforts of the farming systems programme in the Bangladesh Rice Research Institutes (BRRI).

Some of the many issues that have surfaced at the AFSA-organized symposia are clearly policy related. Case studies have also shown that models of integrated farming systems have promising results. Experiments to make credit available on a whole-farm basis, rather than focused on a particular enterprise, are beginning in Sri Lanka. Upland agriculture poses particular policy problems, including sustainability and downstream effects and it is difficult to develop alternatives to the current low input–low yield equilibrium and systems degradation in the acid uplands of south-east Asia.

Farmer adoption of innovations can be expected to be low if rates of return are low, if technologies force a choice between near-term survival and long-term personal or public benefits, or if technologies are difficult to master. Policies regarding upland resources and institutional effectiveness are usually weak, but stakeholder analysis and better micro-level information can improve policy formulation, a new dimension for farming systems practitioners and for the AFSA.

The recommendations of the first symposium providing guidance for the development of FSR have been repeated following subsequent symposia. No great change of direction is needed, but the same foci demand more attention. Sustaining the effort in the face of changing donor perspectives is more problematic. During the initial stages of the development of FSR and that of the ARFSN, donor support came primarily from IDRC. Subsequently it was taken up by a large number of other donors, particularly USAID. IDRC also funded cropping and, later, farming systems programmes across the region, particularly in India, Indonesia, Nepal, the Philippines, Sri Lanka, Thailand and Viet Nam. The Association itself benefited from significant donor support in its formative years. The first symposium relied on support from the Rockefeller and Ford Foundations and IDRC, as well as from USAID and a large number of countries' specific programmes. This support has dwindled and current donor support both to FSR in Asia, and to the Association, is low.

Chapter 10

FSR and the Professional Disciplines

10.1 FARM MANAGEMENT AND THE FARMING SYSTEMS APPROACH

David Norman

Philosophically, the 'current' farming systems approach, as practised mainly in low-income countries, has much more in common with the 'earlier' farm management approach in the high-income countries.

10.1.1 Introduction

Farm management has made important contributions to the idea and the development of the farming systems approach[1]. However, farm management itself has changed over the years and this assessment starts with a short history of the farm management approach. This helps evaluate its contribution to both the early and current characteristics of the farming systems approach. The discussion finishes with a brief evaluation of what the farming systems approach has contributed, and what it can contribute, to farm management in the future.

10.1.2 A short history of the farm management approach

The characteristics of the farm management approach as it was practised during the early part of the century were analogous to what the farming systems approach has become in the 1990s. Experiences and logic at the turn of the century led technical agricultural scientists to start farm management as a field of study[2]. Thus, the early proponents of farm management were biological and technical scientists who knew little economic theory, but were very concerned about the sociological and management dimensions of farm management. In its early days, farm management was multidisciplinary, including the entire range of factors affecting the running of a farm. For example, Warren's classic text of 1913 on farm management included farm accounts, soil types, many agronomic topics and discussions on the factors of production such as land, labour and capital[3]. The text even covered political issues relevant to farming and philosophical matters including those relating to family. One section is titled; 'Some Thoughts for the Farm Boy'!

Because the early leadership in farm management research came from technical scientists with a positivist background, it is perhaps not surprising that normative and prescriptive research was neglected. Although an article by Spillman in 1902 viewed farm management as a merging of the principles of agriculture and economics[4], it wasn't until the 1920s that the balance began to move strongly in the direction of economic analysis[5]. However, this trend continued to such an extent that in the US Land Grant university system, for example, farm management was transferred from departments of agronomy to departments of agricultural economics. As a result, according to Johnson in 1980, 'by the late 1950s, much modern US farm management began to look more and more like a sub-field of production economics which is itself a part of economics, without the multidisciplinary breadth required to handle the problems which arise for farm managers out of technical, institutional and human change'[6]. Perhaps because farm management has become identified increasingly with production economics, in recent decades, greater emphasis has been placed on normative and prescriptive issues through application of techniques such as budgeting and applied decision analysis.

This trend helped entrench an implicit assumption in farm management research: successful farmers had to be thrifty, hard working and were driven by profit. Such farmers would prosper and expand and should, therefore, be emulated. Farm management was thus defined as 'the act of judiciously and skilfully managing the farm' according to Boss and Pond in 1947[7]. As a result, much of the farm management literature has emphasized what farmers should do to be successful, rather than trying to understand the logic of the farming practices that most farmers were using.

Farm management research as it exists today in mainstream agricultural institutes is very different from the original form found about a century ago. However, in high-income countries such as the USA, some vestiges remain of farm management as it was originally conceived, primarily outside the research arena in extension and in implementation agencies such as the Tennessee Valley Authority[8]. Also, the recent interest in sustainable agriculture embraces many principles and much of the multidisciplinary breadth of the early farm management approach.

This introduction to the history of farm management shows that it has meant different things at different times in its development. At no stage in its development has it fully reflected what it ideally should be, if it is to be relevant and helpful to farmers. Thus, the current farming systems approach embraces a mixture of principles that were prominent at different stages in the evolution of farm management. Yang in 1965 suggested that farm management can be viewed as a pure science, because it involves the collection, analysis and explanation of facts and the determination of underlying principles[9]. However, he asserts that it is also an applied science because it involves identifying farm problems and finding solutions to those problems. In essence, therefore, farm management and the farming systems approach should contain positive, diagnostic, and normative elements.

10.1.3 Laying the foundations of a farming systems approach

In the 1960s a large number of formal, structured, cost-route, farm management type surveys were implemented in various countries in Africa, Asia and Latin America. These usually involved interviewing farmers frequently – perhaps once or twice a week – throughout a 1-year period. Most of these studies were carried out by agricultural economists with skills in farm management or production economics. Most were associated with developmental/governmental organizations, academic institutions and/or agricultural research institutions. It is interesting that most of these studies did not involve an interdisciplinary approach, perhaps because they originated with individuals trained in the mode that has characterized more recent, conventional, farm management research. Huge amounts of data were collected, most of which were never completely analysed. Much of the initial impetus for such studies came from a desire for better understanding of the weaknesses of the current farming systems to help plan initiatives for their improvement. Development planners were interested in using the data for social cost–benefit analysis and policy formulation. Many of those from academic institutions, mostly located in high income countries, had a development theory orientation and were interested in applying sophisticated analytical techniques, such as linear programming, in 'traditional' agriculture situations. Finally those individuals associated with agricultural research institutions were primarily interested in using such information to address issues relating to technology adoption.

Two very significant common findings emerged from these studies carried out in a large variety of production environments[10]. These were that the vast majority of farmers:

- Were rational (i.e. sensible) in the methods they used. For example, some such studies revealed that mixed cropping was very rational and compatible with their production constraints and goals.
- Had an intimate understanding of their production environments in which they often

operate complex farming systems consisting of crops, livestock and off-farm enterprises. Such farm systems were often more complex than the commercialized farming systems in high-income countries. Farmers in low-income countries have to work within the constraints they face, rather than break or avoid them by seeking and receiving institutional support.

Unexpectedly, the overwhelming feeling arising from these farm management studies was considerable respect for limited-resource, small-scale farmers, the major focus of most of the studies. This naturally led to the next phase in which agricultural economists, particularly those associated with agricultural research stations in Africa, Asia and Latin America, started evaluating recommended technologies, usually packages. Before the mid 1960s very few station-based experiments were subject to any economic analysis. Consequently it was not surprising that little or no testing was required to indicate that many of the existing recommendations were irrelevant, especially when criteria relevant to farmers were applied in this evaluation. Insights on their criteria came from encouraging farmer cooperators to test conventional recommendations, and three specific insights resulted:

- Farmers were natural experimenters, although the methods they used were informal in nature and were not readily susceptible to statistical analysis.
- Although the recommended technological packages were sometimes compatible with the biophysical environments under which farmers were operating, they often could not be adopted by the farmers because they were incompatible with the socioeconomic circumstances they had to manage. For example, labour was often not sufficient when packages required it, and there were limitations in the policy and support systems, such as non-availability of the recommended inputs and deficiencies in the product-marketing system.
- It was fallacious to assume that the production environments of farmers were homogeneous. Assuming that one technological package would be suitable for all farmers was incorrect, because of variations not only in the biophysical, but also in the socioeconomic environment.

The dominant feeling of those who were engaged in these technology evaluation exercises, most of them agricultural economists, was dissatisfaction with the process that prevailed at that time for developing and evaluating technologies. This was particularly true in less favourable and more heterogeneous production environments. One very significant conclusion was that conventional economic criteria did not ensure identification of a relevant technology. For example, capital scarcity, the value of a balanced and preferred food supply and the risk factors were not handled by a simple profit criterion. Progress in identifying technologies that were relevant only occurred when trials were first implemented on farmers' fields with the non-experimental variables reflecting farmers' practices, and farmers' criteria were used in the evaluation process. However, experiences at that period involved much more interaction between researchers and farmers on the choice of technology already developed than on the design of technology. The latter has now become a focal point of the farming systems approach and marks one important difference from the conventional farm management approach, certainly as it was practised in academic institutions in the 1960s.

10.1.4 Early days of the farming systems approach

Given this evolutionary sequence, it is perhaps not surprising that early manifestations of the farming systems approach generally were linked very closely to technological issues. As a result the farming systems approach, called farming systems research (FSR), was often located in international and nationally sponsored agricultural research institutes in low-income countries. The FSR approach that evolved was based on the premise that the problems of farmers first had to be understood. Consequently, solutions to their problems had to be based on a proper understanding of their production environments, including both biophysical and socioeconomic dimensions. In the light of earlier experiences with farmers, a central principle of this approach was that farmers themselves had a right to be involved in the technology development and evaluation process, indeed, that they could contribute productively to the process. Another central feature was that scientists involved in the process should represent both technical and social disciplines. Thus, in a sense, some, but not all, of the characteristics of the early farm management approach started to reappear. Examples of these were adopting an interdisciplinary method, nurturing interactive relationships with farmers and using an iterative procedure for FSR-related work.

Many of those responsible for developing the FSR approach were trained in the late 1950s and early 1960s. As a result, we were perhaps vulnerable to the observation that we should have had more of a historical perspective and learnt from what had happened in the past. Not surprisingly, Johnson, writing in 1981, observed that 'there has been much reinventing of the wheel in developmental thinking as ... agricultural development economists have discovered farms and (international agricultural research) administrators have rediscovered farm management. Lessons gleaned from historical development of the field of farm management and agricultural economics have been neglected by members of these groups ... Tragically, some important wheels have not yet been reinvented or rediscovered while some wheels demonstrated historically to have flat sides and faulty bearings are being reinvented.'[11] Still, the farm management approach as it existed in the late 1950s and early 1960s did not serve the needs of limited-resource farmers in low-income countries – therefore, the return to an earlier incarnation! However, as the next section will show, the criticism that we were not broad enough in our orientation has some validity, whereas holism was a distinctive feature of early farm management.

10.1.5 Evolution of the farming systems approach

To operationalize the early FSR approach, a couple of compromises were made which involved issues explicitly embraced in the very early days of farm management:

- Although the link between farm and family was explicitly recognized, the family or household unit was treated as a single entity. Only later were intra-household relationships recognized as important.
- Factors relating to the policy/support system were treated as parameters within which the search for improved technologies took place.

Johnson, again in 1981, was particularly critical of the failure to consider family-related issues. Subsequently, many of the methodologies developed to address family relationships came through gender initiatives[12]. The main factor contributing to acceptance of the existing policy/support system was the limited mandates of the technology development institutions in which most farming systems related work was, and still is, found. Such acceptance severely constrained the types of technologies that could be viewed as relevant. The fact is that policy, support and technology related issues are an interacting system and treating them as such facilitates the search for efficient ways to solve farmers' problems. This has become increasingly clear to many practitioners. Recognition of this interaction paved the way for an evolution of FSR *per se* into the farming systems development (FSD) approach advocated by the Food and Agriculture Organization of the United Nations (FAO)[13]. Unfortunately, practical examples of the influence of the farming system approach on policy and support systems are still very limited, and the approach for addressing that interface still needs refinement[14]. In principle, however, there has been increasing recognition of the potential power of the farming systems approach to address farmer problems in a dynamic production environment as well as in issues of equity both within and between generations.

Although a more in-depth examination of the early farm management approach might have sensitized early FSR practitioners to have a more holistic approach, it would not have provided many insights into methods for its operation in the circumstances of developing countries. That said, the traditional methods of farm management have been extremely important in other areas such as ideas on what data to collect, how to collect data via cost route surveys, how to analyse data via budgeting and programming techniques and the importance of incorporating the analysis of risk and uncertainty.

However, the evolution of the farming systems approach has, in turn, stimulated and/or benefited from the development of new methodologies. Three particularly important areas are:

- The tremendous explosion in methods for eliciting farmers' attitudes, opinions and contributions in an inexpensive and systematic manner.
- The learning associated with designing, implementing and evaluating on-farm trials involving farmers themselves.
- The recent greatly increased interest in sustainability issues which has prompted the farming systems approach to incorporate such considerations more explicitly.

New approaches to problem diagnosis are embraced by what is popularly called farmer participatory research (FPR)[15]. These have helped the current trend towards greater empowerment of farmers in the development process. They have also helped legitimize the important role that sociologists and anthropologists can play in improving the effectiveness of the farming systems approach. On the experimental side new techniques have evolved to cope with the much more variable situation on farmers' fields than on fully controlled, usually station-based, locations. For example, a major concern of farming systems practitioners has been the development of relevant evaluative methods for trials carried out by farmers, which may violate *ceteris paribus* conditions and are often replicated across a number of farms. Adaptability analysis[16] is one example of the large number of changes made in trial design and evaluation to improve the credibility and respectability of on-farm experimentation (OFE). The vulnerability of most farmers in low-income countries to severe poverty generally requires that measures to ensure ecological sustainability do not threaten current levels of living, by using time and other essential resources to ensure short-run food needs, for example. One major strategy to attack these dual goals is to piggyback strategies for ensuring ecological sustainability[17] onto those for increasing labour productivity, the inevitable preoccupation of poor farmers and their families. An important component of this is to ensure that policy and support systems to encourage increased productivity are compatible with those designed to encourage ecological sustainability. Bad agricultural practices cause environmental degradation, good agricultural practices are those that can play this dual role – a fact increasingly emphasized by agricultural researchers.

10.1.6 Potential contribution of the farming systems approach to farm management

So what does the current farming systems approach have to contribute to the farm management approach as it currently appears in high-income countries? 'Currently' is an important conditional word. Philosophically, the 'current' farming systems approach, as practised mainly in low-income countries, has much more in common with the 'earlier' farm management approach in the high-income countries. In fact, while the conventional, contemporary, farm management approach has become narrower in the areas it deals with, the farming systems approach has become broader, incorporating more variables[18]. However, what particularly distinguishes the modern farming systems approach from the earlier farm management approach is the considerable evolution in the methodologies for data collection, processing and evaluation.

So the question becomes, 'do these developments have anything to offer the current type of farm management as found in high-income countries?' Part of the answer depends on the direction farm management takes in the future. This currently is a subject of debate[19]. Johnson noted, in 1994, the bimodalization of farm sizes in the USA[20]. One group involves an increasing number of large commercial farms accounting for a very high proportion of cash farm sales but for a small, decreasing proportion of the rural and total populations. The other group of small farms, often part-time plus residential and recreational, accounts for a large and increasing proportion of rural people, but a decreasing proportion of farm-product sales. Johnson, quite rightly, believes that recognizing the dichotomy in farm sizes is critically important in designing relevant farm management programmes. The strengths of the current farming systems approach have most immediate potential in addressing the needs of small farms in countries like the USA. This particularly applies to those that are not hobby farms, where a strong link exists between farm and family, where a need exists to empower farmers so that they have a voice, and/or where a genuine interest in sustainability issues exists. For example, the University of Florida, under the leadership of Peter Hildebrand, has used a farming systems approach in addressing the problems of disadvantaged farmers in Florida for a number of years. Also, the USA sustainable agriculture movement, usually associated with smaller farmers, has increasingly incorporated lessons learned from, and methodologies developed in, applying the farming systems approach in low-income countries[21], as shown in Chapter 6.1. Whether such applications reflect the farming systems approach or a renewal of farm management similar to that which existed earlier can be debated. However, such a debate would be both unproductive and sterile. A lesson learned is the importance of a historical perspective – what was done in the past can provide some idea of what to do in the future.

Perhaps after all, we have come full circle, with the farm management approach having made critically important contributions to the development of the farming systems approach, and the farming systems approach potentially contributing to helping farm management address the needs of smaller farms, particularly those that are diversified and have an explicit sustainable agriculture orientation. The current farm management approach that is present in mainstream academic institutions, for example in the USA, is likely to be most relevant for large commercial, and usually specialized, farmers who can interact with farm management advisers on a one-to-one basis by looking at different options together. However, many of the strengths of the farming system approach can be incorporated into the farm management approach to help address the needs of the less influential, and often more disadvantaged agricultural producers. Finally, on a more general note, the experiences of the farming systems approach in the *ex ante* use of diagnosis for the choice and design of experiments and technologies, also merit wider consideration.

REFERENCES

1. **Dillon, J.L. & J.B. Hardaker, 1993**. Farm Management Research for Small Farmer Development. AGSP, FAO, Rome.
2. **Johnson, G.L., 1981**. Small farms in a changing world. In: International Agricultural Programs. Small Farms in a Changing World: Prospects for the 1980s. Papers Presented at a Conference at Kansas State University, 11–13 November. Kansas State University, Manhattan.
3. **Warren, G.F., 1913**. Farm Management. Macmillan, New York.
4. **Spillman, W.J., 1902**. Systems of Farm Management in the United States. United States Yearbook of Agriculture. USDA, Washington, DC.
5. **Case, H.C.M. & D.B. Williams, 1957**. Fifty Years of Farm Management. University of Illinois Press, Urbana.
6. **Johnson, G.L., 1981**. (Op. cit.).
   Johnson, G.L., 1988. Farm management inquiry: past and present status and implications for the future. In: Robinson, L.J. (Ed.) Determinants of Farm Structure. Proceedings of a conference organized by the NC-181 Committee on Determinants of Farm Size and Structure in north Central Areas of the United States, 16–19 January, San Anthonio, Texas. Michigan State University, East Lansing.
7. **Boss, A. & G.A. Pond, 1947**. Modern Farm Management. Webb Publishing, St. Paul.
8. **Clapp, G.R., 1955**. The TVA: an Approach to the Development of a Region. Russell, Chicago.
   State of California, 1977. The Family Farm in California: Report of the Small Farm Viability Project. State of California, Sacramento.
9. **Yang, W.Y., 1965**. Methods of Farm Management Investigations for Improving Farm Productivity. Revised Edition. FAO Agricultural Development Paper no. 80. Food and Agriculture Organization of the United Nations, Rome, Italy.
10. **Ruthenberg, H., 1971**. Farming Systems in the Tropics. Clarendon Press, Oxford.
    Collinson, M.P., 1972. Farm Management in Peasant Agriculture. A Handbook for Rural Development Planning in Africa. Praeger, New York.
    Walker, T.S. & Ryan, J.G., 1990. Village and Household Economies in India's Semi-arid Tropics. Johns Hopkins University Press, Baltimore.

Norman, D.W., Newman, M. & I. Ouedraogo, 1981. Farm and Village Production Systems in the Semi-arid Tropics of West Africa: an Interpretive Review of Research. Research Bulletin no. 4, Vol. 2. ICRISAT: Patancheru, A.P., India.

11. **Johnson, G.L., 1981**. (Op. cit.).
12. **Feldstein, H.S. & S.V. Poats (Eds), 1989**. Working Together: Gender Analysis in Agriculture. Volumes 1 and 2. Kumarian Press, West Hartford, CT.
13. **FAO, 1989**. Guidelines for the Conduct of a Training Course in Farming Systems Development. FAO Farm Systems Management Series no. 1. AGSP, FAO, Rome, Italy.
14. **Dixon, J.M., M. Hall & J.B. Hardaker, 1994**. Farm and Community Information Use for Agricultural Programmes and Policies. FAO Farm Systems Management Series no. 8. AGSP, FAO, Rome, Italy.
    Upton, M. & J. M. Dixon (Eds), 1994. Methods of Micro-Analysis for Agricultural Programmes and Policies: a Guideline for Policy Analysts. FAO Farm Systems Management Series no. 9. AGSP, FAO, Rome, Italy.
15. **Chambers, R., 1994**. Participatory rural appraisal (PRA): challenges, potentials and paradigm. World Development, 22(10), 1–6.
    Pretty, J., I. Guijt, J. Thompson & I. Scoones, 1995. A Trainer's Guide for Participatory Learning and Action. International Institute for Environment and Development, London.
16. **Hildebrand, P.E. & J.T. Russell, 1996**. Adaptability Analysis. Iowa State University Press, Ames.
17. **Norman, D.W. & M. Douglas, 1994**. Farming Systems Development and Soil Conservation. FAO Farm Systems Management Series no. 7. AGSP, FAO, Rome, Italy.
18. **Hallam, A., 1994**. Farm Management – Its Potential Contribution to Rural America in the Decade Ahead – The 'New' Farm Management. In: Burton, R.O. & D.L. Watt (Eds) (Op. cit.).
    Norman, D.W., F.W. Worman, J.D. Siebert & E. Modiakgotla, 1995. The Farming Systems Approach to Development and Appropriate Technology Generation. FAO Farm Systems Management Series no. 10. AGSP, FAO, Rome, Italy.
19. **Burton, R.O. & D.L. Watt (Eds), 1994**. Future Priorities and Agenda for Farm Management Research. Proceedings of a Conference organized by North Central Regional Research Committee: Farm and Financial Management (NCR-110), 16–18 May, 1993, St Louis, Missouri. Department of Agricultural Economics, Kansas State University, Manhattan.
20. **Johnson, G.L., 1994**. Farm Management – Its Potential Contribution to Rural America in the Decade Ahead. In: Burton, R.O. & D.L. Watt (Eds) (Op. cit.).
21. **Norman, D.W., T.R. Frankenberger & P.E. Hildebrand, 1994**. Agricultural Research in Developed Countries: Past, Present and Future of Farming Systems Research and Extension. Journal of Production Agriculture, 7(1), 124–131.

10.2 ANTHROPOLOGY, SOCIOLOGY AND FSR

Constance M. McCorkle

Both anthropology and sociology have reinforced all of the Fundamentals and Stages of FSR-E to varying degrees in different arenas and different entry points in the FSR-E process. Indeed, alongside economics, the impact of these two disciplines on FSR-E has been so great that some people have mistakenly defined FSR-E as a social science activity.

10.2.1 Introduction

Farming systems research and extension (FSR-E) is often defined as having nine fundamental characteristics (Table 10.2.1) with five broad stages identified in its conduct (Table 10.2.2). These stages also embrace two complementary facets within the overall farming systems concept: a more micro, agrobiological and on-farm facet focusing on the generation, evaluation and delivery of agricultural technology; and a more macro, socioeconomic and off-farm perspective that centres on such subjects as agricultural infrastructure, support services and institutions, and policies, as described by Waugh *et al.* in 1989[1].

Both anthropology and sociology have reinforced all of the fundamentals and stages of FSR-E to varying degrees in different arenas and different entry points in the FSR-E process.

Table 10.2.1. Nine fundamentals of FSR-E.

1.	Small-farmer orientation
2.	Farmer participation
3.	Location-specificity of technical and human factors
4.	Problem-solving approach
5.	Systems orientation
6.	Interdisciplinary approach
7.	Complementarity with component research
8.	On-farm trials
9.	Feedback to shape future agricultural research and policy

Paraphrased and updated from Byrnes,1990[2].

Indeed, alongside economics, the impact of these two disciplines on FSR-E has been so great that some people have mistakenly defined FSR-E as a social science activity[3].

At least from an American perspective, anthropology has been more active in FSR-E than sociology for a variety of reasons. One is anthropology's tradition of working internationally, especially among so-called primitive, indigenous, tribal or peasant peoples – the vast majority of whom rely upon cropping and stock raising as a vital part of their livelihoods. In contrast, 'real' sociology traditionally busied itself with developed industrialized nations. Although rural sociology has an explicit focus on agriculture, it was born with a mandate to solve social problems unique to rural America during the Depression era. It moved into international work much later and more tentatively than anthropology[4].

This contribution discusses the impact of the two non-economic social sciences to FSR-E in terms of conceptual, substantive and methodological inputs. Given that 'intellectual enrichment is a two-way avenue'[5], the ways in which the two disciplines have in turn been enhanced by FSR-E are also noted.

10.2.2 Conceptual/substantive contributions

Whether empirically or theoretically, anthropology has long been concerned with both the particular and the general/comparative study of the agrotechnoecological adaptations of rural people worldwide.

Table 10.2.2. Stages in conducting FSR-E

A. **Diagnosis or description** of farming constraints and potentials as follows:
   i. search of secondary sources for basic data and descriptive information on the target region(s)
   ii. identification of recommendation domains or target groups of farmers, using agroecological and/or farming systems criteria
   iii. exploratory diagnosis, employing rapid survey or reconnaissance techniques, followed by analysis and synthesis of the resulting findings
   iv. formal verification of these findings and construction of a database for monitoring and evaluation

B. **Design or planning,** in which possible strategies to overcome constraints and build on potential are formulated, taking into account their likely technical feasibility, political-economic viability, and social acceptability by target group in light of: the information from A above, past research findings (such as experiment station trials), and farmers' own knowledge and insights

C. **On-farm testing of, or experimentation with,** strategies identified during B above. Strategies can include changes not only in technologies or in cultural, husbandry or resource management practices, but also in farm planning methods, support services (marketing, input supply, credit, infrastructure) and policies geared to have a positive impact at the farm level

D. **Extension or recommendation** of strategies tested in C above

E. **Monitoring and evaluation** throughout the FSR-E process and in tracking farmer adoption of the strategies tested and extended and their consequent impacts (both positive and negative), for feedback into future agricultural research and development, farm planning and policy making

Cultural ecology

This interest stems from one of the earliest and most robust schools of modern anthropological thought: the cultural ecology propounded in the early to mid 20th century by Julian Steward. Subsequently fashioned into various theories of technoenvironmental and economic determinism or cultural materialism[6], and informed by general systems theory[7] and ecology *per se*, this school of thought focuses on: 'the local ecology, that is, the interaction of environment, exploitative devices, and socio-economic habits ... the exigencies of making a living in a given environment with a specific set of devices and methods for obtaining, transporting, and preparing food and other essential goods'[8].

Indeed, the obtaining of food has formed the central core of cultural ecological research. Coupled with anthropology's traditional focus on primary producers, it is small wonder that ecological and economic anthropologists were already contributing to other researchers' understanding of FSR-E fundamentals even before their direct engagement in FSR-E. They provided projects with valuable diagnostic or description information, for example, in the form of detailed ethnographic studies of specific agrarian peoples.

The cultural-ecological tenets carried into FSR-E by contemporary anthropologists have led to increased FSR-E recognition of the complex relationships between human beings and the totality of biophysical resources upon which cropping, stock raising and so many other rural livelihood activities ultimately depend. Anthropologists have had a particularly salubrious effect on FSR-E's greater attention on forestry and livestock components of the farming system, given the discipline's long-time study of forest-dwelling and stock raising peoples.

Certainly, cultural-ecological perspectives have had a hand in re-contextualizing agriculture in ways that facilitate more comprehensive and sustainable approaches to its development. When agriculture is viewed as one end of a continuum from less to more intensively managed strategies of primary resource utilization, it is easier to detect potential complementarities or conflicts between different uses of the local landscape – as well as pressures on resources – that might arise from interventions proposed in the course of FSR-E. The result is better diagnosis, description, design and planning (stages A and B) to arrive at more environmentally sustainable packages of livelihood activities.

Anthropological and, to a lesser extent, sociological contributions in this vein have perhaps been most visible in FSR-E-oriented social forestry and agroforestry[9]. The extensive research of the two disciplines on natural resource tenure regimes has also added much to FSR-E understanding of what cropping and stock raising strategies might or might not be workable and sustainable under different regimes. In mixed, agropastoral farming systems, for example, both disciplines have made major contributions to the previously understudied topic of ownership and management rules for the multiple use resources of community rangelands[10].

Holism

Of course, all anthropologists come to FSR-E with a set, yet operational, concept of holism – essentially, a systems approach that casts its net even wider than that of conventional FSR-E. Indeed, holism is one of the defining principles of the discipline. As the *Encyclopedia of Anthropology* says:

> This term refers to the inclusive nature of anthropology's study of humanity ... as physical, social, and cultural entities, through time and space. As such, anthropology studies many problems that arise at the interfaces of various other disciplines ... [It is] a 'coordinating science'.[11]

This construct helped prod early FSR-E forward from its almost exclusive focus on plant production to what one agricultural anthropologist, Murphy, writing in 1990, terms a farmers' system approach[12]. This more mature FSR-E stance embraces the whole of the agricultural enterprise, beginning with resource access and management and then going beyond *production*, to product *transformation* (post-harvest handling, storage, intermediate and final processing), *distribution* (marketing and other forms of exchange) and last but by no means least, *consumption* – which is of course the point of human involvement in agriculture in the first place.

FSR-E's development was stimulated, for example, by the addition of a nutrition systems research and extension (NSR-E) subcomponent, derived directly from the well-established subdisciplines of nutritional and medical

Box 10.2.1. Beliefs and veterinary practice.
Stock raisers around the world often speak of the etiology of livestock diseases in terms of evil winds, malevolent spirits, genies and haunted locales. Many researchers are quick to dismiss such idiom as mere superstition or nonsense. Informed anthropological probing, however, reveals that often it encodes astute ethno-veterinary savvy, e.g. of aerial modes of disease transmission, climatic stresses on animal health, the ecology and ethology of parasite hosts and vectors, or the distribution of toxic plants. This knowledge in turn dictates apt herding, husbandry, infrastructure and kin- or other social-cooperation strategies that make perfectly good sense when it comes to disease control and avoidance[16].

anthropology[13]. Indeed, together with these two subdisciplines themselves, NSR-E is a model of the sort of interdisciplinary interfacing mentioned in the *Encyclopedia* and embodied in FSR-E fundamental 6.

To take another example, FSR-E's attention to and understanding of distribution was also significantly enhanced by tapping into economic anthropology's (including archaeology's) vast storehouse of both theoretical and empirical research on forms of agricultural commodities exchange and consumption[14]. FSR-E anthropologists of the substantivist economic school elucidated the volume and special values of non-monetary, extra-market or social exchanges plus expedient and/or socially skewed consumption patterns commonly found within non-industrialized societies. These insights helped correct faulty FSR-E assumptions – and thus research protocols and interventions – concerning such diverse issues as farm surpluses, storage losses, real net income, animal offtakes and meat-eating, and food access within and across both producer and consumer households.

Most FSR-E practitioners agree that anthropology has made at least two important contributions in its role as a coordinating science. The first consists of cultural brokerage, including conceptual as well as literal translation between FSR-E 'technologists' and producers[15]. Anthropologists are perhaps uniquely suited to this task because their hyper-holism entails, as Rhoades *et al.* described in 1982: 'tracing, especially in rural communities, the connections between the mundane, bread-and-butter farming activities and the beliefs, religion, kinship, social institutions, material culture [i.e., technology], and ... ecology and economy'.

There are many translation exercises similar to the example shown in Box 10.2.1[17]. These illustrate the kind of disciplinary or subject-matter interfaces embraced by holism – as between social and veterinary sciences, and between culture and agriculture.

Second, and related, is the anthropologist's 'total view of farming and social activities that can yield a special holistic understanding of farmer decision-making', as Rhoades *et al.* said in 1982[18]. As economist Ruttan wrote in 1989, 'Anthropologists, in particular, have demonstrated a capacity to understand the dynamics of technology choice and impact at the household and village level that is highly complementary to both agronomic and economic research.[19]' Along with some sociologists, anthropologists' decision-making research has spanned numerous FSR-E topics and the findings have been used to support virtually every aspect of FSR-E (Box 10.2.2)[20].

Box 10.2.2. Anthropology and farmer decision.
The decision-making topics studied by anthropologists and sociologists working in FSR-E include producer and consumer choices in:

- the assignment of scarce natural resources and limited labour to competing livelihood activities; the mix of cash-versus-food or plant-versus-animal crops raised
- sell/kill/cure options for ailing livestock – and in the case of cures, choice of traditional versus modern veterinary medicine
- different product processing and disposal or acquisition options in relation to timing and type of commodities available, market conditions, sociocultural obligations and networks
- shifts or substitutions between traditional and non-traditional foodstuffs and preparation techniques
- wage-labour or out-migration versus own-farm work

Box 10.2.3. The value of ethno history, an example.
A Haitian reforestation scheme was transformed into a successful farming systems project when the anthropologist team leader pointed out that Haitian peasants would more readily accept reforestation if tree planting were represented to them not as a conservation effort (the original intent) but rather as the domestication of another, multipurpose crop to be folded into their diversified food and farming system – as they had already done across the centuries with other plant and animal 'crops'. This approach was akin to 'a replay of an ancient anthropological drama, the shift from an extractive to a domesticated mode of resource procurement ... replicating this transition in the domain of wood and wood-based energy'[22].

Evolutionism

This third advantage of anthropological holism deserves its own conceptual category. As the encyclopaedia definition acknowledged, anthropology is concerned with the whole of human history from every possible perspective. This of course includes all human systems to obtain food dating from the first hunter–gatherers some four million years ago through the agricultural revolution *c.* 10,000 BC down to the present. No other discipline in FSR-E enjoys this depth of historical vision. Only anthropology consistently examines its research topics from both global and local ethnohistorical vantage points. Because efforts at agricultural development essentially boil down to applied evolution[21] when coupled with the cultural- and location-specific knowledge (fundamental 3) for which anthropologists are rightly reputed, ethnohistory can usefully inform *ex ante* assessments of agricultural innovations that are likely to be acceptable and viable for a given rural people, given their past adaptations as hunter–gatherers, nomadic pastoralists, slaves, colonists and so forth (Box 10.2.3).

Peters, writing in 1986, neatly summed up the value of such intelligence for FSR-E: 'Like any other research endeavour, farming systems research has to ... decipher how social groups and categories are generated, or "sedimented", by social processes and how the developmental or life cycles of households intersect with historical processes ... for determining appropriate recommendation domains and effective technologies'[23].

Emicism

Closely related to holism is emicism – that is, taking into comparative analytic account the beliefs, opinions, experiences, worldviews and knowledge (or lack thereof) of cultural insiders (the *emic*) as well of those of outsiders (the *etic*). In FSR-E, male and female producers most often comprise the insiders, while the outsiders of immediate interest are typically researchers and extra-community extensionists. To a certain

Box 10.2.4. Emicism: its value in FSR.
On an integrated pest management (IPM) project in the Philippines organized by the International Rice Research Institute (IRRI), entomologists' *ex ante* assumptions about, and initial approaches to, IPM technology design and implementation were completely re-vamped once the team anthropologist introduced them to heretofore unsuspected farmer perceptions, difficulties and concerns. These included:

- local definitions of pest problems in terms of overall crop damage, rather than scientists' counts of discrete pest populations at different life-cycle points – this led scientists to design more user-friendly economic threshold cues
- sharp labour and land-tenure constraints to the cultural methods of pest control initially proposed by scientists – this resulted in a re-focus on field drainage measures (which farmers had identified as a major need) that also favoured uptake of IPM
- widespread producer confusion about and misuse of chemical pesticides due to wildly inconsistent labelling and to merchants' adulteration of commercial products, plus national credit-system incentives to overspraying – these findings stimulated expansion of the research agenda to address information, industry and policy, as well as purely technical, IPM issues
- likewise for farmers' worries about the inadvertent poisoning of vital wildlife resources (snails, frogs, fish, crabs), their livestock, their children and themselves[24]

Box 10.2.5. Emicism and research direction.
On an FSR-E project in Mali, farmers and scientists agreed that a priority problem was animal nutrition – to hasten growth, improve overall animal condition and thus generate faster and larger earnings for stock raisers. A series of participatory on-farm feeding trials using urea-enriched straws was launched. Scientists' measurements (the *etic*) showed small but real weight gains among stock on the improved diet. But participating farmers detected no or negligible differences between experimental and control animals. They thus deemed the urea treatments not worth the trouble and expense. With the team anthropologist to emphasize that this conclusion represented valid *emic* feedback, the animal scientists were persuaded to turn their attention to other nutrition strategies instead[25].

extent, emicism was already implicit in FSR-E fundamentals 1, 2, 3 and 8, and in stages B, C and E. But as a paramount principle of anthropology, this discipline put it into practice much more explicitly and universally, especially in comparing and resolving differences between perceptions, definitions and prioritization of problems and possible solutions (Box 10.2.4).

For obvious reasons, the emic perspective is particularly important in stage B planning. And certainly, it constitutes the final proof of the pudding when it comes to technology uptake and the monitoring and evaluation of benefits in stages D and E (Box 10.2.5).

Much of the greatly strengthened emicism that anthropology brought to FSR-E has become ensconced under the now familiar rubric of local knowledge systems or, less felicitously, indigenous knowledge[26]. The formalized study of such systems dates to 1950s research in cognitive anthropology. Whatever the label, there is consensus that, as Wallace and Jones said in 1986: 'The increased appreciation for indigenous technical [and agroecological] knowledge is a major breakthrough which has come to be associated with FSR'[27]. An immediate outcome for FSR-E has been even more appropriate strategy design in stage B, often by blending local/ethnoscientific with universal/Western-scientific knowledge. There are innumerable cases where FSR-E has had successful recourse to techno-blending, and Box 10.2.6 provides just one illustration for soil and water conservation[28]. According to some estimates, as much as 90% of all International Agricultural Research Centre (IARC)-developed and -recommended technologies and practices originally drew upon local agroecological knowledge, materials and practices.

An inevitable part of the study of local knowledge is of course producer innovation and experimentation, in other words, how such knowledge is generated in the first place. Here again, beginning as early as the 1940s, anthropology led the way. This was due not only to the discipline's evolutionary and emic emphases but also to cognitive anthropology's finding that, like language abilities, the scientific

Box 10.2.6. Emicism and technology adaptation.
The Mali project mentioned in Box 10.2.5 identified tied ridging as a possible solution to some of the soil and water problems faced by farmers in the country's dry, Sahelian areas. This land preparation technique involves building earthen cross-ridges at regular distances along field furrows, giving fields a sort of checkerboard appearance. The ridges combat water and soil erosion; and by trapping and better infiltrating scarce rainwater, they facilitate and prolong its availability to the crops. However, tied ridging can require considerable animal traction, which poorer producers lack. In any case, participating farmers argued that mounding was less labour-intensive when it came to incorporating the necessary manures and the enriching crop and weed residues into the soil. It also made for more efficient, targeted application of these scarce inputs. With the cultural brokerage services of the team anthropologist, farmers and scientists together arrived at a blended option that combined the best of both local and outsider knowledge and experience. It consisted of connecting up the mounds with hand-dug earthen partitions, thereby forming 'tied mounds'. This technique preserved the benefits of mounding while adding those of tied ridging without excluding the majority of farmers who had no access to animal traction[29].

> **Box 10.2.7.** The gender dimension.
> An FSR-oriented livestock project in the Andes found that when it came to on-farm work with one carefully delineated domain group of smallholders, farmers seemed strangely indifferent to their animals' health and productivity. They agreed to participate only if agronomic trials were also conducted. But even then, they failed to attend any of the meetings to organize livestock trials. It took the (female) anthropologist on the team to point out that project researchers had defined their domains without attention to the most basic socio-organizational principle of all: gender. In the farming system in question, stock raising was largely the responsibility of women, while men dealt mainly with cropping. Once livestock trials were redirected to women producers, OFR got underway with great success[33].

method is 'hard-wired' into the human brain in at least some empirical, test–compare–correlate sense. Indeed, teasing apart magic, religion and science is a favourite anthropological task.

To take just a few examples, anthropologists have documented how producers worldwide assiduously seek out, test and cross-breed new crop and livestock varieties/races, often using controlled field trials of their own devising; how they borrow and adapt both indigenous and imported soil and water management techniques from their observation of other farming peoples near and far; how they create their own effective (and often techno-blended) veterinary treatments and feed rations; and how they or local craft-workers like smiths devise and market more workable versions of imported farm machinery, on-farm storage techniques or food-processing implements[30]. Rural sociology has similarly documented producer FSR in the USA, where farmers' tinkering and testing directly or indirectly led to the design of most modern-day agricultural machinery and of innovative techniques such as no-till cultivation. Together, anthropologists and sociologists have highlighted the validity of grassroots innovation and experimentation, and the value of harnessing these processes in service of nearly all FSR-E fundamentals, particularly fundamental 8 and especially stage C. Their studies thus helped bolster confidence in and acceptance of FSR-E's insistence on increasingly producer designed, managed, monitored and evaluated on-farm research (OFR).

FSR-E's appropriation of the study of agro-techno-ecological ethnoscience represents perhaps the most striking contribution from anthropology. Unfortunately, scientists' appreciation of local knowledge and practice usually stops at the technological. They show far less understanding of how local knowledge is actualized and transmitted via local systems or of how these local resources could and should be put to work in agricultural development[31]. Or as the World Bank rural sociologist Michael Cernea said in 1985: 'High-yielding social organizations are not [*sic*] less important for development than high-yielding crop varieties'[32].

Human social organization

Fortunately, however, both anthropology and sociology lent FSR-E their unique conceptual and analytic expertise in human social organization – the myriad and overlapping ways in which societies categorize, bound, interlink and assign roles and behaviours to their membership. This expertise made for more explicit recognition and incorporation of important *non*-producer actor and institutional groups into the FSR-E process. Spanning both micro and macro levels, such groups include processors, traders, input suppliers, rural and urban entrepreneurs, consumers, and members and managers of regional, national or international institutions with a controlling role at any point along the production to consumption continuum. Cross-cutting biosocial groups were highlighted. Depending upon the problem focus, these might be functionally defined by, for example age and/or gender; marital, reproductive or nutritional status, lineage and other kinship positions, caste, ethnicity, religion and so on (Box 10.2.7).

Such added socio-organizational savvy makes for expanded yet refined recommendation domains and target groups (stage A-ii), inserting sociocultural variables into what would otherwise be sterile agroecological or economic definitions of domains. As Box 10.2.7 underscores, recognition of such variables is critical in recruiting appropriate farmer/stock raiser cooperators for stage C trials. But access to the knowledge, opinion and socioprofessional networks of groups other than purely producers can be critical in all FSR-E stages. The agricultural development literature is rife with descriptions of how, although producers were persuaded to raise some new high yield variety or livestock species or breed, researchers and extensionists forgot to consult and work with local and non-local people in the distribution chain, such as service providers, processors, transporters, traders and consumers about their willingness to handle or consume greatly increased quantities or greatly modified types of agricultural products.

Intimately linked to their concern with multiple human groupings has been the 'conscience-minding' role of anthropologists and sociologists in urging FSR-E to give attention to equity and distributional issues at both micro and macro levels. Beginning as early as the 1950s, these two disciplines were the first to point out the errors of pre-FSR-E, Green Revolution style agricultural research and development in focusing solely on increasing total food output, with little notice of negative outcomes for equitable development (much less environmental sustainability). All too often, agricultural 'tech-fixes' created social, economic and ecological problems that greatly offset their immediate food production benefits. Indeed, a driving force behind FSR-E's emergence was a growing realization that the supposedly 'soft' social sciences were needed to tackle what was by far the 'harder' part of agricultural and rural development – appropriately matching technologies to sociologies.

Achieving more equitable, wider-scale benefits from agricultural innovations meant that sociologists and especially anthropologists had to look at how more meso- and macro-level social, political, economic and institutional phenomena conditioned the ability of individuals and groups to 'choose for change'. While 'macro' may have multiple senses[34], in FSR-E it has been largely defined in opposition to fundamental 3 as external, higher-order processes that impinge upon, and are reacted to, by specific groups in specific locations. In particular, governmental policies that undermine incentives and disenfranchise came under scrutiny by FSR-E social scientists, particularly laws on the public/private tenure and management of land and other resources such as water; commodity transport, pricing and marketing controls; food processing and veterinary/public-health regulations; the infrastructure for and organization of national agricultural research and service institutions; and so forth. At a higher order still, the policies of international research and development agencies and donors, plus the activities of transnational corporations involved in agriculture, were also subjected to greater anthropological and sociological investigation. The result was that – as suggested in Box 10.2.4 and as ultimately folded into virtually all of Table 10.2.2's stages – FSR-E benefited from more historically and socially informed policy analysis.

Finally, rural sociology merits special mention for its unique contributions to the 'E' of FSR-E. This discipline provided a rich and unique corpus of data and theory on the human social organization of adoption and diffusion[35] of agricultural innovations, including all the informal, non-formal and formal communication networks through which producers obtain credible information and opinion on new farming, processing and marketing ideas[36]. Aside from some anthropological inputs concerning folk and 'oral media', rural sociology took the conceptual point in FSR-E's stage D.

10.2.3 Methodological contributions

Sociology also benefited stage A work in its early days, bringing its formidable skills in statistical methods and formal survey research to bear on stage A, ii and iv. Like farm management economists, rural sociologists were adept at analysing large-scale secondary data sets to establish recommendation domains and/or to typologize production systems[37]. The best such analyses successfully embodied FSR-E concern with the complexity of smallholder farming systems, going beyond targeting in terms of only crop or livestock commodities and very gross socioeconomic categories to capture more of the diversity of actor, institutional and biosocial groups discussed above.

Survey research and analysis of macrodata alone, however, are not always appropriate – and rarely if ever sufficient – to guide specific technology development among small farmers

Box 10.2.8. Trade-offs: relevance and precision.
The insights gathered and the successful technologies and extension strategies adopted in the IRRI IPM project described in Box 10.2.4 were the direct result of the team anthropologist's decision to abandon research protocols that relied on large-scale survey methods with lengthy one-on-one interviews. Instead, she organized weekly group discussions between scientists and rice farmers, taking advantage of the latter's strong preference for social and face-to-face exchange versus individual or indirect (radio, print) types of contact. This qualitative method made for far richer and more nuanced FSR-E information, feedback and evaluation[38].

in developing countries. Quantitative data that are not informed by well-grounded qualitative understanding can be meaningless or, worse, open to dangerous misinterpretation. It was therefore fortunate that anthropology contributed its defining emphases on qualitative methods to nearly all FSR-E stages.

Unquestionably, both anthropologists' and sociologists' role modelling of the value of frequent, direct and intimate but topically purposive contact with producers in their own fields, homes and community organizations stimulated other FSR-E disciplines to greater efforts in this direction, bolstering nearly all FSR-E fundamentals and stages. As Mike Collinson wrote in 1988: 'Anthropologists and sociologists ... have highlighted the potential contribution of increased client participation to [a] systems-based adaptive research process'[39]. The classic example in the FSR-E literature is the interdisciplinary elaboration and implementation of a highly participatory, farmer-back-to-farmer model described by International Potato Center (CIP) anthropologists, as covered by Rhoades in 1984[40].

Certainly, in arriving at the kinds of emic insights discussed in the preceding section, anthropology made an especially apt contribution to FSR-E in the form of both quantitative and qualitative participatory research and development techniques that are both rapid and reliable[41]. As Horton commented in 1984 on the CIP FSR-E team: 'The ... rapid, effective ... methods employed by the [anthropologist] researchers were extremely useful throughout the research process'[42]. The items from the anthropological toolkit that are brought into play in FSR-E are: discriminate sortings and rankings using cards, tokens, colour photos, indigenous board games, and so forth; sketch-mapping, drawing and social diagramming; producer-defined and/or -directed inventories; and of course, first-hand fieldwork in which researchers take part in local agricultural chores, events and discussions. Termed 'participant observation', this last technique is the mainstay of all anthropological fieldwork. Rural sociologists also employ it, albeit often unconsciously. Within the American land-grant university system, for example, they were expected to interact with producers, grower groups and extensionists (usually farmers) at local meetings and events as well as during formal field research. Moreover, until recently, many rural sociologists were themselves farmers or the children of farmers.

And anthropology and sociology has made other methodological, as well as substantive, contributions to FSR-E. For example, NSR-E naturally utilized long-standing methods of anthropometry to gauge human growth and nutrition, which can in turn provide one measure of FSR-E impacts. Work on producer decision making brought fresh, non-economistic decision modelling methods to FSR-E, ranging from qualitative and quasiquantitative approaches derived from behavioural[43] and cognitive[44] anthropology, later incorporating expert systems technology. These modelling methods were used to elucidate not only producer decisions but also donor-agency decisions about funding for agricultural research and development. In the latter context, such methods thus provided social scientists with another policy analysis tool. FSR-E's increased attention to intra-household dynamics and gendered divisions of labour also caused anthropologists to bring sophisticated but streamlined methods of time allocation research to bear on these topics, as well as on others such as malnutrition[45].

10.2.5 Contributions to anthropology and sociology from FSR-E

When the Rockefeller Foundation instituted its Social Science in Agriculture Fellowship Program in 1974, a ready-made pool of anthropologists eager to participate in FSR-E already existed. They recognized its potential for expanding the frontiers of cultural ecology. With its stated aim of inducing change, FSR-E offered an opportunity to select whatever was best in existing systems of obtaining food, or study brand new methods and strategies from a more dynamic and actively interdisciplinary standpoint than before. In the process, cultural ecology and agricultural development concerns became much more closely linked in both theory and praxis.

This linkage gave rise, in large part, to the hybrid subdiscipline of agricultural anthropology in the late 1970s. Its practitioners created their own society and bulletin under the name 'Culture & Agriculture Group' (C&A). By the mid 1980s, C&A was a fully-fledged unit of the 7000-member-strong American Anthropological Society. At the same time, within the Rural Sociological Society, FSR-E stimulated the creation and/or the marked evolution of the RSS's Sociology of Agriculture Research Group and its International Development Research Group, as well as the long-standing Extension Sociology Special Interest Group. And today, a number of American universities offer graduate specializations in the anthropology or sociology of agriculture.

FSR-E's influence, however, runs deeper than just the proliferation of professional units and subdisciplines. In anthropology as a whole, it accelerated an on-going trend to look beyond the small scale, localized groups that traditionally constituted this discipline's focus, to analyse their social and economic role within broader societal structures. In like vein, FSR-E provided strong impetus for rural sociologists to expand their vision to non-industrialized societies. This challenged them to re-test and refine methods, theories and interventions elaborated on the basis of data from developed countries for application to the contexts of other regions. The results were sometimes pleasantly surprising, and sometimes not (Box 10.2.9).

In rural sociology, but also in anthropology, FSR-E led to a wider recognition of the persistence and dynamism of non-mechanized, diversified peasant farming systems, which in many societies continued to generate the bulk of the food supply. Clearly, such systems had outlived repeated predictions of their demise in both socialist and capitalist formations, contradicting earlier modernization-theory dogma that agricultural development must follow a Euro-American evolutionary pattern. Along with the vast amounts of fresh and far more sophisticated comparative data that anthropology and sociology garnered from their participation in FSR-E, this recognition had immediate implications for the construction of a cross-culturally applicable 'anthropology of agriculture'[47] or a unified 'sociology of agriculture'[48]. Indeed, the latter, as Cernea wrote in 1985:

> represents a new approach to rural sociology, one more theoretically informed, holistic, critical and radical ... [that] is increasing the professional relevance of sociology to rural change programs and improving the ability of sociologists to address the social and organizational relationships in agriculture and their change under the impact of technology, market forces,

Box 10.2.9. FSR: a new vehicle for anthropology.
As part of a global FSR-E initiative, remote farmer stock raisers in Niger were investigated to compare their use of national extension services, rather than non-formal sources of agricultural information, to determine their respective influence on technology adoption decisions. To structure the research, a model was selected that had been elaborated by rural sociologists across 30 years' study in one region of the USA, where it was used to trace out the flow of farm information into and within rural communities. The researcher fully expected that the model would require considerable re-working before it could be applied in Africa. But surprisingly, it turned out to fit the Niger situation *mutatis mutandis*, with little modification other than initial identification of local agrarian social structures plus a cultural-linguistic glossing of such 'farm talk' locales as 'church', 'feed store' and 'sewing circle' into, respectively, 'mosque', 'tea shop' and 'community well'. Not surprising, the research found that alien Western models of extension services did not fit Niger[46].

> industrial development, population pressures, and so on.[49]

Even archaeology has benefited. To enhance its study of long-dead agrarian peoples and the evolution of agriculture, it now mines data on contemporary farming families – data often generated by agricultural anthropology and other disciplines influenced by FSR-E. All such data have fed into theorizing about social change and development in agrarian milieux, bolstering disciplinary concepts and constructs in the anthropology/sociology *of* agriculture and development, as opposed to the various modes of disciplinary participation *in* development-oriented endeavours such as FSR-E. As anthropologists Jones and Wallace explained in 1986:

> The breadth and volume of FSR project experiences make them a 'real world laboratory' to compare individual and social responses to change and development, and at the same time offer a possibility for operationalizing and implementing social science concepts. The changes in world agriculture during the past few decades have radically altered the conditions of third world farmers and an evaluation of experience in the implementation of development projects will necessarily lead to the re-evaluation of, and improvements in, the concepts and assumptions which underlie models of development.[50]

In addition to advances in ivory-tower theorizing, however, FSR-E has also lent impetus to a growing trend in anthropology harking back to its disciplinary genesis in social problem solving. One result was the extra-academic renaissance of a sort of neo-Taxian action anthropology, with its associated emphasis on equitable, bottom-up participatory decision making and community mobilization.

Another result of anthropological and sociological engagement in FSR-E was the advent of entirely new fields of study, or the clearer enunciation of those that were previously rudimentary. Examples that come to mind from the author's own experience are, respectively, ethnoveterinary medicine and crop or livestock sociology, with the latter defined as the study of how the biological and technical requirements of raising plant or animal crops may influence family-farm and even larger scale social processes. In some instances, these new areas of study have led, or are now leading, to radical redefinition of long-standing subdisciplines, and to alternative problem solving approaches within them. A case in point is the impact of ethnoveterinary medicine upon medical anthropology, which has suggested possibilities for more cheaply and efficiently delivering primary health care to greater numbers of rural people by integrating delivery with, or 'piggy-backing' it on, veterinary services[51].

Methodologically, FSR-E triggered increased extradisciplinary demand for anthropology's participatory techniques. This in turn impelled anthropologists to greater innovation and refinement of such techniques, to make them even more rapid, reliable and cross-culturally robust. And in rural sociology – where qualitative methods have long taken a back seat to quantitative ones – FSR-E helped make such techniques more broadly acceptable. Of course, both disciplines benefited from exposure to the methods and tools of other sciences they encountered during interdisciplinary teamwork on FSR-E projects. One example is the enthusiastic adoption of geographic information systems by cultural ecologists.

Taking theory, method and praxis together, however, perhaps the most striking outcome of the participation of anthropologists and sociologists in FSR-E may prove to be a much-needed pragmatic one – their increased recognition of and relevance to concrete issues in international agricultural research and development and, refreshingly, as Wallace and Jones said in 1986: 'an expansion in the numbers of a previously rare professional, the social scientist able to communicate social science concerns to non-social scientists, and capable of responding to technical problems as they arise'[52].

REFERENCES

1. **Waugh, R.K., P.E. Hildebrand & C.O. Andrew, 1989.** Farming systems research and extension. In: J.L. Compton (Ed.) The Transformation of International Agricultural Research & Development. Lynne Rienner Publishers, Boulder, CO.
2. **Byrnes, K., 1990.** A Review of AID Experience With Farming Systems Research and Extension Projects. AID Evaluation Special Study no. 67. US Agency for International Development, Washington, DC.

3. **Gilles, J.L., 1992**. Social sciences and collaborative research: toward an agenda for the social sciences in agriculture. In: Proceedings of the Workshop on Social Science Research and the CRSPs. Collaborative Research Support Programs.
4. **Hibino, B. & D.O. Hansen, 1988**. A profile of the international activities and interests of RSS members. Rural Sociologist, 8(2), 142–49.
5. **Cernea, M.M. & A. Adams, 1994**. Sociology, Anthropology, and Development: an Annotated Bibliography of World Bank Publications 1975–1993. Environmentally Sustainable Development Studies and Monographs Series Paper no. 3. The World Bank, Washington, DC.
6. **Netting, R. McC., 1977**. Cultural Ecology. Benjamin/Cummings Publishing, Menlo Park, CA.
7. **von Bertalanffy, L., 1968**. General Systems Theory (revised edition). George Braziller, New York.
8. **Stewart, cited in: Harris, M. 1968**. The Rise of Anthropological Theory: a History of Theories of Culture. Thomas Y. Crowell, New York.
9. **Jones, J.R. & B.J. Wallace (Eds), 1986a**. Social Sciences and Farming Systems Research: Methodological Perspectives on Agricultural Development. Westview Press, Boulder, CO.
10. **McCorkle, C.M., 1992**. Plants, Animals & People: Agropastoral Systems Research. Westview Press, Boulder, CO.
11. **Hunter, D.E. & P. Whitten, 1976**. Encyclopedia of Anthropology. Harper & Row, New York.
12. **Murphy, J., 1990**. Farmers' systems and technological change in agriculture. In: M.S. Chaiken & A.K. Fleuret (Eds.) Social Change and Applied Anthropology: Essays in Honor of David W. Brokensha. Westview Press, Boulder, CO.
13. **DeWalt, K.M., 1983**. Nutritional Strategies and Agricultural Change. University of Michigan Research Press, Ann Arbor, MI.
    DeWalt, K.M. & B.R. DeWalt, 1989. Social science approaches to including nutrition research in the international sorghum/millet CRSP. In: C. McCorkle 1989b (Op. cit.).
14. **SEA (Society for Economic Anthropology), 1983**. Monographs in Economic Anthropology (1 volume annually). University Press of America, Lanham.
15. **Rhoades, R., R. Booth, R. Shaw & R. Werge, 1982**. Interdisciplinary development and transfer of postharvest technology at the International Potato Center. In: Report of an Exploratory Workshop on the Role of Anthropologists and Other Social Scientists in Interdisciplinary Teams Developing Improved Food Production Technology. IRRI and UNDP, Manila.
16. **McCorkle, C.M., E. Mathias & T.W. Schillhorn van Veen (Eds), 1996**. Ethnoveterinary Research & Development. Intermediate Technology Publications, London.
17. **CRSP, 1992**. Proceedings of the Workshop on Social Science Research and the CRSPs. INTSORMIL Publication no. 93-3. Collaborative Research Support Programs.
    Groenfeldt, D. & J.L. Moock (Eds), 1989. Social Science Perspectives on Managing Agricultural Technology. International Irrigation Management Institute, Colombo, Sri Lanka.
    Jones and Wallace, 1986a. (Op. cit.).
    McCorkle, C.M. (Ed.), 1989b. The Social Sciences in International Agricultural Research: Lessons from the CRSPs. Lynne Rienner Publishers, Boulder, CO.
18. **Rhoades *et al.*, 1982**. (Op. cit.).
19. **Ruttan, V.W., 1989**. The international agricultural research system. In: J.L. Compton (Ed.) The Transformation of International Agricultural Research & Development. Lynne Rienner Publishers, Boulder, CO.
20. **Barlett, P.F. (Ed.), 1980**. Agricultural Decision Making: Anthropological Contributions to Rural Development. Academic Press, New York.
21. **DeWalt, B.R., 1991**. Anthropology, evolution, and agricultural development. In: G.L. Johnson & J.T. Bonnen with D. Fineup, C.L. Quance & N. Schaller (Eds) Social Science Agricultural Agendas and Strategies. Michigan State University Press, East Lansing.
22. (Ibid.).
23. **Peters, P.E., 1986**. Household management in Botswana: cattle, crops, and wage labor. In: J.L. Moock (Ed.) Understanding Africa's Rural Households and Farming Systems. Westview Press, Boulder, CO.
24. **Goodell, G.E., P.E. Kenmore, J.A. Litsinger, J.P. Bandong, C.G. de la Cruz & M.D. Lumaban, 1982**. Rice insect pest management technology and its transfer to small-scale farmers in the Philppines. In: Report of an Exploratory Workshop on the Role of Anthropologists and Other Social Scientists in Interdisciplinary Teams Developing Improved Food Production Technology. IRRI and UNDP, Manila.
25. **McCorkle, C.M., M. Coulibaly, S. Coulibaly, D. McHugh, G.W. Selleck & D. Sidibe, 1993**. USAID/Mali Farming Systems Research and Extension Project: Phase II Evaluation Report. Development Alternatives, Inc., and Institute for Development Anthropology, Bethesda, MD and Binghamton, NY.

26. **McCorkle, C.M., 1989a**. Toward a knowledge of local knowledge and its importance for agricultural RD&E. Agriculture and Human Values, 6(3), 4–12.
27. **Wallace, B.J. & J.R. Jones, 1986**. Social science in FSR: conclusions and future directions. In: J.R. Jones & B.J. Wallace (Eds.) (Op. cit.).
28. **Brokensha, D., D.M. Warren & O. Werner (Eds), 1980**. Indigenous Knowledge Systems and Development. University Press of America, Lanham, MD.
Dupré, G. (Ed.), 1991. Savoirs Paysans et Développement. Editions Karthala and Editions de l' ORSTOM, Paris.
Warren, D.M., L.J. Slikkerveer & D. Brokensha (Eds), 1995. The Cultural Dimension of Development: Indigenous Knowledge Systems. Intermediate Technology Publications, London.
29. **McCorkle *et al.*, 1993**. (Op. cit.).
30. **Chambers, R., A. Pacey & L.A. Thrupp (Eds), 1989**. Farmer First: Farmer Innovation and Agricultural Research. Intermediate Technology Publications, London
McCorkle, C.M., R.H. Brandsetetter & G.D. McClure, 1988. A Case Study on Farmer Innovations and Communication in Niger. Communication for Technology Transfer in Agriculture Project, of the Academy for Educational Development, Washington, DC.
McCorkle *et al.*, 1996. (Op. cit.).
Moock, J.L. & R.E. Rhoades (Eds), 1992. Diversity, Farmer Knowledge & Sustainability. Cornell University Press, Ithaca, NY.
31. **Blount, P. & D.M. Warren (Eds), 1996**. Indigenous Organizations and Development. Intermediate Technology Publications, London.
32. **Cernea, M.M., 1985**. Sociological knowledge for development projects. In: M.M. Cernea (Ed.) Putting People First: Sociological Variables in Rural Development. Oxford University Press, for the World Bank, Ithaca, NY.
33. **Fernández, M.E., 1992**. The social organization of production in community-based agropastoralism in the Andes. In: C.M. McCorkle (Ed.) (Op. cit.).
34. **DeWalt, B.R. & P. J. Pelto (Eds), 1985**. Micro and Macro Levels of Analysis in Anthropology: Issues in Theory and Research. Westview Press, Boulder, CO.
35. **Rogers, E.M., 1983**. Diffusion of Innovations. The Free Press, New York.
36. **Lionberger, H.F. & P. Gwin, 1982**. Communication Strategies: a Guide for Agricultural Change Agents. Interstate Publishers, Danville, IL.
37. **Butler Flora, C., 1985**. Sociologists and the farming systems support project. The Rural Sociologist, 5(4), 274–9.
38. **Goodell, *et al.*, 1982**. (Op. cit.).
39. **Collinson, M., 1988**. FSR in evolution: past and future. In: Contributions of FSR/E Towards Sustainable Agricultural Systems: Proceedings of Farming Systems Research/Extension Symposium 1988. Farming Systems Research Paper Series Paper no. 17. University of Arkansas and Winrock International Institute for Agricultural Development, Fayetteville, AR and Morrilton, AR.
40. **Rhoades, R.E., 1984**. Breaking New Ground: Agricultural Anthropology. International Potato Center, Lima.
41. **Chambers, R., 1997**. Whose Reality Counts? Putting the First Last. Intermediate Technology Publications, London.
42. **Horton, D.E., 1984**. Social Scientists in Agricultural Research: Lessons from the Mantaro Valley Project, Peru. International Development Research Center, Ottawa.
43. **McCorkle, C.M., 1984**. Pride, preference, and practice: farmers' grain disposal decisions in a Burkinabe community. In: The Dynamics of Grain Marketing in Burkina Faso, Vol. II. Center for Research in Economic Development of the University of Michigan, Ouagadougou and Ann Arbor.
44. **Gladwin, C.H., 1982**. The role of a cognitive anthropologist in a farming systems program which has everything. In: Report of an Exploratory Workshop on the Role of Anthropologists and Other Social Scientists in Interdisciplinary Teams Developing Improved Food Production Technology. IRRI and UNDP, Manila.
Gladwin, C.H., 1983. Contributions of a decision tree methodology to a farming systems program. Human Organization, 42(2), 146–57.
45. **Paolisso, M. & M.G. Baksh, 1989**. Anthropological Contributions to the Study of Malnutrition: The Nutrition CRSP Kenya Project. In: C.M. McCorkle (Ed.) 1989b (Op. cit.).
46. **McCorkle *et al.*, 1988**. (Op. cit.).
47. **DeWalt, B.R., 1989**. Halfway There: Social Science in Agricultural Development and the Social Science of Agricultural Development. In: C.M. McCorkle (Ed.) 1989b (Op. cit.).
48. **Bonnano, A., 1989**. Sociology of Agriculture: Labor, Technology, and Development in an International Perspective. Concept Publishing, New Delhi and London.

49. **Cernea, M.M., 1985**. (Op. cit.).
50. **Jones, J.R. & B.J. Wallace, 1986b**. Introduction: Social Science in Farming Systems Research. In: J.R. Jones & B.J. Wallace (Eds) (Op. cit.).
51. **McCorkle, C.M., 1996**. Intersectoral Healthcare Delivery. In: J. Chesworth (Ed.) The Ecology of Health: Identifying Issues and Alternatives. Sage Publications, Thousand Oaks, CA.
52. **Wallace & Jones, 1986**. (Op. cit.).

10.3 AGRONOMY AND FSR – A RELUCTANT MARRIAGE?

Peter Hildebrand and Dennis Keeney

Reconciling statistical rigour and reliability with relevancy in experimentation has been a real area of conflict for agronomists in FSR, and perhaps remains the heart of the 'reluctant marriage' label.

10.3.1 Introduction

There have always been important systems-related themes within agronomy. Rotational experiments were, perhaps, the key early examples. Since the 1930s there have been forays into intercropping to improve small-farmer husbandry, particularly in Africa in the 1950s – the work of Evans, for example (1960)[1]. As Dick Harwood explains in Chapter 2, the cropping systems initiatives of Richard Bradfield evolved into important systems-related themes, and he was one of the first agronomists to recognize the value of working in cooperation with economists[2].

Since the mid 1960s the involvement of agronomists in FSR has added a further systems theme, perhaps with greater repercussions for the discipline than these earlier areas. It has brought agronomists in developing countries closer to their resource-poor clientele; people trying to make a living operating complex farming systems. FSR has moved experiments off the research station, out of a controlled environment, and on to farmers' fields to face the vagaries of nature and be managed by local farmers. While it would be wrong to conclude that the discipline of agronomy has changed because of FSR, there is little doubt that as a result of FSR a significant number of agronomists are now working in an interdisciplinary way and using non-traditional methods for designing and evaluating technology.

As earlier chapters show, FSR was born from professional concern at the low acceptance of research recommendations by small farmers. Yet the accumulated wealth of conventional agronomic skills and methods launched FSR into the development of more appropriate technologies.

While it is difficult to discern how the application of FSR to technologies development has influenced agronomy itself, it is possible to identify FSR's five overall effects on agronomy. All arise from the routines of FSR; working with small farmers on their fields, working with an awareness of the farm as a system, and interacting with social scientists:

- The design of farmer-friendly technologies, and evaluation of these technologies on farms.
- An appreciation of the variable quality of the land and other resources managed by small farmers.
- An understanding of the interactions between one enterprise and the rest of the farm system.
- An understanding of the diversity among small farmers.
- The development of new priorities for formal, station-based, agronomy.

Before examining each of these sources, the interaction with social scientists deserves closer attention.

10.3.2 Working with social scientists

Chapter 2 suggests that the introduction of social scientists into agricultural research programmes in developing countries has been an important result of FSR. Social science research arose from criticism, implicit but often also explicit, of the relevance of the products coming from the existing research establishment. The 'hard' scientists in the research services, with agronomists and breeders at the centre, saw these young newcomers (they were often junior professionals) as 'new boys on the block' to be seen but not heard. Not only were they regarded as new, but they also were considered as representatives of a 'soft' science with little apparent relevance to the conventional paradigms which had dominated agricultural research for the last 50 years. Yet these junior scientists, almost before they had their hands dirty, were criticizing the way in which research was being organized and managed! Understandably there was initial resentment against the introduction, often seen as intrusion, of social science into agricultural research.

Perhaps the first step towards mutual respect was the ability of economists to use money as a common denominator, and their efforts to interpret hard science results in a way familiar to agronomists in their everyday life. There was increasing appreciation that economic evaluation added to relevance of the results. Little was perceived as threatening in these early interactions. The economist remained dependent on the agronomist for data, as was seen in an acceptable 'service' role. However, even that limited role could open up wounds. There were occasions, for example, when a programme regarded as a biological and statistical triumph by the agronomist was recorded as an economic failure by the economist. One example is the interpretation of a typical fertilizer response curve. While the agronomist concentrated on levels realizing the highest yields, the social scientist needed to nett out the costs of purchasing and applying the fertilizer. Furthermore, operating funds are so scarce among resource-poor farmers that highest cash return per unit of cash outlay was a much more appropriate criterion than maximizing profit. The levels of fertilizer application at these 'economic' points were almost always much lower down the response curves than the technical optimum.

Misunderstandings increased when social scientists began to question experimental designs and began working directly with farmers. Inevitably they intruded into the territory of agronomists and breeders by establishing their own field experiments. Representative comments from agronomists about the bedraggled nature of FSR-E trials under real resource-limited conditions included: 'This looks just like a trial being run by social scientists' or 'It's a good thing it is well off the road!' One of the most common and disheartening comments was, 'It is obviously not worthwhile to work under these conditions because nothing can be accomplished'[3]. However, it was in these circumstances, when agronomists were drawn on to farms to interact with farmers and to learn about their production conditions, that the values of social science skills were demonstrated. In eliciting and understanding farmers' goals, priorities, strategies and constraints, social scientists could provide information for the design and evaluation of more relevant experiments.

It is still too early to say that the battle is over and the disciplines are reconciled, and there are still too few research establishments with a social science cadre. Today, however, there are many examples of powerful partnerships between farmers, agronomists and social scientists.

10.3.3 The design and evaluation of experiments in farmers' fields

The move from the artificial environment of the experiment station into the real world, has been and will continue to be a major challenge to the agronomy profession.

The move off the station

The research station gave agronomy a safe haven from the myriad sources of variation that frustrated the statistical measurement of responses. Fertility was managed, a regular water supply was often provided, and machinery allowed precise control of timing and consistency of materials. Even where labour was used on-station it was inevitably provided under the banner of precision. Such facilities did indeed provide an environment in which agronomists and breeders could work with precision. However, not only is such a context insulated from the vagaries of the real world, it also has the potential to produce wholly misleading results. One only has to make the facile comparison of the depth and speed of cultivation with a heavy tractor compared to a hand hoe to see that results on research stations may not be replicable in smallholders' fields.

Fostered by FSR, many agronomists moved

experiments on-farm. This was by no means the first foray off the station. Multilocational trials to capture the effects of wide variations in climate, soil and pest and disease complexes were common. However, the locations chosen for these trials were treated as mini stations to ensure that the effects of spatial differences in climate, soils and pests could be isolated and accurately measured. Such trials, while playing an important role, are a poor basis for recommendations to farmers. The move off station prompted by FSR was different and had a range of repercussions, many of these the result of increasing familiarity with small farmers as partners, and with the conditions under which these farmers must produce and survive.

System interactions

As early as 1980 Henry Nix, an Australian agronomist, was proactively urging a wider systems perspective for the profession. As he wrote in 1980: 'A research strategy based on the systems approach would centre around the development of working models of crop production systems. Such models need to be structured so that they remain operational yet capable of continuous improvement in logical structure and function. Ideally it would be useful to have a hierarchy of models capable of application at a range of scales and offering some choice in the levels of precision and accuracy'[4]. Intercropping and cropping systems research widened the range of interactions to be investigated. Farming systems research took this a step further and superimposed the whole set of socioeconomic constraints for consideration while shaping agronomic improvements for farmers. The consideration of interactions between enterprises in their demands for cash, labour and indeed land, and the complementarity of enterprises, particularly livestock with crops or crop residues for feed, manure and draught power for crops, brought a new dimension to agronomy. For example, the fact that increased cotton production is a national priority will not make farmers plant cotton early. They will not use agronomically perfect husbandry. If their own priority is for an early planted, early available, food crop to fend off starvation, they will plant their food crop first, then plant their cotton. Analysing an interaction such as this tells us that greater efficiency in growing early planted food crops is a research priority to induce better husbandry on cotton. A social alternative is to provide subsidized supplies of food in the local market in the pre-harvest months. Both might be acceptable routes towards better husbandry of the cash-producing cotton crop. An understanding of how farmers use system interactions and set their priorities widens both the range of experimental hypotheses and the evaluation criteria needed. Interactions also imply the importance of compromise solutions that, while sacrificing the optimum from any one enterprise, enhance overall system performance.

Within farm and within field variation in soil and water resources

Their growing familiarity with farmers' fields also led agronomists to an appreciation of in-field variability, and to an understanding that farmers often used variability in managing their resources. One example is the exploitation of local differences in water-holding capacity in different parts of a field. Strong interactions occur between key management factors and resource 'niches' that farmers use in a particular way. In rainfed farming the need for 'niche' technologies, rather than broad adaptability, has emerged.

Farmer diversity

Exposure to farmers, the concerns of the social scientists, and the need to identify representative clients for partnership in OFEs, led agronomists to appreciate the diversity of the farm population. They learned that old farmers have attitudes and capabilities that differ from those of young farmers, and that women-headed households usually have a more limited labour force than those with a full family. This diversity provided further evidence of the futility of a single 'best' answer. It reinforced the need for multiple choices, tailored both to the diversity of households in local communities, and to the variety of resource niches used by them. As discussed later, sustainable agriculture research and outreach in the USA and other developed countries benefited greatly from these observations.

Experimental design

Much of the influence of FSR on agronomy as a discipline has arisen from adapting conventional methods, particularly the analysis of variance, for use in the real conditions which small farmers must manage in order to produce sustainably. The classic agronomic goal of producing 'broadly adaptable' technology fitted well with the practice of controlling non-experimental variables at non-limiting levels. This practice created artificially superior environments that large-scale and/or industrialized farmers could mimic but limited-resource farmers could not achieve[5].

Conventional agronomy identified treatment variables as those management factors most likely to contribute to high yields per unit area under relatively homogeneous conditions of climate and soil. Experiments sought optimum levels in each of these key areas. For precise measurement of the responses to changes in their levels, non-experimental variables were, and are, conventionally held at high, non-limiting levels. As with the false environment of the station this completely isolated the results from small-farmer management. Recommendations framed from the findings on the treatment variables alone implicitly assume that non-treatment variables would be non-limiting when such recommendations were implemented on the fields of small farmers. This was never the case. Small farmers never had the machinery, and rarely had the cash or labour, to implement the treatment recommendations, let alone the resources to implement the non-treatment management at non-limiting levels. Overall it was small wonder that there was limited adoption among resource-poor farmers. Few recommendations were accepted, particularly in rainfed farming where the uncertainties of climate also had to be managed. In FSR-driven OFR, farmer practice became the standard for non-treatment management and responses obtained to treatment factors were much closer to those to be expected when treatments were tested by farmers.

In the early days there was a strong reaction from agronomists to this loss of control and, to an extent, the reaction continues. As a result of their training, agronomists inevitably associated the relatively low yield levels and the limited treatment effects with high coefficients of variation and lost trials. Reconciling statistical rigour and reliability with relevancy in experimentation has been a real area of conflict for agronomists in FSR, and perhaps remains the heart of the 'reluctant marriage' label. In much OFR the balance has shifted to relevancy, with replicability more a function of significant numbers of farmers testing out apparently relevant new materials and methods.

Similar reservations have arisen about conventional breeding strategies. It has long been argued that conventional selection methods have had a negative effect on the relevance of germplasm made available to small farmers[6]. In 1990, Hildebrand listed four factors likely to have caused the rejection of genetic material that would have demonstrated superior yielding abilities in both the poorest and the best farm environments:

- Statistical dependence on analysis of variance leading to the concern with reducing genotype by environment interactions. This in turn leads to the nearly universal practice of evaluating material on experiment stations and farms with real or artificially created superior environments to control this interaction or to permit the material to express its yield potential.
- The capability of many farmers in the developed world, over the last few decades to use their resources to modify unfavourable environments.
- The widespread use of a regression coefficient of unity as a measure of stability.

With FSR at least partially responsible, practice has again overtaken theory following the recent surge in participatory breeding. The widening understanding of the farming system, the fact of niche management, and of small-farmers' use of diversity as a management tool, has brought new selection criteria to bear through direct farmer involvement in varietal choice[7]. The burgeoning interest in diversity as an environmental goal has helped spur efforts to service farmers' needs for a range of plant material.

Experimental evaluation

Working with farmers in their fields raised awareness among agronomists of the real circumstances under which small farmers operate. They became more familiar with the choices small farmers must make, and the criteria they used in making these choices. Yield per unit area was seldom their basis for choice. As mentioned, the return per unit of cash or labour outlay was their key criterion and led to the need to modify conventions in evaluation and eventually in the choice of experimental treatments. However, given the priority for most small farmers to meet their daily household food needs, and given the need to manage risks from uncertain rainfall and uncertain markets and prices, the trade-offs across multiple objectives and a range of evaluation criteria become complex and subtle – so much so that the farmers have proved to be the best evaluators, and many FSR teams now involve them as such. These factors, and a need to shift the ownership of on-farm trials to the farmers and to their communities, have led to wide farmer participation in trial evaluation and, increasingly, in experimental design.

Setting on-station priorities

Finally, agronomic priorities for controlled experiment station research have started to respond to information from diagnosis and from the closer association of agronomists with both farmers and with social scientists. Station trials increasingly investigate the underlying biophysical relationships of the problems thrown up by farming systems research on farmers' fields. It is the beginning of the articulation of demands from the hitherto unorganized small-farmer population and is increasingly influencing priorities in both plant breeding and agronomy.

10.3.4 Effects in professional circles

A dichotomy seems to be developing within agronomy. Some, essentially field agronomists, are now devoted to OFR. For others the increased emphasis on sustainability issues has either made them more reductionist in approach, focusing narrowly on the detailed physical processes governing the relationships between soils, water and plants, or has pushed them higher up the systems hierarchy to the broader fields of ecology and soil and water processes at the watershed level. In some senses the battle for wider systems thinking among agronomists has been won. A significant subset is broadening out into farm-level systems (as opposed to plant growth systems) and has found a professional home in the FSR associations, particularly in Africa, Asia and Latin America. Beyond this, systems agronomy is also well represented in conventional professional circles, at least in the USA.

As early as 1976 the American Society of Agronomy published the first major publication on one aspect of increasing food production beyond increasing cultivated area and increasing yields; that is, harvesting more than one crop from the same piece of land in a year[8]. It resulted from an ASA symposium led by the International Agronomy Division on multiple cropping systems in which two agricultural economists participated. And an anthropologist took part in the 1978 symposium led by the Extension and International Agronomy Divisions on transferring technology for small-scale farming, which resulted in another ASA publication. In this the then ASA President John Pesek argued the importance of small farmers as contributors to the global food supply[9], and Nyle Brady, then Director General of IRRI, lamented that only about one quarter of the rice farmers in the tropics had benefited from the improved (Green Revolution) rice technologies. 'For the remaining rice farmers', he said 'no really superior technologies have been developed that suit their conditions and that financially benefit them'[10]. Brady further argued that 'A major constraint to modification and adoption of new technology to small-scale LDC farmers is the failure of researchers and extension personnel to work with them'. In the same publication, Robert Waugh, Adjunct Director of ICTA in Guatemala, argued that 'the use of multidisciplinary teams of biologists and social scientists has contributed more than any other thing to making it possible for the social scientists [at ICTA] to contribute to agronomic technology. By the same token, this integration has had a beneficial effect on the nature of the work undertaken by the biological scientists'[11].

Many members of the International Agronomy Division have long been active in FSR-E, and the division elected a FSR-E agricultural economist as Chair-Elect in 1990. In 1994 the ASA created the Agricultural

Systems Division with support from the FSR-E community within and outside the ASA. In '*Agronomy News*', December 1995, the ASA announced that its premier publication, the 'Agronomy Journal', would open a new section, Integrated Agricultural Systems, to include FSR-E.

10.3.5 The ongoing honeymoon

Continuing interactions between biophysical and social scientists, many within the context of FSR and FPR, are the honeymoon following the reluctant marriage and recent developments, particularly in the areas of natural resource management and participatory breeding, are bringing successes.

Agronomists have increasingly associated FSR with sustainable agriculture. Many agronomists with FSR backgrounds and experience in developing countries have been at the forefront of sustainable agriculture conceptualization in the USA, undoubtedly because of the systems approach underpinning FSR and the location-specific nature of technologies required by limited resource farmers. Obviously, FSR has not been the only influence on the move towards systems approaches to sustainable agricultural development, but its influence is undeniable. When the Competitive Grants Program of the United States Department of Agriculture's National Research Initiative was initiated, an FSR agricultural economist headed the proposal evaluation team that included several FSR agronomy professionals. The USAID-funded Sustainable Agriculture and Natural Resource Management (SANREM) Collaborative Research Support Program (CRSP), the USDA-funded Sustainable Agriculture Research and Extension (SARE) programme, and the reorganized Soils Management CRSP all require methodologies that were pioneered by FSR practitioners.

Iowa State University (ISU) agronomist and former ASA President John Pesek chaired the Board of Agriculture's National Research Council committee that prepared the 1989 NRC book on Alternative Agriculture[12]. Dennis Keeney, another ISU agronomist and also a former ASA President, currently heads the Leopold Center for Sustainable Agriculture at ISU. But, as he points out, administrative declarations do not change paradigms. After a decade of Leopold Center funding and persuasion there are only two FSR-E type projects at ISU that truly involve agronomists, even though there is, and will continue to be, much research and education in alternative systems. Major team successes at Iowa have included those headed by Forestry (development of watershed-level buffer strips), Animal Science (management intensive grazing) and Agricultural and Biosystems Engineering (manure management). These teams include agronomists, but agronomy has not been the leading discipline. An agronomy-led team on strip intercropping operated for several years and developed excellent strip intercrop practices, but these were not adapted to the large scale-farming systems of the midwestern USA.

10.3.6 Conclusion

There is increasing, though grudging, acceptance – at least among 'cutting edge' members of the discipline – that agronomic technologies are seldom, if ever, 'farmer size neutral'. In other words, small farmers have particular needs. There is accumulating evidence that farmers are able to outperform breeders for use of plant germplasm in 'niche' environments. Niche breeding with farmer participation is rapidly establishing a role, particularly in areas and crops where market penetration is weak. There is increasing specialization amongst agronomists as the field widens, to the farming system and up the hierarchy to microecologies and watersheds, and as it deepens with the soil/water/plant relationships becoming increasingly important under the sustainability banner.

FSR can rightfully claim to have brought agronomy to terms with a set of farmer clients operating under circumstances very different from those of the commercial farms where it grew up. At the same time FSR must acknowledge that without the accumulated skills of conventional agronomy it would have had no vehicle to reach out to farmers in their fields in developing countries. The reluctant marriage promises to flourish over time.

REFERENCES

1. **Evans, A., 1960**. Studies of intercropping. I. East African Agricultural and Forestry Journal, XXVI.
2. **Bradfield, R., 1966**. Toward more and better food for the Filipino people and more income for her farmers. ADC Paper, The Agricultural Development Council, New York.
3. **Hildebrand, P.E., 1979**. Incorporating the social sciences into agricultural research: the formation of a national farm systems research institute. Instituto de Ciencia y Tecnologia Agricolas, Guatemala, and the Rockefeller Foundation, New York.
4. **Nix, H.A., 1980**. Strategies for crop research. Procceedings of the Agronomy Society of New Zealand.
5. **Hildebrand, P.E. & J.T. Russell, 1996**. Adaptability Analysis: a Method for the Design, Analysis and Interpretation of On-Farm Research-Extension. Iowa State University Press, Ames.
6. **Republic of Zambia, 1982**. Plant Breeding for Low Input Conditions. Proceedings of a workshop on plant breeding for 'lousy' conditions held at Mount Makulu Central Research Station, Chilanga, Zambia. Government Printer, Lusaka.
   Hildebrand, P.E., 1990. Modified stability analysis and on-farm research to breed specific adaptability for ecological diversity. In: Manjit S. Kang (Ed.) Genotype-By-Environment Interaction and Plant Breeding. Louisiana State Agricultural Center.
7. **Sperling, L., M.E. Loevinsohn & B. Ntabomvura, 1993**. Rethinking the farmer's role in plant breeding: local bean experts and on-station selection in Rwanda. Experimental Agriculture, 29.
8. **Papendick, R.I., P.A. Sanchez & G.B. Triplett (Eds), 1976**. Multiple cropping. ASA Special Publication 27. ASA, CSSA, SSSA, Madison, WI.
9. **Pesek, J., 1981**. Forward. In: Usherwood (Ed.) Transferring Technology for Small-Scale Farming. ASA Special Publication 41. ASA, CSSA, SSSA, Madison, WI.
10. **Brady, N.C., 1981**. Significance of developing and transferring technology to farmers with limited resources, pp. 1–21. In: Usherwood (Ed.) Transferring Technology for Small-Scale Farming. ASA Special Publication 41. ASA, CSSA, SSSA, Madison, WI.
11. **Waugh, R.K., 1981**. Research and promotion of technology use, pp. 67–88. In: Usherwood (Ed.) Transferring Technology for Small-Scale Farming. ASA Special Publication 41. ASA, CSSA, SSSA, Madison, WI.
12. **Committee on the Role of Alternative Farming Methods in Modern Production Agriculture, Board on Agriculture, NRC, 1989**. Alternative Agriculture. National Academy Press, Washington, DC.

Part V

Cutting Edge Methods, Abiding Issues and the Future for FSR

EDITORIAL INTRODUCTION

Mike Collinson

THE CONTRIBUTIONS

This last part of the book has two chapters. Chapter 11 looks at six cutting edge developments in methods at various levels in the systems hierarchy. First, Toon van Eijk[1] examines farming systems research's (FSR) claim as a holistic approach using his own experiences in its implementation set against the characteristics of the positivist, constructivist and transcendentalist paradigms, concluding that transcendentalism can improve the holistic nature of the approach. Evaristo Miranda sets out an application of geographical information system (GIS) in Brazil, demonstrating the breakthrough that the technique offers in using a variety of digitized databases over a range of hierarchical levels. Also from Latin America, Roberto Quiroz and colleagues at the International Potato Center (CIP), illustrate a range of modelling applications within the context of FSR. Louise Sperling and Jacqueline Ashby review progress in participatory breeding, an application evolving so quickly that they have expressed reservations at the already dated nature of their contribution. Clive Lightfoot links a range of experiences in agroecosystems analysis to foster the evolution of a coherent diagnostic and analytical process for use by field teams. Finally, J.P. Deffontaines and a group of his colleagues illustrate the progress made in applying systems approaches in France, linking applications at different hierarchical levels.

In Chapter 12, together with Clive Lightfoot, I take up the issues identified in the four earlier parts of the book and build a prognosis for the future of FSR. Myself, as a farm economist and strong supporter of anthroplogy and sociology in FSR, and Clive, as an agronomist with a strong ecological bent, try to offer a balanced perspective.

A COMMENTARY

My difficulty with van Eijk's discussion of practice and paradigms is in the distinction between current and best practice. It remains very true that much past field implementation has been weak, and much current field implementation remains that way. The cause of much of this gap lies in the continuing weakness of university education to equip undergraduates and postgraduates with a theoretical and conceptual foundation in systems thinking. Current practitioners have usually received short-term training in an FSR process, often one already overtaken by best practice. Not only do they have poor access to best practice but neither do they have the intellectual underpinnings to critique and adapt what they learn as new concepts and methods to move the process on. The gap between current and best practice widens.

The five cutting edge developments in methods reach across levels of the systems hierarchy and, not for the first time in the book, raise the issue of the scope of FSR. Are all these applications, and the methods they use, appropriately subsumed under the banner of FSR? Are advocates of such an all-embracing FSR empire building? The Association for Farming Systems Research and Extension (AFSRE) 1994 symposium in Montpellier, perhaps influenced by the breadth of French experience, titled its proceedings volume quite generally: 'Systems Oriented Research in Agriculture and Rural Development'[2]. Should we relabel FSR or redefine it as one among a family of systems-based applications that have a contribution to make in agricultural R & D, particularly in the development of developing world agriculture? Other systems exponents have couched their contributions in these more modest terms, including Penning de Vries, Teng and Metselaar in 1993[3].

There is confusion among practitioners at the ever widening array of activities and processes subsumed under the FSR acronym. There seem to be two interlinked dimensions to the confusion. First, how far those factors beyond the farm that influence farmers' decision-making, such as policy and the mandates and procedures of enabling institutions, should be subsumed by FSR. Second, while recognizing the linkages, and seeking to partner together institutions with complementary roles, how far should the focus in FSR be the improvement of the farm, rather than improvement of the economy, or, within the last decade, of the ecology?

The limits of FSR *vis à vis* the economic, ecological and social context of the farm has been an issue raised in many fora. Most of the early discussions revolved around the questions of how far FSR should reach beyond the farm to mitigate farm problems. While it was acknowledged that the original models of the late 1970s embraced off-farm factors[4], it was equally clear that practice was more restrictive. At an Australian Council for International Agricultural Research (ACIAR) symposium in 1985 Norman and Collinson[5], both from research backgrounds, categorized FSR in three ways:

- In the large – in which all parameters, including those exogenous to the farm – were variable.
- In the small – FSR arrives at a focus within the farming system as a result of diagnosis.
- With a predetermined focus – diagnosis aims at understanding the system to allow improvement at a predetermined focus.

All three are valid and perhaps each is appropriate under different circumstances. Norman and Collinson passed over FSR 'in the large' as too complex. This restrictive view of FSR almost certainly arose from the fact that their own FSR programmes operated within the research arm of the agricultural establishment, an arm which rarely had the authority to question either policy or the mandates and procedures of agricultural enabling services. Nevertheless, the conclusion of the 1985 Hawkesbury meeting was that the exogenous constraints on farm activities were clearly so important to development that they needed to be internalized within FSR. The idea of farming systems development (FSD) to which these external factors were central was mooted[6] and, as we have seen, was subsequently promoted by the Food and Agriculture Organization of the United Nations (FAO).

Both Baker in 1993 with particular reference to policy, and Berdegué and Escobar in 1995[7] as a new economic and political context took over in many Latin American countries, recently resurrected this issue, Berdegué and Escobar perhaps the more stridently. Most of their conclusions hinge on the opening of markets and hence the importance of identifying new market opportunities – a recent revolution for Latin America. With a longer view this is perhaps another iteration of a recurring phenomenon. For example, the introduction of export crops to the small farmers in East and southern Africa in the 1950s and 1960s had a very similar impact. One quarter of a million small farmers in north-west Tanzania raised cotton production there from 40,000 bales in 1950 to 405,000 bales in 1970 – a 10-fold increase in 20 years[8]. Coffee, tea, tobacco, pyrethrum, sugar and cashew followed similar patterns in Africa, as did wheat, maize, sorghum, cassava and beans, increasingly grown for cash as domestic markets expanded with rapid urbanization. It is clearly important that FSR reaches beyond new technology on existing crops to all opportunities for farm improvement. It raises the point that how and where FSR is institutionalized strongly influences its scope, a point that justifies elaboration in Chapter 12, given the widening brief for FSR and the changing nature of the FSR process.

The second question remains. How far should FSR 'umbrella' R & D activities at levels of the systems hierarchy beyond the farm? Is it more appropriate that FSR be seen as a source of information for R & D at other levels, even when systems approaches are being used there? At a lower level of the hierarchy, for example in crop modelling, FSR input may be valuable in establishing parameter values based on farmers needs and preferences, but crop modelling remains a systems application in its own right. Similarly at higher levels, in watershed modelling or in aggregate economic modelling for policy analysis, FSR may provide valuable information but cannot justifiably subsume such activities.

I favour a conclusion that the improvement of identified farming systems operated by significant groups of farmers remains the focus for FSR. The word 'identified' is important in stressing an operational role for FSR. With such an emphasis the aggregation question is not properly an FSR issue. All R & D activities at any hierarchical level that contribute to farm system improvement are important targets for FSR information, including policy, even policy decisions not directed to any identified farm system, but which clearly exert influence on decisions there, are valid targets. Even activities at hierarchical levels above and below the farm that do not affect farmers' decisions, but where farmer understanding is important, will benefit from FSR information.

REFERENCES

1. **van Eijk, T., 1998**. Farming systems research and spirituality: an analysis of the foundations of professionalism in developing sustainable farming systems. Thesis, Landbouwuniversiteit, Wageningen, The Netherlands.
2. **CIRAD, 1996**. Systems Oriented Reseach in Agriculture and Rural Development. An International Symposium, Montpellier, France, November 1994. CIRAD-SAR, Montpellier, France.
3. **Penning de Vries, F., P. Teng & K. Metselaar, 1993**. Systems Approaches for Agricultural Development. Kluwer Academic Publishers and IRRI, Dordrecht, The Netherlands.
4. **Baker, D., 1993**. Inability of farming systems research to deal with policy. JFSRE, 4(1), 67–86.
5. **Norman, D. & M. Collinson, 1985**. Farming systems research in theory and in practice. In: J.V. Remenyi (Ed.) Agricultural Systems Research for Developing Countries: Proceedings of an International Workshop. ACIAR Proceedings no. 11, ACIAR, Canberra.
6. **Remenyi, J.V., 1985**. Opportunities for farming systems research collaboration: issues and recommendations. In: J.V. Remenyi (Ed.) Agricultural Systems Research for Developing Countries: Proceedings of an International Workshop. ACIAR Proceedings no. 11, ACIAR, Canberra.
7. **Berdegué, J.A. & G. Escobar, 1995**. New Directions of the Systems Approach for the Modernization of Latin American Peasant Agriculture. JFSRE, 5(1), 1–30.
8. **Lele, U., 1975**. The Design of Rural Development. Johns Hopkins University Press for the World Bank, Washington, DC.

Chapter 11
At the Cutting Edge

11.1 HOLISM AND FSR
Toon van Eijk

To my mind it is evident that the dynamic and emergent nature of interactions taking place between farmers and nature, including forces which lie beyond the interface situation itself, and between numerous other actors and their networks, puts the process of rural development beyond full human control.

11.1.1 Introduction

FSR is part of a larger 'systems movement' which breaks away from narrow, disciplinary thinking about agricultural research and development. It provides a 'market' orientation to agricultural research through focusing on specific client groups[1] and was, at least in part, a response to weak representation of small farmers in research and extension institutions. One can hope it will be a temporary substitute for well-articulated small-farmer demand. In FSR theory the complexity of small farming systems is the interdependence of components which constitute a coherent whole, and the centrality of the farmer – the human factor. While the farming systems perspective must be complete or 'whole', FSR is not a separate science. It is rather, as Stoop said in 1987 'an approach and a scientist's attitude towards agricultural research'[2], a perspective that should permeate the whole agricultural research process, including discipline and commodity-oriented on-station research (OSR).

The interpretation of the term 'holistic' is probably the main source of confusion in FSR. Plucknett *et al.*, in 1987, referred to the 'often fuzzy and seemingly all-embracing nature' of many FSR programmes[3]. In most, the holistic approach requires that the whole farm system serves as a framework for analysis during the diagnostic stage, but in later stages only specific components, subsystems or interactions are targeted for intervention[4]. The selection of intervention points is one of the problematic areas in FSR[5]. Ideally, one works on a selection of technological constraints while maintaining the whole system perspective, which implies that the complex interactions between interdependent components are recognized and taken into account[6]. The question is whether the interactions with other subsystems are really kept in mind: whether one can maintain the farming systems perspective while working within subsystems. Some programmes carry FSR labels but are nothing more than on-farm experiments (OFEs) with no systems analysis evident[7].

In a holistic FSR approach one might expect horizontal as well as vertical integration: between various disciplines at the farm level and between different levels, for example, the farm and the watershed. In the end the concept 'farming systems perspective' implies, as Anandajayasekeram said in 1995: 'seeing things from the farmers' viewpoint'[8]. Although some smallholder production conditions can be simulated at research stations, others, such as system interactions and farmer criteria, can only be properly studied under actual farming conditions, i.e. in farmer managed and farmer implemented on-farm trials[9]. Adapted OSR cannot replace OFE. The perspective of systems, and therefore of the farmers, cannot be suspended[10]. Rhoades noted in 1994 that: 'only farmers can bring realistic "holism" to a research project. ... "technology" is only part of the story. Important political, social, and even religious concerns affect farmers, who must weigh technologies within a broader framework of "life" '[11].

One major problem with the holistic aspect of FSR is the delineation of the system under study: what are the boundaries of the system? The focus on the farm often ignored important structural and macro-economic factors. In the early days 'holistic' meant a break away from a monocropping focus to intercropping, then to cropping systems research. Subsequently, crop/livestock systems and off-farm activities were included. More recently, the agricultural sector, national economy and world economy have been seen by some as parts of a holistic FSR approach. The most broadly conceived FSR programme integrates agricultural research and development strategies. The Francophone approach to FSR is a rare example in which agrarian infrastructure and services are treated as variable[12]. The definition of the boundaries of a system is a fundamental step in systems analysis. The decision on which factors to treat as endogenous variables (under farmers' control) and which as exogenous parameters (not under farmers' control), is often a dilemma for researchers. FSR, which is basically a 'hard' systems theory, neglects the fact that the delimitation of systems' boundaries is a subjective process[13]. Despite lip-service to holism, systems thinking is still the exception rather than the rule[14].

When FSR becomes more holistic, multidisciplinary and location specific, with wider roles included under its umbrella, the analysis becomes more relevant but brings with it far-reaching consequences in terms of methodological complexity, interdisciplinary communication, skill requirements and organization and management. Demands on institutions and personnel applying the concept rapidly increase and this raises serious doubts about its practicality. Simmonds captured this in 1985 when he said: 'In real life ... systems isolated for study are always subsystems arbitrarily defined for the purpose in view. They are never holistic in any serious sense of that rather over-used word. In practice, what is wanted is sufficient understanding to attain the necessary level of FSP [Farming Systems Perspective] and no more. I wish the words holism and holistic were avoided in FSR contexts except when a really deep analysis of a whole-farm system is being attempted. ... For OFR/FSP [On-Farm Research with a FSP] a partial, non-holistic, subsystem knowledge will suffice or, anyway, has to suffice in practice. ... FSP rightly takes a common-sensical rather than a formal view of systems and, only exceptionally, needs to make numerical models. ... A system is what an experienced worker says it is'[15].

The systems approach concentrates on interrelations or interconnections, it is a dynamic approach that considers processes more fundamental than structures. Bawden, in 1995, argued that FSR is more systematic than systemic – 'more concerned with the rigour and linear logic of the process, than with the systemic interconnections of either the object of the research or the process used'[16]. The capacity of FSR practitioners to think and act systemically must be improved. Sustainable development requires systemic competence, expressed in a systemic perspective which portrays, as Bawden wrote: 'the sense of wholeness in all of this' and which promotes the participation of all relevant stakeholders in the rural development process. Successful practitioners of systems analysis pursue it more as a 'craft' than as a science[17].

It must be emphasized that agricultural research, and thus also FSR, is only one component in the mix of conditions that must be catered for in order to facilitate rural development. Other components in the multidimensional process of rural development, such as an adequate infrastructure, input supply, credit, marketing, land tenure and price policy, are often a prerequisite to research and extension making a difference. Which components in this mix should be treated as endogenous variables and which as exogenous parameters? In the Anglophone FSR approach, the common approach in eastern and southern Africa, research is adapted to the external conditions which are seen as largely given. As a practice-oriented field agronomist I support this choice: what can a resource-poor farmer or FSR agronomist do about infrastructural bottlenecks or inappropriate price policies? After all, it is unlikely that the countervailing power of resource-poor farmers will increase very quickly. At the same time, however, it is clear that infrastructural bottlenecks hamper the effectiveness of investments in agricultural research. In recent years the tendency in FSR is to treat more and more institutional factors as potential leverage points. The farming systems perspective is enlarged. Whether the problems of implementation implied by an enlarged perspective can be solved remains to be seen. Coordination of the input of farmers, researchers, extensionists, input suppliers, credit and marketing organizations, private traders, NGOs, planners, donors and politicians is difficult. The whole system perspective is hard to realize, yet it is evident that, in location-specific farming systems, the central position of farmers as 'experts at adaptive management' requires more attention[18].

11.1.2 Four main issues

A list of 15 operational problems in FSR is presented in Table 11.1.1. Virtually all of these problems arose in each of my work experiences in Kenya, Mozambique, Tanzania and Zambia. Not all emerge in each FSR programme but they will, nevertheless, be familiar to FSR practitioners, certainly in eastern and southern Africa. Some are not specific to FSR and have hampered conventional research on a continuous basis, a few are FSR specific. Many are interrelated which makes it difficult to rank them in order of importance.

The initially simple FSR methodology has been under constant revision, many innovations have been introduced: the farmer-first paradigm, participatory rural appraisal (PRA) techniques, gender, informal research and experimentation, and so on. These are no substitute for FSR, but complement conventional procedures which are flexible enough to incorporate new techniques and methods[19]. Practitioners have suggested modifications to solve all the problems listed in Table 11.1.1, but these make the process so elaborate that for most, if not all, FSR teams, problems of practicality arise. Bearing in mind that much FSR is implemented by relatively junior researchers working under difficult conditions radical innovations easily 'overload' FSR teams. Too elaborate definitions of FSR result in an approach that is difficult to implement by current research systems. The dilemma is to balance holistic, interdisciplinary and pragmatic approaches.

This long list of operational problems presented in Table 11.1.1 implies a gap between FSR theory and practice. Apparently holistic theory is difficult to implement. While a dichotomy between theory and practice has been inherent in the FSR approach since the beginning, it has become more pronounced in recent years with the introduction of new roles and new methods. As early as 1982, Byerlee *et al.* noted the following paradox: 'There is a potentially serious inconsistency

Table 11.1.1. List of operational problems in FSR[20].

- Lack of systems perspective
- Lack of client-oriented attitude
- Lack of farmer participation and lack of countervailing power
- Lack of participatory attitudes with researchers and extensionists
- Neglect of indigenous knowledge and gender issues
- Lack of feedback to OSR and weak priority setting there
- Lack of collaboration FSR–OSR and weak institutionalization of FSR
- Lack of involvement of extension and NGOs
- Lack of ecological sustainability
- Neglect of variation in time and space
- Neglect of role of intuition
- Lack of quality in field experimentation
- Lack of balance in breadth and depth of research
- Lack of incentives and resources
- Lack of interdisciplinarity

between our advocacy of a farming systems perspective as a holistic view of an often complex farming system and the use of research methods which are cost effective and emphasize rapid results'[21].

It seems to me that incessant improvement of methods has sought to bridge the gap between theory and practice but the impact on the well-being of resource-poor farmers has remained limited. Although each innovation makes sense, individual team members and the team as a whole cannot cope with the added complexity. Formal training has not succeeded in providing scientists who can handle increasingly elaborate FSR methodology adequately. Multidisciplinary teams are not a sufficient answer since each member must master the farming systems perspective and gain a perception of the whole, before effective interdisciplinary communication and collaboration can occur. I have attempted to cluster the operational problems into four main issues: holism, interdisciplinarity, attitudinal factors and lack of countervailing power of resource-poor farmers. These four are closely interrelated. If, for example, resource-poor farmers had more power, they could, in theory, 'enforce' a more client-oriented, interdisciplinary approach to agricultural research. Holism and interdisciplinarity are key features of FSR theory and, thus, logical points to cluster operational problems.

The problematic character of holism in FSR has already been discussed. While interdisciplinary collaboration among natural scientists in FSR teams may sometimes be problematic, the key failing is weak collaboration between natural and social scientists. Social scientists in developing countries tend to be young and inexperienced and this makes it difficult for them to function as equal partners in multidisciplinary teams[22]. Few trained professionals in these disciplines choose work at farm level, preferring to gravitate to planning or to academia[23]. These main issues of holism and interdisciplinarity are 'white spots' in FSR theory and practice. They demand fundamental conceptual innovation at a level above the fragmented agricultural sciences, and new problem-solving methods.

Attitudinal factors and lack of countervailing power are, at first sight, less obvious choices as key issues in balancing FSR theory and practice. Although regularly mentioned in recent FSR literature, they are to my mind insufficiently explicated. In 1994, Pretty and Chambers advocated a new agricultural professionalism about which they remark: 'Personal behavior and attitudes remain the great blind spot of agricultural research and extension. The quality and sensitivity of personal interactions are critical. ... Methodologically, a major frontier for institutional change is how first to enable individuals to change, for personal change will often have to precede as well as accompany changes in the cultures of organizations'[24]. Unfortunately, changing attitudes is another 'white spot' in FSR practice, conventional training is just not up to it. On the fourth issue the central question is how to balance the power of change agents, such as researchers and extensionists, and the power of resource-poor farmers.

As long as the four issues remain problematic, the gap between FSR theory and practice cannot be bridged. But the question arises of 'why' these issues exist. Indeed, understanding 'why' is the key to their solution. More training and new methods are unlikely to make a great difference, eventually a more thorough examination of the causes is needed. In my view the key operational problems originate from erroneous theoretical assumptions. Although the FSR principle is commendable, it is, unfortunately, based on a faulty theoretical paradigm.

11.1.3 Emergent properties and synergy

Three different paradigms can be distinguished in contemporary science: the positivist, the constructivist and the transcendentalist. The positivist paradigm has, historically, underpinned most agricultural research, although the constructivist paradigm gains influence there through the introduction of participatory methods. The three paradigms can be characterized by numerous criteria, some of which are shown in Table 11.1.2. Within the positivist paradigm we can distinguish two different belief positions: holism and reductionism[25]. Holism refers to the belief that the world is structured in the form of coherent whole entities (systems) with each subsystem, system and suprasystem having unique characteristics or emergent properties. These emergent properties are a key

Table 11.1.2. Characterization of the positivist, constructivist and transcendentalist paradigms.

Paradigm	Positivisit	Constructivist	Transcendentalist
Keyword	Matter	Mind	Spirit
Methodology	Experimental testing	Debate, interaction, communication, actor-oriented approaches	Methods for consciousness development, e.g. meditation
Nature and role of science	Natural sciences	Hybrid of natural and social sciences	Hybrid of sciences and methods for consciousness development
	Science is source of innovation	Communicative interaction is source of innovation	Consciousness development facilitates innovation
	Keywords: explain, control, prediction; solve problems	Keywords: understanding, interpretation, participation, negotiation, facilitation of joint learning, improve situations	Keywords: participatory attitudes, facilitation of positivist and constructivist-oriented methodologies
	Scientist is problem solver	Scientist is one active partner in the social construction of reality, equal partner, co-learner facilitator	Scientist is equal participant, co-learner, facilitator
	Studies consequences of human action	Studies human activity itself	Studies the underlying basis for human activity
	Reductionist and holistic position	More holistic position	Holistic position
	Conventional station research, most FSR	Some FSR which incorporated participatory methodologies	Research with a farming systems perspective, combination of science and spirituality
Nature and role of extension	Transfer of technology (TOT) teaching	Facilitation of participatory learning processes	Integral human development
	Transfer of data and information	Sharing interpretation and transformation of data and information	Transformation of attitudes
	Do to, do for	Do with	Do with, do themselves

concept in systems thinking: they are properties which emerge at the system level and which cannot be fully understood or predicted by studying each component separately nor by simply taking the sum total of the properties of the components[26]. The whole is more than the sum of the parts – the basic tenet of holism. Reductionism, on the other hand, refers to the belief that one must analyse and understand the parts in order to understand the whole.

Conventional OSR in East Africa operates within the reductionist perspective of the positivist paradigm, while most FSR attempts to operate within the holistic perspective of the same paradigm. However, holism's central tenet is rarely fulfilled in FSR. The FSR approach fits in with the historical and philosophical underpinnings of the development paradigm that, during the last 40 to 50 years, has guided the way development is conceptualized, planned and implemented[27]. Two predecessor programmes of FSR, community development and integrated rural development, suffered the same internal contradiction as FSR – the dichotomy between theory and practice, the dilemma of how to strike a balance between a holistic and a pragmatic approach. Past integrated rural development projects attempted to be so all-encompassing that they became unmanageable[28]. Conway, in 1985, said the same about communal self-help projects, which he described as exercises in social engineering[29].

As noted earlier, most FSR attempts to operate from a holistic position within the positivist paradigm. With the recent incorporation of participatory research approaches some FSR has moved towards the constructivist paradigm. Therefore, contemporary FSR must be positioned somewhere at the point of overlap between the positivist and constructivist paradigms. In resource-poor farming, however, there is no simple 'techno-fix' nor a simple 'participation-fix'[30]. Nevertheless, the gradual shift from the positivist to the constructivist paradigm is a positive development in FSR. Paradigm shifts are not a matter of replacement, but of addition and extension. The prevailing paradigm becomes a subset of the new one[31]. Notwithstanding the widespread belief in the reductionist scientific method, it is evident that a wide range of problems in the 'real world' are beyond the grasp of a complete scientific analysis[32]. The starting point for a holistic ecology must be that nature is always more complex than we, to the best of our understanding, can know, and that changes in our association with nature always will have unpredictable consequences[33]. This parallels Bawden's observation that from a holistic perspective 'surprise is anticipated, but never predictable'. The disappointing impact of agricultural science on farming systems of resource-poor farmers indicates that the reductionist scientific method has not been very effective in improving these farming systems. The long list of operational problems points to a similar bleak conclusion for FSR.

In 1990, Brouwer and Jansen argued that interdisciplinary collaboration in multidisciplinary teams is based on the presupposition that disciplinary knowledge is complementary and collaboration will result in a more complete view of reality[34]. This, they say, conflicts with the basic principle in systems theory, namely that 'emergent properties' exist: the system is more than the sum of its parts. It is unclear how multidisciplinary teams handle these emergent properties. Savory, in 1991, held a similar view when he said that 'the fact that wholes have qualities not present in their parts causes the interdisciplinary approach to fail'[35]. Only by having seen the whole, can one ask the right questions about the parts. In multidisciplinary teams with various single-discipline trained specialists, or in interdisciplinary teams with generalists trained in several disciplines, the researchers look from the outside to a whole, in our case a farming system. Approaching matters from this direction leads to confusion because the whole can never be seen from the perspective of the disciplines. We must reverse the arrows, and look outward from the perspective of the whole at all available knowledge from the various disciplines. Only the persons who are directly involved in, and manage, the whole, command the outward-looking perspective vital to their particular management needs. This puts the farm-household members centre stage, and underscores the importance of participatory approaches and indigenous knowledge. Nevertheless, it remains important, in my view, that researchers (specialists and generalists) master, to the largest extent possible, the farming systems perspective. If researchers are not able to operate within a systems perspective, requests from farmers and other stakeholders for sound advice that serves the holistic view cannot be met.

The emergent properties of farming systems only emerge when the components of a system interact. The synergetic effect of these interactions makes the farming system more than the sum of its parts. Schiere in 1995, remarks that the word holism does not necessarily imply a mystical sense, but he simultaneously speaks of the deeper sense of the word 'system' as a unit, i.e. an 'organism' with an irreducible integrity[36]. Positivist researchers do not speak about farms as 'organisms' but use the word 'systems' – a technical term for a complex biological whole[37]. The holistic argument that 'the whole is more than the sum of its parts' has a certain 'elusive' connotation. The emerging synergetic effect of interaction can be puzzling. To my mind it is evident that the dynamic and emergent nature of interactions taking place between farmers and nature, including forces which lie beyond the interface situation itself, and between numerous other actors and their networks, puts the process of rural development beyond full human control. The question is how to reduce complexity to manageable proportions without ending up in a unidimensional, positivist rationalization, which violates multiple cause and effect relationships in ecosystems. In an attempt to formulate a tentative answer I hypothesize that when the transcendentalist paradigm is brought to bear on the 'manageability' of complex situations, the far too narrow, positivist and constructivist views of human agency can be extended.

With regard to the irreducible integrity of organisms, the Gaia hypothesis holds that life itself creates the conditions for its own existence. The regulation of, for example, the earth's temperature and atmospheric composition are, in this theory, emergent properties of the system 'earth', which emerge, automatically and without any teleological plan, as a consequence of cyclical feedback mechanisms between organisms and their environment. In the end all living beings are composed of atoms and molecules, but they are not 'nothing other than' atoms and molecules. Only the non-material and irreducible organizational pattern makes them alive[38]. Most reductionist researchers do not grasp the importance of this pattern, which is lost when an organism is dissected. The non-material organizing principles that were, in the past, attributed to, for example, 'souls' are now thought of in terms of 'systems properties' or 'emergent principles of organization' or 'patterns which connect' or 'organizing fields'[39]. The question remains – what exactly are these elusive principles of organization, and do people have access to these non-material organizing principles?

The agronomic principle of input interaction can serve as a practical example of a synergetic effect: the combined effect on yield of applying several inputs jointly is greater than the sum of the effects of each applied separately. On the one hand, there is little mysterious about such interaction effects: the outcomes of these mutually reinforcing interactions can be logically explained. On the other hand, however, unexpected synergisms often occur: the outcomes of interactions are not always predictable, because of the large number of factors that can be involved. Evans, in 1993, for example, pointed out that the yield improvements from the last few decades were due to often unexpected synergisms between agronomy, plant breeding, fertilizers, pesticides, fungicides and herbicides[40]. Another (non-agricultural) example is a football team: sometimes a team performs well and the whole is more than the sum of the parts, another time performance is moderate and the team is 'just' the sum of the parts. What causes the 'magic' of holistic performance, how does effective and well-timed interaction occur, what causes the synergetic effect of interaction among players, what makes a collection of 11 individuals an 11-headed unit, a true team, rather than an aggregate of 11 individuals? Similarly, interdisciplinarity in multidisciplinary FSR teams can emerge with the synergetic effect of interaction among team members: the team functions then as a synergic whole. The question is how to create synergy in a systematic way?

Schroevers, in 1984, and Van Asseldonk, in 1987, were among those emphasizing the importance of holism as a scientific paradigm and methodology, and not only as a general philosophy of life or an article of faith[41]. Van Asseldonk distinguished between holistic and reductionist generality. Holistic generality is an approach in which agriculture is seen as a 'whole' and problems are tackled in an integrated way without splitting them up in subproblems to be covered by specialized disciplines. Reductionist generality is the multidisciplinary integration in retrospect of subsolutions developed by specialists. It is important to know whether knowledge about the cohesion of a system can be obtained by means of integration in retrospect or by means of an integral approach: when both approaches can yield this knowledge, then the choice between reductionism and holism is no longer a fundamental issue.

Koningsveld, in 1986, distinguishes two types of problems in his analysis of conventional agricultural science: problems as anomalies and problems as crisis situations[42]. A problem is an anomaly when it can be solved with available, time-tested conceptual means and with the standard technical approach, although sometimes new instruments must first be developed. A problem is a 'borderline' problem or crisis situation when it cannot be solved with available conceptual means, but requires a fundamental theoretical innovation in the conceptual framework of agricultural research, and a new problem solving method. In a crisis

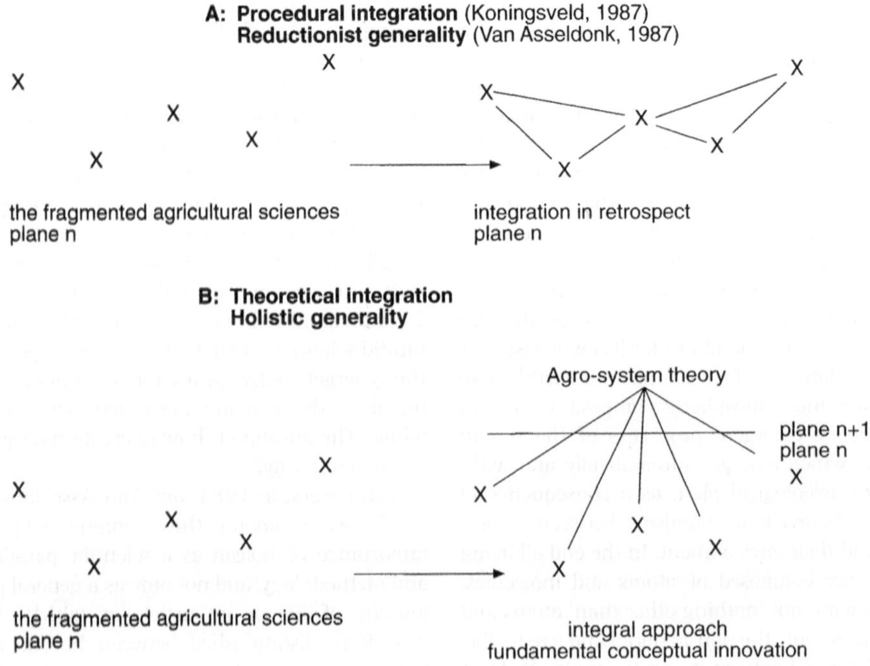

Fig. 11.1.1. Holism in agricultural science.

situation the limits of the old theory become more and more visible: the problem is exceeding these limits. In 1987, Koningsveld spoke of procedural integration, when problems in agriculture are interpreted as anomalies, and a systems approach is used as a formal methodology (often a mathematical instrumentarium) to integrate the fragmented agricultural sciences[43]. Such a systems approach does not add much to conceptual innovation: it is just a methodological tool to integrate parts of a whole. This is the case in most positivist 'hard' systems research, in which so far mainly technical disciplines and economics play a role. Also most FSR attempts to operate within the holistic perspective of the positivist paradigm (albeit mainly without formal modelling exercises). In practice, however, most FSR boils down to a mere linking up of a limited number of disciplines, mainly agronomy and agricultural economics: it is at best integration in retrospect of subsolutions developed by discipline-oriented scientists, which does not yield more knowledge than the sum of the parts. Joint reports by FSR teams are often just the sum of the parts.

In situation A of Fig. 11.1.1, at best, integration in retrospect of subsolutions developed by discipline-oriented specialists takes place. In situation B a new theory about the system agriculture, at a level above the fragmented agricultural sciences, is developed.

When fundamental conceptual innovation is at stake, Koningsveld speaks theoretical integration. It entails the formation of a theoretical framework at a level above the fragmented agricultural sciences in order to encompass agriculture in its totality; allowing for an integration of the contents of the fragmented agricultural sciences. The agricultural system in its totality is more than the sum of the parts studied by the fragmented agricultural sciences, so that for an adequate study of the whole, concepts of a higher level must also be developed.

The reductionist and holistic generality, as formulated by Van Asseldonk, and Koningsveld's procedural and theoretical integration, are clarified in Fig. 11.1.1. In my view the long list of operational problems and the hitherto limited impact of FSR are signs of, what Koningsveld calls, phenomena signalling a crisis: a crisis which demands fundamental conceptual innovation and a new problem solving method.

11.1.4 The transcendentalist paradigm

The frequently difficult collaboration between natural and social scientists in FSR is due to the fact that most natural scientists work from the positivist perspective, while many social scientists are based in the constructivist paradigm. Each paradigm occupies its own niche, but integration of the two paradigms requires an understanding of their philosophical base, and an open-minded attitude. An emerging third paradigm, the transcendentalist paradigm, can facilitate the integration of the previous two. This encompasses the earlier paradigms in the sense that it is a hybrid of the natural and social sciences, and techniques for consciousness development. It focuses on the underlying basis of human activity, i.e. consciousness, and on a transformation of attitudes, 'the great blind spot of agricultural research and extension' according to Pretty & Chambers in 1994[44].

The FSR experience shows a gross underestimate of how difficult it is to change attitudes of scientists and extensionists. When more than two decades of fostering interdisciplinary collaboration in relatively small and permanent multidisciplinary FSR teams have been rather unsuccessful, how do we then 'enforce' group synergy on social platforms with a multitude of actors of different walks of life? At higher levels of social aggregation the task will become increasingly difficult.

The holistic aspect of farming systems, their irreducible integrity, is 'intangible' in the sense that it is incomprehensible: it is beyond the discursive intellect. The number of variables that are at play in agriculture makes it difficult to grasp the complexities of farming systems at the intellectual level. Conventional FSR suffers from the 'illusion of intellectual holism'[45]. Although the development process is beyond full human control, we might increase our 'steering capacity' by a new paradigm of development that pays attention to the underlying base of the multitude of interfaces and interactions.

Earlier on I spoke of a non-material, organizational pattern that underlies organisms. The Indian philosopher Maharishi Mahesh Yogi called this organizational pattern 'the field of creative intelligence', a field that underlies all nature, including people[46]. People have access to this field through their own consciousness, when they experience 'the field of transcendental consciousness' they are at home in the field of creative intelligence. In the view of Maharishi a level of pure or transcendental consciousness, a consciousness-as-such without any content of consciousness, exists. Through meditation techniques the mind can be trained to 'transcend' the subtlest stage of thinking until one reaches this level of pure consciousness. The 'field of transcendental consciousness' and the 'field of creative intelligence' are identical, this field is the source of 'subjective' as well as 'objective' existence. It is the basis of all creation and evolution. In the course of human history this field has been given numerous names: in theistic traditions one refers to God(s), while in non-theistic traditions one postulates, for example, a non-local 'Tao'.

The emergent properties of organisms that emerge as a consequence of cyclical feedback mechanisms between organisms and their environment, the autonomous self-organization of organisms, the autonomy of natural processes and their triggers and feedback, are, in my view, 'produced' by the underlying 'field of creative intelligence'. The underlying base of the multitude of interfaces and interactions among social actors, and between these actors and nature, is the field of creative intelligence or transcendental consciousness.

My hypothesis is that regular access to the field of transcendental consciousness guides attitudes and behaviour in a societally and environmentally friendly direction. Regular access to this field can be obtained through, for example, meditation techniques. Extensive scientific research on the transcendental meditation technique shows that the individual and collective effects of transcendental meditation are beneficial and societally favourable. The ultimate objective of participatory approaches is synergistic performance of a multitude of actors. Synergy emerges 'when certain conditions prevail', but hitherto these conditions have not been sufficiently specified. In the perspective of the transcendentalist paradigm it is the agency of the field of transcendental consciousness that facilitates the management of the multiple aspects of sustainable development. Language-mediated interaction must be supported by consciousness-mediated interaction.

The process in which one systematically trains the receptivity to gain regular access to transcendental consciousness can be labelled spirituality. In order to create sustainable farming systems I recommend a sustained use of the critical intellect in combination with an experiential, non-dogmatic spirituality. A spirituality that highlights personal transformation through do-it-yourself techniques. A spirituality that refers to the original meaning of religion, i.e. *religare, religio*: to (re)connect (to the field of transcendental consciousness). In addition to the outward-oriented approaches of the positivist and constructivist paradigms, I recommend an inward-oriented approach which focuses on consciousness development.

REFERENCES

1. **Fresco, L.O., 1986**. Cassava in Shifting Cultivation. A Systems Approach to Agricultural Technology Development in Africa. Royal Tropical Institute, Amsterdam.
   Röling, N., 1988. Extension Science. Information Systems in Agricultural Development. Cambridge University Press, Cambridge.
2. **Stoop, W.A., 1987**. Issues in implementing research with a farming systems perspective in NARS. Working Paper no. 6. ISNAR, The Hague.
3. **Plucknett, D.L., J.L. Dillon & G.J. Vallaeys, 1987**. Review of concepts of farming systems research: the what, why and how. In: International Agricultural Research Centres. Proceedings of the Workshop on Farming Systems Research, Feb. 1986, ICRISAT, pp. 8–15. ICRISAT, India.
4. **Merril-Sands, D., 1986**. Farming systems research: clarification of terms and concepts. Expl Agric., 22, 87–104.
   Norman, D., J.D. Siebert, E. Modiakgotla & F.D. Worman, 1994. Farming Systems Research Approach. A Primer for Eastern and Southern Africa. FAO/SIDA Farming Systems Programme, Gaborone, Botswana.
5. **Van Eijk, A.M., 1998**. Farming systems research and spirituality. An analysis of the foundations of professionalism in developing sustainable farming systems. PhD thesis, Wageningen Agricultural University, The Netherlands.
6. **Woolley, J. & R. Tripp, 1994**. Priority setting tools for research groups. In: Meindertsma J.D. (Ed.) Setting Research Priorities. Towards Effective Farmer-oriented Research. Royal Tropical Institute, Amsterdam, pp. 47–64.
   Anandajayasekeram, P., 1995. Farming systems research: concepts, procedures and challenges. Key note address presented at the Fourth Regional Conference of the Southern African Association for FSR-E, 2–4 October 1995, Harare, Zimbabwe.
7. **Brouwer, R. & K. Jansen, 1989**. Critical introductory notes on farming systems research in developing Third World agriculture. Systems Practice, 2 (4).
8. **Anandajayasekeram, P., 1995**. (Op. cit.).
9. **Van Leeuwen, J., 1988**. On-farm trials in farming systems research. In: An Introduction to Farming Systems Analysis. Department of Tropical Crop Science, Wageningen, Agricultural University.
10. **Norman, D., 1994**. The research-research interface. In: Meindertsma, J.D. (Ed.) Setting Research Priorities. Towards Effective Farmer-oriented Research. Royal Tropical Institute, Amsterdam, pp. 32–46.
11. **Rhoades, R.E., 1994**. Farmer participation in priority setting. In: Meindertsma, J.D. (Ed.) Setting Research Priorities. Towards Effective Farmer-oriented Research. Royal Tropical Institute, Amsterdam, pp. 65–76.
12. **Fresco, L.O., 1986**. (Op. cit.).
    Huijsman, A. & A. Budelman, 1996. A crossroads for agricultural R & D: merging approaches on behalf of sustainable development. In: Budelman, A. (Ed.) Agricultural R & D at the Crossroads. Merging Systems Research and Social Actor Approaches. Royal Tropical Institute, Amsterdam, pp. 9–17.
13. **Brouwer, R. & K. Jansen, 1990**. (Op. cit.).
14. **Sevilla Guzmán, E. & ISEC Team, 1994**. The role of farming systems research/extension in guiding low input systems towards sustainability: an agroecological approach for Andalusia. In: Dent, J.B. & M.J.

McGregor (Eds) Rural and Farming Systems Analysis. European Perspectives. CAB International, Wallingford.

15. **Simmonds, N.W., 1985**. Farming Systems Research. A review. World Bank Technical Paper no. 43. The World Bank, Washington, DC.
16. **Bawden, R., 1995**. On the systems dimension in FSR. Journal for Farming Systems Research- Extension, 5 (2), 1–18.
17. **Maini, J.S., 1987**. Review. Session 5: Integration of crop/livestock/agroforestry and other land-use systems. In: International Agricultural Research Centers. Proceedings of the Workshop on Farming Systems Research, Feb.1986, ICRISAT, pp. 100–2. ICRISAT, India.
18. **Holling, C.S., 1995**. What barriers? What bridges? In: Gunderson *et al.* (Eds) Barriers and Bridges to the Renewal of Ecosystems and Institutions. Colombia Press, New York, pp. 3–37.
19. **Anandajayasekeram, P., 1995**. (Op. cit.).
20. **Van Eijk, A.M., 1998**. (Op. cit.).
21. **Byerlee, D., L. Harrington & D.L. Winkelmann, 1982**. Farming Systems Research: issues in research strategy and technology design. American Journal of Agricultural Economics, 64, 897–904.
22. **Horton, D.E., 1984**. Social scientists in agricultural research: lessons from the Mantaro Valley Project, Peru. IDRC-219e, Ottawa.
23. **Stoop, W.A., 1987a**. Organization and managerial implications of a farming systems approach for national agricultural programs. In: International Agricultural Research Centres. Proceedings of the Workshop on Farming Systems Research, Feb. 1986, ICRISAT, pp. 72–6. ICRISAT, India.
24. **Pretty, J. & R. Chambers, 1994**. Towards a learning paradigm: new professionalism and institutions for a sustainable agriculture. In: Scoones, I. & J. Thompson (Eds) Beyond Farmer First. Rural People's Knowledge, Agricultural Research and Extension Practice. Intermediate Technology Publications, London, pp. 182–202.
25. **Bawden, R. 1995**. (Op. cit.).
26. (Ibid.).
    Checkland, P.B., 1981. Systems Thinking, Systems Practice. J. Wiley, Chichester.
    Röling, N., 1994. Platforms for decision making about eco-systems. In: L.O. Fresco *et al.* (Eds) Future of the Land: Mobilising and Integrating Knowledge for Land Use Options. John Wiley, Chichester, pp. 386–93.
    Engel, P.G.H., 1995. Facilitating innovation: an action-oriented approach and participatory methodology to improve innovative social practice in agriculture. PhD dissertation, Wageningen Agricultural University.
27. **Jamieson, N., 1987**. The paradigmatic significance of rapid rural appraisal. In: Khon Kaen University. Proceedings of the 1985 International Conference on Rapid Rural Appraisal. Rural Systems Research and Farming Systems Research Projects, Khon Kaen, Thailand.
28. **Horwith, B.J., P.N. Windle, E.F. MacDonald, J.K. Parker, A.M. Ruby & C. Elfring, 1989**. The role of technology in enhancing low resource agriculture in Africa. Agriculture and Human Values, VI (3), 68–84.
29. **Conway, G.R., 1985**. Agricultural ecology and farming systems research. In: Remenyi, J.V. (Ed.) Agricultural Systems Research for Developing Countries. Proceedings of the International Workshop, Hawkesbury Agriculture College, Richmond, Australia, pp. 43–59.
30. **Scoones, I. & J. Thompson, 1994**. Knowledge, power and agriculture – towards a theoretical understanding. In: Scoones, I. & J. Thompson (Eds) Beyond Farmer First. Rural People's Knowledge, Agricultural Research and Extension Practice. Intermediate Technology Publications, London, pp. 16–32.
31. **Röling, N., 1995**. Naar een interactieve landbouwwetenschap. Inaugurele Rede, Landbouwuniversiteit Wageningen.
32. **Chalmers, A., 1990**. Science and its Fabrication. Open University Press, Ballmoor, Buckingham.
    Funtowicz, S.O. & J.R. Ravetz, 1994. The worth of a songbird; ecological economics as a post-normal science. Ecological Economics, 10, 197–207.
33. **Schroevers, P., 1984**. Inhoud en betekenis van een holistisch natuurbeeld. In: Achterberg, W. & W. Zweers (Eds) Milieucrisis en Filosofie. Westers Bewustzijn en Vervreemde Natuur. Ekologische Uitgeverij, Amsterdam.
34. **Brouwer, R. & K. Jansen, 1990**. FSR: model en werkelijkheid. In: Brouwer & Jansen (Eds) Het Systeem, de Som en de Delen: 5 Opstellen Over FSR. Studium Generale-LUW. SG-paper 115, Wageningen.
35. **Savory, A., 1991**. Holistic Resource Management. Southern Africa Edition. Gilmour Publishing, Harare, Zimbabwe (first published in 1988 by Island Press, USA).
36. **Schiere, J.B., 1995**. Cattle, straw and system control. A study of straw feeding systems. PhD thesis, Wageningen Agricultural University, KIT Press, Amsterdam.

37. **Koepf, H.H., 1989.** The Biodynamic Farm. Anthroposophic Press, Hudson, New York.
38. **Capra, F., 1996.** Het Levensweb. Levende Organismen en Systemen: Verbluffend Nieuw Inzicht in de Grote Samenhang. Kosmos-Z & K Uitgevers, Utrecht, The Netherlands (Dutch translation of 'The Web of Life').
39. **Sheldrake, R., 1989.** The Presence of the Past. Morphic Resonance and the Habits of Nature. Vintage Books, New York.
40. **Evans, L.T., 1993.** Crop Evolution, Adaptation and Yield. Cambridge University Press.
41. **Schroevers, P., 1984.** (Op. cit.).
Van Asseldonk, J.S.O., 1987. Holisme en reductionisme in de landbouwwetenschap: Samen meer dan de som der delen. Landbouwkundig Tijdschrift, 99 (9).
42. **Koningsveld, H., 1986.** Landbouwwetenschap als systeemtheorie? Landbouwkundig Tijdschrift, 98 (12).
43. **Koningsveld, H., 1987.** Fragmentatie en systeembenadering in de landbouwwetenschappen. In: Koningsveld, *et al.* Landbouw, Landbouwwetenschap en Samenleving: Filosofische Opstellen. Mededelingen van de Vakgroepen Voor Sociologie, 20. Landbouwuniversiteit, Wageningen.
44. **Pretty, J. & R. Chambers, 1994.** (Op. cit.).
45. **Van Eijk, A.M., 1988.** (Op. cit.).
46. **Maharishi Mahesh Yogi, 1968.** Transcendental Meditation. Signet Book, New York (Original title: The Science of Being and Art of Living, first published in 1963).
Maharishi Mahesh Yogi, 1969. On the Bhagavad-Gita. A new translation and commentary, Chapters 1–6. Penguin, UK (first published: 1967).

11.2 THE GIS AND REMOTE SENSING CONTRIBUTION TO THE ELABORATION OF SYSTEM HIERARCHIES IN FSR

Evaristo Miranda

Increasingly, the challenge of relating agriculture and farming to different land uses or landscape categories can be attacked using GIS and RS in both research and management.

11.2.1 Introduction

Land use and changes in land use are critical elements in FSR. Rural areas are markedly heterogeneous and many factors drive land use changes in both time and space[1]. In the agricultural frontier areas of the developing countries change is particularly intense and, in addition, one type of land use (annual crops, permanent crops, grazing lands, forestry) may be a feature of several different production systems within the landscape. This spatial and temporal diversity often precludes the application of the same technologies across the land-use type as a whole. To add a further dimension, technologies which are useful on an individual farm may be harmful when used more widely.

As a result, three important questions arise for FSR:

- How to establish the spatial distribution of production systems' at different scales?
- How to characterize the link between individual farming systems and different uses of land?
- How to evaluate the sustainability of the diversity of farming systems and the interactions between them in the landscape?

Historically, FSR has concentrated at the farm level of what is in fact a hierarchy of systems. Analysis and evaluation have widened recently, helped by new research tools. GIS and remote sensing (RS) techniques are proving valuable in establishing and monitoring systems hierarchies in which the farm is one important level. This contribution summarizes recent developments in GIS and RS methods, using examples from applications in the FSR-based systems hierarchy in Brazil. The methods were developed in a research programme executed by the NGO ECOFORCE (Research and Development @ http://www.ecof.org.br) with the technical and scientific collaboration of the

Environmental Monitoring Center (NMA-EMBRAPA, http://www.nma.embrapa.br) in the county of Campinas, in São Paulo State, Brazil.

11.2.2 Complexity and systems hierarchies

It has been acknowledged, historically, that farms are complex systems. FSR has tried to manage this complexity through its conceptual frameworks and models. This understanding of complexity is related to many factors; the non-linearity and asymmetry of the relationship among the components and its structure; the different levels of organization and constraints; the simultaneous existence of functional and structural boundaries; the level of uncertainty of systems indicators; the permanent state of evolution in systems' components and interactions between them; varied sources of perturbations which destabilize a range of parameters; the spatial diversity of the landscape and the interactions between the social, economic and ecological systems.

While the concept of systems hierarchies is helpful in understanding these sources of complexity, the establishment of hierarchical levels in FSR cannot be an arbitrary process. Research and rural development programmes[2] demonstrate a continuum of levels, their interactions and linkages. Scale is a central concern to those modelling dynamic multivariate structures that respond to the interactions between many levels of organization[3]. Generally speaking, FSR hierarchical levels are associated with spatial scales or organizational levels in agricultural production[4]. In FSR the early notion of hierarchy involved discrete levels: field, farm, watershed, valley, county. Hart, in 1985, presented a series of concepts on agroecosystems, based on this kind of hierarchical analysis[5]. His study led to many FSR applications in Latin America. Many researchers have considered the farming system hierarchy as nested, which requires that upper levels contain lower levels in a continuum of structures. However, research results can rarely be generalized, aggregated or disaggregated, from one level to another. In fact, in general, hierarchies which include farm systems are non-nested, with strong interactive tendencies. The levels are frequently a more convenient focus for research than the hierarchy, despite its explanatory power.

Computers are able to display structures in hierarchical levels without human value judgements[6]. In exploring typologies of farming systems, a process using clusters and multivariate analysis, employing only numerical criteria given by data sets is now well known. Some authors, such as Simon in 1962, suggest that hierarchical structure itself is a consequence of human observations[7]. Without raising questions of ontological reality, it seems fundamental for given levels in FSR to take account of hierarchical continuity and cohesiveness. Scale is one continuously varying function that can describe the continuum of levels and their interactions[8]. The available methods try to elucidate the process and the critical parameters for the different spatial scales and landscape units (land, property, hydrographic basin, community microregions, regions and county) using GIS and RS tools[9].

This contribution shows how GIS and RS are increasingly used in a complementary way to simulate farming systems strategies at the micro level; technologies for agriculture production systems, and at macro levels; public policy, politics and land use. Ongoing developments are improving the resolution available at several hierarchical levels. Increasingly the challenge of relating agriculture and farming to different land uses or landscape categories can be attacked using GIS and RS in both research and management[10].

11.2.3 GIS and RS: new techniques in FSR

GIS and RS techniques are proving their worth in helping to establish and monitor systems hierarchies in FSR. GIS offers sophisticated spatial analyses of the numerical descriptors of the farming system. Whereas FSR developed complex numerical models, GIS spatial analysis of production systems is limited to farm fields. The spatialization of productivity variables, of farm system typology parameters, or of the system's environmental impact on a given resource at several hierarchical levels (e.g. field, farm, groups of farms or region) widens our understanding and opens new horizons for FSR. GIS allows area, perimeter and volume calculations and a series of basic operations for quantifying the spatial expression of variables. GIS also allows qualitative spatial analysis, such as diversity, proper or improper land use, the simulation of alternative uses, the interactions between different uses and the probable impact of new agricultural technologies on the environment[11].

Spatial analysis of systems can occur at different hierarchical levels, however, it often demands information that is not readily available and sometimes does not even exist. Recently, the evolution of RS has made available series of spatial data on existing production systems and land use which have helped fill many of these data gaps. In the developing countries RS is frequently the only way to get these data due to the lack of census data[12] and the difficulty in reaching some rural areas[13]. Satellite imagery gives the researcher the means to evaluate land use and changes in use[14]. It also allows the researcher to detect the use of some technologies, particularly in soil conservation[15] and, importantly, to relate land uses and vegetation behaviour[16].

The terrestrial monitoring satellites represent an efficient instrument to characterize land use, measuring the spatial distribution of farms and land use in a very precise way[17]. Some farming systems can be identified from orbital images and the uses of RS in FSR have been increasing with the development of new imaging softwares and more sensitive sensors and satellites. In 1996 satellites can already observe detail smaller than $50m^2$.

11.2.4 Remote sensing and FSR

The first LANDSAT satellite, originally called ERTS-1, was developed and launched by NASA in July 1972. Today about 300 satellites are available to monitor terrestrial ecosystems, agriculture and changes in land use. The interest in the use of RS in FSR is linked to three properties of orbital images: spatial resolution, temporal resolution and radiometric or spectral resolution.

Spatial resolution

This is important to the study of farming system based hierarchies. The orbital digital data's plasticity allows works at different spatial scales[18]. Agriculture can be analysed in different perception or hierarchical levels (local, microregional, regional, national), and each perception level can be at least partially associated with a cartographic scale in spatial terms. Local studies range from 1 : 1000 to 1 : 10,000, microregional studies from 1 : 25,000 to 1 : 100,000. Regional studies generally work with scales ranging from 1 : 100,000 to 1 : 250,000 and national or macroregional studies sometimes use spatial scales smaller than 1 : 1,000,000. The same image can be analysed from 1 : 1,000,000 to 1 : 50,000. Recently there has been remarkable development in the spatial resolution and scales of 1 : 25,000 and 1 : 10,000 can now be obtained, for example, from the IRS-C (India), SPOT 4 and 5 (France) and ORBVIEW (USA) satellites.

The LANDSAT TM image, used since 1985, covers an area of 34,000 km^2 approximately and has a 30-m-pixel resolution. Agriculture can be studied hierarchically from scales based on the LANDSAT images which extend from 1 : 1,000,000 to 1 : 50,000. The French satellite SPOT 3 has a 10-m resolution and the Indian satellite IRS-1C has a 6-m resolution. Both these satellites offer stereoscopic views and their images are already available. The next generation of satellites will be even more accurate, providing resolution between 10 and 100 times better than the existing commercial satellites, formerly available merely as expensive aerial photos. As an example, the panchromatic sensor of the satellites QuickBird and OrbView will have a 1-m resolution at nadir and the multi-colour sensor will have a 4-m resolution[19].

The improving spatial resolution of the imagery allows ever better sampling plans in FSR. The distribution of land owners, the land uses and their localization can be mapped *a priori*. This is particularly important in areas where censuses are insufficient or non-existent. It is also vital in expanding agricultural frontiers such as the Amazon, or in areas where agriculture has a strong spatial dynamic, frequently expanding and contracting[20].

Temporal resolution

This defines the frequency of repetition in the image's coverage at a same point: 16 days for the LANDSAT, 23 for the SPOT. Remote sensing satellites are able to provide a monthly monitoring of agricultural and land use systems. If different orbital systems are combined, for example, LANDSAT, SPOT and IRS, a weekly monitoring can be obtained. At a finer level the NOAA/AVHRR satellites provide information on temperature[21], drought, fires and burnings[22], soils moisture[23] and vegetation activity at least four times a day. Thus orbital images can help monitor nutritional stress in vegetation, irrigation efficiency and even pest attacks.

In the last few years, the time between image acquisition by the satellite and availability to the user has been reduced to 1 or 2 months. The next generation of commercial high-resolution satellites will reduce this time still further as the images will be made available through electronic networks within hours. Several orbital systems have been working since the 1970s. The images obtained are preserved on files and made available through networks. This allows the reconstitution of land use evolution and the monitoring of dynamic phenomena, like deforestation[24], erosion, the expansion of the cultivated area and salinization.

Spectral resolution

This is defined by on-board instrument bands. The instruments' spectral range includes the panchromatic (PAN), the visible and the near infrared (VNIR), the short wave infrared (SWIR), the multiband thermal infrared (TIR) and the synthetic aperture radar (SAR). Satellites do not take pictures, but generate images. Those images are digital and can be processed digitally. Each part of the spectral range 'recognizes' different surface elements such as soil, humidity, vegetation, dust, etc. The combination of the several spectral bands through mathematical and statistical algorithms allows the identification and qualification of diverse cultures and the different kinds of land use[25].

The discriminatory power of the images is greatly superior to that of the human eye. While the eye distinguishes an average of 20 grey tones, hundreds of grey tones can be identified on a satellite image. This, for example, makes it possible to identify irregularities in photosynthetical activity that would be invisible to the human eye. Several vegetative stages can be identified on the same kind of plantation. Phytomass and productivity levels can be evaluated, on pastures, sugar cane and cereal fields. Different soybean varieties have been distinguished on orbital images, due to their differences in height and the insertion angle of the leaves. In the microwave field, the radar sensors allow imaging during the night and under any weather conditions[26]. In humid tropical regions with frequent cloud cover the radar images are of great help.

11.2.5 The use of GIS in FSR

The agricultural production cycle rarely corresponds to the time scales for the evaluation of environmental phenomena (pedogenesis, morphogenesis, loss of fertility, land compactness, acidification, biodiversity reduction, river obstruction). There has been little integration of environmental phenomena in the reconstitution of the history of production systems, or in the modelling of new ones. Strict numerical models are inadequate to show spatial realities. The spatial and temporal dynamics of land use, either on the farm or at regional level, are good examples of crucial hierarchical issues that are difficult to resolve without the use of cartographic methods. In FSR a spatial view can be acquired through the use of GIS.

GIS was born as a way to digitize cartography. Linked to the numeric data bank, a GIS is an efficient tool to characterize the spatial division of a great number of phenomena and their dynamics. GIS allows spatial analysis to be linked to maps according to rules archives/files, equations or logical sequences[27] and can then create new maps showing agricultural production cycles and the environmental impacts related to those cycles[28].

GIS have a different mathematical structure from the satellite images treatment systems, but there are interfaces between both. Cartographic data can be digitally confronted with the orbital images and vice versa. A great deal of software uses GIS in a variety of applications, the thematic and cartographic precisions vary for different studies. Many GIS are in the public domain and some versions can be very expensive to use. A GIS surrounding is convenient enough to analyse land-use maps of different periods and articulate them with the production systems.

Several routines of agricultural zoning are currently operational in GIS[29]. Land use planners are increasingly using it and, in FSR, it is contributing to sample planning, extrapolation of data and the multivariate and multilocational analysis of production[30].

11.2.6 GIS, RS and system hierarchies: an example

The work described in this contribution was carried out in an area of approximately 800 km^2 in Campinas county, São Paulo State, Brazil. The data were obtained during a multidisciplinary research project which included the survey of a sample of 100 small farms. The environmental and land use characterization was supported by RS and the integration of these cartographic results with the FSR survey data on the production systems was accomplished through GIS. The project used the 2.4 version of the GIS (GIS 2.4) of the National Institute of Spatial Researches (INPE). Four main methods based on GIS and RS use were developed and validated: mapping land-use capacity, characterization of present land-use systems, relationships between farming systems and land use and a hierarchical evaluation of agriculture environmental impacts.

Over 100 thematic and synthetic maps and orbital images were analysed and treated. Several themes were analysed at different hierarchic levels: farm, farm groups, watershed and county. It is impossible to reproduce these maps and images here, but the main methodological and operational results are discussed. These results were the object of several publications and are available on the Internet (at the URL, http://www.nma.embrapa/projetos/cmp/gis.html/).

Mapping land use capacity

There are several analogue methods of calculating land use capacity. The method used by FAO was adapted for GIS. The main steps developed and validated in the process were:

- Digitalization of the county limit.
- Digitalization of the contour curve map.
- Digital generation of the hypsometric map by the GIS.
- Digital elevation model (DEM) generation.
- Declivity map digital generation, by the GIS.
- Hydrographic map digitalization.
- Basins and sub-basins map generation and digitalization.
- Generation and adjusts of the pedological map, by fieldwork.
- Pedological map digitalization.
- Digital generation of the erodibility map, by the GIS.
- Digital generation of the hydric availability map, by the GIS.
- Digital generation of soils' phosphorus fixation capacity, by the GIS.
- Digital generation of free aluminum toxicity, using the pedology.
- Digital generation of the interchangeable bases availability map, by GIS.
- Digital generation of a chemical fertility map, using the pedology base.
- Constitution of an integration programme for generation of a land use capacity map.
- Digital generation of the agricultural land use capacity map.

Characterization of present land use

The characterization of present land use was carried out using satellite multispectral images (LANDSAT TM 5 and SPOT) and IBGE's (Brazilian Institute of Geography and Statistics) topographic charts in combination with fieldwork. The categories of land use were defined in the county. In the Campinas case 17 categories were identified.

1. Urban areas.
2. Water (lakes, irrigation dams and rivers).
3. Natural forests and woodlands.
4. Riparian forests.
5. Savannas.
6. *Capoeiras* (deforested areas, secondary vegetation).
7. *Pinus* plantations.
8. Eucalyptus plantations.
9. Sugar cane.
10. Citrus.
11. Coffee.
12. Fruit plantations.
13. Annual crops.
14. Natural pastures.
15. Artificial pastures.
16. Vegetable gardens.
17. Others (roads, rock, mines).

Digital classification methods were used to interpret the satellite images and the results incorporated in GIS. This information can be extracted at several hierarchic levels: farm, farm groups, basin or sub-basin, region or county. The main methodological steps developed and validated were:

- Preliminary definitions of the agricultural and non-agricultural land use categories in Campinas county.
- Images acquisition of the SPOT and LANDSAT TM satellites.
- Digitalization of Campinas county boundaries in GIS, and extraction of its area from the images.
- Migration of the corresponding digital records.
- Preliminary digital treatment of the satellite images to differentiate and limit the main land use categories.
- Terrestrial verifications in a diffused and concentrated way.
- Map making of agricultural land uses in a definite way.
- Vectoring the land use map and entry to GIS.

The relationships between farming systems and land use

Each type of land use can contain one or more production or farming systems. The existing production systems were identified and their technical coefficients quantified and the relationships between the use and production systems were established at a variety of hierarchical levels: camp, farm, farms groups, basin, region and so on. Every farm was geocoded in GIS and a data bank was created, articulating the land use and the production systems. The main methodological steps were:

- Land use inventory and elaboration of hypothesis about the existing variability between land uses and production systems.
- Preliminary identification of the main production systems, in relation to agricultural land use.
- Acquisition of a 100-farm sample using aleatory stratified sampling techniques.
- Hypotheses verification about the relationships between the production systems and the uses from the camp survey.
- For applications including more than one production system, definition of a complementary farm sample.
- Preliminary map making of spatial division of the main production systems interaction with land use.
- Farming systems technical coefficient quantification from the 100-farm survey.
- Final evaluation of the variability of main land uses across production systems.
- Technical coefficient quantification and assessment of the possible environmental impacts of the production systems.
- Data bank constitution, for 1 ha of each type of use identified in the fieldwork complemented with bibliographical data and discussions with researchers.
- Creation of a data bank of technical coefficients, production systems and possible current environmental impacts.
- Adequate polygon labelling of the present land use map to allow their association with the data bank (GIS).

Hierarchical evaluation of agricultural environmental impacts

The environmental impact map of the agricultural activities, based on GIS, was made by linking the survey data from the farming systems and the cartographic bases. The goal was to evaluate the impact of agricultural activities on the land, air, surface waters, fauna and natural vegetation. The impact of agricultural inputs was also evaluated. A final synthesis was drawn up for the ecosystems. The more recent data linking routines via GIS enables the evaluations for several hierarchic levels: farm, farm groups, basin, sub-basins, regions, counties or any other desired area. First, through GIS, the following maps are produced at several hierarchical levels:

- Nitrogen use (kg N ha^{-1} $year^{-1}$).
- Herbicide use (l ha^{-1} $year^{-1}$).
- Pesticide (l ha^{-1} $year^{-1}$).
- Synthesis map about chemical inputs impact.
- The use of burning for agriculture.
- Soil compactness.
- Run-off map.
- Soil loss.
- Land use stability.
- Land exposure period per year or bareness.

Second, the following synthesis maps are produced at several hierarchic levels:

- Farm system environmental impact on the land.
- Farm system environmental impact on the water.
- Farm system environmental impact on the vegetation.
- Farm system environmental impact on the air quality.
- Farm system environmental impact on the non-biotic systems.
- Farm system environmental impact on the biotic systems.

11.2.7 Conclusion

FSR's recent history shows new research themes and methods emerging. System hierarchies are a useful operational device to help deal with the complexity of farming systems. They can be addressed through GIS and RS which can be used independently or together in FSR. The results of the ECOFORCE and NMA/EMBRAPA application in Brazil demonstrate that GIS and RS can contribute to the improved elaboration of system hierarchies in FSR, reducing the costs and time required. The nature of the RS data allows the scale theme to be treated as a continuum across several hierarchical levels. At the same time thematic complexity gets adequate spatial treatment through the GIS. GIS and RS can contribute to FSR, from sampling to the extrapolation of results across space. New sensors are providing unprecedented data, with detail to 1 m and wide spectral resolution, and will widen GIS's application to FSR. The great challenge is not in GIS and RS development anymore but perhaps on the researchers' willingness to incorporate these techniques into their FSR toolbag.

REFERENCES

1. **Nassauer, J.I., 1995**. Culture and changing landscape structure. Landscape Ecology, 10(4), 229–37.
2. **Berdegué, J.A. & E. Ramírez, 1995**. Investigación con Enfoque de Sistemas en la Agricultura y el Desarrollo Rural. RIMISP, Santiago, Chile.
   Berdegué, J.A. & E.E. de Miranda, 1991. Assesment of sustainable land-use systems research in South America. In: Sustainable Land Use Systems Research, 12–16 Feb. 1990, New Delhi, India. Proceedings. India: Rodale Institute-USDA-India Council of Agricultural Research, pp. 201–18.
3. **Allen, T.F.H. & T.B. Starr, 1982**. Hierarchy: Perspectives for Ecological Complexity. University of Chicago Press, Chicago.
4. **de Miranda, E.E., 1985**. De Diferenciação camponesa e tipologias de produtores (Município de Euclides da Cunha). EMATER-BA/Embrapa-CPATSA, Salvador, BA.
5. **Hart, R.D., 1985**. Agroecosistemas; Conceptos básicos, 2nd edn. CATIE, Turrialba, Costa Rica.
6. **Gauch, H.G. & R.N. Whittaker, 1972**. Comparison of ordination techniques. Ecology, 53, 868–75.
7. **Simon, H.A., 1962**. The architecture of complexity. Proceedings of the American Philosophical Society, 106, 467–82.
8. **Forman, R.T.T. & M. Godron, 1986**. Landscape Ecology. John Wiley, New York.
9. **Davidson, D.A., S.P. Theocharopoulos & R.J. Bloksma, 1996**. A land evaluation project in Greece using GIS based on Boolean and fuzzi set methodologies. International Journal of Geographical Information Systems, 10(8), 369–84.
10. **de Miranda, E.E., C.O. Mattos & J.A.C. Mangabeira, 1995**. Na força das idéias. Indicadores de sustentabilidade agrícola na Amazônia. Embrapa-NMA, Campinas, SP, Brasil, p. 86.
11. **Mitasova, H., J. Hofierka, M. Zlocha & L.R. Iverson, 1996**. Modelling topographic potential for erosion and deposition using GIS. International Journal of Geographical Information Systems, 10(5), 629–41.
    Xiang, W-N. & W.L. Stratton, 1996. The *b*-function and variable stream buffer mapping: a note on 'A GIS method for riparian water quality buffer generation'. International Journal of Geographical Information Systems, 10(4), 499–510.
12. **GIS Aids EL Salvador Land Redistribution, 1996**. GIS World, 9(8), 13.
13. **Petrie, G. & A.E. Niweiri, 1996**. The applicability of space imagery to the small-scale topographic mapping of developing countries: a case study – the Sudan. ISPRIS Journal of Photogrammetry and Remote Sensing, 51(4), 209–15.
14. **Lenney, P.M., C.E. Woodcock, J.B. Collins & H. Hamdl, 1996**. The status of agricultural lands in Egypt: the use of multitemporal NDVI features derived from LANDSAT TM. Remote Sense of Enviroment, 56(1), 8–20.

15. **Rogowisk, A.S., 1996.** Quantifying soil variability in GIS applications: II. Spatial distribution of soil properties. International Journal of Geographical Information Systems, 10(4), 455–75.
16. **Chuvieco, E. & J. Salas, 1996.** Mapping the spatial distribution of forest danger using GIS. International Journal of Geographical Information Systems, 10(3), 333–45.
17. **Benson, B.J. & M.D. Mackenzie, 1995.** Effects of sensor spatial resolution on landscape struture parameters. Landscape Ecology, 10(2), 113–20.
18. **Moody, A. & C.E. Woodcock, 1995.** The influence of scale and the spatial characteristics of landscapes on land-cover mapping using remote sensing. Landscape Ecology, 10(6), 363–79.
19. **Corbley, K.P., 1996.** One-meter satellites: practical applications by spatial data users. Part 3. Geo Info Systems, 6 (5), 39–43.
20. **de Miranda, E.E. & C. Mattos, 1992.** Brazilian rain forest colonization and biodiversity. In: Agriculture Ecosystems and Environment. Elsevier, Amsterdam, 40, 275–96.
21. **Petkov, L., M. Pieri, F. Maselli & G. Maracchi, 1992.** Study and modelling of temperature spatial variability by NOAA-AVHRR thermal imagery. ISPRIS Journal of Photogrammetry and Remote Sensing, 47(1), 1–42.
22. **de Miranda, E.E., A. Setzer & A. Takeda, 1994.** Monitoramento Orbital das Queimadas no Brasil. Embrapa-NMA, Campinas, SP, Brasil, p. 149.

    Pereira, J.R. & A.W. Setzer, 1996. Comparison of the fire detection in savannas using AVHRR's channel 3 and TM images. International Journal of Remote Sensing, 17(10), 1925–37.
23. **Tateishi, R. & C.H. Ahn, 1996.** Mapping evapotranspiration and water balance for global land surfaces. ISPRIS Journal of Photogrammetry and Remote Sensing, 51(4), 209–15.
24. **Foody, G.M., G. Palubinskas, R.M. Lucas, P.J. Curran & M. Honzak, 1996.** Identifying terrestrial carbon sinks: classification of successional stage in regenerating tropical forest from LANDSAT TM Data. Remote Sense of Enviroment, 55(3), 205–16.
25. **Cherrill, A. & C. MacClean, 1995.** An investigation of uncertainty in field habitat mapping and the implications for detecting land cover change. Landscape Ecology, 10(1), 5–21.
26. **Taconet, O., C. Vidal-madjar, C.H. Emblanch & M. Normand, 1995.** Taking into account vegetation effects to estimate soil moisture from C-band radar measurements. Remote Sense of Enviroment, 56(1), 52–6.
27. **Zheng, D., E.R. Hunt & S.W. Running, 1996.** Comparison of available soil water capacity estimated from topography and soil series information. Landscape Ecology, 11(1), 3–14.
28. **Hallett, S.H., R.J.A. Jones & C.A. Keay, 1996.** Environmental information systems developments for planning sustainable land use. International Journal of Geographical Information Systems, 10(1), 47–64.
29. **Arze, J., 1992.** Sistemas expertos y fase de diseño de la metodología para la investigación y desarrollo de alternativas tecnológicas. In: International Meeting of RIMISP 5. RIMISP, Proceedings Santiago, Chile, 20.

    de Miranda, E.E., 1992. Posibles aportes de los Sistemas de Información Geográfica al diseño de sistemas de producción. In: International meeting of RIMISP, 5. RIMISP, Santiago, Chile, p. 13.

    Merten, G.H., J.H. Kaviglioni, C.C. Giacomini, R.L. Rufino, D. Saintrand, C.A. Kessler, G.C. Ribas & R.O. Dedecek, 1995. Uso de SIG e do modelo USLE para determinar a erosão potencial e atual nas microbacias pilotos de Água Grande e Córrego do Pensamento, Maborê, Paraná. In: Encontro da Sociedade Brasileira de Sistemas de Produção, 2Anais. SBS-IAPAR, Londrina, pp. 273–74.
30. **Zonneveld, I.S. & R.T.T. Forman (Eds), 1990.** Changing Landscape; an Ecological Perspective. Springer-Verlag, New York.

11.3 FSR FROM A MODELLING PERSPECTIVE: EXPERIENCES IN LATIN AMERICA

Roberto A. Quiroz, Carlos León-Velarde and Walter Bowen

... modelling is only an analytical tool and not an end in itself. Models are, at best, simple representations of reality. Thus they should be seen only as decision support tools. The question to be asked is, what is a suitable (valid) model for experimentation?

11.3.1 Introduction

Agricultural researchers apply the scientific method in search of solutions to the factors limiting agricultural production. The lack of appropriate technology is a key limiting factor, especially in resource-poor farms. A closer look at what is defined as the scientific method makes us wonder if this method *per se* may be applied to solve technological problems. In a restrictive sense, the scientific method can be seen as a process that uses existing knowledge to generate new knowledge[1]. Problem solving requires adaptation of knowledge to surmount limiting factors. The successful use of technology to overcome major limitations to agricultural production depends upon an adequate acquaintance with the problems within a specified context and a good application of available knowledge. When this interface is used to solve agricultural problems of small farmers with their active participation, we say we are applying FSR methodology.

Preaching FSR is a lot easier than actually solving small-farmer's problems, as shown elsewhere in this book. This contribution discusses the role of modelling in knowledge integration and management for a more effective application of FSR. It is not our intention to present a comprehensive review of the state of the art in simulation and modelling, rather to show some examples of the application of this tool by FSR teams in Latin America.

One of the lessons learned from the application of FSR is that farmers do not adopt complete technological packages, rather components of them[2]. Changes in productivity of a small farm from a low level to its potential level, might be pictured as an ascending spiral with the actual production level located at the lower end and the potential at the upper end. The number of turns required to go from the actual to the potential level differs among farming systems (FS). It can be envisioned that a complete turn of a cycle requires the application of several methodological steps: diagnosis, experimentation, validation and diffusion. These steps are commonly applied by most of the research teams following the FSR methodology and are worth closer examination, with application examples showing how models might be used as tools.

11.3.2 Modelling in FSR methodology: the diagnosis phase

The objective of this phase is to select target areas, divide the frame of farming families into target groups, and to ascertain the major constraints on farming in the area and also the degree of flexibility that exists for modifying the farming systems. In-depth knowledge is desirable in order to ascertain the major constraints of FS as well as the degree of flexibility[3]. Objectives of target FS should be well defined, important components and their interactions should be understood, there should also be a good hypothesis of how the FS functions, and the information (hopefully including quantitative information) must be organized and analysed in a systematic way.

Thus the core of the diagnosis phase is nothing more than making a model of the target FS. As stated by Pandey and Hardaker in 1995, no research study of a farming system is possible without a model[4]. The main issue here is what kind of model is required to do the best job possible.

An alpaca FS from the Andes could be used as an example. This system is characterized by mixed camelid flocks (alpacas and llamas) raised in locations higher than 4000 m above sea level; low production; poor quality rainfed pastures, limited amounts of natural irrigated rangelands, low temperatures to −16°C and frost throughout the year; and low-income

Table 11.3.1. Voluntary intake, live weight and fibre production in the high plateau of Peru, compared with model results.

Output variables	Real world	Simulated
Voluntary intake (kg DM d^{-1})		
Reiner *et al.*, 1986	0.98	1.07
Proyecto Alpaca	1.00	1.09
Clavo and Perez, 1986	1.19	1.20
Clavo and Ravillet, 1987	1.40	1.30
Liveweight (kg) at different stages		
at birth	6.5	6.2
1 year	27.0	26.0
2 years	35.0	34.0
3 years	44.0	43.0
4 years	49.0	49.0
Fibre production of different animal categories (kg^{-1} $year^{-1}$)		
young animals	1.1	1.1
males	0.8	0.8
dry females	1.3	1.4
lactating females	0.8	0.8

Source: Arce *et al.*, 1994[5].

farmers. Despite this harsh environment, many Andean farmers inhabit these areas and are dependent upon camelid production as their main source of income. As a result, research on alpaca production is a main topic for animal science research teams in Bolivia and Peru. The diagnosis phase was improved by the construction of a mathematical model of the alpaca system. A combination of on-farm surveys, literature searches, field measurements and OSR led to a definition of main constraints and possibilities of the FS. These findings were systematized in a mathematical model based on four dynamic components: herbage production and quality; flock categories and weights; energy intake and utilization for maintenance and production; and management decisions. Comparisons between collected data from small farms and research results with those generated by simulation are shown in Table 11.3.1.

11.3.3 Modelling in FSR methodology: the experimentation phase

This phase constitutes the backbone of the FSR methodology. Without experimentation there will not be a continuous generation of the knowledge required to solve the diverse limiting factors faced by small farmers. It is in this phase that the scientific method should be rigorously applied to solve real problems. Experimentation is required at several steps of the FSR methodology (Fig. 11.3.1). Experimentation in the real world, although absolutely necessary, is in some circumstances an expensive and time-consuming process, facing severe difficulties in controlling variables exogenous to the experiments[6]. In computer modelling, by contrast, experimentation is easy, cheap and speedy, once a suitable model has been developed[7]. A word of caution is important at this stage; modelling is only an analytical tool and not an end in itself. Models are at best simple representations of reality. Thus they should be seen only as decision support tools. The question to be asked is, what is a suitable (valid) model for experimentation? A robust statistical test for model specification does not exist[8] and it may not be that important. Models, as imperfect representations of reality, should be used to generate a hypothesis of how a component of a system or a whole system functions under a given set of conditions. Comparison of trends of simulated and real responses is more meaningful than a statistical test of model adequacy.

A valid model is useful for evaluating scenarios that are difficult to test in the real world.

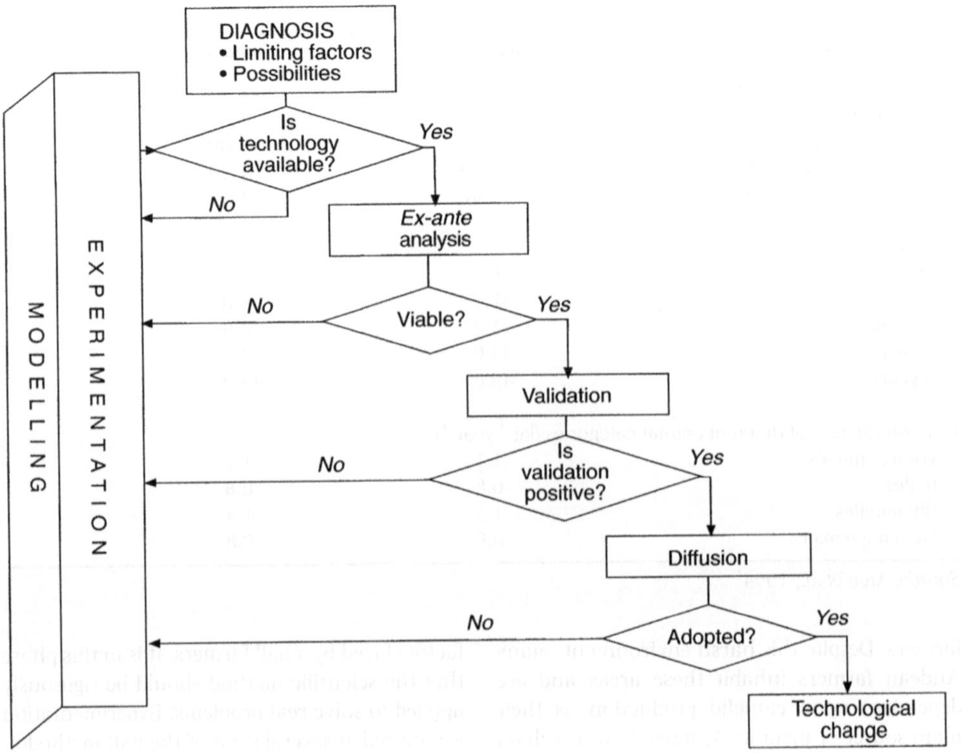

Fig. 11.3.1. The role of research in FSR.

This is especially true when assessing the long-term consequences of different management practices or the possible impact of climate change. In such cases, comprehensive models are extremely valuable for formulating and testing hypotheses that otherwise require many years of data. The following two examples illustrate how models may be used for hypothesis testing, and how simulated results might be useful for guiding further field research.

11.3.4 A livestock system[9]

Problem

The high plateau of the central Andes, located near lake Titicaca on the border of Bolivia and Peru 3900 m above sea level, is home to a large portion of the rural families of these two countries. As altitude increases, cropping becomes impractical due to high frost and drought risks. Camelid (alpaca and llama) production is the most important farming system for a large population of rural families. Fibre production is attractive as it is transformed and exported, thus guaranteeing good prices to farmers. Meat, a less valued good, is a complementary product.

Animals graze year round in two types of grasslands; rainfed rangelands and naturally irrigated rangelands of higher quality and productivity called 'bofedales'[10]. On average, most families have access to 80 ha, 70 ha of rainfed rangelands and 10 ha of *bofedales*. Actual gross income is around US$1100 per family per year. Biomass production as well as the quality of forage on offer is limited. Carrying capacity is decreasing, exacerbating overgrazing and its consequences on land productivity, carbon sequestration, soil protection and fertility and the hydrological cycle.

Actual experiments to evaluate management options for optimizing profit as well as minimizing the impact on the environment are impossible to conduct due to the high cost, the time required to evaluate responses, and the difficulties of access during the rainy season. Most important variables determining the system's output may be examined in computer simulation studies.

Procedure

The first question addressed in the simulation study was what combination of stocking rate (X1), pasture growth rate (X2) and digestibility of selected forage (X3) will maximize profit. The second question focused on the time course of changes in productivity/income by the adoption of the technology that would maximize profit. The simulation model described by Arce *et al.* in 1994[11] was used to evaluate the combination of X1, X2 and X3. A central composite rotatable design[12] was used to evaluate five levels of each factor. There were 23 factorial treatments, 23 axial treatments and one central treatment repeated six times. Other management variables were held constant and the simulation time was 15 years:

- Pasture type: *bofedal* and rainfed rangelands.
- Area: 80 ha divided into 70 ha rainfed rangelands and 10 ha *bofedal.*
- Initial biomass availability: rainfed rangelands, 1000 kg DM ha^{-1}; *bofedal*, 1500 kg DM ha^{-1}.
- Grazing: 3 months in rainfed rangelands (January/March) 9 months in *bofedal* (April/December).
- Residual biomass: rainfed rangelands, 500 kg DM ha^{-1}; *bofedal*, 700 kg DM ha^{-1}.
- Herd management: mating in January; weaning in September; shearing in August; sales in December.

Treatment allocation ($2k + 2^*k + 1(n)$) with their respective code is shown in Table 11.3.2. The rotatability of the design is given by a value alpha = $(2k)1/4$. In this example alpha is equal to 1.682. When $k = 3$, as it is in this case, the highest efficiency of a central composite rotatable design is attained (0.67). Once the combination of treatments that maximized profits was established, the time course of gross income changes due to implementation of technology in a farm was simulated. Several mathematical models were used to analyse the obtained response, the logistic function offering the best fit ($Y = b_0/1 + b_1{}^*e^{-ct}$).

Results

Average gross income for simulated treatments, over a 10-year period, ranged from \$331 to \$3421. The analysis of the data using the central composite rotatable design resulted in the equations used to generate a three-dimensional response surface. Maximum profit was obtained with a stocking rate of 0.54 heads ha^{-1}, in a grassland producing 8.14 kg DM ha^{-1} d^{-1}, and a digestibility of 65.9%. This combination of studied variables enables a gross income of \$2859 $farm^{-1}$ $year^{-1}$. The stocking rate required to attain maximal profit is higher than the 0.4 heads ha^{-1} recommended by local researchers and extension agents and smaller than actual stocking rates used by farmers (0.9 heads ha^{-1}); forage growth rate derived from the model is similar to actual growth rates and derived digestibility is found in the field during the first few days that animals graze *bofedales*.

Table 11.3.2. Treatment allocation to generate a simulated response surface of alpaca production in the Andes as a function of stocking rate, forage growth rate and forage digestibility, using a central composite rotatable design.

			Treatment levels code				
Variable	Treatment code	Ecology	−1.682	−1	0	+1	1.682
Stocking rate	X1	Both	0.56	0.70	0.79	1.10	1.24
Forage growth rate	X2	Rainfed range	0.57	1.25	2.25	3.25	3.93
		Bofedals	8.12	8.80	9.80	10.80	11.40
Forage digestibility	X3	Rainfed range	53.64	55.00	57.00	59.00	60.36
		Bofedals	67.36	66.00	63.00	70.00	71.36

Source: Leon-Velarde and Quiroz, 1994a.

By implementing recommended technology (increasing the area with *bofedales*, introduction of white clover with strategic irrigation, and improved pasture management) farmers may increase gross income up to \$3891 $year^{-1}$. Going from actual gross income to the estimated potential may require 10 years (Fig. 11.3.2).

11.3.5 A cropping system[13]

Problem

Nitrogen supply is a major factor governing the sustainability of many cropping systems. Without an adequate supply of nitrogen, crop yields eventually decline, making the system unsustainable. Just as important, an excessive supply of nitrogen may make a system unsustainable in environmental terms because too much nitrogen enters the water supply. Thus, nitrogen management is an important component of sustainable land management.

The long-term consequences of some alternatives for managing nitrogen may be examined in computer simulated studies. As an example, consider a hypothetical 50-year simulation study conducted to determine the feasibility of growing a short-duration leguminous green manure each year as a nitrogen source for maize in the wet–dry tropics (savanna) of central Brazil. In this area, farmers normally grow one crop of maize each year during the wet season, which starts in October and runs through April.

Procedure

The questions addressed in the simulation study were whether or not the wet season would be of sufficient length for growing a short-duration (2–3 months) leguminous green manure followed by maize and whether or not the nitrogen supply by the decomposing green manure would be adequate for maintaining maize yields. For purposes of comparison, maize growth assuming no additional N other than that supplied by the soil organic matter was added. As a measure of potential yields, maize growth was also simulated assuming non-limiting nitrogen and water supplies. The simulation study was done using the crop models distributed with the Decision Support System For Agrotechnology Transfer (DSSAT), version 3[14]. Soil and weather data for the simulation site were provided by the Brazilian Corporation for Agricultural Research (EMBRAPA).

Automatic planting decision modules in the

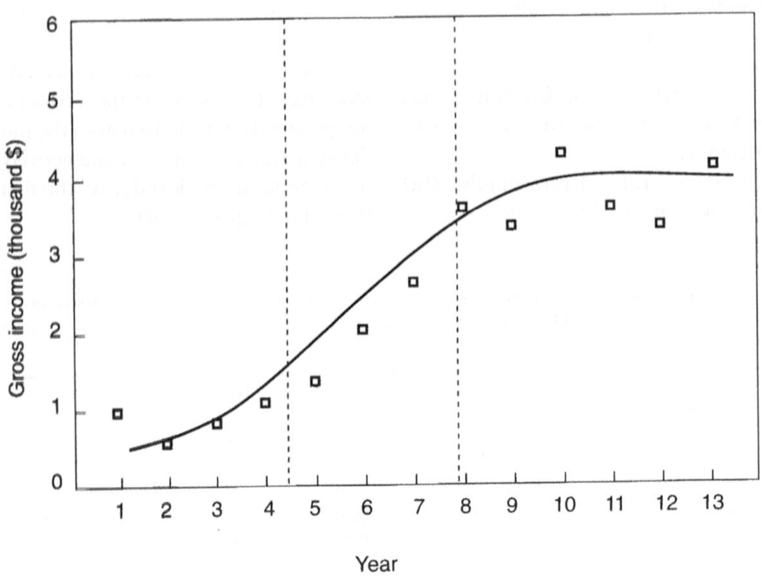

Fig. 11.3.2. Simulated gross income accrued by farmers in time, by adopting new pasture management techniques.

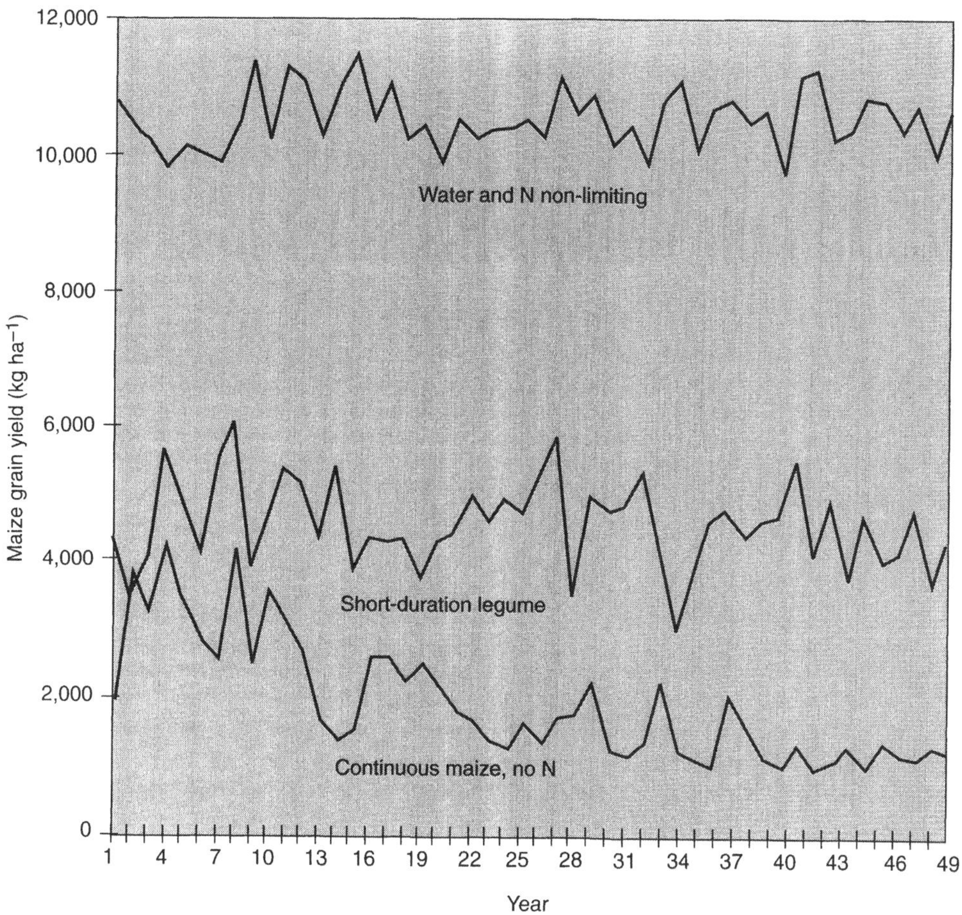

Fig. 11.3.3. Simulated variation in grain yield for maize under different management systems during a 50-year period in central Brazil.

DSSAT were used so that sowing of either the legume or maize would not occur until the soil profile was sufficiently moist within a specified planting window. To simulate short-duration legume green manure growth, an automatic harvesting decision module was used to automatically incorporate the legume on the day the crop reached its flowering stage. The short-duration leguminous green manure was inserted into the cropping system at the beginning of the wet season, with the planting window set to begin in mid October. After sowing, legume growth and nitrogen fixation were simulated until the plant was incorporated at flowering, which usually occurred by early to mid December and reached harvest maturity by late April. This same rotation of a short-duration legume followed by maize was simulated continuously for 50 years. To account for the effect of potential variability in the weather, the DSSAT weather generator was used to provide 10 replications of 50-year weather sequences.

Results

The mean variation in grain yield (average of the weather sequence replications) for each management alternative during a 50-year period is shown in Fig. 11.3.3. If no nitrogen was applied to the system, there was a downward trend in yields particularly notable after 10 years. Average yields of about 3.5 t ha^{-1} during the first 10 years were almost identical to yields obtained with no nitrogen (3.3 t ha^{-1}) during a 10-year experiment near the

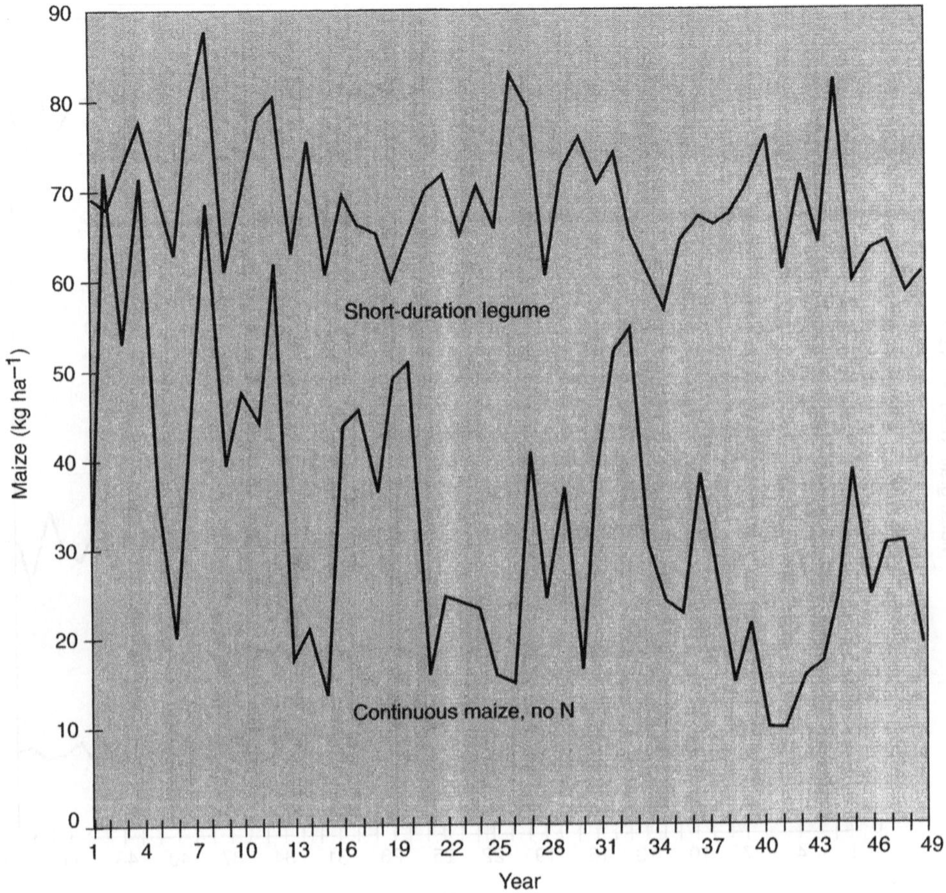

Fig. 11.3.4. Simulated variation in nitrog en uptake by maize under different management systems during a 50-year period in central Brazil.

simulated site[15]. Nitrogen from decomposition of the short-duration legume resulted in greater yields averaging 6.0 t ha^{-1} during the first 25 years before showing a slight trend downward during the last 25 years. The highest maize yields were of course obtained under non-limiting water and nitrogen conditions (Fig. 11.3.3). These yields show the maximum potential production thus providing a standard for comparing alternative management practices.

Indicators other than grain yield can be examined also with model output, e.g. nitrogen uptake by maize (Fig. 11.3.4) or legume dry matter production, amount of nitrogen fixed, total nitrogen in the legume at the time of incorporation (Fig. 11.3.5).

Nitrogen uptake by maize followed much the same pattern as grain yield. Legume dry matter production was usually in the range of 1.5–2.0 t ha^{-1}, and the amount of legume nitrogen at the time of incorporation was usually between 60 and 80 kg ha^{-1} (Fig. 11.3.5). The simulation results showed that the wet season at this site is probably of sufficient length for growing a short-duration green manure followed by maize in the same season. Such a system would be expected to provide a sustainable source of nitrogen for maize, although maize would likely respond to more nitrogen.

11.3.6 The validation phase

These brief simulation studies demonstrate how alternative management systems may be evaluated first on the computer. In other words, the computer may be thought of as a screening tool. Once systems with the greatest potential for success have been identified through simulation, components of such systems may be evaluated more closely in field studies

In the validation phase the technology is given to the farmer and the researcher plays a passive role of monitoring the outcomes. The critical issue in this step is to maintain the farmer's confidence in research results to solve their problem thus increasing profit. The closer a researcher can assess the outcome of a new technology under the conditions of a target farmers farm, the closer the FSR team is to producing positive technological changes. Simulation may be of value in tailoring technology for different farmers, as it is relatively easy to evaluate different scenarios. One must bear in mind that this judgment applies for biophysical and economic scenarios. Social aspects are not yet adequately modelled[16] and social assessments are, therefore, incorporated through interviews with participating farmers, discussing model results[17]. Recent advances in incorporating social aspects into simulation models in the Andes will be discussed below. The following example shows the application of simulation to predict the outcome of a portfolio of feeding technology for dual purpose cattle at lake Titicaca basin in the border area of Bolivia and Peru.

Problem

Dual purpose cattle production is a main income generating activity of resource-poor farmers of the area (*c.* 3950 m above sea level). Breeds adapted to environmental conditions produce less than their potential, especially during the winter time, due to restricted nutritional plane, a limitation that is amplified by the energy required to bring the animals' temperature up

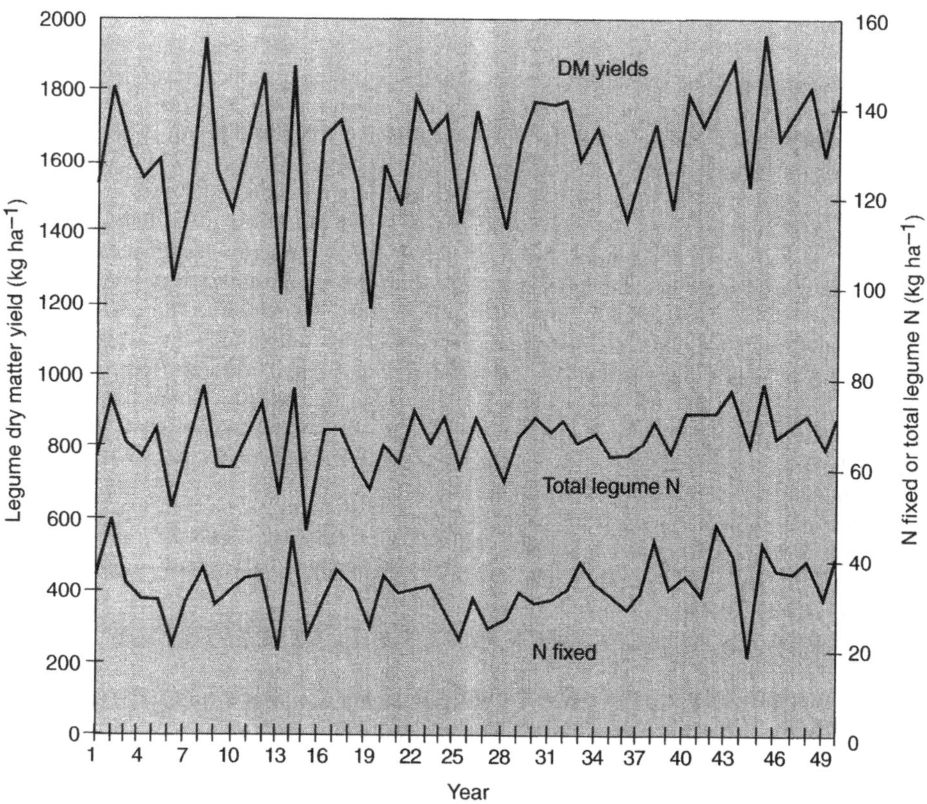

Fig. 11.3.5. Simulated variation in legume dry matter production, amount of nitrogen fixed, and total nitrogen in the legume at the time of incorporation during a 50-year period in central Brazil.

Table 11.3.3. Milk and meat production in the high plateau of the Andes with animals fed lliachu and totora.

Output variables	Real world	Simulated
	Milk (kg d^{-1})	Milk (kg d^{-1})
Forage ratio		
100% llachu	3.04 (4–7)*	4.69
75% llachu + 25% totora	4.17 (5–8)	6.02
50% llachu + 50% totora	3.99 (5–8)	5.68
100% totora	4.01 (5)*	6.07
	ADG (kg)	
Scenario		
fresh (as harvested) llachu without shelter	0.31	0.3
fresh llachu with shelter	0.47	0.6
pre-dried (c. 40% DM) without shelter	0.45	0.4
pre-dried with shelter	0.98	1.5
Farmers' assessment of validated technology (0 = poor, 5 = good)		
acceptability	5	
profitability	5	
viability	4	
time needed to produce results (years)	< 1	
productivity	5	

* Estimated milk production range, considering calves in dual purpose dairies consume between 1 and 3 l d^{-1}.
Source: Osuji and Smith 1988[22].

to their thermoneutral zone. Despite the general belief that local feed resources are low quality roughage, hydrophyte plants from the lake constitute an alternative since their energy[18] and protein[19] contents are high (DM digestibility above 85% and and crude protein above 20%).

Even though hydrophyte plants (*E. potamogeton*, llachu and *S. tatora*, totora) have been used for many years in traditional feeding systems, they are offered as harvested with *c.* 10% DM. The output levels are the same as those obtained with other local forages and crop residues. Evidently the feeding system for the hydrophyte plants does not utilize their higher quality. Research showed that the water content in the forage decreased the residence time in the rumen thus the ability of the animal to use available nutrients[20]. Research findings needed to be translated into technology capable of increasing a farm's productivity. In addition, an acceptable estimate of the expected outcome at farm level was desirable.

Procedure

Two questions were addressed in the simulated study. The first one referred to establishing the combination of the two types of plants (llachu and totora) that would maximize profit from dual purpose cattle systems; and the second considered the possibility of using the highest quality forage (llachu) as the unique feed resource in a feedlot, and the evaluation of the use of rustic shelters. A simulation model driven by energy, protein, calcium and phosphorus consumption and metabolism and considering the impact of environmental variables such as temperature, humidity, and wind speed, on animal performance, was constructed, tested and then used for these simulation studies[21].

The first study considered five forage combinations (pre-dried for obtaining *c.* 40% DM).

1. 100% llachu.
2. 75% llachu + 25% totora.
3. 50% llachu + 50% totora.
4. 25% llachu + 75% totora.
5. 100% totora.

Response variables were 4% fat corrected milk production and net income. The second study included four scenarios:

1. Fresh (as harvested) llachu without shelter.
2. Fresh llachu with shelter.
3. Pre-dried (*c.* 40% DM) without shelter.
4. Pre-dried with shelter.

Response variables were average daily gain and net income.

The results from the simulation were shown to farmers in order to find those willing to try out the technologies in their farms. A field day was organized where collaborating farmers presented the results to other farmers. A follow-up survey was then conducted to appraise farmers opinion on proposed technologies and to obtain feedback.

Results

Ex ante assessment of milk production was very close to that found by validating farmers (Table 11.3.3). It is important to note that the model predicted total milk production, but registered productions at the farms do not consider the amount consumed by calves. Given the genetic potential for milk production of local breeds, it is not economical to use llachu, as the costs of the labour required to pre-dry it are not covered by the increment in milk production. If animals with higher milk production potential are not acquired, farmers should use totora combined with other feed resources. Pre-dried llachu could either be used for fattening steers or be sold to local dairy farmers.

When llachu was fed to fattening steers, its nutritional quality was reflected in an up to threefold increment in average daily gain (Table 11.3.3), compared to the traditional way of using the forage. Model predictions were good, except for the last scenario, where the model overestimated the outcome. Farmers' evaluation of this technology was very positive, and an NGO is now providing credits to farmers to construct rough shelters and pre-dry llachu.

11.3.7 Diffusion phase

The use of simulation at this step is similar to the ones described above. The main difference might be the heterogeneity of conditions where technology is expected to be adopted. Models may play a key role assessing expected outcomes for different environments and in defining minimal conditions required for a good response of the technology, such as management practices, weather and markets. Models are also useful in *ex post* analysis of diffusion programmes. A typical question could be why expected increments in productivity are seldom attained. Different scenarios looking at environmental and managerial constraints may be simulated to examine what went wrong and what lessons can be learned.

11.3.8 Model integration

The models explained above are particularly useful for researchers and extension agents. They lack a sociocultural component, vital for emulating real-world farms. Representing biophysical and economical aspects of real life in a model, despite the fact that many can be translated into mathematical equations, is already difficult. Modeling sociocultural attitudes imposes a greater challenge. The degree of difficulty not only relates to the heuristic nature of the process, but is further entangled by the differences that might exist among farmers, in a similar environment, making decisions about the same problem.

Social scientists play a key role in determining, ranking and defining probability of occurrence of the most important rules in the farmer's decision-making process. Applications of available methods for eliciting subjective probabilities[23] might be useful in this endeavour. Once the FSR team is satisfied with the degree of knowledge on sociocultural aspects and decision-making process, rules can be incorporated into an expert system as a knowledge base.

The Pumani model, developed by French scientific cooperation (ORSTOM) and the Bolivian Institute of Agricultural technology (IBTA), is an example of this type of model integration[24]. The model utilizes object programming to integrate different components typically found in FS in the central high plateau of Bolivia. The core component of the model is an expert system at a farm level where the rules for the organization of the work on the farm are assembled in a knowledge base, and integrated by an inference motor. This expert system is composed of rules about agronomy, animal production and sociology. Whenever required, the core model con-

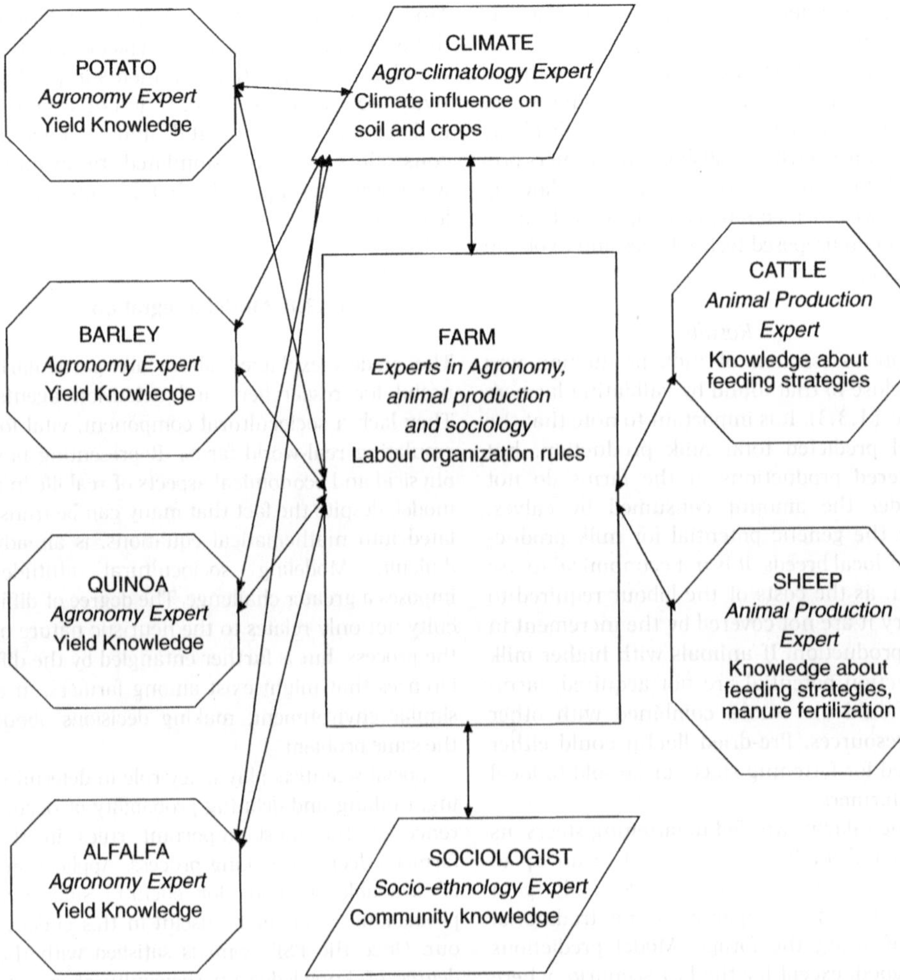

Fig. 11.3.6. An integrated model: in which the core model consults other expert systems when specific knowledge is required.

sults other expert systems (objects) where specific knowledge about climate, different crops and pastures, animal species and sociology is stored and processed (Fig. 11.3.6). The knowledge base is a combination of farmers' intuitive knowledge, research results as well as simulated results. These databases are combined to better assess farm problems and to evaluate feasible solutions.

Other ways of integrating model results exist. Results from process-based models can be used as coefficients for optimization models[25]. Outputs from component models can be used as inputs for less precise models built for assessing outcomes at regional levels. One example was presented by Quiroz *et al.* in 1995[26]. Another tool for integrating model results is the geographical information system[27].

11.3.9 Training

The training of new practitioners of FSR is vital to the process. The best way of learning the process is *in situ*, but exposure to a sound theoretical framework is helpful. Farm simulation models, where environmental conditions, capital, stocks, prices, taxes, market demand and other variables are easily changed, might be used to familiarize new scientists with the type of problems they will face in the real world, forcing them to make decisions similar to those being taken by farmers, considering environmental and market restrictions. A combination of a good field experience and modelling will allow FSR scientists to be more efficient in identifying constraints and proposing appropriate technological alternatives.

11.3.10 Concluding remarks

Practitioners of FSR methodology are increasingly being pressed to deliver technological options for overcoming small-farmers' problems. Farmers' demands for solutions require short-term answers, a high degree of specificity and a minimal negative impact on the environment. Researchers are, on the other hand, limited by lower financial resources. Under these circumstances, models and simulation techniques may become very handy in improving our efficiency in promoting a sustainable agricultural development among resource-poor farmers, most of them living in fragile environments. Fortunately, computer technology is becoming more accessible, both in price and performance, and new professionals are better trained in its use. Just do not let the computer think for you.

REFERENCES

1. **Cañas, R. & J. Lavados, 1989.** Tecnología, Gestión y Desarrollo: Aspectos Básicos Generales. Series manuales I&D. CINDA, Chile.
2. **Quiroz, R., 1994.** Metodología de Sistemas: una herramienta para la solución de problemas productivos de pequeños agricultores. In: Iñiguez, L. & Tejada, E. (Eds) Producción de Rumiantes Menores en los Valles Interandinos de Sudamérica. RERUMEN, La Paz, Bolivia, pp. 153–70.
3. **Norman, D. & M. Collinson, 1986.** Farming systems research in theory and practice. In: Remenyi, J.V. (Ed.) Agricultural Systems Research for Developing Countries: Proceedings of an International Workshop Held at Hawkesbury Agricultural College, Richmond, N.S.W., Australia, 12–15 May 1985, pp. 16–30.
4. **Pandey, S. & B. Hardaker, 1995.** The role of modelling in the quest for sustainable farming systems. Agricultural Systems, 47, 439–50.
5. **Arce, B., C. Aguilar, R. Cañas & R. Quiroz, 1994.** A simulation model of an alpaca system in the dry Puna of the Andes. Agricultural Systems, 46, 205–25.
6. **Dent, J.B., 1993.** Potential for systems simulation in farming systems. In: Penning de Vries, F., P. Teng & K. Metselaar (Eds) Systems Approaches for Sustainable Agricultural Development. Kluwer Academic Publishers and International Rice Research Institute, Dordrecht, pp. 325–39.
7. **Pandey, S. & B. Hardaker, 1995.** (Op. cit.).
8. **Jones, P.N. & P.S. Carberry, 1994.** A technique to develop and validate simulation models. Agricultural Systems, 46, 427–42.
9. **Leon-Velarde, C. & R. Quiroz, 1994.** Análisis de Sistemas Agropecuarios: Uso de Métodos Biomatemáticos. CIRNMA, La Paz, Bolivia.
10. **Quiroz, R., G. Mamani, R. Revilla, C. Guerra, J. Sanchez, M. Gonzales & G. Pari, 1991.** Perspectivas de Investigacion Pecuaria para el Desarrollo de la Comunidades de Puno. In: Arguelles, L. & R.D. Estrada (Eds) Perspectivas de la Investigación Agropecuaria para el Altiplano. Lima, Peru, pp. 357–406.
11. **Arce, B. *et al.*, 1994.** (Op. cit.).
12. **Montgomery, D.C., 1984.** Design and Analysis of Experiments, 2nd edition. John Wiley, New York.
13. **Bowen, W., J.W. Jones & P. Thornton, 1993.** Crop simulation as a potential tool for evaluating sustainable land management. In: Kimble, J.M. (Ed.) Proceedings of the Eight International Soil Management Workshop: Utilization for Sustainable Land Use. Oregon, California and Nevada. Soil Conservation Service USDA.
14. **Tsuji, G.Y., G. Uehara & S. Balas (Eds), 1994.** The Decision Support System for Agrotechnology Transfer (DSSAT) Version 3. University of Hawaii, Hawaii, USA.
15. **Suhet, A.R., J.R.R. Peres & M.A.T. Vargas, 1985.** Nitrogenio. In: Moedert, W.J. (Ed.) Solos do Cerrados: Tecnologias e Estrategias de Manjo. EMBRAPA/CPAC, Planaltina, pp. 167–202.
16. **Dent, J.B., 1993.** (Op. cit.).
17. **PISA, 1993.** Informe Final de Proyecto. Puno, Peru.
18. **Quiroz, R., 1992.** Apuntes sobre modelos matematicos en problemas ganaderos. In: Ruiz, M.E. (Ed.)

Simulacion de Sistemas Pecuarios. Serie de Ponencias, Resultados y Recomendaciones de Eventos Tecnicos. IICA, San Jose, Costa Rica, pp. 157–87.
19. **Bautista, J.L., 1992**. Fraccionamiento y degradabilidad ruminal del llachu (*E. potamogeton*) y totora (*S. tatora*) en comparación con heno de avena (*A. sativa*) en bovinos. Tesis Magister Scientiae, Universidad Nacional Agraria La Molina, Lima, Perú.
20. (Ibid.).
21. **Quiroz, R., B. Arce, R. Canas & C. Aguilar, 1994**. Development and use of simulation models in animal production systems research. In: M.E. Ruiz (Ed.) Animal Production Systems Research: Methodological and Analytical Guidelines. CR. IDRC-IICA-RISPAL, San Jose.
22. **Osuji, P.O. & J.W. Smith, 1988**. Proyecto de sistemas de produccion de leche, Guyana. In: Ruiz, M.E. & A. Vargas (Eds) Informe de la VII Reunion de RISPAL. Lima, Peru 22–27 de marzo de 1987. Serie de Ponencias, Resultados y Recomendaciones de Eventos Tecnicos. IICA, San Jose, Costa Rica, pp. 89–101.
23. **Anderson, J.R. & J.L. Dillon, 1992**. Risk analysis in dryland farming systems. Food and Agricultural Organization of the United Nations. Farm Systems Management Series no. 2.
24. **Muigueis, J., 1995**. PUMANI: Un Modelo de Representacion de la Actividad Agropecuaria en los Andes. ORSTOM-IBTA, Bolivia.
25. **ESTRADA, 1996**. Incidencias de las Políticas Económicas en la Conservación de los Recursos Naturales de la Zona Andina. REPAAN-CONDESAN.
26. **Quiroz, R.A., R.D. Estrada, C. Leon-Velarde & H. Zandstra, 1995**. Facing the challenge of the Andean zone: the role of modeling in developing sustainable management of natural resources. In: Eco-Regional Approaches to Develop Sustainable Land Use and Food Production. Kluwer Academic Publishers, The Netherlands, pp. 13–31.
27. **Goodchild, M.F., B.O. Parks & L.T. Steyaert, 1993**. Environmental Modeling with GIS. Oxford University Press, Oxford.

11.4 MOVING PARTICIPATORY PLANT BREEDING FORWARD: THE NEXT STEPS

Louise Sperling and Jacqueline A. Ashby

Breeders no longer need to ask whether farmers have knowledge important for formal breeding: we know they do. Moreover, breeders should be moving beyond the stage of simply 'consulting farmers' to face today's challenges: they should be defining the appropriate division of labour between farmers and breeders for a given objective in a breeding programme and consciously developing new organizational approaches suitable for participatory plant breeding.

11.4.1 Introduction

Participatory plant breeding (PPB) involves farmers and other users such as consumers, vendors and industry, in formal breeding research. It is 'participatory' because users can have a role in all major stages of the plant breeding and varietal selection process: they help set priorities, make crosses, screen germplasm entries in the pre-adaptive phases of research and usually take charge of adaptive testing. While both farmers and scientists breed and select varieties, the division of labour between the two is undergoing dynamic change. Better understanding is emerging of how each partner can take the lead at different stages, according to their respective skills, experience and available resources.

Two broad approaches can be defined under the general rubric of PPB: farmers joining breeding experiments which are strongly shaped by formal breeding programmes; and scientists supporting farmers' own systems of breeding, varietal selection and seed maintenance. The two approaches represent a continuum, with users (e.g. farmers, processors and consumers), development workers and scientists actively involved in both. The seminal differences hinge on who ultimately controls the breeding process and seed systems, whether researchers or farmers are the driving force, and the scale on which the work is undertaken. So far, farmer-driven PPB tends to be clustered in a few communities, while formal breeding/PPB aims for wider geographical coverage.

Table 11.4.1. Schema of conventional and PPB approaches.

Stage of involvement	Conventional	Participatory
Definition of objectives and breeding strategy	☆	□
Selection of parent's crosses	☆	□
Selection from segregating populations	☆	□
Screening advanced lines on-station	☆	□
Adaptive on-farm testing	☆	○
Validation	□	○

Choices including choices of materials made by: ☆, breeders; ○, farmers; □, jointly.

PPB has achieved recognition as a crop improvement strategy over the last 10 years in response to the need for impact in non-commercial crops or in highly diversified, segmented markets. Centralized, researcher-driven breeding (or supply-driven research) has been extremely effective in high potential, uniform environments and for commercial farmers who can afford inputs and have access to institutions for credit and insurance to help manage variable production conditions. Conventional breeding has been less effective in difficult environments, in reaching farmers with few resources, and, in general, in reaching users with specialized concerns, such as those with rigorous product quality requirements[1]. Studies of farmers' own knowledge of varieties, and their plant-breeding or seed systems, have also encouraged the development of PPB, demonstrating that local expertise in germplasm management can be very precise – particularly, in regions with high varietal diversity[2]. Moreover, plant breeding is a familiar activity in farm communities: for generations, some farmers have been selecting and promoting the better adapted or quality entries and matching cultivars to particular production niches. This farmer-based experimentation is still lively and widespread in most rural communities, whether in low-income or wealthy countries. PPB builds on farmers' unique capacity to articulate precise preferences and to match varietal traits with specific niches, and their ability to lead the way in site-specific selection and testing.

11.4.2 PPB and conventional approaches compared

Today, PPB is being used in a wide range of crops and locales: for example, pearl millet in India, barley in Syria, common beans in Brazil, rice in Nepal and cassava in Colombia. One of the most important features of the evolution of PPB is the progressively earlier involvement of farmers in the formal breeding process. Early farmer involvement has taken many forms: farmers have been taught to handle crosses[3], they have been involved in screening segregating populations[4], and farmers have been brought directly onto experimental stations[5] and OFR sites[6] set up for screening pre-release lines.

Another important feature of PPB strategies is decentralization, which involves disaggregating the breeding population into subsets. In contrast to conventional plant breeding strategies, decentralization is designed to identify materials with adaptation to specific environments, within a given agroecosystem. Breeding for broad adaptation minimizes variability of the gene pool, whereas decentralization maximizes variability of the gene pool resulting from selection. While the organization and methodologies used in PPB are evolving rapidly, most PPB efforts have several features in common.

In conventional breeding, researchers make the major decisions on germplasm creation and promotion, from the initial stages – when a desired plant ideotype is defined, the specific traits to be incorporated are targeted, and a breeding strategy (e.g. pedigree) is identified – through to the stage of on-farm testing. Varietal screening criteria usually include yield and broad adaptation to multiple environments measured in controlled experimental plots, including tolerance to important diseases. Other non-yield related varietal traits of interest to farmers are seldom considered.

Client input is obtained right before varieties are to be released for diffusion – if it is sought at all. At this late stage farmers have only two

Box 11.4.1. Beans in Rwanda.
Farmers in Rwanda have considerable experience in managing local bean diversity: some 550 varieties were in use across the country in the early 1990s. In contrast, the selection sequence of L'Institut des Sciences Agronomiques du Rwanda (ISAR), using conventional breeding, sharply narrowed the diversity of bean cultivars in the national breeding programme. Some 200 entries were initially screened, but between two and five were eventually selected to enter on-farm trials. An experimental programme of the International Center for Tropical Agriculture (CIAT) and ISAR from 1988 to 1993, sought to draw on farmers' experience early in the selection process, when varietal options were still extensive. During a first phase, women experts evaluated 15 pre-released lines in on-station trials two to four seasons before normal adaptive testing. Subsequent on-farm trials showed farmers' ability to extrapolate from station fields to their own home plots. Farmer selections outperformed their checks with average production gains of up to 38%, while breeder choices in the same region showed insignificant gains. During a second phase, participants screened a broader range of cultivars even earlier: 80 to 100 entries in on-station trials five to seven seasons before conventional on-farm testing. Longer-term results were promising. The number adopted from the first 2-year period alone, 21, matched the total number of varieties released by ISAR in the previous 25 years[7]. The experiment shows select benefits of pre-adaptive screening: enhanced and diversified production on-farm (at least in heterogeneous environments) and significant savings in OSR time.

options: to accept or reject some two or three finished cultivars. Finally, conventional breeding most often works with industry or with individual farmers, on the assumption that once the variety is validated according to breeders' criteria, farmers can be persuaded to plant it by an extension system, and seed can be provided by an independent seed service.

In contrast, a PPB approach is based on a firm breeder–farmer partnership in research, validation and seed multiplication, from the earliest stages of the breeding project. Objectives are identified with farmers and the initial germplasm pool is directly shaped with strong client input. Farmers' criteria for varietal selection are identified promptly, such as post-harvest quality, maturity cycle or plant architecture. Farmers contribute to developing the germplasm pool at this pre-adaptive stage, and so work with a diverse range of germplasm. Depending on the crop, the varieties may be stabilized or still segregating. Adaptive testing in PPB is handed over to groups of farmers and integrated into farmers' ongoing experimentation. To meet different client needs and to target effectively for different zones, PPB tends to involve work in niches. Clients have a deciding vote on the acceptability of germplasm at each screening stage, just as researchers may decide to withhold or advance a given material based on their own criteria. As a result, PPB strategies involve decentralization at a relatively early stage in the process. Generic traits of broad importance to most farmers are identified in a

Box 11.4.2. Rice in Nepal.
Since the early 1960s, when the first rice variety was released by the Department of Agriculture, researchers have been having difficulty finding rice varieties for the high hills (> 1500 m asl). Only three of 43 varieties released have been recommended for these areas. Two of these are from a recently launched participatory breeding initiative.

Starting in 1993, the Lumle Agricultural Research Centre (LARC) began a major innovation: putting segregating lines (F5 bulk seed) directly into farmers' fields to incorporate farmer selection and evaluation into the breeding effort. To date, two populations, selected independently by farmers in two sites, are showing unusually high yields, under both farmer and researcher management. The entries have very good resistance to the major stresses of sheath brown rot and chilling injury, and the straw yield is judged by farmers as superior to the available local varieties. Both populations are spreading quickly from farmer to farmer. In response, the lines have been entered into the formal testing system to obtain their official release. These results were achieved in only 2 years. Researchers elaborate that 'in a conventional breeding system ... (such material) would have been still in the very preliminary stage of varietal screening ... and at least 7 years away from being given to farmers'[8].

Box 11.4.3. Colombia: CIALs – farmers' committees.
Farmers' committees in Colombia (Comites de Investigacion Agroppecuria Local, or CIALs) managed the testing of some 1000 varieties of beans, maize, peas, groundnut, fruits, vegetables and forages in the period 1990 to 1994 through the Participatory Research in Agriculture (IPRA) project of CIAT. The early involvement of farmers (starting in on-station trials) led to the identification of farmer-acceptable varieties. In addition, there have been two major spin-offs. The CIALs themselves have initiated small seed enterprises: more than 10,000 farmers have purchased CIAL seed which generated an estimated gross income of over US$2.5 million over one season. Shifting adaptive testing to the CIALs has not only extended the coverage of adaptive testing, but has reduced expenses: CIAL managed experiment costs 60% less in labour than the same trial run by a government extension agent[9].

breeding population early on, and then niche testing (in contrast to multilocational testing) of the breeding population is initiated leading to rapid differentiation of diverse subsets of materials screened with farmer involvement, for specific adaptation.

PPB has developed to improve the impact of breeding in difficult environments, but there remain several major important methodological and organizational challenges, which we shall explore.

11.4.3 Advances in PPB

PPB is being used in a variety of institutional settings, including International Agricultural Research Centres (IARCs), National Agricultural Research Services (NARS), universities and NGOs. During the past decade, a relatively short period in terms of varietal improvement, the accomplishments of the approach have been promising. The following examples illustrate the different kinds of gains made using this approach (Boxes 11.4.1–11.4.3).

Such PPB experiments as described in Boxes 11.4.1–11.4.5 suggest some of the potential gains which can be made by using participatory approaches: important production increases, enhanced varietal diversity on farm, cost savings in both formal research and support services. Two recently initiated programmes point to other novel dimensions of the PPB frontier (Boxes 11.4.4 and 11.4.5).

As PPB programmes gain currency in Latin America, Asia and Africa, the prospects widen for institutionalizing the approach, but practitioners need to address a series of immediate concerns.

11.4.4 Towards a new division of labour in plant breeding: three challenges

As the above examples show, PPB involves more than just consulting with farmers during the breeding process. To institutionalize active farmer breeding and to decentralize research, the breeding process has to be reorganized, with breeders and farmers taking on new roles. This need for a new division of labour in plant breeding raises several issues in need of more systematic analysis: What is the role of each partner at different stages of germplasm development, how might these roles vary across crops and across environments? There are three main challenges:

Box 11.4.4. Barley in Syria.
In Syria, where there has been relatively little breeding impact for barley in the marginal low rainfall environments (350–200 mm), the International Center for Agricultural Research in the Dry Areas (ICARDA) is experimenting with moving its entire screening programme on to farmers' fields, in some cases as early as the F2. The rationale is straightforward: site-specific adaptation needed for stress areas can only be achieved by breeding on-site[10], and farmers know their specific environments and needs better than breeders do. Staring in late 1996, the ICARDA programme is bringing over 200 barley lines to farmers' fields to be jointly selected by breeders and farmers. The lines are highly diverse, including landraces, élite lines, mixed varieties with and without good straw, small and large heads, in order to explore the bounds of what farmers consider acceptable and what will actually produce.

Box 11.4.5. Rainfed farming in India.
In India, a KRIBHCO Rainfed Indo-British Project, funded by the ODA and the Government of India, has had remarkable success in marginal areas of Gujarat, Madhya Pradesh and Rajasthan in selecting among pre-released and released lines with farmers who have generally never been exposed to materials developed for neighbouring Indian states. Between 1992 and 1996, one variety of maize, three of chickpea and two of blackgram were identified as being markedly preferred by farmers[11]. The project is now asking whether such an approach is relevant to more favourable areas – particularly where diversity may need to be expanded.

- How should breeding be 'sociotechnically divided' among farmer and formal researchers?
- What are the organizational options for the new breeding paradigm?
- What is the cost-effectiveness of the technical and organizational alternatives?

Defining the comparative advantage of scientists/farmers at different stages of the breeding process

The results of PPB suggest that the involvement of farmers in decentralized breeding can target germplasm selectively for specific environments more effectively than centralized breeding. This precision results not just from farmers knowing their own preferences, but also from having a keen awareness of the agroenvironmental factors that determine a variety's performance. Will it be intercropped? On what kind of soils will be it planted? However, it is not possible at present to say how any given crop improvement effort should define the optimal balance between farmer and breeder input. It is unlikely that a 'recipe' for determining this balance will ever be forthcoming.

Instead, plant breeding for different environments needs to differentiate routinely, with client participation, the various user groups and agroecological niches to be targeted. Numerous techniques for participatory 'stakeholder' identification and needs assessment (diagnosis) are available. The results supply a profile of the social and biophysical environment, that incorporates local knowledge and priorities. In PPB, this is an essential starting point for setting breeding objectives, in combination with information about the genetic variability available, as well as the biotic and abiotic stress in the target environment(s).

In the case of Rwanda (Box 11.4.1), the discovery that farmers are well equipped to target a range of germplasm led researchers to rethink their own roles. Rather than focusing on refining the development of finished varieties, scientists could devote their efforts to identifying a diversity of 'exotic' varietal options (including farmer and improved materials) and screening these varieties for traits not readily amenable to farmers' own observation (Table 11.4.2). For example, identifying genes tagged to specific disease-susceptible or antinutritional traits which will be 'invisible' to farmers.

The division of labour that developed in Rwanda, was corroborated by PPB for beans in Colombia. There, an experiment began with joint farmer–breeder selection of parents and F1 progeny. The resultant breeding population was selected independently by breeders and farmers. The farmer-developed lines tended to have more attractive seed colours, patterns and sizes, selected for a very demanding urban market. However, several of the lines bred by farmers also turned out to be susceptible to anthracnose, a seed-borne disease which can be devastating in cool wet seasons. Breeder-developed lines were more anthracnose tolerant and higher yielding, but of inferior grain quality

Table 11.4.2. Conceptualizing a new division of breeding labour.

Breeders	Farmers
• Create new genetic variability • Make accessible a wide range of germplasm (local and exotic) • Screen for key stresses 'invisible' to farmers	• Help to create and select from initial genetic variability • Target for agronomic conditions • Target for socioeconomic circumstances • Screen large amounts of material for 'minimum' criteria

and lower priced. For future work, breeders concluded that they should screen for a minimum set of requirements to maintain varietal viability, while farmers should screen for quality characteristics[12].

The question of whether the division of labour developing in some crops is appropriate for other crops requires further analysis of work in progress. As a first step, the technical expertise and limitations of each partner must be better defined by sites and crops. For example, in Rwanda, where bean farmers were exposed to perhaps a 100 different varieties in a lifetime, and where they regularly adjusted varietal mixtures for different soils, seasons and intercropping sites, their ability to extrapolate from one site to another was not particularly surprising. Yet in neighbouring Uganda, where most women have screened only 10 or so varieties during their farm career and where markets demand a very narrow product, farmer screening of diverse lines on-station has yielded few insights in initial PPB trials. So the question 'What traits should breeders screen for, and what can farmers best define?' has to be considered with reference to the knowledge and experience of the client population.

Similarly, production constraints, farmer expertise and relative success of conventional breeding, should determine the stage at which farmers are involved. Participation earlier in the germplasm sequence does not necessarily mean better participation. That is, crossing of materials with farmers, rather than screening, for example, at the F5 generation, does not necessarily assure more farmer-acceptable material. Intense participation can be very time-consuming – and most farm women (women being the varietal and seed experts in many cultures) don't need additional activities to fill up their time. PPB practitioners need to develop a better sense of the relative gains in breeding as they relate to the timing of participatory involvement. Only then can guidelines be sketched on 'stage of involvement' in this technical division of labour.

In this respect, experience suggests that selection for varietal traits that are highly sensitive to G × E interaction may be more likely to succeed in decentralized breeding with farmers taking the lead. For example, a PPB study carried out in three locations in Colombia showed that adaptation of resistant materials to local races of anthracnose in beans is highly variable from site to site. Consequently, ranking in terms of yield performance of the same set of varieties differed significantly from site to site, as did farmers preferences among these varieties. Farmers in each site selected different materials, instead of the one material which obtained the highest yields on the average. In contrast, there are varietal traits which are relatively stable across different environments and of broad-based or generic interest to resource-poor farmers, examples are varietal earliness and plant architecture. Early maturing varieties are of enduring interest to resource-poor farmers who need greater flexibility to manage their circumstances:

- To obtain rapid turnaround on scarce land and capital.
- For early food supply in the new season after a poor harvest.
- For late planting when lack of rain decimated the early crop.
- To prolong the availability of fresh food supplies (of green maize) when used in combination with regular maturing materials.

Maize breeding in eastern Africa, for example, made gains in adoption by resource-poor farmers, once on-farm researchers identified their preference for early maturing varieties[13].

A potential division of labour in PPB may be one which brings farmers into the identification of breeding objectives; then assigns to breeders leadership in identifying those traits that are amenable to centralized breeding (i.e. G × E insensitive); and to farmers, those which require decentralized breeding (i.e. highly G × E sensitive).

The questions of 'which traits?' and 'which stage of the breeding process?' and 'which farmers should participate?' must be analysed in devising any PPB programme. Not all farmers may be equally suited to join in, for a number of reasons. Communities themselves recognize that farmers may have different varietal expertise – with these differences especially highlighted in areas of varietal diversity. While women may be generally known as seed selectors and storers, many communities recognize important differences among women selectors. For instance, in Rwanda, 'seed experts' tended to be those with good land, access to manures and perhaps the extra labour needed to carefully tend the plants. In contrast, 'varietal experts' emerge even from among the poorest farmers and those fascinated by varietal differences tend to be farmers who constantly experiment. Both these 'expert' categories are distinguished locally from the general female cohort.

Beyond expertise, different farmers have also different needs. For instance, the poor may heavily lean towards short cycle cultivars and those that tolerate intercropping; the rich may go for the highest cooking quality – regardless of yield or time to maturity. A PPB programme may have to represent these varied beneficiary interests. So in choosing farmers, both expertise and representativeness are issues in PPB programmes, and it is likely that there will be trade-offs in trying to meet both aims.

Defining organizational options for decentralization

Site-specific divisions of labour between farmers and breeders affect the organization of PPB programmes, but the experience of most case studies shows that testing has to be decentralized early on. As even routine on-farm testing is being curtailed in many national agricultural research systems, the question of 'how to achieve decentralization' must be placed at the core of any PPB programme.

Relatively few PPB programmes have tried to decentralize through existing formal research or government structures. Government agronomists, while often technically skilled, may have little experience in working with farmers as partners, particularly women farmers, and they are rarely rewarded for listening and feeding back information, rather than issuing concrete research and extension recommendations. Moreover, they are frequently biased towards so-called progressive farmers who are easily accessible. Scarce resources may discourage them even more from venturing to communities located further afield.

To date, decentralization has worked most effectively when adaptive testing is devolved in two situations: when researchers have linked with existing community focused groups, such as NGOs[14] and when researchers themselves have catalysed the development of local farmer research groups (such as the Colombia CIAL case, described above). Several issues have to be clarified when adaptive testing is devolved to other organizations. The first occurs when multisectoral partnerships are established in any kind of joint enterprise, and different organizations, including community groups, have various goals and needs. A research programme has to clarify diverse expectations from the outset: for example, who will contribute what

Table 11.4.3. Participatory breeding programmes: potential evaluation criteria.

Functional perspectives	
Production/impact enhancement	Genetic diversity
• Number of farmer acceptable varieties	• Genetic profile of farmer acceptable varieties
• Number of disease-resistant varieties	• Incidence of landrace parents
• Absolute production gains	• Number of farmer-acceptable varieties per site
• Rates of adoption	• Extent to which new varieties are compatible with old varieties
• Longevity – number of years farmer-selected varieties remain in use	
• Extent to which different types of users are reached: men/women, commercial/subsistence	
Control/empowerment perspectives	
Degree to which: farmers' skills are enhanced to more effectively cross/select themselves farmers gain fuller access to wide pools of germplasm farmers control local testing farmers are involved in decisions on varietal release farmers receive royalties on farmer-scientist developed varieties	

labour and resources; by what parameters will the programme be assessed; and who will have rights to the germplasm identified as promising? Such negotiations are always crucial but particularly so in the breeding arena where Intellectual Property Rights (IPRs) for farmer-breeders remain controversial.

The second issue that has to be clarified if adaptive varietal testing is to be successfully devolved, is quality control. What type of data is achievable with farmer participation led by community groups? What type of data is actually necessary to understand if a variety is farmer-acceptable? Some researchers have suggested that farmers need to be able to internalize Western principles of experimentation if the results of their research are to be credible to the research establishment[15]. Others argue that qualitative farmer assessments, if carried out rigorously, may be more predictive of 'good adaptation' than precisely measured yield data. Given such rationale, farmer evaluations should serve as usable data for varietal release committees.

The third key issue to effective decentralization focuses on the service support sector. If breeding is going to be decentralized, the structure of related delivery systems (seed, extension, credit) must also be decentralized. Partnership in decentralization within the support sector will probably have to be sought both within and outside existing public sector structures in many countries. With respect to seed support *per se*, many PPB practitioners see local seed systems as better equipped to handle PPB products than the formal sector: varieties identified may be very site-specific and heterogeneous, and most current seed regulatory frameworks have tightly defined ownership rules – it is not clear where farmer breeding efforts could fit in.

Assessing the benefits and costs of PPB approaches

The adoption of PPB will ultimately depend on its benefits and costs. This contribution ends by raising issues related to assessment of the possible gains. Much rests on the following:

- How the goals and benefits of a PPB are defined.
- The costs of the approach: to different stakeholders.

At least three types of goal guide participatory breeding: some practitioners focus on production achievements; some work mainly to enhance varietal diversity among user groups; and others put the greatest weight on shifting control of the breeding process to communities and other grassroots organizations. While these goals of production, diversity enhancement and empowerment are not mutually exclusive (and some PPB approaches do try to combine several at once), the design of a PPB programme will reflect which goal is paramount. For instance, groups focusing on community empowerment might take the farmer breeding (not the formal system) as their starting point and promote breeding and selection protocols which are transparent foremost for farmer-breeders. Those with primarily a production focus might look for widely adapted farmer-selected varieties, rather than work toward a varied genetic profile.

Table 11.4.3 illustrates some parameters with which different PPB approaches might be evaluated. In practice, successful participatory breeding will show positive indicators in all three categories, with the relative emphasis varying according to the primary goal.

Direct cost comparisons of participatory approaches are not available at present, nor has participatory breeding been costed against conventional breeding. The need to set up multiple case studies to make cost comparisons was agreed on in 1996 and is being carried out through an international research effort. Cost comparisons of PPB approaches in at least three continents should be available in the next few years. PPB is expected to reduce the costs of the development and diffusion of new varieties. This should happen in several ways: breeding programme goals will be on target from the initial stages, unacceptable varieties will be quickly identified and eliminated; a greater number of more useful varieties will emerge as farmers are exposed to a wider range of germplasm. Some practitioners also anticipate that initial diffusion begins from the moment good material is recognized: farmers take it and run.

There may be a cost–benefit 'downside' to PPB which also needs to be examined. One might ask whether breeding will have to be 'doubly efficient', in order to develop materials for specific as well as wide adaptation. Even within specific locales, varieties may be appropriate only for one specific user group, such as women; but there may be a number of other clients in need of new germplasm. In thinking about how to measure the efficiency of PPB, the various divisions of labour and of organizational linkages (described above) show that there may be different benefits and costs for those participating in a PPB. Different approaches may also deliver various gains to non-participating beneficiaries. PPB may be 'participatory' but this doesn't necessarily make it equitable, for example, 'gender-neutral' or 'wealth neutral'. It is necessary to differentiate the cost and benefits of PPB and conventional breeding for different participants and end-users, such as rural women and low-income urban consumers.

11.4.5 The future

PPB embraces many of the vogue concepts in agricultural research: enhanced impact in marginal environments, cost-saving, use and conservation of biodiversity, and benefits provided to a greater range of user groups, especially the poor. The initial decade of experimentation has been extremely positive. As the first steps are made towards institutionalization, PPB has to become more targeted, and has to encompass organizational issues as well as the breeding process. Breeders no longer need to ask whether farmers have knowledge important for formal breeding: we know they do. Moreover, breeders should be moving beyond the stage of simply 'consulting farmers' to face today's challenges: they should be defining the appropriate division of labour between farmers and breeders for a given objective in a breeding programme and consciously developing new organizational approaches suitable for participatory plant breeding. Finally, an essential task is to define how to involve farmers so as maximize the benefits for a broad range of users. Such substantial challenges may well realize substantial gains.

REFERENCES

1. **Byerlee, D. & T. Husain, 1993**. Agricultural research strategies for favoured and marginal areas: the experience of farming system research in Pakistan. Experimental Agriculture, 29, 155–71.
   Ceccarelli, S., 1994. Specific adaptation and breeding for marginal conditions. Euphytica, 77, 209–15.
2. **Richards, P., 1985**. Indigenous Agricultural Revolution. Ecology and Food Production in West Africa. Hutchinson and Westview Press, London and Boulder.
   Sperling, L., 1992. Farmer participation and the development of bean varieties in Rwanda. In: J. Moock & R. Rhoades (Eds) Diversity, Farmer Knowledge, and Sustainability. Cornell University Press, Ithaca.
   van der Heide, W.M., R. Tripp & W.S. de Boef, 1996. Compilers; Local Crop Development: an Annotated Bibliography. IPGRI, Rome Italy/CPRO-DLO(CGN), Wageningen, The Netherlands /ODI, London, UK.
   Voss, J., 1992. Conserving and increasing on-farm genetic diversity: farmer management of varietal bean mixtures in Central Africa. In: J.L. Moock & R.E. Rhoades (Eds) Diversity, Farmer Knowledge and Sustainability. Cornell University Press, Ithaca.
3. **Kornegay, J., J.A. Beltran & J. Ashby, 1996**. Farmer selection within segregating populations of common bean in Colombia. In: Participatory Plant Breeding: Proceedings of a Workshop on Participatory Plant Breeding, 26–29 July, 1995, Wageningen, The Netherlands. IPGRI, Rome.
4. **Sthapit, B.R., K.D. Joshi & J.R. Witcombe, 1996**. Farmers' participatory crop improvement III: participatory plant breeding, a case study for rice in Nepal. Experimental Agriculture, 32, 479–96.
5. **Sperling, L., M. Loevinsohn & B. Ntabomvura, 1993**. Rethinking the farmer's role in plant breeding: local bean experts and on-station selection in Rwanda. Experimental Agriculture, 29, 509–19.
6. **Weltzien, R.E., M.L. Whitaker & M.M. Anders, 1996**. Farmer participation in pearl millet breeding for marginal environments. In: Participatory Plant Breeding: Proceedings of a Workshop on Participatory Plant Breeding, 26–29 July, 1995, Wageningen, The Netherlands. IPGRI, Rome.
7. **Sperling, L., *et al.*, 1993**. (Op. cit.).
   Sperling, L. & U. Scheidegger, 1995. Participatory selection of beans in Rwanda: results, methods and institutional issues. Gatekeeper Series, no. 51. International Institute for Environment and Development, London.
   Sperling, L. & P. Berkowitz, 1994. Partners in Selection: Bean Breeders and Women Bean Experts in Rwanda. CGIAR Gender Program. Consultative Group on International Agricultural Research, Washington, DC.

8. **Sthapit, B.R., *et al.*, 1996.** (Op. cit.).
9. **Ashby, J.A., T. Gracia, M.P. Guerrero, C.A. Quir¢s, J.I. Roa & J.A. Beltran, 1996.** Innovation in the organization of participatory plant breeding. In: Participatory Plant Breeding: Proceedings of a Workshop on Participatory Plant Breeding, 26–29 July, 1995, Wageningen, The Netherlands. IPGRI, Rome.
10. **Ceccarelli, S., 1989.** Wide adaptation. How wide? Euphytica, 40, 197–205.
11. **Witcombe, J., A. Joshi, K.D. Joshi & B.R. Sthapit, 1996.** Farmer Participatory Crop Improvement I. Varietal Selection and Breeding Methods and their Impact on Biodiversity. Experimental Agriculture, 32, 445–60.

    Witcombe, J. & A. Joshi, 1996. Farmer participatory approaches for varietal breeding and selection and linkages to the formal seed sector. In: Participatory Plant Breeding: Proceedings of a Workshop on Participatory Plant Breeding, 26–29 July, 1995, Wageningen, The Netherlands. IPGRI, Rome.
12. **Kornegay, J., *et al.*, 1996.** (Op. cit.).
13. **Haugerud, A. & M.P. Collinson, 1990.** Plants, genes and people: improving the relevance of plant breeding in Africa. Experimental Agriculture, 26, 341–62.
14. **Choudhary, M.K., E. Weltzien, R. & M.M. Sharma, 1996.** Evaluating pearl millet varieties with farmers in Barmer district. In: L. Sperling & M. Loevinsohn (Eds) Using Diversity: Enhancing and Maintaining Genetic Resources On-farm. Proceedings of a Workshop held on 19–21 June, 1995. IDRC/SARO, New Delhi.
    Witcombe, J. & A. Joshi, 1996. (Op. cit.).
15. **Ashby, J.A., T. Gracia, M.P. Guerrero, C.A. Quir¢s, J.L. Roa & J.A. Beltran, 1995.** Institutionalizing farmer participation in adaptive technology testing with the CIAL. Agricultural Administration (Research and Extension) Network paper 49. Overseas Development Institute, London.

11.5 AGROECOSYSTEMS ANALYSIS: A SYSTEMS APPLICATION WITH A FUTURE?

Clive Lightfoot

Detailed descriptions of agroecosystems in the form of maps and transects are not enough. Future visions and experimental models of more sustainable farming systems are not enough. Similarly, Venn diagrams and linkage maps of stakeholders are not enough. But, linking stakeholders to a particular vision of improved agroecosystem management and negotiating roles for each one might be enough if performance indicators both for stakeholders and the farming system are regularly monitored.

11.5.1 Introduction

The spectacles through which I view the last 15 or so years of research in the field of agroecosystem analysis are purely personal. Though the views expressed here have been shaped by many others, I have largely drawn from my own experience to provide examples, focusing on the farm and community levels, and emphasizing participatory methods, where my interests lie. A disciplinary background as an agronomist puts a biophysical coating on my lenses.

I draw upon the work of others to establish why the management of agroecological niches within larger agroecosystems provides an entry point for more ecologically sound agriculture. The first part of this contribution demonstrates the need for agroecosystem analysis, seeking to show that one ignores agroecosystems at the peril of natural resource degradation and pollution, of losing traditional knowledge, and of missing opportunities for innovation. I then explain what agroecosystem analysis is. Brief reviews of the scientific traditions that formed agroecosystem analysis are followed by an outline of the achievements of its pioneers, a discussion of the problems they encountered and recent advances to overcome these. Finally, looking into the future, I see agroecosystem analysis integrated with other research tools, particularly stakeholder analysis, to extend its utility. It will evolve from a data gathering tool within a FSR context to a learning tool within an action research context. Tools for identifying stakeholders, eliciting multiple perspectives of the agroecological resource base, brainstorming alternative management options, negotiating concerted action and tracking change are illustrated here, but can only hint at what is possible.

Other important applications for agroecosystems analysis are omitted. I expect strong developments in the field of GIS and agroecological zoning, and in the field of ecological modelling. Moreover, I believe advances in what is now called 'precision farming' will continue to explore linkages between these two. None of these are elaborated here. However, even within the narrower focus of an emerging action research context, improved sustainability in farming systems, the original rationale for agroecosystems analysis, may be significantly enhanced. The magnitude of the opportunities will be influenced by the way local institutions work together, and by the way local people are able to involve themselves in policy dialogue.

11.5.2 Why agroecological niches are important

The active management of agroecological niches means making the best use of the variation in natural resource endowment within an agroecosystem. We are all familiar with the small garden plots that surround homesteads; the trees, hedges and grass strips that bound fields, the watercourses and swamp areas that dissect the landscape, the terraced plots on hillsides. These are the niches that bring diversity to the natural resource endowment of any landscape, and their destruction can lead to resource degradation and pollution. Farmers of traditional systems understand this. Agroecological niche management was an inherited strategy for ecologically sound agriculture. A number of recent successes described in this contribution are built on these strategies. Recent but limited research experience in Ghana, Malawi and the Philippines suggests that agroecological niche management does offer an entry point to more ecologically sound and sustainable farming.

The destruction of agroecological niches

People destroy agroecological niches in both favourable and harsh environments. In harsh environments the destruction happens because people are poor: in the face of increasing populations and limited resources, sheer survival forces them to 'mine' their soils long after serious erosion has set in. Fallow periods are reduced, rotations and crop mixtures give way to continuous monocropping. Survival forces them to deforest sacred groves, once protected by tradition and social norms, for fuelwood. A vicious cycle sets in as eroded soils silt up streams and deforested spring lines dry up. Almost unnoticed, parallelling the destruction of agroecological niches, is the breakdown of community cohesion and social capital. As degradation deepens and rehabilitation costs rise, it becomes more difficult for local people to break this vicious cycle.

Breaking 'poverty cycles' is not just a matter of investments for agroecological rehabilitation – bringing back the streams and forests – or even investments in modern agricultural technology. Nor is escape just a matter of developing markets for a more diverse array of biological products. Reversal requires more. I am not only thinking of the rebuilding of social organization, institutions for local empowerment and community spirit, but also the careful development of markets and technologies.

I say careful because much of the destruction of agroecological niches in favourable environments has been driven by modern technology and market opportunities. When used inappropriately, pesticides and fertilizers pollute environments such as irrigated rice fields, and damage human health[1]. Modern technology also reduces agroecosystem diversity. The landscape is homogenized to exploit the economies of scale from mechanization and monocropping, to satisfy markets that favour large volume suppliers. Farmers clear, plough and level large numbers of fields, fallows and wetlands, to plant rows and rows of a single crop. How many highly diverse wetlands have been drained for cropping? More often than not high external input monocropping is the kind of agriculture promoted by government extension agencies and supported by government policies.

Farmers in the Philippines have transformed diverse agroecological niches including ricefields, fish ponds, vegetable gardens and orchards into intensive high external input vegetable or pig operations[2]. Again it was their own survival that motivated this destruction. Farmers knew that in a few years they must sell their farms for industrial development. With no future in farming the pressure was there to generate cash as quickly as possible.

The salaried labourers on nearby industrial sites provided a ready market for high priced meat and vegetables. With no future, and the need to make money quickly, it made sense to seize the immediate market opportunity and use modern technology to maximize production. Fortunately, the short duration and small scale of these operations meant that there was little pollution or damage to human health. The older and larger pig, poultry, or vegetable operations surrounding most Asian cities are more persistent and may be a long-term threat to the environment.

Learning from tradition

Much of the anthropological record of traditional farming illustrates how well farmers understood the importance of agroecological niches. Small plots were micro-managed for food production and maintenance of natural resources. Crops were planted into mounds of ash in central African 'Chitemene' systems[3]. Crops were planted into beds in Indonesian 'Sorjan' systems. Terraces were built for cropping on Andean and Asian hillsides. We all know that many of these systems could not cope with the increasing demand for food from rapidly growing populations. Shifting cultivation systems that had sustained small populations for centuries, now degrade many forest areas. Too many poor people move in after the 'loggers' and crop for too long for any significant forest regeneration to occur. While traditional systems have no place in today's agriculture, agroecological niches do. Farmers in the densely populated district of Machakos, Kenya have restored lands badly degraded 50 years ago by terracing sloping plots and integrating stall-fed livestock into their farming systems[4]. As in Machakos, increased diversity in agroecological niches underpins the success of farmers in China operating mulberry dyke–pond systems, farmers in Viet Nam operating *vuon, ao, chuong* (garden, pond, livestock pen) systems and farmers in Indonesia operating Sawah Tambak rice–shrimp systems.

Just like farmers, researchers have learned from traditional agroecological niche management. A traditional technique in parts of Papua New Guinea involves the planting of food crops on terraces formed between close planting of *Cordyline fruiticosa* shrubs[5]. Many researchers designing agroforestry systems and agricultural technologies for sloping lands learned from this tradition[6]. This is not to say that farmers have nothing to learn from researchers; far from it. Many farmers are operating in environments estranged from their traditions. Degradation may have changed the landscape, poverty may have forced them on to land once considered too poor for agriculture, or they may be immigrants from resettlement schemes or refugees. Beyond these factors many biophysical processes can only be understood using modern scientific techniques and laboratory equipment. Farmers have no knowledge of many of these processes, especially those that they cannot see. Indonesian farmers involved in FAO's Integrated Pest Management Programme did not know that their ricefields harboured many beneficial insects[7]. Once they learned this in 'Farmers' Field Schools' they went on to reduce pesticide application by 50% without yield loss. Farmers in their thousands in Bangladesh, China and Sri Lanka have emulated the success of IPM in Indonesia. Farmers learning from researchers and researchers learning from farmers about agroecosystems have provided a route to more sustainable rice farming.

Sustainability through agroecological niche management

Attempting to improve the management of more than one agroecological niche – the ricefield in the case of IPM – was the challenge for those working on designs of whole farm systems. Most of these entailed the integration of many species in a diverse array of agroecological niches[8]. One particularly complex form was developed by Bill Mollison in what he called 'permaculture' systems[9]. Another involved the integration of aquaculture into mixed farming systems. Integrated aquaculture-agriculture farming included animals, crops, vegetables, trees and fish[10]. A modest attempt to quantify the sustainability of integrated aquaculture-agriculture farming systems involved case study farmers for a period of 2 years in Ghana, Malawi and the Philippines[11]. Modest though this attempt was it provided new information on the importance of agroecological niche management for sustainability.

Rehabilitation of water resources in the Philippines required the digging of ponds and

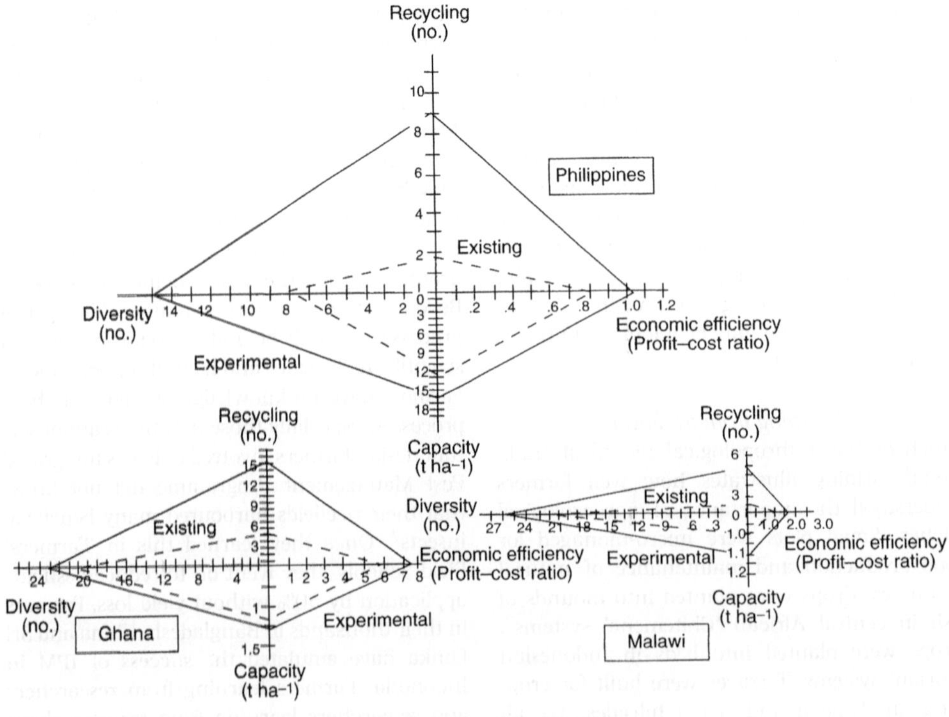

Fig. 11.5.1. Ghana, Malawi and the Philippines: sustainability kite plots.

trenches in ricefields to improve water control. Fish and ducks were introduced into the farming system. Ducks were fed on ipil-ipil leaves and rice bran while the fish were fed household wastes and vegetable wastes; and the pond fertilized with cattle and goat manure. In Ghana and Malawi the farmers rehabilitated spring lines and 'dambo' wetland agroecological niches, introduced fish, vegetables and animals, and recycled cattle manure, household waste, weeds, tree leaves and crop by-products like maize bran and stover. In each case sustainability was measured using four indicators:

- Economic efficiency being a simple cost–profit ratio.
- Agroecological niche capacity being the amount of biomass produced.
- Species diversity being the number of species utilized.
- Bioresource recycling being the number of flows of wastes recycled.

Each indicator was measured before the experiment began, to get a value for the existing farming system, and again 1 year later, to get a preliminary value for the experimental integrated farming system.

Figure 11.5.1 shows the changes in sustainability indicators after 1 year of experimentation in a four-way 'kite plot'. In all the case studies resource rehabilitation, increases in recycling and wider species diversity pushed the value of each sustainability indicator beyond that of the existing farming system. Although still unreported, subsequent participants in these experiments recorded similar results. It appears that rehabilitating water resources and increasing recycling and diversity not only improves ecological soundness but may also improve economic efficiency. More detailed ecological research conducted on the Philippines case-study farm confirmed that, compared to rice monocropping, the integrated farm delivers higher scores in key ecological attributes like species richness, functional agricultural diversity (Shannon's index), nutrient cycling (Finn's index) and harvest indexes. The integrated farm scores lower, however, when it comes to net system yield[12] – you can harvest more from a well-fertilized monocrop.

We all know that farmers will not change their farming systems for purely ecological reasons. Just as we know they will not change their farms purely to maximize yield. Farmers, particularly resource-poor farmers, will change their farming systems to enhance food security. This is what happened in Malawi. Cash was saved to buy food because composts and pond mud substituted for purchased fertilizer in the production of vegetables. More cash was earned because vegetables grown with water from the pond could be sold at high prices during the 'drought'. And, more impressive still, households did not lose their savings during the 1992–93 drought. Sales from vegetables grown in their drying ponds secured their food supplies[13].

The rehabilitation of 'wetland' agroecological niches to conserve water resources has had similar impact in other parts of Africa. In Zimbabwe's 1992 drought, households with 'dambo' gardens did not go hungry. Cash from vegetable sales amounting to some Z$70.00 week^{-1} was used to buy maize. These households did not suffer malnutrition and so maintained high labour productivity. They also held on to their capital[14]. One farmer reported that 'Dambos and streams can mean an improvement in the food security and self reliance of communal farmers'[15]. Water conservation through terracing and stream diversion in Machakos, Kenya, prevented complete crop loss where the second rains failed[16].

A need for agroecosystems analysis

Regardless of favourable or harsh environments, farming that degrades or pollutes natural resources and threatens human health is no longer acceptable to society at large. However, although it may seem obvious that the diversity in agroecological niches conserves both structural and species biodiversity, objective evidence for it remains limited. Similarly, the notion that species and agroecological diversity will increase income opportunities, food security and sustainability is poorly supported. More and better evidence is needed. While we know something about degradation processes and techniques to rehabilitate degraded agroecological niches, we know little of the human and institutional management requirements for change. It was the concern for the ecological dimension of farming systems and their sustainability that stimulated the development of agroecosystem analysis.

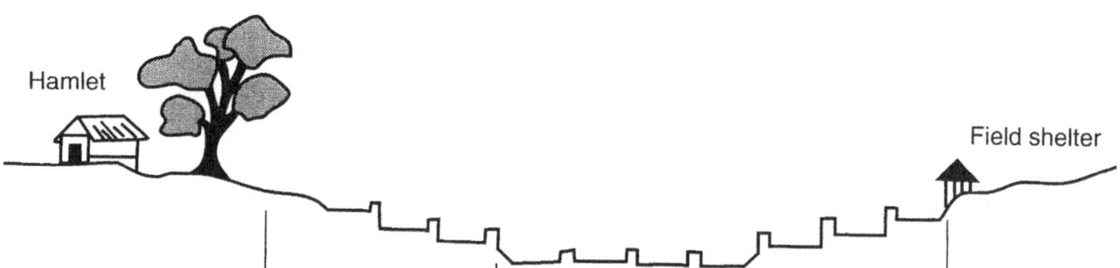

	Upland	Upper Paddy	Lower Paddy	Upper Paddy	Upland
Soils	Paleustult	Paleaquult/ Paleustult	Paleaquult		
Crops	Cassava Kenaf Sugarcane Water melon	Cassava Rice	Rice followed by vegetables		
Problems	Drought Erosion	Insufficient water	Occasional flooding		

Fig. 11.5.2. Conway transect.

11.5.3 Agroecosystem analysis: past and present

It was the search for sustainability which drove the development of agroecosystems analysis in the early 1980s. The case for it rests on the hypothesis that agroecological niche management contributes to sustainable farming.

Scientific traditions

About 15 years ago one could find the term agroecosystem in three different scientific traditions. The first is a 'classification' tradition in which agroecosystem analysis was a technique for characterizing and classifying agroecological environments. Perhaps the most ambitious expression of this tradition is the agroecological zoning and mapping exercise undertaken by FAO[17]. Modern tools like GIS and advances in crop modelling offer increasing scope in this classification. The second is the 'ecological' tradition in which scientists focused 'agroecosystem analysis' on the crop field. While they recognized Bob Hart's hierarchy of systems[18] (1982 and Chapter 3.1) and indeed may have worked at other hierarchical levels, their success has been in understanding the ecology of a field for the improved management of pests. IPM strategies emerged from Peter Kenmore's and Miguel Altieri's work at the level of the crop field[19]. The third is the 'farming systems' tradition, and I focus on this here.

Within the farming systems tradition agroecosystem analysis has concentrated on the farm and watershed level. The analysis has brought a much needed 'ecological' perspective to farming systems research, helping researchers understand the wider ecological setting – again a recognition of the hierarchy of systems. This understanding was achieved by breaking down wider agricultural landscapes into smaller agroecological niches – the 'land types' of the now familiar agroecosystems transect shown in Fig. 11.5.2. Management of these niches was further analysed through seasonal calendars and causal linkages between the problems encountered. A brief account of the achievements of this tradition in agroecosystems analysis covers the work of Gordon Conway, Percy Sajise and Terry Rambo[20].

Fifteen years of achievement

For the first time agroecosystems analysis provided farming systems researchers with tools to help them understand the ecological dimensions of farming. It has helped scientists look beyond their disciplinary interest in the crop field or the livestock unit and has extended system boundaries to include a wider context of natural resources – streams, rivers, forests and grazing lands. The maps, transects, seasonal calendars, flow diagrams and decision trees, brought a new richness to the description of farming systems. Unfortunately moving from 'rich description' to research questions and the determination of systems properties of productivity, stability, sustainability and equitability has proved more difficult.

Even if moving from description to action has proved difficult, the descriptive tools did achieve another important goal by improving communication between researchers and farmers. The maps, transects and seasonal calendars were easy for farmers to understand. Indeed, as the practice of working with farmers improved, and RRAs evolved into PRAs, farmers began to draw the maps and calendars themselves. Moreover, the use of local terms helped build a common understanding between farmers and researchers. Today, agroecosystem maps, transects and calendars can be found in the reports of many community-based agriculture research and development projects[21]. This achievement is due, in no small measure, to the work of the South-east Asian Universities Agroecosystem Network, SUAN[22]. Established in 1982 by six institutions from Indonesia, the Philippines and Thailand with the support of the East West Center in Hawaii and the University of London's Imperial College of Science and Technology, the network now involves scientists from Bangladesh, Cambodia, China, Laos, Nepal and Viet Nam. In recent years the International Institute for Environment and Development in London and Institute for Development Studies at the University of Sussex, England, have spread the use of agroecosystems tools in Africa and Latin America. While agroecosystems analysis has many achievements, its story is not one of undiluted success.

Three problems

Even though agroecosystems analysis improved researchers' understanding of how farming systems worked they found it difficult to use it to identify improvements. Typically, PRA exercises generate lots of visuals – maps, transects and calendars – which describe how agroecosystems are managed, but fail to identify directions for experimentation or development. Lots of information describing the farming system and so what? I call this the 'so-what' syndrome. In the absence of an informed link between description and action, diagnostic procedures have often been channelled by the commodity or disciplinary interests of the researchers themselves. The technologies tested in subsequent OFEs have tended to be a function of researchers' prior interests rather than products of the agroecosystems analysis.

A second problem for practitioners I call 'blinkers and blindfolds'. Many find it difficult to see ways to improve farming systems within the complexity of agroecosystems, they are effectively 'blindfolded'. Others fail to address the whole agroecosystem. Their 'blinkers' only show them a component like a forest, a ricefield or a pond that is of special interest. As a result impacts, good or bad, across the rest of the system go undetected. Worse still, viable improvements go unexplored because they impact parts of the agroecosystem beyond the blinkers.

One further area of difficulty is 'who participates' in agroecosystems analysis. Often analyses, even in so-called participatory appraisals, are extractive with too little attention to their use within the community itself. In the generation, interpretation and use of information gathered in PRA exercises practitioners pay too little attention to stakeholder representation. This includes the 'gender-benders'. Even though most practitioners do recognize the importance of gender they often fail to capture the separate perspectives of men and women. They condense stakeholder information, 'bending' perspectives into one type, usually that of the man. Once differences are lost it is hard to identify and reassert them when interventions are being developed.

Recent advances for better practice

The last 5 years or so have seen some significant advances in the application of participatory research methods. Participatory appraisals are giving way to participatory learning exercises[23]. Many local stakeholders in these, including researchers, engage in experiential or action learning activities in which they learn together. We have moved away from extractive exercises by researchers to joint brainstorming tools and investigation and analysis by local people themselves.

Much of the impetus for this move to action learning approaches comes from the utilization of 'systems thinking' techniques in agriculture[24]. Experience shows that methods are context specific and must be adapted or re-invented to fit each situation[25]. Participatory learning approaches work best when groups of stakeholders work together. Learning embraces the different perspectives of different stakeholders and innovation and change emerge from an appreciation of multiple perspectives and from negotiation between different stakeholders on what to change and what innovations to test. In recent years considerable effort has gone into understanding stakeholder interactions and the social organization of innovation[26]. This is in part due to our realization that efforts failed because we paid too little attention to the different interests of the various stakeholders involved[27]. Within a learning approach it is the stakeholders themselves who use the tools to better understand each other's views and values. With greater understanding comes the potential to act together.

Agroecosystems analysis has advanced through the use of analytical diagramming or modelling tools for exploring interventions to promote sustainability. Groups of stakeholders, including researchers, brainstorm ways to improve the sustainability of farming systems using 'bioresource flow models'. These are conceptual diagrams depicting agroecological niches, species captured or cultivated, and the bioresources such as fodders, feeds, fuels, composts and manures that flow between species and agroecological niches. One such model is shown in Fig. 11.5.3. It is a visual representation of an existing situation that the stakeholders, including researchers, involved in its construction, understand. Bioresource flow models provide a useful tool for all to accumulate options to increase the sustainability of the farming system. All ideas for future action and

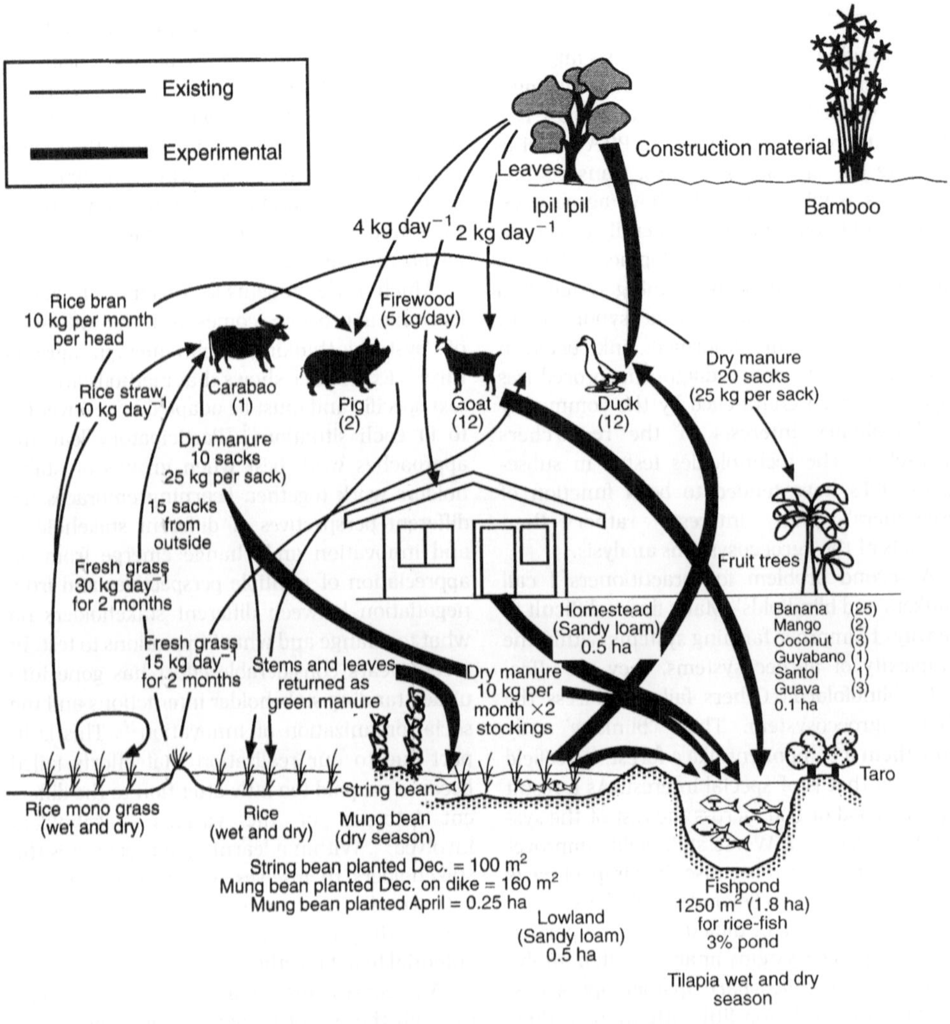

Fig. 11.5.3. Philippine bioresource flow model.

experimentation in agroecological niche rehabilitation, diversification of species and increasing recycling are visualized on the bioresource flow diagram. Thus a vision of a more sustainable farming system emerges, and is common to all stakeholders. How this is achieved is discussed in more detail in the next section. Bioresource flow modelling can provide a link between understanding agroecosystems and subsequent action to improve the system's sustainability. This is illustrated in the following examples.

In Cavite in the Philippines, brainstorming using bioresource flow diagrams with farmers, researchers and extensionists produced the model shown in Fig. 11.5.3. Superimposed on the existing situation, in thicker lines, are the areas for experimentation. Thus a lowland agroecological niche, sketched with thicker lines, will be rehabilitated through the construction of deep trenches and a small pond in the rice paddy. The impounded water will provide dry season irrigation for the introduction of taro and string beans and enough water to grow fish (tilapia) in the rice field. New bioresource flows, again shown in thicker arrows, will see rice bran, household waste and livestock manure used to feed the fish; green manure to fertilize the ricefield; and rice bran and ipil-ipil leaves to feed the newly acquired ducks[28]. The impact on the sustainability of the farming system when these changes were implemented was shown in Fig. 11.5.1. The kite plots there also showed similar impacts on sustainability in Ghana and Malawi. These impacts were achieved using models to brainstorm ways to increase sustainability[29]. Brainstorming with bioresource flow models in Mali on ways to improve soil fertility management gave similar results. Farmers started contour farming to rehabilitate degraded croplands. Recycling flows were increased through composting of crop residues rather than burning crop residues. Cattle were penned so that more manure could be collected for use on crops. New species – fodder maize and dolichos – were introduced to feed the growing number of stall-fed cattle[30].

Action followed understanding in the examples presented, because changes in agroecosystem management were linked to negotiated and agreed roles for the different stakeholders. The models allowed the perspectives of each stakeholder to be seen and understood by all. Brainstorming improvements to the agroecological niches allowed stakeholders, including researchers, to learn from each other and negotiate roles so that concerted actions could follow.

The importance of coupling agroecosystems analysis with stakeholder analysis for the achievement of sustainable farming has been raised in several symposia of the AFSRE. Susan Poats stressed this need in her summary of papers on agroecosystems analysis at AFSRE's 13th Symposium in Montpellier, France[31]. In their paper to the Second European Symposium, Niels Roling and Janice Jiggins argued that without such a linkage 'ecological knowledge systems', a necessary condition for sustainable agriculture, could not be established[32]. A similar plea was made by Ray Ison and colleagues who called for iteration between biophysical systems and human activity systems[33]. They state that more research in the use of systems learning and process monitoring methods is needed. Much of this discussion has, however, been of a rather theoretical nature and the next section proposes a practical course of action for weaving together agroecosystems analysis and stakeholder analysis.

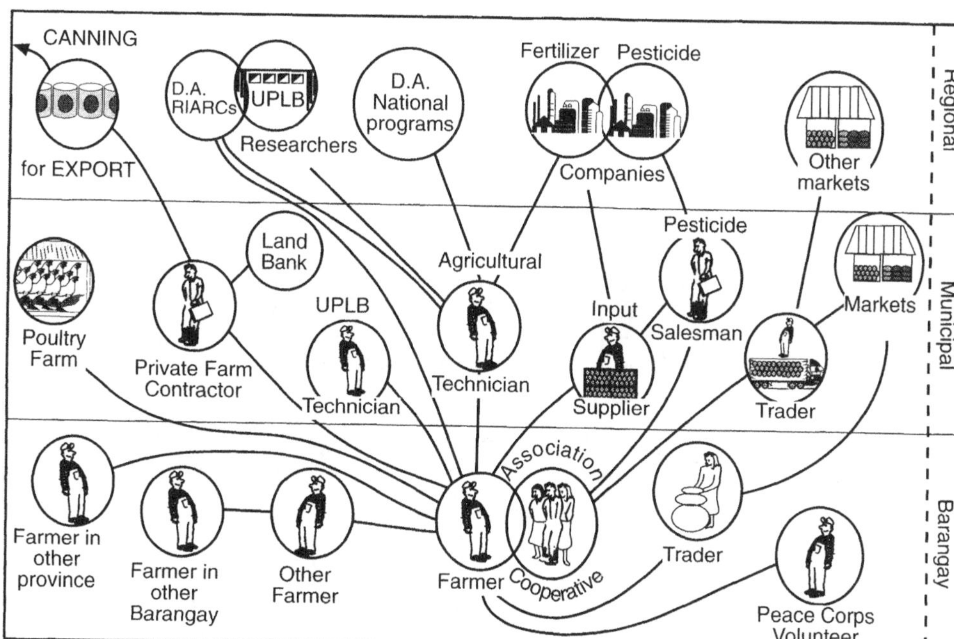

Fig. 11.5.4. Philippine stakeholder linkage map.

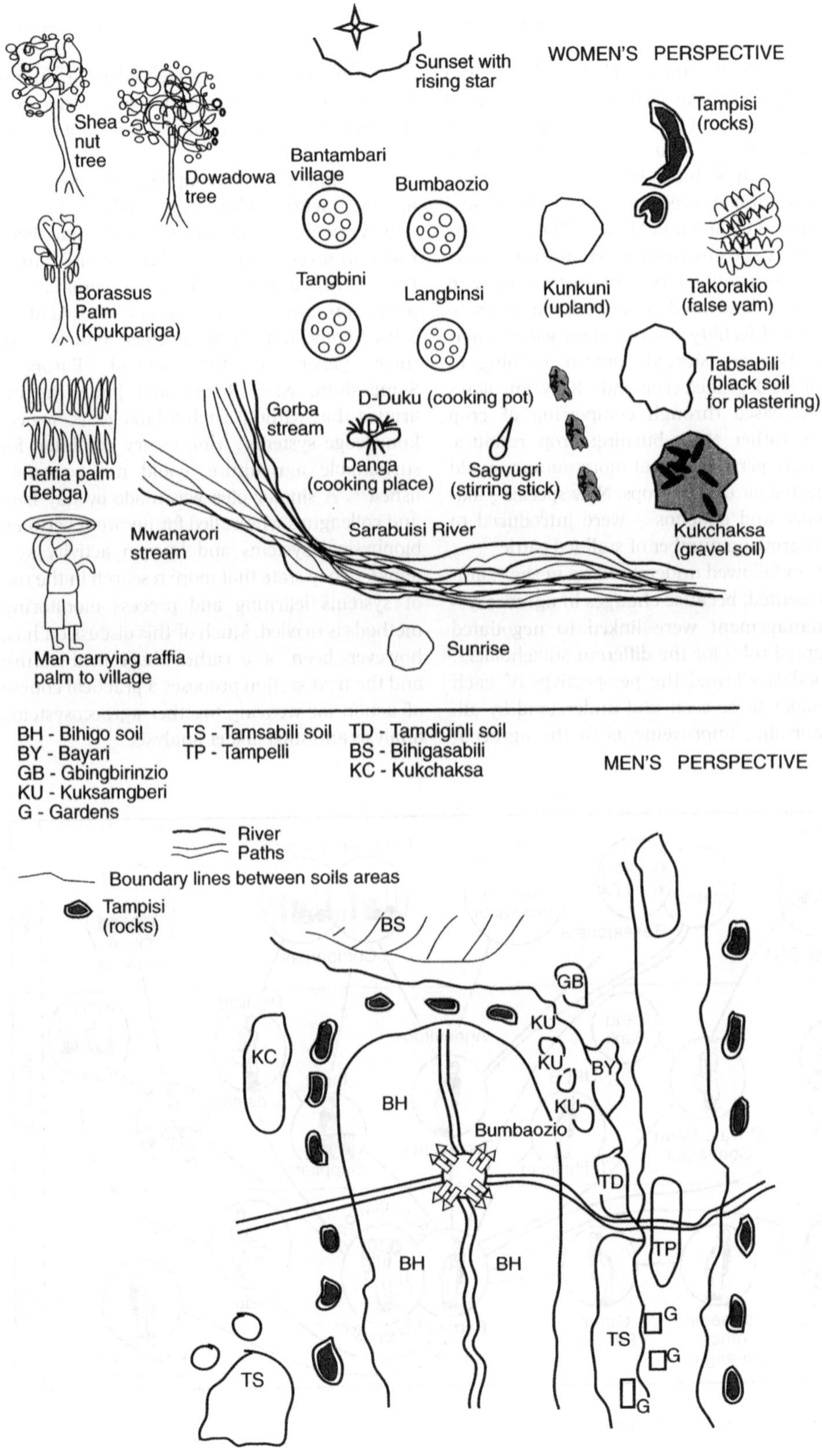

Fig. 11.5.5. Ghana: gender perspectives on agroecological resources.

11.5.4 Agroecosystem analysis in a learning and action research process

This part of the contribution uses recent advances in action learning, agroecosystems analysis and stakeholder analysis as a springboard for outlining a hypothetical process. The proposed process enables multiple stakeholders to engage in joint learning and action research to guide the management of agroecological niches towards more sustainable farming systems. It iterates between agroecosystems and stakeholder analyses. To my knowledge such a process has not been implemented as a whole, so I cannot say it works, nor can I describe how to weave the various tools together into a coherent process. I can only illustrate each tool using examples from very disparate sources. My illustration will cover tools for:

- Identifying stakeholders.
- Eliciting multiple perspectives of agroecological resources.
- Brainstorming alternative agroecosystem management options.
- Negotiating concerted actions for alternative management.
- Tracking change using indicators.

Identifying stakeholders

Identifying stakeholders uses a tool drawn from the methods of rapid appraisal of agricultural knowledge systems[34]. The stakeholder linkage map shows all the key actors in the agricultural knowledge system and the linkages between them. In the case of agroecosystem management such a map would show all the key actors using, or controlling the use of, agroecosystems and the linkages between them. Once a first map is prepared each of the stakeholder groups can be interviewed to obtain their perceptions of the linkage map and to better understand their motivations and interests. A deeper understanding of stakeholders and their interests also informs the process of gathering multiple perspectives on agroecological resources. As we learn who the different stakeholders are, so we learn whose perceptions on agroecological resources are needed. Thus we link the use of stakeholder and agroecosystem analyses tools.

The stakeholder linkage map illustrated in Fig. 11.5.4 comes from a case study in the Philippines[35]. Here farmers were asked to identify the main actors who provided them with information about agricultural innovations. As one would expect, farmers get most of their information from other farmers and even from farmers in neighbouring villages. Farmers also hear from government and the private sector technicians: in this case the pesticide industry and a pineapple canning company looking to contract farmers to grow pineapples. Knowledge sources are extended through formal association with other farmers who are linked to traders and salesmen. In their turn these local level actors are linked to national level institutions in the private and public sectors.

Eliciting multiple perspectives of agroecological resources

Enabling local stakeholders to communicate and record their own categories of agroecological resources involves the use of maps and transects. First the stakeholders themselves need to feel comfortable with each other, with the outside facilitators, and with the task at hand. Second, stakeholders need to feel at ease with the medium for recording the categories, in this case a map. Finally, the idea of 'agroecological resources' must be clarified. One way to get started with this is to ask different stakeholders to show each other places of interest and importance to them. Once on the spot it is easier to elicit names of plants, animals, soils and water resources. After a series of visits a local vocabulary builds up of place names, land use names, and soils names from which names for local categories of agroecological niches emerge. After several visits to the field the amount of information shared exceeds most peoples' memories. This is the point where visualization techniques like mapping become useful. These local maps communicate best with local audiences. Communicating with wider audiences such as policy makers or research planners requires more formal representations of local knowledge in maps and transects.

Figure 11.5.5 compares the perspectives of men and women on a common set of agroecological resources. The example comes from Bumbaozio village in northern Ghana. Here separate groups of men and women were asked to point out places of interest and importance to them. The women walked to shea nut trees and raffia palms – women's crops used in cooking and basket making. They visited the stream that supplies water for the household and an area of black soils which the women use for plastering the house. On returning to the village the women visualized what they had shown us in a 'map'. In contrast to the women who showed us plant species the men walked us to areas of land. 'In this area we have such and such soils on which we grow such and such' they said. Because men are concerned with the control and management of land it is not surprising that their drawing of agroecological resources looks more like a conventional map than that of the women.

Brainstorming alternative agroecosystem management options

Bioresource flow models, an agroecosystems analysis tool, can help local stakeholders, including researchers, learn about more ecologically sound management options for agroecosystems. These 'conceptual' models show not only the local categories of land and water resources or agroecological niches and species in the farming system, but also the flows of biological resources (fodders, fuel, waste and by-products) between agroecological niches and species. These then are models drawn by farmers of how they conceive their existing farming systems – the niches they work, the species they capture, collect or cultivate, and the wastes and by-products they recycle. An illustration of a bioresource flow model has been provided in Fig. 11.5.3.

Learning occurs between the stakeholders when they challenge each other to think of ways to increase species diversity, increase nutrient recycling flows, and improve soil and water resource quality. It is easy to see in the diagrams what species are present and discuss what new ones could be integrated into the system. All can discuss what biological resources they have and how they can increase their use through recycling. A more ecologically benign balance of internal and external inputs can be struck. All can discuss which agroecological niches are degraded and how to rehabilitate soil and water resources. In these 'brainstorming' sessions farmers can see how to save money through the substitution of external inputs like inorganic fertilizer with recycled internal inputs like manure and composts. Farmers often challenge researchers to tell them how a particular piece of land or water resource could be rehabilitated. Researchers are asked for their ideas on what new species could be cultivated or what commodities are in the market that they might cultivate. The new ideas are draw in on top of the existing bioresource flow model to produce an 'experimental' model: a common vision of a more ecologically sound farming system. Then, moving from vision to action requires that we switch from agroecosystems to stakeholder analysis.

Negotiating concerted actions for alternative management

The stakeholder analysis tool for negotiating roles and responsibilities between the different stakeholders or actors, is called the analysis of linkage performance[36]. In this analysis different stakeholders agree to provide each other with different services. For example a researcher may agree to provide germplasm and diagnostic 'help' services; or a farmer may agree to share his labour, experiences or germplasm with other farmers; or an extension worker may agree to provide training and equipment; and so on. Agreeing on who will provide what service and how each actor is linked to another dominates the negotiation process. Linking actors and developing criteria through which that linkage can be assessed, provides an incentive for people to work together. Beyond an awareness of each other's contribution, key criteria will be the relevance of each actors service, the accessibility of their service and timeliness in their fulfilment of their role.

The illustration of analysing linkage performance comes from a rapid appraisal exercise in the Philippines[37]. The exercise focused on linkages between farmers and private farm contractors. Their analysis of linkage performance is shown in Fig. 11.5.6. Here farmers and private contractors assess each other's roles. Each actor makes the other aware of their role or service. The farmer finds the proposal of contract growing of pineapples promising, but it is the contractor who chooses the farm to participate. Arrangements are made to make

CRITERIA FOR ASSESSING PERFORMANCE							
Linkage	**Awareness of other actor's service**	**Relevance of other actor's service**	**Timeliness of other actor's service**	**Accessibility to other actor's service**	**Communication medium through which link is mediated**	**Linkage control**	**Remarks**
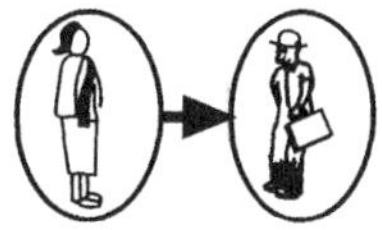 Farmer 'TPM' private farm contractor group	F aware of TPM commercial interest, F appreciates the TPM's resources invested in the linkage (without understanding where the hidden costs are)	F considers TPM's proposal very promising	Regular, following crop-growing cycle	F able to access TPM package as it includes credit + marketing	Interpersonal communication	F control limited to negotiating terms of contract	F is very optimistic about economic prospect and not aware/ concerned with risk
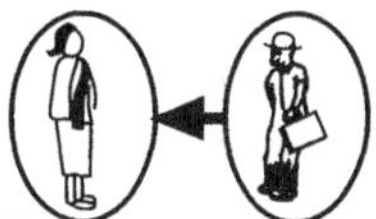 (linkage outcome: pineapple contract production)	TPM aware of F's needs for economically viable attention	TPM chooses farms with the necessary conditions to guarantee production	(as above)	TPM has means of transportation for regular visits to F	Good quality brochures in tagalog; business-like negotiation	TPM initiated + controlled via contractual agreement links set by TPM	TPM relationship with F considerably superior to GEA as it includes credit + market commitments and regular visits

Fig. 11.5.6. Philippines: matrix for analysing linkage performance. GEA, government extension agency.

time for each other and for accessibility of the service. The outcome of the examination is very positive. Farmers report that private contractors are superior to government extension agents.

Tracking change with indicators

Learning and change are kept moving by tracking with sets of performance indicators. Regular assessment of performance enables stakeholders to change what they do, both in terms of the roles they play and the agroecosystem management options they test. People too often trap themselves in roles they cannot perform satisfactorily. People too often keep using technologies that degrade natural resources. Indicators should at first address key questions regarding performance targets or expectations. Second, targets should be chosen so that they provide information that guides management decisions. Indicators for tracking change in agroecosystem performance and stakeholder performance are illustrated here.

The key questions raised by farmers in the assessment of sustainability were: how much more profit does the farm make for every dollar invested? How much more productive is the farm? How much more diverse is the farm? How much recycling is going on? From these questions four simple indicators emerged to track changes in performance of the whole farming system. Economic performance was indicated through changes in profit–cost ratios. Improvements in physical productivity were indicated through changes in weight of biomass produced from all enterprises. Ecological performance was indicated through changes in the number of recycling flows, and number of species cultivated. Benchmark values can be estimated for the existing farming system, recall can be used to assess performance of past farming systems, and potential values can be estimated for future experimental farming systems. With the passage of time changes in existing farming systems are tracked.

Estimates of these four sustainability indicators from the Philippines illustrate what kind of answers one can get[38]. These data, shown in Fig. 11.5.7, are plotted for a time series of 4 years. A trend emerges over the years against the background of wet–dry seasonal variation. With few exceptions the farmer has steadily increased the recycling and species diversity of the farm. This, along with work on water rehabilitation has produced a steady rise, the drought in the dry season of 1993 excepted, in the productive capacity of the farming system. Ecological soundness has improved constantly and more often that not, contrary to conventional wisdom, with increases in economic efficiency.

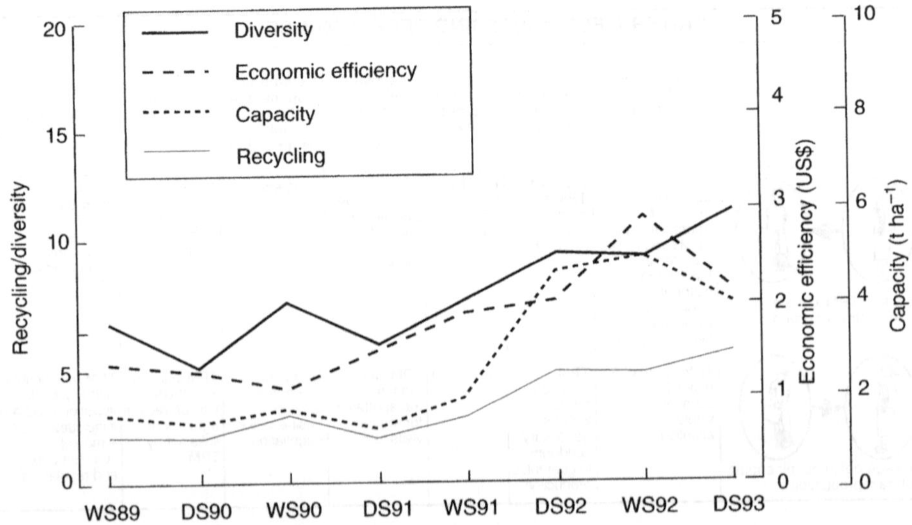

Fig. 11.5.7. Philippines time series plot.

Using indicators to track performance also carries over to learning about working together. Indicators for stakeholder performance are derived from the criteria chosen to assess roles defined for each actor linkage. So, from the matrix for analysing linkage performance we find relevance of the service. The use to which the information was put, timeliness of the delivery of services and access to meet other stakeholders, might be three indicators to start off with. The assessment of stakeholder performance is undertaken to learn, not condemn. Poor performance offers opportunities to adjust roles. An ability to learn and adjust roles quickly will be an important asset in the rapidly changing institutional environment.

The following example of the stakeholder analysis from the Philippines can be used to illustrate how scoring of indicators for stakeholder performance might work. Taking the views of farmers about the government extension agent and the private sector scoring of relevance, access and timeliness indicators might end in a result as shown in Fig. 11.5.8. Farmers thought the relevance of the contractor's service was very high and the timeliness of the delivery of that service was also high but that getting access to them was difficult. The government extension agent scored rather poorly, with little of the technical information they had to give being relevant and usually arriving too late. Although farmers rarely met the extension agents, they trusted the agents and valued their ability to link them up to other sources of information. In a learning mode, this creates opportunities for re-examining roles and making changes.

In this section I have presented some tools and some suggestions as to how they might link together. I have not described a *de facto* learning and action research process. In so far as we understand such a process it would start with the identification of stakeholders and their gaining a mutual understanding of each other's motivations and interests. The process would seek a deeper understanding of stakeholders' perspectives on agroecological resources. We must recognize that it cannot be a linear process. During the exploration of agroecological resources new stakeholders may be identified. Often, seasonal users of agroecological resources, such as nomadic herders, are not identified until one explores these resources. Building on multiple stakeholder perceptions the learning process moves on to brainstorming alternative agroecosystem management options. Again, which stakeholders are around the table determines what kind of options will emerge. New stakeholders are again likely to emerge during the phase of negotiating concerted actions on alternative management options. Iteration between stakeholder and agroecosystems analysis tools becomes essential. The learning action process is kept moving through tracking change. It is crucial that indicators for both farming system performance and stakeholder performance are monitored. What we learn through monitoring indicators changes our ideas about the roles each stakeholder plays. Recognizing that we should change what we do or invite other stakeholders in are useful results of monitoring. Similarly, realizing that we should change the technologies or practices we use to manage agroecological niches is a useful result. Thus a continuous learning and action research process may evolve that contributes to the goal of sustainable farming.

11.5.5 Conclusion: making a contribution

In this contribution I have argued that agroecological niche management makes an important contribution to sustainable farming. Not only has it sustained some long-lived traditional systems, it also forms the basis of modern intensive integrated farming systems which give good performance on at least some indicators of sustainability. Agroecosystems analysis, particularly recent advances in bioresource flow modelling, can contribute to the development of intensive integrated farming systems. Its contribution, however, would be greater if agroecosystems analysis were linked with stakeholder analysis and the two embedded within a learning and action research process. Detailed descriptions of agroecosystems in the form of maps and transects are not enough. Future visions and experimental models of more sus-

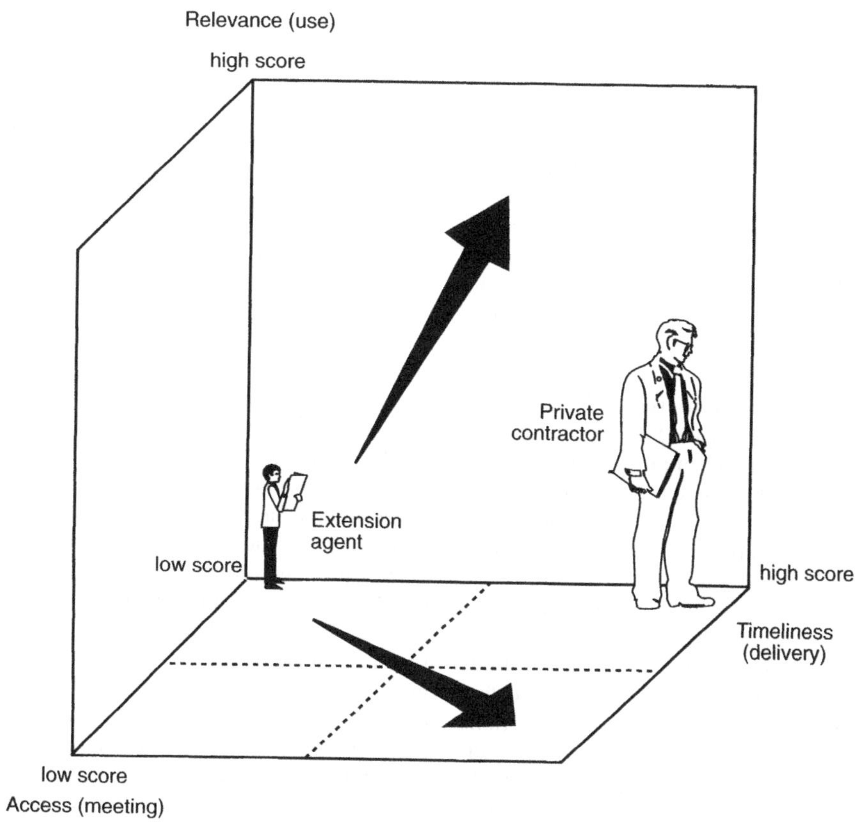

Fig. 11.5.8. Philippines performance indicator plot.

tainable farming systems are not enough. Similarly Venn diagrams and linkage maps of stakeholders are not enough. But, linking stakeholders to a particular vision of improved agroecosystem management and negotiating roles for each one might be enough if performance indicators both for stakeholders and the farming system are regularly monitored. Linking agroecosystems and stakeholder analysis within a learning and action research process permits both dynamism and flexibility in the transformation of existing farming towards more sustainable systems. The proposed process cannot get far without favourable institutional settings and policies. The way local institutions work together, and the way local people involve themselves in policy dialogue will be crucial.

Local learning for institutions and policy

Learning to work together requires more than a change in the skills and attitudes of individuals. It also challenges institutions to adjust their functions and mandates to fit with others. Involving wider sets of stakeholders in developing common visions and in negotiating agroecological niche management will create new opportunities for farmers, NGOs and government research and extension services to work together, and to link with other local stakeholders like local authorities and banks. In a future characterized by greater government decentralization and privatization of agricultural support services, skills in learning and action research will be a great asset.

Learning and action research processes, particularly bioresource flow modelling for brainstorming on the improvement of farming systems, will widen the search for new ideas and new sources of knowledge. The process of negotiating who does what to realize future visions of more ecologically sound farming will bring a more diverse set of stakeholders into play. But deciding who does what will not happen automatically. Negotiations facilitated by stakeholder analysis must define the complementarity of skills and functions across participants. Using performance indicators like timeliness of service, accessibility of service and relevance of service, for stakeholders to assess each other would ease the difficult task of forging local interinstitutional linkages. The observation that such stakeholder groups may encourage farmers to make multiple simultaneous interventions suggests that farmer adoption is not always stepwise component by component. This questions both the conventional wisdom of farmer adoption, and the widespread T & V extension process. Get the right stakeholders together and change proceeds at a far greater pace.

Speeding up learning and the action research process, and engaging more farmers in it, requires a larger and more diverse array of stakeholders operating outside the 'project' mode. Staff from local universities, government research and extension agencies, and local NGOs need to link up with farmers in a 'non-project' learning group. Learning groups created and run by local stakeholders allow farmers' own experiences and outside information to be exchanged. It is doubtful, however, that wide-scale expansion could occur without greater organization among the farmers themselves. A local learning group would better interact with community organizations than with individual farmers or even with small groups. Creating opportunities for stakeholders to come together will challenge them to sort out their various roles. Survival will depend on near-term tangible benefits. That farmers can benefit is clear. Working together allows them to spread labour requirements, especially for rehabilitation activities, across members. One farmer may never be able to dig bench terraces but groups can. Groups can leverage access to resources and markets. Importantly, groups can enhance social cohesion as the following examples illustrate.

In the process of negotiating experimental interventions, farmers in the Philippines had to visit farmers of the neighbouring village as pesticide runoff was killing the fish in their rice paddies. As a group they managed to successfully negotiate with their neighbours a reduction in pesticide runoff. Agroecosystem maps prepared by women's groups in Ghana when compared with those prepared by men's groups exposed a potential conflict. The men expressed their intention to plant cassava in a 'fallow' area which the women, according to their map, used to gather grasses for weaving baskets. Exposing potential resource use conflicts allows them to be resolved before they threaten social cohesion.

Decentralization and privatization may provide some motivation for government research and extension agencies, universities and NGOs to participate in non-project local learning groups. A 1997 article by Steven Biggs suggests that professionals in these institutions who have formed the right alliances and coalitions will have a considerable advantage over others[39]. Local learning groups may have a better chance at leveraging policy as coalitions than farmers or their lobbyists. Such groups generate many ideas about policy when brainstorming future visions of more sustainable farming systems. These ideas could inform policy makers on policies that work and perhaps more importantly on those policies that do not work. One much talked about instance is that of land tenure policies acting as disincentives for owners and tenants to plant trees or build other soil conservation structures.

Local learning groups, as coalitions involving government and university researchers, could have better leverage on research policy than NGO lobbyists. Crucial gaps in the knowledge of researchers are exposed in the brainstorming sessions. These often involve rehabilitating obscure agroecological niches and minor crops that researchers have no mandate to address. Women farmers in Shoshong, Botswana, who exposed knowledge gaps in the rehabilitation of 'molapo' water resources, in the selection of multipurpose trees for planting in crop headlands, and in leguminous forages for overseeding hillside grazing areas remain ignored – policy insists that scarce research money must be spent on staple food crops[40]. Engaging local people in policy dialogue may be of special value in decentralized government settings.

Sustainable farming systems

I have shared my personal views on the importance of agroecological niche management for sustainable farming within complex, diverse, integrated systems. I hold this view not only because of the evidence presented here; but also because micro management of many agroecological niches is something that resource-poor farmers do better than operators of large modern 'farms'. Moreover, it could provide an instrument for the alleviation of rural poverty. The agroecological niche management route to sustainable farming also makes sense when one remembers that dismantling of complex diverse integrated farming systems into simple high external input single enterprise operations threatens sustainability. Such simplification threatens not only species and habitat biodiversity and flexibility; but also the knowledge base that makes both risk management and system evolution possible. Oversimplification of farming reaches beyond cropland to threaten fallow land, forests and water bodies as well. Wetlands have been drained for cropping and mangroves turned into shrimp ponds to the detriment of both resource-poor farmers and the environment. But that was in the past. In future it will be less easy for profit and production to come at the expense of poverty and environmental goals.

Understanding why farmers cannot pursue 'sustainable' farming systems will become increasingly important[41]. We already know that farmers will not pursue species and habitat biodiversity strategies for their own sakes. These strategies have to pay. We know that the benefits from diversity in terms of income and food security only go so far. We see all around us that they do not go far enough for resource-poor farmers struggling to survive. As governments realize the cost of poverty and environmental destruction to national economies, ways will be found to make investments. Unfortunately, I think we still do not know enough about sustainable farming systems to make the right investments. The narrow base of research on complex, diverse integrated farming systems means that we can conclude little with confidence. This is disappointing. The sustainability indicator work in Ghana, Malawi and the Philippines suggested that there is no necessary trade-off between ecological and economic objectives. If such findings are repeated, getting resource-poor farmers into sustainable agriculture will not be a matter of welfare. Agroecosystems analysis, within a larger learning and action research process, can make a contribution. Through my spectacles it is a systems application with a future.

REFERENCES

1. **Conway, G.R. & J.N. Pretty, 1991**. Unwelcome Harvest: Agriculture and Pollution. Earthscan Publications, London.
   Pingali, P.L. & P.A. Roger (Eds), 1995. Impact of Pesticides on Farmer Health and the Rice Environment. IRRI, Los Baños, Philippines.
2. **Lightfoot, C., J.P. Dalsgaard, M. Bimbao & F. Fermin, 1993**. Farmer participatory procedures for managing and monitoring sustainable farming systems. Journal of the Asian Farming Systems Association, 2(1), 67–87.
3. **Richards, A., 1939**. Land, Labour and Diet in Northern Rhodesia. CUP, Cambridge.
4. **Tiffen, M. & M. Mortimer, 1992**. Environment, population growth and production intensity: a case of Machakos, Kenya. Development Policy Review, 10, 359–87.
5. **Brookfield, H. & G.S. Humphreys, 1991**. The use of unstable steeplands in the mountains of Papua New Guinea. Mountain Research and Development, 11, 295–318.
6. **Raintree, J.B., 1984**. Designing agroforestry systems for rural development: ICRAFs D & D approach. International Council for Research in Agroforestry (ICRAF), Nairobi, Kenya.
   Watson, H.R., 1983. Developing a hillside farming technique for Mindanao hill farmers – SALT (Sloping Agricultural Land Technology), pp. 107–19. In: E.C. Quisumbing & J.G. Marasigan (Eds) Proceeding of the Hilly Land Development Workshop. MAF, Manila, Philippines.
7. **Kenmore, P.E., 1991**. Indonesia's Integrated Pest Management: a Model for Asian Rice Farmers' Environmental Conservation, Production, and Profits. FAO, Manila, Philippines.
8. **Altieri, M.A., 1989**. Agro-ecology: the Science of Sustainable Agriculture. Westview Press, Boulder, CO.
9. **Mollison, B., 1993**. Introduction to Permaculture. Tagari Publications, Tyalgum, Australia.
10. **Edwards, P., R.S.V. Pullin & J.A. Gartner, 1988**. Research and education for the development of integrated crop-livestock-fish farming systems in the tropics. ICLARM Studies Review, 16, Manila, Philippines.
11. **Lightfoot, C. & R. Noble, 1993**. A participatory experiment in sustainable agriculture. Journal for Farming Systems Research and Extension, 4(1), 11–34.
    Prein, M., J.K. Ofori & C. Lightfoot, 1995. Research for the Future Development of Aquaculture in Ghana. ICLARM Conference Proceedings 42. Manila, Philippines.
12. **Dalsgaard, J.P., 1996**. An ecological modeling approach towards the determination of sustainability in farming system. PhD thesis, Royal Agricultural and Veterinary University, Copenhagen, Denmark.
13. **Brummett, R.E. & R. Noble, 1995**. Aquaculture for African Small-holders. ICLARM Technical Report 46.
14. **Kundhlande, G., J. Govereh & O. Muchena, 1994**. Socio-economic constraints to increased utilization of dambos in selected communal areas, pp. 87–96. In: R.Owens, K. Verbeek, J. Jackson & T. Steenhuis (Eds) Dambo Farming in Zimbabwe. CIFAD, Cornell University, Ithaca, New York.
15. **Maseko, Z. & M. Bussink, 1994**. Wetlands and water harvesting as resources in communal farmers' food security, pp. 33–8. In: R. Owens, K. Verbeek, J. Jackson & T. Steenhuis (Eds) Dambo Farming in Zimbabwe. CIFAD, Cornell University, Ithaca, New York.
16. **Tiffen, M. & M. Mortimer, 1992**. (Op. cit.).
17. **Bunting, A.H., 1987**. Agricultural Environments. Characterization, Classification and Mapping. CABI, Wallingford.
18. **Hart, R.D., 1982**. An ecological conceptual framework for agricultural research and development, pp. 44–58. In: W. Shaner, P. Philipp & W. Schmehl (Eds) Readings in Farming Systems Research and Development. Westview Press, Boulder, CO.
19. **Altieri, M.A., 1989**. (Op. cit.).
20. **Conway, G.R., 1982**. A guide to agroecosystem analysis. In: Tambon and Village Agricultural Systems in Northeast Thailand. KKU Ford Cropping Systems Project, Khon Kaen University, Thailand.
    Conway, G.R., 1985. Agroecosystem Analysis. Agricultural Administration, 20, 31–55.
    Rambo, T.A., 1984. Human ecology research by social scientists on tropical agroecosystems, pp. 39–60. In: T.A. Rambo & P.E. Sajise (Eds) An Introduction to Human Ecology Research on Agricultural Systems in Southeast Asia. University of the Philippines, Los Baños, Philippines.
    Sajise, P.E., 1986. Agroecosystem Analysis: The SUAN Approach. Proceedings 3rd SUAN-EAPI Regional Symposium on Agroecosystem Research. Chaing Mai University, Thailand, October 1996. East West Environment & Policy Institute, Hawaii, USA.
21. **IDS, 1996**. PRA Annotated Bibliographies. IDS, University of Sussex.
22. **Rambo, T.A., 1984**. (Op. cit.).

Sajise, P.E. & T.A. Rambo (Eds), 1985. Agroecosystem Research in Rural Resources Management and Development: Proceedings 2nd SUAN-EAPI Regional Symposium on Agroecosystem Research. Baguio City Philippines. March 1985. PESAM, University of the Philippines, Los Baños, Philippines.
Sajise, P.E., 1986. (Op. cit.).
23. **Chambers, R., 1997**. Whose Reality Counts? Putting the First Last. IT Publications, London.
24. **Bawden, R.J., 1991**. Systems thinking and practice in agriculture. Journal of Dairy Science, 74, 2362–73.
25. **Pretty, J., I. Guijt, J. Thompson & I. Scoones, 1995**. Participatory Learning and Action: a Trainer's Guide. IIED Participatory Methodology Series. IIED, London.
26. **Engel, P., 1997**. The social organization of innovation. A focus on stakeholder interaction. In: Engel, P. & M. Salomon (Eds) Facilitating Innovations for Development. KIT, STOAS and CTA, Amsterdam, Netherlands.
27. **Grimble, R. & K. Wellard, 1996**. Stakeholder methodologies in natural resource management: a review of principles, contexts, experiences and opportunities. Agricultural Systems, 65(2), 173–93.
28. **Lightfoot, C. & R.S.V. Pullin, 1995**. (Op. cit.).
29. **Brummet, R.E. & R. Nobel, 1995**. (Op. cit.).
Prein, M., *et al.*, 1995. (Op. cit.).
30. **Defoer, T., S. Kante, T. Hilhorst & H. DeGroote, 1996**. Towards More Sustainable Soil Fertility Management. ODI Agriculture Research and Extension Network Paper no. 63. ODI, London.
31. **Poats, S., 1994**. The place of the farmer in agroecosystems analysis, pp. 235–49. In: M. Sebillotte (Ed.) Systems-Oriented Research in Agriculture and Rural Development: Lectures and Debates. CIRAD, Montpellier, France.
32. **Roling, N. & J. Jiggins, 1996**. The ecological knowledge system. In: Proceedings of the 2nd European Symposium on Rural and Farming Systems Research. Granada, Spain.
33. **Ison, R.L., P.T. Maiteny & S. Carr, 1997**. Systems methodologies for sustainable natural resources research and development. Agricultural Systems, 55(2), 257–72.
34. **Engel, P.G.H. & M. Salomon, 1994**. RAAKS a participatory action research approach to facilitating social learning for sustainable development. In: Systems-Oriented Research in Agriculture and Rural Development. Papers of ASFRE International Symposium, Montpellier, France, November 1994.
35. **Ramirez, R., 1997**. Understanding Farmers' Communication Networks: Combining PRA with Agricultural Knowledge Systems Analysis. IIED Gatekeeper Series no. 66. IIED, London.
36. (Ibid.).
37. **FAO, 1995**. Understanding Farmers Communication Networks: an Experience in the Philippines. Communications for Development Case Study 14. FAO, Rome, Italy.
38. **Lightfoot, C., *et al.*, 1993**. (Op. cit.).
39. **Biggs, S., 1997**. Livelihood, coping and influencing strategies of rural development personnel. Project Appraisal, 12(2), 101–6.
40. **Norman, D.W. & C. Lightfoot, 1992**. A Formative Evaluation of the FAO/SIDA Farming Systems Program. FAO, Rome, Italy.
41. **Bunting, A.H. & B. Pickersgill, 1996**. What is a plant genetic resource? Biologist, 43(5), 227–30.

11.6 WATER QUALITY, AGRICULTURAL PRACTICES AND CHANGES IN FARMING AND AGRARIAN SYSTEMS

J.P. Deffontaines, J. Brossier, M. Barbier, M. Benoit, E. Chia, J.L. Fiorelli, M. Gafsi, F. Gras, H. Lemery and M. Roux.

In this case, one major peculiarity is that while the request for changes (and therefore the funds needed for them), was made by a company, the changes involved other parties who were not attracted to them, possibly did not even want them, and who were obviously not interested in funding the research.

11.6.1 Setting the scene, research background and research programme

Setting the scene

A private mineral water company wanted to avoid increases in the level of nitrates in its water by bringing in preventive measures on the farming practices in its catchment area – an area entirely dominated by agriculture. Recent changes in farming systems (cultivation of pastures, development of maize crops, increased fertilizer inputs) seemed to be the major cause for an increase in nitrates. Pollution did not yet exist as the nitrate levels were still far below the European minimum standards, but the economic activity of the water company could be prejudiced if water quality deteriorated due to increasingly intensive cropping practices.

First some facts. The catchment was a relatively clearly bounded area of about 5000 ha with diverse soils. There were some 40 farmers involved; dairy and cereal companies and farmers' professional organizations with annual outputs of 60,000 hectolitres of milk and 8000 metric tons of cereals. The mineral water company is an important employer with 1500 staff, the main employer in this rural area and a leader in its market with sales of 1 million bottles each year. The case is unusual compared to many classic environmental problems, due to:

- The constraint arising from the targeted nitrates level (10–15 mg l^{-1}).
- The restricted area.
- A 'polluted' party prepared to help the 'polluters' curb nitrate leaching.

In order to stop increases in nitrates the company asked the French National Institute for Agricultural Research (INRA) and the local Chamber of Agriculture to identify solutions.

Research background

In 1988, the company producing mineral water (Vittel) formulated a request that was to be the starting point for a research project[1]. The question put to the INRA research team by the company was: Which changes in the farming systems can slow up the increasing level of nitrates in water draining through the upper layer of the soil, and how can these changes be introduced? There were strict constraints from the outset: no pesticides were to be used, and a nitrates level lower than 10 mg l^{-1} in the soil solution under the roots was to be achieved.

The problem was reformulated by researchers from the INRA department working on Agrarian Systems and Development (SAD): under what conditions can local agriculture develop when subject to new constraints regarding groundwater quality? The question was addressed by a programme using action-research principles. The scientific organization of the programme and its basic hypotheses are summarized here.

The research programme

The research undertaken by INRA (Fig. 11.6.1) is based on three principles taken from systems modelling and constructivist theory:

- The holistic aspect of complex systems.
- Identification of interrelations between actor-groups.
- The action-research approach, with actor-groups and researchers integrated in the process of research and change.

The project involved two branches of knowledge which need to be associated. The first branch concerns the biotechnical and economic mechanisms responsible for nitrogen output from local agriculture. The simulation models generated were based on these mechanisms. The second branch of knowledge is the result of a collective approach to action-research involving all actor-groups, including the researchers themselves.

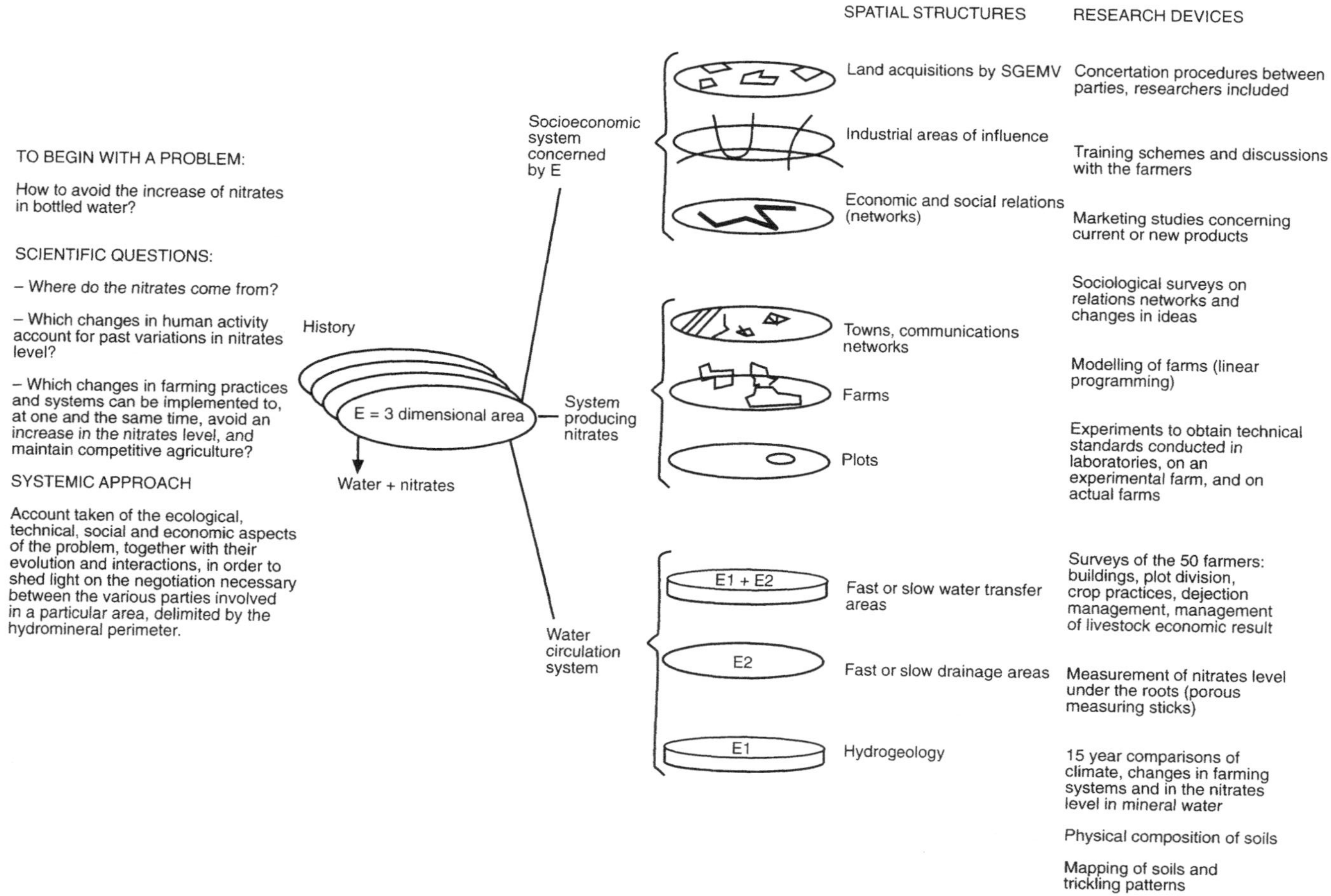

Fig. 11.6.1. Structure of the interdisciplinary research programme.

Table 11.6.1. Number and results of nitrates leaching measurements 1988–92 on main plant cover types in Vittel catchment.

Plant cover	Number measured	Average (mg NO_3 l^{-1})	Standard deviation
Forest	5	2	–
Hay field	9	19	14
Permanent pasture	18	31	25
Temporary pasture	3	28	–
Lucerne	13	23	8
Winter wheat	27	46	25
Winter barley	27	46	25
Oil seed rape	8	120	52
Spring cereal	8	28	20
Maize	28	126	77

The general structure of the programme is based on an agrarian systems model, linking the land with the farm operators and their activities. The hydrogeological, technical and socioeconomic aspects of the problem, together with their changes and interactions are considered, and the negotiation to be established between the various actor-groups involved is clarified. A diagram identifies these main components (Fig. 11.6.1).

Conclusion

Since several scientific publications present the research project and the main findings, the purpose of this contribution is to emphasize two aspects of this research. First we will highlight the key role of spatial organization in systems research, especially for environmental questions. Second, we will present a participatory approach (action-research) as a means of reaching a compromise between conflicting parties.

11.6.2 The spatial organization of farming activities

The scientific objectives of the SAD department of INRA can be described as the analysis of development processes through study of the relationships between the historical, technical and social factors involved, on different organizational levels, in activities connected with agriculture and the rural areas. Spatial analysis is integral to meeting these objectives. In this context, a rural area is a geographical entity; a complex and changing structure of which the farmer is an integral part. Spatial issues have changed over time, as have the analytical methods and tools for examining them. There are two key roles for spatial factors:

- Their introduction into the analysis of the technical and economic functioning of activity systems on the farm level.
- The description and comprehension of the spatial organization of both farm activity and other activities carried out in an identified geographical area, by connecting up spatial structures and activity functions.

These two roles were adapted to the research programme on water quality.

An area for managing water quality was identified, the water catchment area

In connection with the monitoring of changes in water quality incurred by changes in cropping systems, agronomists and soil scientists raised the following questions:

- What changes in cropping systems and farming practices can be observed?
- Can nitrate transfers be modelled in a relatively precise way, and can the agronomic input parameters of these models be specified?
- Do these models allow the effect of the changes in farming practices on groundwater quality to be simulated?

Two spatial scales were selected, the crop field and the water catchment area

The crop field is a portion of unbroken land which has the same plant presence over a production cycle and on which the farmer practises the same crop management sequence. It is the best definition of a basic unit of the biotechnical system for nitrates production. However, studies of the soil have shown that the circulation of water, and thus of nitrates, does not respect field limits, especially when the soil and climate conditions are favourable to lateral draining. The catchment area of local water sources is the geographical entity in which water quality is determined. This enables verification that cropping patterns and management sequences do actually have an effect on the quality of groundwater. The nitrate level of these sources is solely dependent on the leaching mechanisms produced in the soils.

A model for determining water quality within a catchment area was built

Several methods were used to determine the relationships between cropping systems and the nitrate levels beneath the roots (samples came from 65 representative crop fields). Balances between nitrate inputs and outputs were calculated for several farm field patterns using the BASCULE indicator[2] (Spatialized Nitrogen Balance of Farm Cropping Systems). Porous ceramic cores allowed changes in the quality of the water beneath various plant covers to be monitored.

A study of 17 sources[3] demonstrated that it was possible to model the degree of concentration of nitrates levels in each crop field within the water catchment area, by taking on one hand the average rate of the water drained from the preceding crop of each crop field, and on the other hand the surface allotted to this crop within the catchment area. The concentration of nitrates in the soil solution beneath the root layer depends greatly on how the soil is used. The highest rates can be observed beneath maize and oil seed rape, the lowest rates beneath lucerne, meadow or grazed pastures and beneath forests (Table 11.6.1).

Inclusion of types of soil in the model

A GIS was used to model the change in crop systems, enabling the data furnished by soil scientists and agronomists to be reconciled. The permanent pastures are localized on impermeable clay soils and on soils that are stony on the surface and impermeable deep down (Lettenkohle and Ceratites layers) and near villages. These pastures are less common and have made way for rotating crops especially maize and oil seed rape. For example, we noted that in one of the catchment areas covering 144 ha, the percentages of land use in 1970 and 1991 generally indicate an intensifying trend which has resulted in an increase in the level of NO_3 at the source from 30 to 50 mg l^{-1} (Table 11.6.2).

Inclusion of hydrological functioning of the catchment area

Surface hydrological functioning is highly dependent on the topography of the soils, their hydric and hydrodynamic properties, and where they are located. The combination of these different aspects furnished a simplified diagram of surface draining patterns.

Choice of cropping systems within the catchment area

Analysis of the positioning of cropping systems highlighted the effect of two crucial factors in the choices farmers make as to soil use:

- Type of soil.
- Distance between crop fields and farm buildings.

The proportion of the total farmland managed by each farmer that fell within the catchment area (in this case 11 farmers) identified those farmers responsible for the largest portions of

Table 11.6.2. Areas (in %) occupied by main plant covers in 1970 and 1991.

Plant covers	1970	1991
Permanent pasture	38	10
Lucerne	12	8
Cereals	39	48
Peas and spring crops	11	–
Oil seed rape	–	14
Maize	–	20

the area and allowed the calculation of an 'involvement rate' for farmers expressed by the ratio of farmland within the catchment area to the total area of the farm.

Making the case for a water 'terroir'

Changes in the quality of water from a catchment area are a result of the crop pattern dynamics and farming practices used in that area. Spatial correlation between detrimental cropping systems and the water catchment areas is relevant for tackling the problems of regional groundwater pollution[4]. Management of the cropping systems used within a water catchment area is the key to water quality. How can individual farmer decisions, rational on the farm level, such as ploughing in a lucerne crop, be reconciled with the logic of catchment management which must limit this detrimental practice to a very small proportion of the catchment area? Can the water catchment area not be considered as a water 'terroir'? In a given physical environment (soil and climate), farmers are using farming practices in fields (cropping systems) which define the quality of the water resource in the catchment area. The farmers feel involved in the catchment area to a greater or lesser extent according to the percentage of their land which lies within the catchment area (involvement rate). Improving the quality of water in these water 'terroirs' implies collective management and probably recognition of a new function for farmers: as co-producers of water quality.

11.6.3 Modelling the water catchment area for its coherent management

The aim of modelling was to apply knowledge acquired in the various scientific domains to create land organization models which maintain sustainable agriculture while reducing nitrates leaching as much as possible. The basic idea was to ignore the current boundaries of the land belonging to each farm and to consider large hydrological units of the catchment as management units. While this abstract approach distanced us from the real constraints involved in managing the land, it did provide a benchmark for comparison with real situations that come up during the successive stages of *ad hoc* land reorganization (land consolidation, exchanges, land concentration and restructuring). This benchmark should also pinpoint the aspects needing more detailed work and suggest land forms, layouts and structures, all of which contribute to programme objectives.

A methodological sequence was proposed in order to construct this model. Its underpinnings were the spatial entities significant for the problem in hand. A large hydrological unit in the perimeter was studied, together with the distribution of soil types, substratum fracturing and morphological characteristics, which define soil use. A production system is designed which realistically weighs the trends affecting the current systems; increases in land used, the extensive management of portions of land, together with effective management of the dairy herd. The system designed also incorporated the recommended technical specifications. The aim is the organization of the land contained in a hydrographical catchment area that corresponds to a hypothetical farm, with headquarters located in a village, which implements the chosen production system elements, and the functioning of which follows a set of rules which result from knowledge acquired during the research programme.

This abstract approach to spatial organization should not be construed as a technocratic procedure for restructuring activities. For the actor-groups or stakeholders in land management, it facilitates discussion of the coordination of activities under the environmental constraints which are new to them.

11.6.4 Modelling, a participatory approach and action-research

Action-research

Action-research combines an intention to change and a plan for research. Each action-research project is specific and it is difficult to define a standard model. In this case, one major peculiarity is that while the request for changes (and therefore the funds needed for them) was made by a company, the changes involved other parties who were not attracted to them, possibly did not even want them and who were obviously not interested in funding the research. The Vittel case of action-research is the result of a research plan formulated by a team and the desire of one stakeholder to change the activities of others[5]. A crucial point was to involve all stakeholders in the action-research.

Various operators were involved in this action-research. Three institutions commissioned and financed the research and two, the project managers – INRA and a subsidiary of the Mineral Water Company (Agrivair) – defined the framework for the programme, and were directly involved in it. Various committees and teams were set up during the research programme. These allowed negotiation and arbitration, as well as the management and improvement of research performance:

- Research Management Committee: to determine responsibilities and activate research groups, fix priorities, manage the contacts and negotiate with various partners.
- Contract Steering Committee: to bring clients and project managers together to ensure that the programme was carried out in accordance with the contracts.
- Scientific Committee: composed of scholars not involved in the research programme, for scientific protection and openness.

Scientific protection ensures that pressures from the clients are not an obstacle to either the scientific process or scientific progress. Scientific openness is there to diversify the disciplines of the team, to specify scientific issues and find new ways of achieving scientific collaboration.

The ongoing negotiation between the various actors and partners covered:

- The programme: content, deadlines, partners.
- Pollution limits and the identification of the geographical area of the research.
- The conditions and specifications.
- Pollution limits (constraints and pesticide constraints).

The role of models in systems research work and action-research

Le Moigne pointed out, in 1990, that modelling can be analytical (cartesian) or systemic[6]. It is systemic if the actions and interactions are intentionally modelled as a project, the teleological project of the person responsible for the model being essential to it. In 1996, this idea was taken up by Dent, for whom the main function of soft systems models is to create debate and discussion about an agreed area in development, by providing indicators that point to possible outcomes of alternative actions[7].

The Vittel research required collaboration between agronomists, economists, an extensionist, several farmers and the company, for building and analysing farm models. It involved simulating the nitrates constraint in the farms, defining new and adapted production systems, studying the consequences of the changes proposed for farm functioning and profitability, and suggesting possible terms of negotiation between the mineral water (MW) company and the farmers. The linear programming models used incorporate the data available on the harmful effects of certain farming practices, the programming output indicates the practices that should be used and gives information about their performance. The models enabled pertinent questions to be posed on the nature of the nitrates constraint and what this means for the farm (role of dairy-cow grazing for example).

Management economists elaborated marginal productivity curves for nitrates based on milligramme increase or decrease in relation to the nitrates constraint (see Fig. 11.6.1). Such a curve is the typical fruit of collaboration between agronomists and economists, and is based on the connections between fertilization practices, N balance readings per hectare and nitrates loss under the roots measured by porous ceramic cores. At the threshold level requested by MW (10 mg l^{-1}), each milligramme drop in nitrates is extremely costly to obtain. There are possible solutions, but they require considerable improvements in management: drastic changes in production systems, strict crop management and the improved development of products.

These models, elaborated jointly by researchers and farmers, have been used in a participatory approach to open up negotiations between the different stakeholders. Using scenarios from the current situation and possible future situations that are sufficiently credible, discussion elaborated the sensitivities and the challenges involved and aspects requiring clarification. From a concrete point of view, the debate between the farmers, MW and the researchers, stimulated by the marginal cost curve of the nitrates and the constraint threshold, underlined the interests at stake in the negotiations between the farmers and the company: the scope of change expected of the farmers, the level of support that the company must/can contribute, and so on. In this case, since these models were used to promote dialogue, the preparatory work in model specification, stakeholder participation and the discussion the results stimulated are all more important than the model results themsleves.

The difficulties inherent to this time-consuming approach should not be minimized, because the means proposed by research may be inopportune. In this case not all the farmers involved were willing to participate in the process.

11.6.5 The results of interdisciplinary research on agriculture and water quality

The research as an aid in the water quality negotiation

The project is a good example of the transposing of research knowledge required to cope with a specific question not usually encountered by farmers: the pollution of groundwater by fertilizer residues. The issue was the consequences of their practices on a criterion – water quality – which is beyond the normal knowledge of farmers. They are not involved in water production and do not feel responsible for its quality, nor do they spread nitrates but rather what they perceive as organic and mineral fertilizers for their own objectives.

The notion of the marginal productivity of nitrates introduced by agricultural researchers and economists working together, in contact with their partners in the field, enabled two worlds to meet up: one using the criteria of water quality (particularly nitrates content) and the other more usually involved in production factors, particularly costly inputs such as mineral fertilizers or undesirable intermediate products such as liquid manures. Negotiations between these different categories of actor-groups were made relevant for all the stakeholders through a modelled simulation of the situation.

The research team designed methods and models which are in fact procedures, in this case for the arbitration of mineral water quality, which facilitated the collective invention of solutions. The method can be generalized to other situations involving the management of natural resources.

Table 11.6.3. Research tools used in the three systems under study.

	1988	1989	1990	1991	1992	1993	1994	1995	1996
Water quality monitoring		—	—	—	—	—	—	—	—
Soil map		-	—	-	-	—			
Tectonic studies						—	—		
Water circulation in the soil			-	—	—	—	—		
Monitoring cropping	-	—	—	—	—	—	—	—	—
Porous ceramic cores	-	—	—	—	—	—	—	—	—
Laboratory work; mineralization			-	—	-				
Measuring residual N in fields	-	--	--	--	--	--	--	--	--
Composting experiments	-	—	—	—	—	—	—		
Green fertilizer trials		-	-						
Monitoring dairy cattle grazing					-	—	-		
Forage monitoring						-	—	—	-
Farm interviews		-	-						
Farm analysis and monitoring			—	—	—	—	—	—	—
Information system			-	—	-				
Economic simulations	-	—	-						
Sociological interviews		-	-				-	—	—

—, Full year of involvement; -, half a year of involvement; --, intermittent involvement.

Systems research: goal-oriented researchers and stakeholders interacting

This mineral water example demonstrates the relevance of the model building hypothesis of systems science. To deal with the complex issue raised by MW and the farmers, the research team proposed a system to reduce nitrates while developing efficient agriculture. Study and monitoring of this elaboration process require tools and methods which, although not unusual in research, imply the interconnection of several disciplinary points of view and of several levels of investigation: from porous ceramic cores to the development of 'clean' farm products and action-research that involves actor-groups and researchers in the process of knowledge elaboration, the MW research project has numerous interrelated facets. Given the often contradictory interests of the various partners involved, this implies the flexibility to adapt the research as it evolves.

The contribution of research is not to solve an incompletely identified problem – to prevent the increase of nitrates in water bodies, even though this has been requested, but to help properly identify and formulate the problem: where do the nitrates come from, who is responsible for the increase, which practices need modifying?[8] The solution cannot be 'invented' by the research, it is produced collectively by the creation of new relationships and organizational innovations connecting the stakeholders and by drawing them together. Such innovations are not easy to realize because of the influence of a multitude of endogenous and dynamic factors.

An outdoor laboratory linking fields of knowledge

Setting up a research field of enquiry led to the definition of a land area as an outdoor laboratory. In addition to the knowledge acquired through the research, the very presence of scientists helped to deal with a water quality problem with many ramifications. The situation proved difficult for the scientists at times due to the novelty of the tasks they had to handle. Indeed, their work had to produce knowledge at various levels of observation using a variety of disciplinary inputs to describe, measure and understand the phenomena involved, and, practically in parallel, the work had to define new practices that could be proposed to farmers to produce a level of nitrates not higher than 10 mg l^{-1} and using no pesticides.

The first lessons that can be drawn are on the ways in which a hydrographical unit can be defined; in this case a land area supplied with water from a spring (the water basin) to be protected by the practices which farmers use to manage their fields. Researchers had to find ways to link two fields of knowledge. First, this meant the detailed understanding of the ways in which their hydric, chemical and biochemical properties could affect the nitrates levels in the soils, requiring the use of many complementary techniques that were themselves often completely novel. The heterogeneity of the geological substratum and the soil cover, and the need to measure the effects of farming practices at the scales of the crop field, the farm and the small water basin, led the researchers to multiply the sites and observation levels in order to connect up biophysical phenomena with the farming practices. The technical, economic and social dimensions of these practices had to be defined and analysed in detail. Techniques used ranged from conventional interviews and agronomic measurements supplied by farmers to participatory research.

11.6.6 Conclusions: analytical and systems sciences: no longer at loggerheads

A large number of disciplines were used in the research, corresponding to the three subsystems presented in the diagram showing the interdisciplinary structure of the research programme (Fig. 11.6.1).

Each discipline used its own tools and analytical models: Table 11.6.3 shows the diverse set of research techniques used, most of which study one aspect of the problem and are strongly analytical. It is clear that the systems approach did not dispense with the tools furnished by the hard sciences such as hydrogeology, soil science and agronomy. These different techniques serve to develop knowledge that is authoritative in the individual disciplinary fields. This authoritative knowledge is backed up by five postgraduate theses that have been defended in different disciplinary fields, and by 20 articles in peer reviewed publications. The majority of these deal with scientific fields other than systems science. It is contrary to our way of thinking to oppose systems science and analytical science, often caricatured as reductionist. Opposing holistic and reductionist approaches does not bring progress since the holistic approach is necessarily reductionist itself. A phrase of Pascal might be our motto: 'I cannot know the whole if I do not know the different parts, neither the parts if I do not know the whole'. The interdisciplinary approach is the key point – the desire to interconnect these different sciences in an approach which addresses the varied and interconnected aspects raised by the overall problem. The interaction across disciplines benefits from a vision in terms of a system and from addressing the interfaces. In a nutshell, this is epistemology – a state of mind, and it underlines the importance of the integration of research and development with the stakeholders involved.

REFERENCES

1. **Brossier, J. & E. Chia, 1994**. Participatory research: water quality and changes in farming systems. In: Dent, B. & M.J. McGregor (Eds) Rural and Farming Systems Analysis. European Perspectives. CAB International, Wallingford.
   Brossier, J. & E. Chia, 1990. Pratiques Agricoles et Qualité de l'Eau. Construction d'une Recherche-Développement dans le cas d'un périmètre hydrominéral. Economie Rurale, 199 (Sept–Oct), 6–13.
   Deffontaines, J.P., *et al.*, 1993. Agriculture et Qualité des Eaux. Diagnostic et propositions pour un périmètre de protection. INRA SAD, Paris.
   Deffontaines, J.P. & J. Brossier, *et al.*, 1995. Agricultural practice and water quality. A research and development project. In: Brossier, J., L. de Bonneval & E. Landais (Eds) Systems Studies in Agriculture and Rural Development. INRA, Paris.
2. **Benoit, M., 1992**. In indicateur des risques de pollution nommé 'BASCULE' (Balance Azotée Spatialisée des Systèmes de Culture de l'Exploitation). Fourrages, 129, 95–110.
3. **Gaury, F., 1992**. Système de culture et teneurs en nitrates des eaux souterraines. Dynamique passée et actuelle en région de polyculture-élevage, sur le périmètre d'un gîte hydromineral. Thèse, ENSA Rennes, INRA, SAD, Paris.
4. **Benoit, M., 1992**. (Op. cit.).
5. **Liu, M., 1992**. Présentation de la recherche action; définition, déroulement et résultats. Revue Internationale de Systémique, 6(4), 293–311.
6. **Le Moigne, J.L., 1990**. La théorie du système général, théorie de la modèlisation. PUF, Paris.
7. **Dent, B., 1996**. Theory and practice in FSR/E: consideration of the role of modeling. In: Sebillotte (Ed.) Systems Oriented Research in Agriculture and Rural Development. CIRAD, Montpellier, France.
8. **Simon, H., 1972**. In: Newell & Simon (Ed.) Human Problem Solving. Prentice Hall, Englewood, New Jersey.

Chapter 12

The Future of Farming Systems Research

Mike Collinson and Clive Lightfoot

12.1 THE FARMING SYSTEMS RESEARCH (FSR) PROCESS – THE ORIGINAL APPROACH

Research efficiency is an important idea and the difficulties of linking research to the market have long been acknowledged, even outside agriculture. The origins of FSR in the late 1960s and early 1970s coincided with a major study of 1000 manufacturing companies, the study report stated: 'For each product that succeeds in the market, a typical manufacturing company generates 58 new product ideas. After business analysis seven of these generally reach the development stage. Of these seven, six are eliminated during development, testing or commercialization. Almost 75 percent of new product expenses (and thus the work of eight out of ten developmental scientists and engineers) are devoted to projects that will not be justified in terms of commercial success'[1]. The early purpose of FSR was to improve efficiency by increasing the relevance of research programming and research findings to smallholder farming. It was a purpose prompted, at least in part, by independence from colonial rule for an increasing number of developing countries. And, in part, by popular predictions of food shortages and famine in many developing countries.

Much early FSR was the result of the addition of a social scientist, usually a farm economist, to the research station complement of natural scientists. It was an addition which became associated with a shift to on-farm experimentation (OFE) by station agronomists, in part in response to the clear contrasts between farmers' circumstances and research station circumstances spelled out by the social scientists. By the mid 1970s on-farm research (OFR), incorporating FSR and OFE and almost always mounted from research stations, had crystallized an approach with two roles:

- To identify how research station recommendations could be reshaped to better fit local farming systems.
- To identify key problems limiting production that could be addressed by informed agricultural research to help set a more relevant station agenda.

Early FSR followed a consistent operational sequence: (a) description: to describe and understand the system; (b) diagnosis: to identify key problems; (c) design of experiments; (d) testing through implementing experiments; and (e) evaluating the results: either to reformulate station-based recommendations or to plan a further cycle of experiments or transfer recommendations to extension. The diagnosis used farmer surveys, some formal but increasingly informal, and generated hypotheses about how recommendations might be adjusted to fit the farmers' system. Diagnosis also identified general leverage points in the system, usually labour bottlenecks, some of which could be addressed by agricultural research. In the early OFEs, as Stroud and Kirkby described in Chapter 4, agronomists carried their designs, management and analytical methods off the research station, essentially using farmers' fields as sites more representative of climate and soil conditions under which they operated. Soon farmer practice was adopted for the non-experimental variables, initially these were 'typical' practices to avoid uncontrolled non-experimental variables creating a new source of variation. At the same time researchers' understanding of farmers' evaluation criteria began to be used for the analysis of results.

This early paradigm was a sequence of activities essentially within the aegis of the research services. Early efforts were usually implemented on an *ad hoc* basis, often when influential individuals were persuaded to have a look at the process, rarely as a commitment to institutionalization. Organized programmes to disseminate and institutionalize FSR in the context of OFR also began in the mid 1970s. Since these early days practice has evolved radically and with it the role, scope and potential of FSR.

12.2 MILESTONES IN CONCEPTS AND PRACTICE FROM 1970

Nine improvements and potential improvements stand out as milestones in the evolution of the FSR process and in its application. These have, perhaps, been the most important factors in the reshaping of best practice. They underpin a contemporary FSR process that offers greater flexibility in both organizational options and promotional strategies, important to widening its use. One or two of these milestones, however, have also confounded the role for FSR, confusing both practitioners and development managers, and perhaps inhibiting wider support for its adoption.

12.2.1 A wider conceptual framework: the systems hierarchy

The hierarchy concept has greatly enhanced awareness of different levels of activity, particularly above the farm, and the need for cohesion across them. Though not yet commonly used, the hierarchical framework has a great potential to improve the orientation of development efforts. Shared ownership of a common hierarchical framework can improve partnerships across international, national and local bodies, guiding subsidiarity and helping to reconcile the perspectives of decision makers operating at different levels and in different but interacting hierarchies.

12.2.2 Merging physical and human factors in characterization

Within such a framework characterization, and indeed the full range of systems methods, can be clearly linked to hierarchical levels and can address both natural and economic systems. The proponents of human groupings have long been at loggerheads with the traditional zoning school focused on climate, soil and production potential[2]. The fact that global databases on human characteristics are less developed than on climate and soils has inhibited reconciliation. However, perhaps accelerated by growing confidence in the link between poverty and environmental degradation, human characteristics are now drawing increasing attention. Aiding reconciliation is GIS, a revolutionary technique. The superimposition of digitized spatial databases allows the pattern of existing human activities to be juxtaposed on the pattern of physical potential. Such juxtaposition conjures up novel ideas on the spatial dimensions of development strategy and associated policy measures. For example the easier identification of populations which can best be helped by new infrastructure and marketing initiatives, or, at the other extreme, which should be encouraged into urban areas because they occupy fragile environments of low agricultural potential. There is an ongoing effort to reconcile natural and socioeconomic parameters at the international level, with a focus on natural resource management[3].

Yet internationally derived solutions to developing country problems are increasingly rejected as patronizing. They contradict the partnership ethic now widely advocated. The principles of participation can be applied at all levels of the hierarchical framework. Experience teaches that ownership by all stakeholders is a prerequisite if a new intervention is to be acceptable and successful. Although broad socioeconomic patterns can be developed from available or easily constructed global databases they are essentially rough proxies for the more detailed human groupings required for operational effectiveness. Moving down the hierarchical scale is best achieved through partnerships with interested countries that provide the input needed to define local domains. Such a partnership strategy would allow an iterative approach to reconciliation between physical and human characters at an operational level, and a sequential approach to the coverage of developing countries.

12.2.3 Insights into diagnosis of problems

Simon Maxwell's 1986 article[4] questioned the value of developing technology for a farming system likely to have changed before the targeted technology was available. His criticism carried particular weight when the focus in FSR was on programming the local research station. The new emphases on multiple sources of technology and menus of options for farmers have muted his point. Yet the idea of system evolution remains important. Low research efficiency has largely been caused by prematurely thrusting land intensification practices at farmers with a need for improved labour productivity. Even now 'low yield' as a problem is a common output from superficial diagnosis. This clumsy determination of relevance has lowered the credibility of government research and extension services in the eyes of rural people. Despite agricultural researchers, and more recently environmentalists, stridently anticipating increasingly scarce land, the economic threshold which shifts resource-poor farmers from extensive to intensive methods is dictated by their own judgements, not by outside rhetoric. That threshold generally occurs at much higher population densities than commentators believe, particularly where market access is weak. Recent work, for example at the International Institute of Tropical Agriculture (IITA)[5], has reinforced both market access and population density as drivers of intensification and has taken an important step towards more insightful choices of technology.

12.2.4 Rapid and participatory rural appraisal

Illiteracy and the lack of farm records preclude easy data capture on small peasant farms. The escape from the inflexible, expensive, formal collection of questionnaire-based quantitative survey data for modelling farm households was a tremendous breakthrough in process. At the same time the costs of informal diagnostic surveys, eventually called rapid rural appraisal (RRA), were relatively low and coverage potential encouragingly high. Qualitative understanding has been legitimized in anthropology for many years and more recently in soft systems analysis circles[6]. The final step in the evolution of informal methods for interacting with local communities came with the advocacy and practice of participation under the leadership of Chambers[7]. It has been widely adopted in the 1990s, particularly by NGOs in their work with communities. Biggs'[8] classification of 'extractive', 'consultative', 'collaborative' and 'collegial' methods helped clarify the degree of participation. Early informal diagnoses, carried on in the survey tradition, were perceived as 'extractive', or at best 'consultative', by participatory advocates. As interest surged, participatory rural appraisal (PRA) joined RRA and participatory methods ranging from consultative to collegial became important diagnostic and evaluative tools in the portfolio of FSR practitioners. Participation itself has evolved[9] over the last decade as it incorporated more and more from the experience of participatory development[10]. This evolution, however, has taken off within NGOs and international research leaving national researchers behind.

12.2.5 Multiple sources of technology

The early convention that OFR was essentially an interaction between the research station and local farming systems it served, was overthrown by Biggs and Clay in 1981[11]. They pointed to a variety of alternative sources of technology that could be tapped for local farming systems. These included results from other research stations in similar ecologies, and communities further along the development continuum which had already managed the pressures now engaging the local system in question. Multiple sources opened up best practice in FSR, potentially revolutionizing its role in development.

12.2.6 Wider sources of farm improvement

Crops dominated the improvement strategies of early FSR work. A livestock component and the study of crop–livestock integration was the first widening of improvement for smallholder farming. This was soon followed by the addition of aquaculture into so-called integrated systems. Concerns over fuelwood and deforestation expanded the sources of farm improvement to include agroforestry. With the re-emergence of market access as a key factor for development, the introduction of cash crops onto small farms as a development tool is regaining popularity. Assessing the repercussions of a new enterprise on an existing system requires the same understanding as assessing the repercussions of a new technology. The farmers' approach to adoption will usually follow the familiar pattern; a small-scale trial to minimize disruption to an established means of livelihood, and then expansion if judged successful. Here again is a widening of the role and scope of FSR, bringing it closer to the role of farm management in Western agriculture, but working at the system rather than the farm level.

12.2.7 Embracing 'exogenous' variables

As we saw in Part II of the book, even the early conceptual frameworks for FSR acknowledged the key role of policy, and the need for appropriate mandates in the major enabling services such as banking, credit, marketing, input supply and extension, for influencing farmer behaviour. They also acknowledged the value of FSR output for policy formulation and for guiding strategy in enabling services. However, the typical FSR niche within the research services was too 'thin' a platform for influencing either policy or service mandates. In terms of public service, research is usually one division of an executive ministry with few links to policy makers, often even policy makers within the ministry itself. Research fights its own corner for resources, a corner often under considerable political pressure for poor performance. FSR remains a novelty, its niche has typically been as a project, often with *ad hoc* sites in a few selected locations, and often driven by the donor community. Finally, and against this background, an increasingly broad scope is claimed for a farming systems approach, sometimes embracing the whole agricultural development process. This accumulation of factors, and the contradiction between claims and practice, has threatened FSR's credibility and inhibited its progress.

A strong, recognized and acceptable in-country platform is the *sine qua non* for FSR to move beyond its original role in research into wider farm improvement, policy and the orientation of the enabling services. The hierarchical framework has clarified the huge 'distance' between decision makers at the national and national institution levels, and farmers. In this context the farming systems development initiative from the Food and Agricultural Organization of the United Nations (FAO) has laudably sought to broaden the FSR platform. The liberation of 'best FSR practice' from its role as an extra stage in the formal research process does offer new organizational opportunities. But promotional strategies used to broaden its applications must carefully weigh claims on scope and the current status of FSR in the country, and address those opportunities for widening its platform which can also strengthen its credibility.

12.2.8 Research at aggregated hierarchical levels

Economic impacts have long been dealt with at an aggregate level, using production and price information. To date, however, economists have not consistently examined higher level impacts by aggregation from farm-level data. This failure is reciprocated by the lack of success in formulating policy at the national level to accommodate the diversity in local circumstances. There is a major ongoing effort to solve the mathematics of the aggregation problem. When solved it would allow better disaggregation of policy effects and better estimation of aggregated effects of local change. Further stimulus to research at the wider scales of the hierarchy has come from the increased interest in the sustainability issue. There is a high demand for research on the effects of changed activity on farms and in communities at the level of the watershed, the landscape or ecosystem, as well as the impact on biodiversity. As mentioned by Stroud and Kirkby this has drawn in new disciplines; particularly ecology with a focus on diversity and natural resource management, and geography with both a physical dimension in terms of soil and water, and a human dimension in population and social dynamics. Both these disciplines work on process at higher levels of the hierarchy and, like economics, value a link with FSR as a source of grassroots information. They should not, however, be considered as within the scope of FSR itself.

12.2.9 Stakeholder analysis

Both the hierarchical framework and the aggregation issue have helped raise awareness that different stakeholders exist and they often have different perspectives on agricultural development. Stakeholder analysis is a tool to identify stakeholders and understand their varying perspectives. Gender is an important dimension here, often analysis indicates that men hold very different perspectives than women. Stakeholder analysis also provides an entry point for reconciling conflicting perspectives or negotiating a common position on farm improvement. Its future value is assured by the way the contemporary literature witnesses the growth of local conflict as pressures increase on access to land and water.

Each of these nine milestone innovations of concept and process is an important contribution, or more accurately, a potentially important contribution to contemporary best practice in FSR. Their adoption is as yet modest and at best piecemeal. This is explored more fully here in our examination of four outstanding issues in FSR and its application.

12.3 THE SCOPE OF FSR

Chapter 1 earmarked the scope of FSR as an issue and it has resurfaced several times throughout this book. It has been claimed that confusion over scope has blurred FSR's role and distracted from its application and institutionalization. Indeed, one hears research managers and new staff struggling to articulate what FSR actually is. A significant part of the FSR constituency has difficulty in relating to issues that appear far removed from what concerns them on the ground. Four factors seem to have combined to make scope an issue:

- Rapid evolution in concept, process and method in FSR.
- Proliferation of constituencies for a range of systems applications in agricultural R & D, some of which are closely identified with FSR.
- Confusion from professionals, often outsiders, wanting either to contribute to development theory or to test new, often sophisticated, methods, using FSR as a vehicle to the household level.
- Institutionalization that has occurred in FSR has usually been within the narrow mandate of the research services.

Experienced people in FSR feel that the movement has lost its momentum. We believe this is due to a coincidence of two factors; criticism of the weak performance of FSR capacity created for adaptive research and the simultaneous promotion of FSR to a wider role in development. Weak performance in the narrow, closely focused role of adaptive research is certainly a poor recommendation for an enhanced role and this coincidence threatens the credibility of FSR as a whole. As the nine milestones suggest, there has been rapid evolution in concept, process and method in FSR, and this has outpaced the ability of practitioners to refresh their training and absorb new ideas. This failing has been exacerbated by their lack of grounding in systems concepts and methods during formal professional education, perhaps the Achilles' heel of FSR. Many other reasons contribute to the loss of momentum: negative promotional strategies have often sought to discredit, rather than build on, what exists. Early FSR itself fell into this trap. Condemning existing research recommendations created barriers to its acceptability in the research establishment. Many participation advocates were similarly scathing about FSR. They perceived, and continue to perceive, participation as a replacement for, rather than an improvement of, the FSR process.

While FSR originated in the field it has, over the last 15 years, been distracted by professionals and academics with their own horses to ride[12], one of the many attempts to impose innovations in development processes from outside. False trails, many created by outside academics wanting to demonstrate what is theoretically desirable and methodologically possible, have exacerbated the problem, widening the gap between theory and what is operationally practical. There are historical examples of the promotion of over-sophisticated processes such as the 'have tool will travel' mentality of American universities with linear programming in the 1960s and 1970s and the touting of full-blown economic surplus models for priority setting in national agricultural research systems in the 1990s. Outsiders often know 'what is best', but with little understanding of the inside circumstances 'what is best' may remain irrelevant – like classic agricultural research recommendations evaluated on the wrong criteria.

Early FSR methods were essentially a response in the other direction, agricultural professionals recognized that their 'Western' conceived tools did not cope with three characteristic circumstances of developing country smallholder agriculture:

- Large numbers of farmers.
- Small absolute benefit levels, even from significant improvements on the very small farm units.
- Low numbers of professional agriculturalists.

Our concerns about FSR stepping beyond the boundaries of practicality in the context of relatively weak developing agricultural institutions are elaborated in discussing the quality of contemporary FSR practice below.

Perhaps the most confounding influence on FSR has been the growing demand for research at wider scales in the hierarchy. Much of this has arisen from the long-standing effort, particularly among academic development economists, to aggregate economic models from the household to national levels. This research on methods has associated itself with the FSR movement because of its potential to contribute at the household level. More recently it has been paralleled by similar interest from the disciplines of geography and ecology also seeking roles at wider scales in the course of the assault on natural resource degradation. A problem identified at a higher level in the hierarchy will only occasionally be endogenous to a single farming system. The Vittel case in Chapter 11 is a good environmental example. Deteriorating water quality was identified at the level of the water-catchment area, the cause was the externalities from farming practice across the catchment. In this case the solution sought was a compromise on farm performance for the benefit of a stakeholder with catchment wide interests.

Policy, particularly at the national level and usually even at the local level, has a very diverse impact. Changes influence many different farming systems often in different ways, and the perceived national interest, as well as the interests of stakeholders in commerce and the enabling services, will carry considerable weight in the outcomes. In a hard operational context it becomes almost cavalier to claim that FSR can ever *drive* policy. *Informing* policy is another matter, and here of course FSR can make an extremely valuable contribution. Links through the hierarchy feed FSR practitioners with information to aid farm improvement, and, in return, feed development managers with local information to improve the trade-offs in decisions on policies and institutional strategies which themselves seek to influence farmer and community behaviour. The fundamental reorientation of policy and institutional strategy, as the introduction of FSR has itself demonstrated, will usually be a long-term process. That should not compromise short-term farm improvement within the existing environment. Policy and strategy change will help immediate farm improvement when it can be achieved in the same timeframe. This will usually be limited to small changes in institutional operating procedures which facilitate the introduction of improvements identified for a specific farm system.

Research higher in the hierarchy involving applications over wider scales must deal with spatial aggregation, wider stakeholder diversity and the national interest, none of which are within the operational scope of FSR. Jiggins has published two interesting statements, which, taken at face value, conflict. First; 'Engagement with method and process in turn has led us to recognize that we cannot, as agricultural professionals, limit ourselves anymore to the crop system, or to the farm system. We are having to deal with systems on a larger scale'[13]. Second; 'When moving up from the farm in aggregation we must be careful not to be swept away by the relatively painless generalization that this allows. Remember, just as the farming system aggregates farmers' perspectives losing any one of them, so further aggregation loses perspective of any one farming system'. These two statements can be reconciled through the linked levels of the local/global hierarchy. It is important to have competent agricultural professionals making decisions at all levels, but it would be dangerous to assume that a farmers' perspective, and indeed a farming systems perspective, can survive aggregation. FSR is an essentially operational process with a focus on the farming system and community levels in a systems hierarchy, and with useful by-products for decision makers on policy, and in enabling institutions, at higher levels.

Even within the operational boundaries of a focus on the farm system the scope of FSR has expanded significantly with far reaching implications. The multiple sources of technology concept broke the link between the local research station and local farming systems and shifted emphasis to the identification, rather than the development, of new technologies. It has potentially liberated the FSR process both from its niche in research institutions, and from its controversy with the scientific establishment. More recently the sources of farm improvement embraced by FSR have widened beyond technology in the strict sense, to include new enterprises with a market, and opportunities to add value through local processing or by local group action. Conventional farm management has always handled these wider sources. Including these in FSR suggests that it can be seen as a farm management approach in circumstances where the masses of very small farmers preclude cost-effective face-to-face advice. Here improvements must be identified for significant groups of farmers at the system level. Again, its early (and largely continuing) niche in research institutions has inhibited the broadening of FSR horizons in farm improvement. Both multiple sources of technology and new enthusiasm for market linkages through new cash earning opportunities reduce dependency on research for a home and open other options for institutionalization.

The increasing complexity of the conceptual framework adopted in FSR and the proliferation of academic interest, for one reason or other, has blurred operational practicalities for the relatively unsophisticated institutions available for its implementation. FSR has lost momentum. Field practitioners feel lost in the vast scope of the local to global systems hierarchy. Development managers look sceptically at its performance and question its credibility in adaptive research let alone wider roles. Nevertheless, if FSR is to survive in the future, ways to meet demands for new and better farming systems will have to be found. Governments and donors aim to improve the management of natural resources, aim to improve rural livelihoods and FSR practitioners will have to decide what contribution they want to make towards these aims. In thinking about their contribution FSR will be challenged to redefine its scope. We must look to those practitioners with a foot in the theoretical camp to articulate new concepts and methods to meet these.

12.4 POOR FSR PERFORMANCE – THE QUALITY ISSUE

The second crucial issue identified in the opening chapter was slow institutionalization. Though this is analysed later in this section it is important to a discussion of the low quality of FSR applied in the field. Supported by the donor community FSR made inroads into public research services in the late 1970s and 1980s. During this process institutional characteristics emerged which called for changes well beyond the horizons of FSR. These characteristics are responsible both for the slow adoption of FSR as a process in public institutions, and for its poor performance in the field.

- The science-driven culture in research institutions.
- A top-down, control culture in enabling services.
- Alien curricula in higher agricultural education.

Efforts to change these began, again almost exclusively through donor support. However, much of the learning in these projects occurred among the expatriate technical advisors and not among their local counterparts. So, the knowledge gained from hard lessons learned left at the end of the project leaving little behind. A valuable and increasingly prominent role was, and is, being played by the participation movement. Its advocacy and practice of empowerment of the poor is a direct challenge to all three of these circumstances, but also, in many countries, has opened up a wide gap between NGOs and government services.

The contribution of the participatory movement has not, however, been without its shortcomings. Demand for PRA training soon outstripped the supply of experienced practitioners. The gap was happily filled by 'consultants' prepared to train people in techniques they had learned from books or picked up from a week of training. While many people learned to go through the PRA steps few grasped the concepts, attitudes and behaviour necessary to do it properly. Beyond the friction between government and NGOs, poor performance in implementing FSR is most often a function of the general malaise in the public services of many developing countries. Poor performance is by no means unique to FSR, it generally permeates public service functions. Even capable and conscientious individuals are demoralized by the frustrations of timid leadership and stagnant, under-resourced institutions.

The universities are often no exception but, beyond the poor conditions for staff, there are shortcomings of the curricula based on Western agriculture and a disciplinary culture giving allegiance to American and European journals also rooted in the West. Graduates are poorly equipped to understand the circumstances of farming in their own countries and often prefer to insulate themselves from the realities of small farmers. Even those who move into fieldwork do not have the motivation or methods to understand smallholder farming, the conceptual foundations to absorb innovations in techniques or indeed the access to the publications where such innovations are to be found. Anandajayasekeram's account of the progress made with bringing farming systems concepts and methods into university curricula in eastern and southern Africa over the last decade in Chapter 8 suggests patchy and uneven progress. Much more effort is required here if the necessary revolution in culture, perspectives and practice is to begin. A better balance of effort between the development of new concepts and methods and capacity building for good practice in the field is urgently needed. It is an area where developed-country universities can assist when they are prepared to meet the real needs of the institutions they work with, rather than pursue their own predispositions.

12.5 PARTICIPATION AND FSR

There is no doubt that 'participation' and 'empowerment' are currently much more attractive ideas in the eyes of the donor community than FSR. As noted in the introduction in Chapter 1, donor funding, still searching for a cost-effective approach to smallholder development, has shifted from FSR, a hero of the 1980s, to participation and empowerment, new gods of the 1990s. As with FSR, participation is represented by an increasing number of acronyms: FPR, PRA, PTD; most of these are flag flying, some re-labelling, others emphasize different dimensions in process. What is often forgotten is that farmer participation was one of the founding principles of FSR. Indeed, 'giving voice to farmers' was the early slogan of FSR. Moreover, we should not forget that most of the development in participatory methods occurred within FSR projects trying to enhance the participation of farmers in their work. For some FSR practitioners participatory research is purely a natural evolution of their pursuit of farmer participation within FSR. For others, notably donors and participatory development advocates, the goal of participation is empowerment. While this goal is much more fundamental than improving farming systems, it is also more abstract and more political. As with the community development movement in the 1950s and 1960s, being abstract it needs vehicles to move from principle to practice. Agriculture has provided one vehicle among others, like health, to which the participatory principle has been applied. However, regardless of labels, for best practice in farm improvement FSR needs to embrace farmer participation more fully.

In his contribution John Farrington expresses the view that: 'FPR-E is an approach to the development of technologies (embracing diagnosis, screening, testing, verification) that

Box 12.5.1. Contributions of FSR and participatory approaches.

Contributions from FSR

Systems hierarchy framework
- links to outside to identify improvement opportunities
- links for informing and influencing higher level decisions

Farming system typology
- framework for priority setting, programming and resource allocation

Farm level framework
- understanding through a replicable analytical process farm families' priorities
- resource constraints and evaluation criteria
- whole farm system modelling and analysis
- more relevant and appropriate menus for farm improvement; identification of relevant, appropriate solutions; guidance in shaping solutions for acceptability

Contributions from participatory approaches

Community trust
- building trust between communities and outsiders
- identification of articulated (rather than analysed) problems

Methods
- participatory methods, often using diagrams
- bring new dimensions to diagnosis and evaluation
- add ownership for community stakeholders

Empowerment
- adds scope to improvement, through community action and organization in marketing and local processing, input purchase, technology testing, infrastructure investment

Widens partnerships
- mobilizes socially conscious external institutions for local action

meet farmers needs. As such it is equivalent to FSR-E but utilizing a wider range of methods and relying on a wider range of institutional linkages'. FPR-E represents a natural improvement and progression for the FSR paradigm and is a 'participatory umbrella' under which FSR can stand. While FSR has much to learn from progress made in participatory research we feel that contributions flow in both directions.

12.5.1 Mutual Learning

FSR and participatory approaches both make contributions to the improvement of farming. Some of the major contributions are listed in Box 12.5.1. The implementation of interventions on issues suggested by farmers themselves is an important vanguard for reinforcing the credibility of outsiders. The trust it builds becomes, in turn, a key vehicle for 'outsider' functions, including FSR, to make their contribution to local development. Beyond this, participatory approaches make a major contribution to the efficiency of the FSR process itself. Many opportunities for mutual learning exist. FSR, having emerged from a rigorous research background, provides frameworks that underpin the organization and orientation of participation. The phrase 'replicable analytical process' is particularly important in describing FSR's contribution. The systems perspective at its heart provides the understanding of the constraints in, and opportunities of, the farming system. At the same time its 'outsider' perspective on available technologies, new market opportunities, processing possibilities and policy influence draws on resources not normally available to local farmers and communities. The discussion of scope in FSR earlier in the chapter demonstrates the potential to move beyond technology adaptation into wider sources for farm improvement. In turn the strength of participation in community organization, already widely evident in practice on the ground, has also widened the scope of farming systems development.

As with FSR, there are also weaknesses in the implementation of participatory approaches. Weakness reaches into concept when practitioners, carried away by the emotive appeal of empowerment, ignore the fact that it needs purpose, and that it is the dynamics beyond the community that provide the development opportunities. Baker[14], reacting to Chamber's writing and seeking reorientation of FSR not a reversal from it, caught an important flavour: 'There is little evidence however, that a farmer-first paradigm, when carried to an extreme, will produce substantial benefits. To the contrary, some of the methods and behaviour advocated ... could lead to decreased attention to technology supply options, overinvestment of public resources in small numbers of farmers, and reduced overall effectiveness in assisting farmers'.

Much participation places great store on eliciting and acting on articulated farmer and community priorities. In this it opens itself to an early criticism of FSR made by hard scientists in the research establishment – that traditional small farmers often do not know how to solve their problems, they are unaware of the wider environment. The participation movement sometimes appears to deny, rather like the extreme advocates of indigenous knowledge, that 'outside' knowledge will necessarily play a part in providing sustainable livelihoods. It is an outlook reminiscent of academic anthropology and the culture of 'study them as they are'. Recent cracks have appeared between rhetoric and reality in participation, some are listed by Rhoades[15] who wrote; 'the social scientists who attempt to raise analytical points about stratification, differential access to power and resources, and other social shaping dynamics are accused of being top-down and then are marginalized by turf guarding NGOs'. There is a vast difference between letting farmers articulate their perceived problems and acting on these; and an analysis of the farming system they operate and the problems and opportunities it presents. As Stroud and Kirkby point out in Chapter 4, farmers' perceptions of problems are confounded by life: labour peaks, gender specialization and social obligations. Their perceptions of opportunities are confined by the scope of their world-view. Only an analytical framework that penetrates below the surface to the problems of the system they operate has the full potential to reach into their lives. The importance of an interface with both an inside and an outside understanding (neither of course perfect) cannot be overemphasized. It is outside opportunities that are the key to the improvement of the local situation. This is not to underestimate the importance of acting on farmers' perceived and articulated problems, particularly to build community confidence in outsiders, in turn, helping them access the more fundamental issues surrounding smallholder livelihoods.

Over the last few years the coverage issue has gained attention in the participation movement. There is a growing awareness among practitioners that, in working with small groups, they are spending time and resources on relatively few farmers. The question of how the benefits of a participatory approach can reach the vast numbers of rural resource-poor households is of increasing concern. FSR on the other hand evolved with the coverage issue in mind. Facets of the FSR process, the farming systems typology and the identification of representative communities and farmers were developed as sampling devices to achieve coverage. Similarly, qualitative, low cost and rapid diagnostic methods sought the appropriate trade-off between coverage and the intensity of professional effort required. FSR has much to offer participation on the coverage issue.

Because of its political nature effective local empowerment is not easily achieved. Reconciling stakeholder interests, even at the very local level, implies an often significant shift in the distribution of power. Beyond the local level the process of articulating local needs has to avoid capture by inherently more powerful and experienced stakeholders. Local empowerment is a first step, it must be complemented by responsive public and private institutions. Part III of this book demonstrates how difficult this has been with FSR over the last 20 years.

Nor should it ever become a full reversal. There is a need to balance local actions with the wider interests of other stakeholders, not least with the national interest benignly defined. Local farmers' aspirations must be weighed with the aspirations of other local farmers and other stakeholders more broadly, and with the interests of the country as a whole, in determining local programme priorities. For some countries final empowerment will remain a dream for a long time, such political maturity may never arrive. Van Eijk's[16] 'interim measure' may be permanent, it is certainly better to have an understanding, informed interface where the

Box 12.5.2. Embracing FSR.

Process and methods

Continue to use the stages of FSR as the core of process
Further develop collegial methods for each stage
Ensure the use of the replicable FSR analytical framework to provide understanding

Hierarchical links

Accept that good understanding of the local system by outsiders is important
Use the understanding as a link to influence higher hierarchical levels
Seek partnerships with government that provide upward linkages to local and national levels

Coverage

Identify with a farming systems typology as a platform for rural development
Accept that not all communities can be covered by a full participatory programme
Accept that those that are favoured serve as 'models' for others operating similar farming system
Facilitate farmer-to-farmer and community-to-community networking to spread changes
Seek partnerships with government institutions that
    prioritize development initiatives in the national interest
    enable networking and provide support for networked changes in adopting communities

empowerment road is a long one. Hall[17] identified FSR as an interface to marry the local skills and enthusiasms of NGOs with the policy and science base in government. NGOs need to recognize their limitations and appreciate the political need for governments to be seen to improve the lot of all their diverse rural and urban constituencies. Empowerment and NGO/GO partnerships will be increasingly tested as donors seek better governance and an increased voice for the resource-poor. FSR, its beneficiary allegiance strengthened by its partnership with participation and with its widened scope for farm improvement, has a strong interface role to play as a public service and offers a robust vehicle for moving towards empowerment. The participation movement, if it is to demonstrate success where it counts, needs to embrace FSR more completely, including the practical implications of a hierarchical conceptual framework. The aspects set out in Box 12.5.2 seem the most important.

12.6 FSR AND THE RESEARCH AND DEVELOPMENT PROCESS

With the participation partnership in mind three further issues on the role of FSR in the R & D process merit discussion. Perhaps the most controversial is the promotion and use of a farming system typology as an improved framework for agricultural R & D.

12.6.1 Uses of a farming systems typology as a development framework

Typology is an issue that is discussed in length in Chapter 3. There are two particularly important development interfaces: we have discussed the interface at the grassroots level between farmers, their communities and 'outsiders'. The second is at the farming systems' level, where the grouping of farmers by the system they operate represents a compromise between the (prohibitive) cost of improving small farmers individually, and, at the other extreme, treating the small-farm sector as uniform. The groups of individual decision makers in diverse systems will react differently to decisions made higher up in the hierarchy that seek to influence local actions both on farms and in communities. The farming system is the most rational intermediary unit in an aggregating, or disaggregating, sequence between the farm and the nation, and indeed between the farm and international resources for development.

One challenge that uses the full scale of the system hierarchy, is to develop a practical way for local teams working with communities to draw on worldwide experience of interventions at farm, community and policy level to improve local livelihoods. Two issues seem to underlie this challenge:

- An effective marriage of biophysical and human variables in characterization to foster the transfer of development experience.

- Partnership between global and local skills, both in making such a resource available, and in its application.

An effective global technology database will be wholly dependent on the input from thousands of local situations and will be shaped by their needs in evaluating input from other local situations for relevance to their own. Widening acceptance of the link between poverty and environmental degradation, and the widening recognition of human decisions as the cause of degradation, are forcing a much needed reconciliation between traditional physically based definitions of zones in terms of climate and soil, and people-based definitions. Looking ahead, geographical information systems (GIS) will allow the matching of farming system profiles in similar agroecologies worldwide. Local professionals could draw on databases of technologies, and of community and policy actions, found effective in similar situations elsewhere. The superimposition of farming systems on agroecology through GIS could be a common planning tool for institutions at all levels of the hierarchy. Stroud and Kirkby note the inadequate profiling of experimental sites in Chapter 4. If global databases to provide local options are to work successfully, profiling protocols would need to be system based, thorough and common to all. The act of choosing potential improvements from such a database is indicative of the profiling needs that might take the following steps:

- Compare physical profiles at the level of agroecology for a technical 'first fit'.
- Compare socioeconomic profiles on market access and population density for human 'first fit'.
- If for a crop already grown, then
  - socioeconomic pre-screening from local system understanding
  - decision on whether the technology needs proving technically locally.
- If for a crop not yet grown, then
  - market evaluation as either a cash opportunity or an alternative to current subsistence crops, if positive
  - socioeconomic pre-screening from local system understanding
  - decision on whether the crop needs proving technically locally.
- Adoption or not as one in the menu of options to be put to local farmers.

The database idea recalls 'think globally act locally', the well-used quote from John Naisbett's vision of the future. Too frequently, perhaps, development efforts at the global level have taken on their own life, flags flown by international agencies for those perceived as beneficiaries, though out of touch with them and their needs. Schemes developed at a global level and promoted to developing countries have often fallen short; the use of agroclimatic zoning for policy formulation, and T & V for extension are two examples. Such 'flag flying' has frequently led in a circle. One such circle is the way 'scaling up', a new preoccupation for participation practitioners, was a core consideration in the original shift from intensive measurement to qualitative understanding in the FSR process,

Box 12.6.1. Definition of research levels.
Strategic – to understanding natural processes surrounding an identified problem – perhaps the supply of nitrogen from the soil. Some understanding will be generic, much will be ecology specific. In a resource scarce situation the strategic research portfolio should be dictated by the need for information to facilitate the development of a new, powerful prototype solution. For example one might justify the long search for nodulation in grasses by the potential offered if grasses could fix atmospheric N

Applied – using the available understanding of natural processes to define prototype solutions to an identified problem. Soil nitrogen can be maintained by following one or more of a number of strategies. A prototype solution defines the technically ideal way to use chemical nitrogen, manure, compost, green manure, rotation, rhyzobial bacteria or leguminous trees, and will usually be crop and ecology specific

Adaptive – selecting a range of prototype interventions for farmers with an identified need or problem. Working with them to choose between alternatives and adapt the chosen options to their local circumstances itself subsequently adopted by the participation movement.

Beyond its hierarchical linkage role a farming system typology also offers more informed management of geographical space, an underused but, given accumulating environmental concerns, an increasingly important dimension in policy formulation. An example might be for areas where degradation of fragile land threatens urban water supplies. If farming potential is poor due to the marginal resource base, an alternative strategy to agricultural improvement would require incentives to encourage migration to urban livelihoods. Such choices further illustrate the importance of reconciling national and local interests.

12.6.2 A modified paradigm for applied research

FSR has, to date, most frequently found its government home in the research services, cheek by jowl with the more traditional disciplinary and commodity scientists, as an approach to adaptive research. To keep a consistent distinction between the different levels of agricultural research in the discussion which follows we offer some definitions (Box 12.6.1).

More broadly, complementarity across the strategic, applied and adaptive levels provides integrity to the research process as a whole. This requires changes in the classical applied research paradigm in order to reconcile more effectively the role of OFEs with FSR as a key component in an adaptive research process. Historically, applied research has sought a technically ideal management system for a commodity in a biophysical environment[18]. The ideal is selected by the physical productivity of land as a measure of performance. Each commodity has warranted a discrete research effort for the major ecologies in which it grows. There are at least three failings in this paradigm:

- It fails to recognize that physical productivity is never the evaluation criterion used by farmers. Improved labour and capital productivity are their primary criteria, they recognize higher yield as a possible means to these ends.
- It fails to recognize that economic and cultural diversity also creates discrete environments and requires accommodation in technology choice and design.
- The search for an ideal technical model and the dominance of statistical precision in experimentation discounts relevance and flexibility and fosters prescription.

FSR has already addressed the issues of evaluation, through farmer assessment, and of economic and cultural diversity, through the farm typology concept. The final issue is one for applied researchers who continue to use a paradigm that discounts both relevance and flexibility, the latter perhaps the *sine qua non* for effective farm management, particularly in dryland agriculture. Applied research needs to recognize that its outputs – recommendations for the technically ideal system for growing a commodity in an agroecology – are inadequate to equip adaptive researchers with the information for shaping innovations to fit farmers' circumstances. Three sets of factors create the need to fully understand not only that ideal but beyond that its *plasticity*. These sets highlight the flexibility needed in management. It is the same flexibility that has stimulated recent writing on 'agriculture as a performance'[19]. It is the same flexibility that adaptive management has sought in the field of environmental management[20].

- Within and between season variations in climate; small farmers manage this directly by their cropping pattern and practices. Variability in rainfall pattern is of particular importance.
- Farm systems produce a range of products to meet a complex of family objectives, all compete for the labour, land and cash farmers have, no one product will be given 'ideal' management.
- Farmers often do not know whether a particular management practice will work. Thus variations are created so that they can learn their way to appropriate practices.

Bringing new crops, new varieties and new management practices into farmers' systems demands compromises on the ideal. Adaptive researchers working with farmers need field procedures to learn the level of compromise that can be made on the main components of the technically ideal management model in order to help farmers shape it to their system. In such a scenario the 'ideal technical management' forms a first step for applied researchers, not a final output. Beyond the ideal, applied research needs to learn about the loss in productivity or of conservation potential, or both, that whole farming system management forces on the ideal. When fitting interventions to farming systems deviations from the ideal occur, compromises are inevitable. Two illustrations are provided in Box 12.6.2.

12.6.3 Multiple interventions: back to system transformation?

Conventional wisdom characterizes the adoption of changes by small farmers as component by component, perhaps more than one where these are clearly interdependent and interacting, initially on a small scale, expanding the scale of adoption as experience is gained. Historically, at the other end of the spectrum, failed settlement and re-settlement schemes have demonstrated that small farmers cannot be successfully introduced to completely new farming systems. 'Designer' systems rarely meet their family priorities and they do not have the abilities to manage wholesale transformation unless very heavily supervised over several seasons.

Recently some ecologists working to improve natural resources management have argued that combining agroecosystems analysis with participatory approaches allows more comprehensive change. Using participatory approaches that allow farmers to become the 'designers' of the systems addresses a major source of error in the past; imposition by 'outside' designers. Moving from changing one to many components comes through agroecosystem analysis that allows farmers to identify a wide range of interventions because it explores each and every agroecological niche available to the farm family. In this work agroecosystems analysis introduces change in three areas:

- Rehabilitation of degraded natural resources.
- Increasing biodiversity of cultivated species.
- Increasing internal recycling of biological wastes or by-products.

What changes should be made are determined by farmers through a participatory process in which they 'model' existing farming systems and vision 'future' farming systems. Farmers do not approach the task of realizing their future

Box 12.6.2. Examples of system circumstances comprising ideal technical practice.

During the 1950s tied-ridging was identified as a yield-enhancing, soil-conserving practice for cotton in Tanzania[21]. The 'ideal technical system' showed that 6-foot intervals between ties made just before or just after planting gave the highest increments in yield, and this formed the extension recommendation. However, 6-foot intervals also required an extra 6 days per acre of labour at a time in the calendar when farmers were cultivating and planting new land or weeding their established crops. In the average year losses from reducing the extra area cultivated, from delays in planting it or from delays in weeding, more than offset the yield gains from 6-foot tie-ridges. Twelve-foot intervals between ridges compromised the 'ideal', reducing the yield increment, but halving the labour requirement, offering a more competitive reward for the use of scarce labour. Alternatively, delaying tie-ridging until the second weeding also compromised the yield increment, but it also moved the labour requirement 'off-peak', again offering a more competitive reward for the use of labour.

The need for the measurement and comparison of plasticity extends beyond the technically ideal system for commodity management, it should also influence the choice of cultivar[22]. An alternative cultivar, or even an alternative species, might be more resilient in the face of local weather variations or labour shortages, despite a lower physical potential. Forage crops are a good example; their introduction into smallholder systems is more likely to be successful if their management regime draws on land and labour resources not being used to produce vital family food and cash. Forages can rarely compete with small farmers' priorities for enterprises of direct benefit to the household. Adaptive researchers need to compare species not only on the physical productivity and nutritional quality offered by an 'ideal' technical system, but also on criteria of flexibility for 'system fit'. The outputs from applied forage research need to demonstrate the responses to changes in criteria commonly influencing 'system fit', such as time of planting, maturity period and the levels of labour and cash needed for establishment, maintenance, harvest and processing under alternative management regimes.

visions by experimenting with changing single components; rather they simultaneously change several components across the system. Similarly, farmers assess the performance of the changes not component by component but by the effects on the system as a whole. Lightfoot's account of agroecosystem analysis in Chapter 11 details the participatory process that moves from the 'bioresourceflow models' of existing systems to 'visions' of future systems.

This way of introducing interventions into the farming system was found to have a number of other advantages. First, the impact on farming system performance from the household's point of view was greater than that achieved through changes in single components. Second, brainstorming future visions generated more human energy and commitment to local action than diagnosing often intractable problems. Third, brainstorming between farmers and scientists identifies what is not known, thus a farmer demand-driven agenda for applied research emerges. Fourth, the use of performance indicators allows farmers and researchers to adjust the future vision in the light of their expanding knowledge. 'Multiple simultaneous change' certainly has its disadvantages, but most stem from the way research is currently organized, rewarded and funded. Although farm households may be impressed with big changes in the performance of the whole farming system, research institutions and the donors who support them are only impressed by improvements in the performance of their target commodity. If the commodity in question is for example fish and in a drought year all the fish die, researchers see failure; even if the resulting water resource improvements allowed the household to grow vegetables in their fish ponds and feed themselves during the drought without spending their savings. Only by resource-poor families themselves articulating the significance of surviving drought without losing capital[23] are the full, system-wide benefits of such changes identified. The conventional requirements for rigour and the need for standardization of treatments and controls make this work unacceptable by traditional standards. Each farmer starts with their own unique farm, envisions their own farm of the future, and incorporates different interventions along the way. Because nothing at the component or enterprise level is controlled, or can be compared across farms, conventional researchers are unconvinced. Similarly, few conventional scientists are interested in tracking and measuring changes in performance of whole farming systems over time. These experiments suggest that the degree and speed of the transformation of their farming systems is not held back by farmers' capacity to experiment with many ideas. It is their poor access to inputs, particularly germplasm of new species, to markets for new products, to financial resources, to knowledge and information, to power which endows the authority for access to resources, in short their institutional environment, which is restrictive.

The 'multiple simultaneous intervention' approach offers a halfway house between single component improvement and new farming system imposition. It offers a range of changes to the existing system, shaped by negotiation based on understanding. The immediate questions it raises are whether the intensive level of professional interaction at the farm level can be reduced, and whether farmer-to-farmer diffusion will be effective in scaling up adoption to provide a healthy benefit stream to that heavy on-farm investment of professional time. As agroecosystems analysis develops it will be important to record the ideas generated, their uptake by farmers in participating and non-participating communities, and also the 'outsider' partnerships required to support the realization of the visionary models.

12.7 THE FSR PROCESS – CONTEMPORARY BEST PRACTICE

Contemporary best practice is itself a vision. It is never yet found on the ground and thus remains essentially hypothetical. Best practice is no longer solely technology focused, and even in technology choice, is no longer dependent on the local station research. The scope of FSR has widened over the last 20 years to an outward looking search for development opportunities, for application in farm improvement, through knowledge networking, more appropriate organization in the enabling services, input supply, credit, processing and marketing, and better policy formulation. Strategies for each application are shaped by the existing circumstances of the farming system when these cannot readily be modified by policy or actions of the enabling services. This wider scope also widens the options for successful institutionalization. Within contemporary best practice an improved understanding of the causes of spatial diversity and local variability across farming systems, and the growing reality of small farmers as empowered partners with their own experimental abilities, have usurped the need for final, precise recommendations.

Our ideas on contemporary best practice were governed by the realization that farms change continuously over time responding to co-evolving social, economic, market and ecological contexts. Thus any idea of best-bet innovations must change with these co-evolving contexts. Moreover, not only do we have multiple sources of innovation but also multiple viewpoints from different stakeholders on farm management. These all have to be integrated to find a way to move forward. Such conditions call for 'learn-by-doing'. Because one cannot have perfect knowledge of either socioeconomic or ecological processes within non-equilibrium systems the sensible course of action is to learn our way to improved farming. Moreover, we do not forget that our understanding of ecological soundness changes as knowledge expands. Thus over several learning cycles future visions will change with expanding knowledge. Here we do not see learning as a passive process of teaching or transferring information for farmers to apply but as an active process of learning-by-doing. Sharing or disseminating the lessons learned in our view should focus on social and organizational processes involved in creating information networks rather than concentrating on technology transfer processes.

With the above observations in mind our contemporary best practice in FSR would include the following five key elements.

12.7.1 Characterization of farming systems

Characterization creates a farming systems typology based on spatial, cultural and resource endowment differences, including agroecological resources, between groups of farmers as basic units for agricultural development. Profiles each farming system allow matching with systems elsewhere as possible sources of improvements, and the thus construction of information networks.

12.7.2 Multi-stakeholder visioning

Within those farming systems which are a national priority teams identify those to be involved in the learning process, and assemble the material to be learned. Participatory techniques are used to identify and build trust between stakeholders. Once the stakeholders are identified work starts on understanding the past and present situation, including who brought change to the farming system. This is followed by brainstorming visions of future farming systems. Brainstorming embraces special techniques for gender analysis and for agroecosystems analysis for the improved management of natural resources. This work goes beyond the farm system to identify both exogenous factors critical to farm performance, including other off-farm means to rural livelihoods, and the endogenous factors creating externalities. There follows a process of negotiation over common learning points and ways and means to realize future visions.

12.7.3 Partnerships for concerted action

Working from the future vision, farmers, with the assistance of other stakeholders and particularly researchers, identify future information linkages needed to realize their visions Partnerships are then negotiated to actuate the needed linkages. Partnerships will include access to relevant new technologies from all available sources, both to enhance productivity and natural resource management. Partnership, especially among farmer groups, would secure opportunities for new enterprises in marketable products, new opportunities for group action in marketing, processing, input purchase and conservation.

12.7.4 Action learning and tracking change

Part of the negotiation process for building partnerships would include the development of performance indicators for those partnerships. Uncertainty concerning the technical performance of chosen options is resolved through development of performance indicators for technical interventions. Learning-by-doing proceeds through experiments and trials. Performance in both technology and partnerships is assessed using the indicators developed. Debate and dialogue on performance indicators tracks learning in both improved farming systems and improved roles of researchers and other enabling institutions. Learning rejects the failures, secures the successes and stimulates further adaptation.

12.7.5 Building knowledge networks

The process of partnership building also lays the foundations for knowledge networks. All stakeholders involved learn who are providers and users of what information. This allows them to facilitate lateral networking: farmer-to-farmer, researcher-to-researcher and so on. Information is passed to policy makers and to enabling institutions, first to highlight changes to encourage further support for wide-scale adoption of preferred options, and second to provide better understanding of the way local communities can be expected to respond to policy or institutional initiatives. Feedback also helps prioritize and programme applied and strategic research.

In our attempt to construct best practice we are mindful that practice must accommodate large numbers of small farmers. This is achieved through the engagement of many stakeholders at the community level. Indeed, farmers themselves will largely control who and how many among the farming population can participate in the learning. Many more stakeholders are brought into the learning process than just researchers and extension workers. Stakeholder analysis indicates that change is brought to farming systems by a wide array of actors and not just extension agents. Engaging this wide array of actors overcomes the problem of low numbers of agricultural research and extension staff and can promote farmer-to-farmer and community-to-community dissemination of information and diffusion of changes.

12.8 FSR ORGANIZATION

Cost has long been held up as a major barrier to the introduction of FSR. In one sense it is a non-issue, as it is cost-effectiveness which is important. However, there is no doubt that a reputation for high operating costs has been a deterrent to the adoption of FSR in a world of falling revenues and rising salary bills. Elon Gilbert presents in Chapter 7.3 cost data from country cases reported in ISNAR's study of the late 1980s. Gilbert comments: 'The OFR cost structure is almost identical to the average structure for the NARI. This is particularly surprising in view of the common impression that OFR is more dependent upon operating costs than the NARI in general. In absolute terms, average operational expenses for OFR programmes are less than two-thirds those for the NARI. Although OFR travel costs per researcher may be equal or greater that those of the NARI, on-station research can be expected to be more expensive in virtually every other category, including equipment, supplies, structures and administrative overheads. This, together with the lower salaries for the relatively junior professionals involved in OFR (62% of NARI average salaries), and a larger support staff for on-station researchers, leads to a general conclusion that the NARI as a whole has higher costs per researcher than OFR'. The cost of travel has been a particular *bête noire* for FSR, perhaps this is attributable more to its high demand for transport in institutions where, in the 1980s and beyond, working vehicles have been scarce. Clearly as on-farm professionals age and salary bills rise, the difference will be eroded. Nevertheless the data and analysis offers no evidence that the costs of OFR are higher than those of on-station research and the study, although now over 10 years old, does much to lay this ghost of FSR.

As yet few countries have achieved full coverage with FSR field teams, though the evidence suggests that, using rapid survey and participatory techniques, it is not manpower expensive. The following rough estimate of the workload of an FSR team is based on Ewell's[24] assessment that the 'minimal pair' – a social scientist and agronomist, supported by other disciplines on a needs basis – is the most efficient team of some 2.5 professional person years. Such a team will cope with initial diagnosis, continuous monitoring, and with an experimental programme in 3 or 4 out of every 7 years, collaborating with farmers of three logistically adjacent farming systems. With the expanding use of participatory techniques at the group level the social scientist will increasingly be an anthropologist or rural sociologist, better equipped than the economist to facilitate the group processes involved, and usually familiar with the simple economics required. Thus coverage of 100 discrete farming systems would, therefore, require some 80–90 professionals – a modest total in the context of the research and extension cadres of many developing countries.

FSR is a process for understanding smallholder farmers. The nature of the process generates allegiance to local beneficiaries and brings this unique perspective to the range of applications for which such understanding is useful. The broader scope of contemporary FSR allows its application in research, in development programming and in policy formulation, offering a variety of institutional entry points for the process. The country's characteristics, and the existing organization for agricultural development, will play a key role in choosing the most appropriate promotional and institutional strategy.

Ideally, FSR links rural communities with both public and private institutions promoting agricultural development. Elaborating on Hall[25] who highlighted the complementarity of NGO local skills and governmental research and policy skills, we believe there is a strong case for the FSR process to be a publicly sponsored interface with three important functions as follows:

- To bring the beneficiaries point of view to local development planning.
- To improve coherence between local and national development perspectives.
- To improve coherence between development initiatives from government, NGOs, donor and private-sector sources.

These functions are probably best performed through local development committees, local community development institutions or through extension as operating platforms. The rapid spread of government decentralization in developing countries now makes a general model of district-based teams mandated for FSR suitable for most. Only such a deployment will allow convincing input into the newly established district development programming and policy formulation.

The broader scope of contemporary FSR suggests a switch away from the research services as an operating platform. The district-based organization of extension is much easier to reconcile with FSR than the commodity and disciplinary organization of centralized research institutions. The diffusion function is increasingly being undertaken through district level enhancement of farmer-to-farmer processes. A district-based FSR team with access to multiple sources of farm improvement will increase the power and authority of farmer-to-farmer diffusion. Where farmer-to-farmer dissemination is adopted, extension services would facilitate the construction and operation of knowledge networks and help broker partnerships with input suppliers, credit and other services needed to mobilize innovations. The FSR process can put the essential flesh on newly formed district structures for agricultural development. The FSR practitioners can act as the essential link between rural communities and district agricultural planners and policy makers. They can play a key role in linking communities to a full range of services from both public and private institutions.

FSR has been seen, historically, as part of the research establishment, and promotional efforts for a wider role will often need to take this as a starting point. It is our view, expanded below, that, despite structural adjustment, the public sector in most developing countries will continue to make huge investments in agricultural R & D. By improving the effectiveness and efficiency of these investments FSR can also improve the political credibility of the R & D process.

12.9 THE CONTEXT FOR THE FUTURE APPLICATION OF FSR

In many developing countries the use of FSR as a process to facilitate agricultural R & D is dependent more on the 'rehabilitation' of degraded economies and institutions, than on the improvement of the process itself. Two questions seem particularly important. Will the organization and management of economies and public institutions improve? Will R & D in agriculture continue to be public funded? Among the reasons for weak institutionalization of FSR perhaps two stand out. First, many public institutions in developing countries, particularly in Africa, have been reduced to a standstill through management failure, both in government generally and in the institutions themselves. Second, public institutions have become heavily dependent on donor funding. This has been subject to the vicissitudes of developed-country politics with divergent views on priorities and a lack of coherence across bilateral donors. This caused a boom in both central FSR units and in outreach FSR teams followed by a withering of these structures where donors withdrew their support. NARI responses have been characterized by bringing their outreach teams back to the research stations. Central units have often survived but in a much reduced and demoted form. Cutbacks and downsizing does not, however, mean that governments think less of FSR. Indeed many now ask all their research staff to utilize participatory approaches and engage systems perspectives in their work. Incentives for scientists to develop FSR skills and conduct their research on farm are further increased through competitive research grants that call for participatory research with systems perspectives.

Structural adjustment programmes imposed through multilateral institutions of the international community have had some devastating short-term consequences, particularly where subsidies had been cushioning the production of marketed commodities. The long-term effects of adjustment are expected to be more favourable to agriculture, removing burdens imposed by governments in exploiting an easy source of revenue and in favouring burgeoning urban populations. Beyond this the emerging donor preoccupation with democracy, good governance and accountability augur well for future standards in government generally, and for the management of public institutions in particular.

The adjustment programmes of the 1980s coincided with the privatization of agricultural research in Western countries and the fall of Communism opening more countries to privatization. This, an acknowledgement that public R & D in many developing countries was in need of reform, and success with the privatization of R & D in some Latin American countries, led to a perception that the private sector would provide an R & D solution for Africa and Asia. It seems clear, however, that for countries heavily dependent on smallholder agriculture to drive development, many product markets are too narrow and too uncertain to attract the private sector. More than this, even for crops with a well-defined market, the political and economic uncertainties continue to discourage more than short-term private investment in adaptive research and extension. For many countries the continuing importance of non-market food crops, and the burgeoning need for better conservation of land and water resources will require continuing public investment in research and extension. Beyond this, as a reaction to the effects of adjustment on R & D budgets and the widening realization of the imbalance between salaries and operating funds, particularly in research, there is immense current interest in new forms of funding to improve research efficiency in public institutions. By encouraging the use of competitive grants[26] the international community will again provide leverage to lift the management skills and rehabilitate degraded institutions. At the same time their emphasis on targeting resource-poor beneficiaries will encourage new reward systems for scientists.

These factors suggest a long-term future for publicly funded R & D and imply that private sector research will remain the exception rather than the rule, particularly in those developing countries heavily dependent on smallholder agriculture. It is important to emphasize, however, that such factors in and of themselves will not revolutionize R & D – that requires an enlightened process. Standing on the axiom that 'understanding farmers is a prerequisite for improving their farming systems' all this augurs well for participatory based FSR. Added to this, an increasing number of development economists are basing their policy research on a better understanding of household-level decision making. Tims[27], a prominent economist from The Netherlands, has recently called for a household-based approach to development. FSR is one (and one of the few) practical vehicle to allow this on a national scale.

12.10 COUNTRY CHARACTERISTICS AND THE PLACE OF FSR IN R & D

Three important factors in the organization of R & D are size, diversity in both agroecology and socioeconomic circumstances, and political culture. Size and diversity are critical to the way countries organize their agricultural R & D. We define size by population: the size of the total population determines the base for raising revenues for public activities including R & D, and the size of the rural population is a measure of the extent of the market for new agricultural technologies. Diversity in agroecological and socioeconomic circumstances decides the number of discrete farming systems – the basic units for rural development. This affects the level of resources needed in agricultural R & D and, at the same time, dictates the size of the market for particular agricultural technologies.

Four extreme country cases are identified here. No attempt is made to define what is a large or small population, high or low revenue and wide or narrow market – but all these are important economic questions in applying the principles outlined.

Case A: a large population, a single agroecology, a single social type and good market access. Such homogeneity allows a focused research effort, the large population offers high revenue potential and a wide market for research results.

Case B: similar homogeneity but a very small population. A focused research effort, but low revenue potential and a narrow market for research results.

Case C: a large population, diverse agroecology, diverse social types and variable market access. Offers high revenue potential but a diverse research effort is required and there are narrow fragmented markets for research results.

Case D: a small population, diverse agroecology, diverse social types and variable market access. Offers low revenue potential but requires diverse research effort and only has narrow, fragmented markets for research results.

Cost-effective strategies for the organization of agricultural R & D clearly vary across these examples. Again, a look at the extremes helps clarify the message. High revenue countries with a single agroecology and a single social type (case A) are ideally placed. Such countries can mount a single strategic, applied and adaptive research agenda targeting farmers in that agroecology. A low revenue country with a diverse agroecology (case D) is the worst-case scenario. Here one should look to agroecological homologues in other countries for strategic and applied research results. They should concentrate their limited funds on obtaining this information, and on OFR guided by FSR, to identify technologies and other farm improvements suitable to the agroecological and socioeconomic circumstances of their rural people. Many countries would fall between these extremes (cases B and C). Most of these should organize full research agendas for their own major agroecology, but seek results from outside homologues for agroecology of lower national priority.

The important point for a book on FSR is that, for all these example scenarios, whether the strategic and applied research is done in country or outside, FSR capability is required in-country to identify and deploy appropriate farm improvements. Strategically even very small countries need resources in FSR and OFR but should seek technical research results from agroecological homologues in other countries. We firmly believe it is strategic and applied research which is the more questionable investment for many countries, particularly those with diverse agroecology. These require heavy investment and yet the markets for the diffusion of results will often be too narrow to support such overheads. Strategic and applied research, wholly biophysical in orientation and content, are most cost-effectively organized by agroecology and seen as an overhead cost to OFR organized locally by farming systems. Agroecological niches and the economic crops grown in them are best addressed by a single strategic and applied research effort organized across countries where that agroecological niche is important.

Finally our third factor: political culture is another important dimension in considering the promotion and institutionalization of FSR. Strongly hierarchical centralized political structures will seek to maintain control of the vehicles for development. Public institutions will always play an important role and voluntary agencies, if tolerated at all, will be closely guided. The scope for small-farmer empowerment may be limited. Promotional strategies will need to weigh the politics associated with the methods employed to operate the FSR process. This caution may apply more widely. In many less formal and in some degraded societies the voluntary agencies have had a great deal of initiative over the last decade, much of it due to disillusion with public institutions on the part of donors. The long-term effects of structural adjustment and shifts in donor policy towards governance as a criteria for lending suggest that this ascendancy will be short lived. Effective development demands appropriate reconciliation between local and national interests which, in turn, demands a degree of coherence among the multitude of voluntary bodies loose in many countrysides. Under the trends stimulated by structural adjustment and the donor community the freedom of the voluntary agencies will be reduced. Fragile nation states, under heavy population pressure, or with a significant proportion of their populations operating a fragile resource base, will be unwilling to devolve too much responsibility for R & D from public institutions. Some countries will not manage their population dynamics unless their public institutions are reoriented and revitalized. The dangers here are clearly reflected in the increased conflicts over land and water use. If they do not get a responsive institutional structure large numbers of poorer people will suffer. Political culture will be an important influence in selecting an appropriate FSR process and in choosing the strategy for its promotion.

12.11 ENABLING TRENDS AND POTENTIAL BARRIERS

A number of ongoing trends favour the introduction of FSR as a process to provide understanding of rural communities, their production environment, and ways to develop them.

12.11.1 Structural adjustment

We have already noted the potentially positive long-term effects of structural adjustment in opening markets and reorienting policy to favour a productive agriculture. We assume safeguards will be in place to enhance smallholder competitive efficiency and ensure environmentally sound practices and equitable development for resource-poor communities in remote marginal areas.

12.11.2 New donor preoccupations

We have also noted the new emphasis in the international community on democracy, good governance and accountability in improving economic management and the management of public institutions. Both these should improve institutional performance and professional morale in the long term. Other donor emphases on poverty, the environment and concern for measurable impact on the beneficiaries will encourage institutions to reorient their programmes and place a premium on understanding the beneficiaries as a foundation for effective action.

12.11.3 Decentralization, community organization and empowerment

The local/global hierarchy offers a vehicle for the rational application of the principle of subsidiarity. Decentralization of authority, recently pursued in some Latin American countries, is demonstrating success in improving the management of local affairs, particularly when accompanied by the organization and empowerment of communities. Line management through the executive ministries of a centralized government has been, and continues to be, a nightmare in local terms. Independent plans, for example in agriculture, urban expansion, tourism and the environment have led to overlapping, often contradictory actions, and local conflicts of interest. With decisions taken locally a more balanced, holistic view of local needs is possible and action can be planned coherently. Given the local articulation of environmental problems and the importance of reconciling the interests of local stakeholders in solving these, and in planning the development of their communities, decentralization and empowerment are trends that can only enhance the value of FSR as a beneficiary interface.

12.11.4 Regional collaboration in research

Subsidiarity works both ways in the global/local hierarchy. We have discussed country diversity and the high cost of strategic and applied research. There is no need for this to be replicated in each country, particularly for staple food crops and for the management of natural resources. There is immense scope for organizing technical research on the basis of each agroecology, with each country where that ecology is important contributing to and drawing from the effort. The international community has been working for many years to encourage such regional collaboration. The Special Programme for African Agricultural Research (SPAAR) implemented through the World Bank and supported by many donors has stimulated regional collaboration in Africa. The Southern African Development Community (SADC) has enjoyed some success in organizing research on a regional basis. The Association for Agricultural Research of Eastern and Central Africa (ASARECA) is seeking to emulate this for East and southern Africa. Similarly, the donor community is also supporting the revival of the long-established regional fora for agricultural research in Asia (APAARI, Asian Pacific Association of Agricultural Research Institutions), the Middle East and Latin America. The trust needed for such collaboration takes time to build but it is a trend that, by reducing research costs, will release domestic resources for locally focused adaptive research.

12.11.5 Globalization of information

A final trend that complements both decentralization and regional collaboration is the globalization of information. The ability to review global information sources from a local situation via the internet offers a facilitating mechanism with the potential to make regional, indeed global collaboration a reality.

Enabling trends apart we can, however, expect to meet a number of barriers. Change and especially institutional change always meets opposition. For a number of years now institutional cultures in R & D have worked against participatory approaches. We see this continuing. Many scientists will continue to discount non-scientific sources of knowledge; a considerable constraint to FSR that looks for multiple sources of intervention. Many scientists cannot let go of the notion that they know what is best for farmers. Asking such scientists to undertake research that supports farmers' future visions usually earns the response that such futures will not work or are plain wrong and my way is best. A further dimension to institutional inertia is the belief that asking commodity scientists to use participatory approaches and systems perspectives will result in farm improvement. It will not. It will only add to the long list of ideal management practices that smallholders cannot use. While nobody is given the mandate to look at whole farm systems too much technology will remain on the shelf to the discredit of research services. We do not suggest a return to centralized FSR units; what we suggest is that some people, probably at the district level, engage in the five key elements of contemporary best practice we outlined earlier.

A further barrier to FSR will be the continuing reliance on transfer-of-technology (TOT) models in R & D. Research managers and policy makers will stick with it because it works for commercial large-scale farming interested in commodity production technologies. Moreover, TOT will get a further boost as governments seek to modernize their agricultural sector through farm specialization to capture contract-farming opportunities for export and other large urban markets. Modern farming to many is still about the use of high external inputs of fertilizer and pesticides on large monocropped areas. Arguments for the continuance of TOT will be further strengthened by weaknesses in FSR practice and early failures in a knowledge system-based model. Fragmented information systems and knowledge systems will undoubtedly result in early failure for those attempting to build the kind of knowledge networks we have described. One source of difficulty we foresee is the lack of strong institutions that have sufficient legitimacy to convene parties across sectors at district and national levels. A further difficulty is the perception that time and resources spent negotiating partnerships for learning are not seen as budget worthy items.

By and large, despite these barriers, we feel

little need for pessimism about the future. The FSR constituency can feel confident taking an active role in promoting trends towards decentralization, community empowerment, the regional coordination of research and the globalization of information, all of which will enhance the use and institutionalization of the FSR process.

12.12 FUTURE IMPACT

We have examined the wider context within which FSR will have to develop and seen that many of the necessary improvements are beyond the scope of FSR itself, although FSR professionals can encourage such improvements through their activities and opinions. We turn here to future promotion of the process itself. What are the priorities for action if its constituency is to widen the adoption of FSR as a routine part of the research and development process? We look first at where its future impact on the process will come from.

Wider acceptance of systems hierarchies as conceptual and operational frameworks common to institutions in development would be an important step forward. Such frameworks allow better definition of functional roles, better understanding of the functional roles of others, the identification of essential linkages between roles and thus the forging of logical operational partnerships. Amongst these benefits the understanding of the roles of others stands out. It raises awareness that success in one's own work depends on the success of others in theirs and forms the basis for collegial rather than competitive relationships, both up and down hierarchies and at levels across hierarchies. Hence its value in the forging of partnerships. Paralleling systems hierarchies, in a mainly (but not entirely) spatial dimension, is the wider use of the farming system rather than commodity as the basic unit for development planning and action in smallholder dominated agricultural sectors. This requires wider acceptance of the economic need to group resource-poor farmers operating the same system in order to provide cost-effective support through enabling services, public or private. It is a point which economic studies could do much to evidence.

The further evolution of some original FSR principles will be an important source of development. The original aim of FSR was to draw small farmers into the R & D process. Best practice has progressed from their involvement and collaboration to full farmer participation and is moving towards farmer empowerment. However, the process should not be made the victim of local politics. In countries that are politically sensitive to the degree and means of empowerment the beneficiaries may be better helped, certainly in the medium term, by an FSR process which respects this situation. Otherwise it is a 'baby with the bath water', 'cut off the nose to spite the face' syndrome. Ideological campaigning should be led from other platforms. In the same way this applies to the relief of poverty, currently high on donors' agendas. Just as with empowerment it is important FSR is not swept up by poverty as a cause. The ability of the FSR process to understand rural poverty, and to identify types of improvements that the poorest of the poor can absorb, is a valuable promotional focus for donors with strong poverty commitments. However, there are a wide variety of strategies for addressing poverty. In communities where most families are earning say US$5–10 day^{-1} all are poor by most standards. Choosing to address only the poorest in such communities may not be the best use of funds and personnel. Lifting the whole community in a way that increases employment opportunities for the very poor may be more effective. We appreciate that the latter course, indeed all courses, are easier said than done, and when they fail the poor are left even further behind.

With its origins in research services, innovation through FSR was restricted to the 'technical fix'. Best practice now expands this to the integration of innovations impacting the farm directly, such as new technologies, new enterprises and new processing and marketing opportunities, with changes impacting the farm indirectly, including infrastructure development, new policies and revised institutional mandates and operating procedures. Both greatly enhance the scope and impact potential of FSR, as well as the options for its institutionalization. Given the commodity and disciplinary organization of conventional research the original 'technical fix' was inevitably within an commodity, often a single technical component, at best two or three. This also should give way to multiple interventions across a range of enterprises, including resource rehabilitation strategies and enterprise integration providing synergy to the farm system as a whole, or indeed to off-farm activities, further widening the scope for impact from better understanding.

Institutionally FSR was labelled as a new vehicle for research-extension linkage, an area traditionally weakened by the separation of the two services. A shift to decentralized government and district decision making should pave the way for cross-institutional learning groups seeking coherence across programmes at both district and local levels. Resources and efforts applied to building knowledge networks provide the infrastructure for such cross-institutional learning. Such a development will begin to erode the competitiveness of line management in the economy, reduce spending on conflicting priorities and make more effective use of revenue. Such a development will begin to rationalize the many overlapping and conflicting roles evident in newly decentralized services.

Finally, impact will also come from new information technology. The internet offers a cheap and easy way to exchange experience and learning between the very many small isolated and scattered innovators in all sorts of R & D projects. Advances are being made all the time by people who do not publish in the formal 'white' literature. They share their innovations in conferences and seminars to the few who can afford to attend them. We see the internet as a new tool that will in future bring these innovations to many homes. The tremendous boost to farmers' involvement from the introduction of participatory techniques can be repeated. While validation of participatory tools is important for credibility the acid test should be farmer utility not scientific rigour as defined in a positivist paradigm. It is very easy to render a good participatory research tool useless by overburdening it with more and more data requirements. Other new tools including the use of GIS, simulation modelling and the accumulation of global databases, will be of vital importance to the development of FSR. All will contribute to improved characterization and knowledge networking.

12.13 PROMOTIONAL STRATEGIES

Experience shows that stimulating the changes in organization to operate an FSR-based interface requires the collaboration of as wide a range of stakeholder institutions as possible, including those line ministries, enabling institutions and NGOs which will benefit from the bottom-up flow of information. It demands an institutional culture that recognizes adoption and uptake by resource-poor farmers as the goal of the R & D process. Transition to an FSR-based interface needs to be nurtured by parallel strategies; on the one hand to build up capacity and coverage of the FSR cadre, and on the other to train all stakeholders, research scientists, extension staff, policy makers and institutional managers to understand:

- Small farmer circumstances, the challenges they face and the way they take decisions on change.
- The full farmer-to-farmer R & D process, their own role in it and their dependence on a range of institutional partners for success.
- The implications of a demand-driven R & D agenda.

There has, as yet, been no concerted effort to achieve such changes. We believe this to be a great omission on the part of the international research community and their donors, particularly those supporting the development of national agricultural research organizations. The root of much of the problem is bilateral donor behaviour. Independent action on the part of a multitude of donor countries inhibits appropriate institutional change. In part this is due to a very poor understanding of the research and development process and in part to wholly unrealistic ideas of the timescale required to achieve such change in the circumstances of developing countries. The disparate nature of project-driven development, dominated by bilateral aid, inhibits coherence in approach and learning from wider experiences. Donor strategies are short term, often from ignorance of the time required to achieve institutional goals, often due to staff turnover and the need for new people to make their mark and fly their flag. Vested institutional interests also address their requests for funds to donors whose policy is non-threatening and finally, as the Stroud contribution in Chapter 7.1

demonstrates, domestic funds have rarely been made available to maintain FSR initiatives once donor funding has been withdrawn. Similar behaviour is beginning to characterize the International Agricultural Research Centres, which, in striving to show impact to maintain funding, have become more self-promotional of their own technologies and achievements. The need for this has confounded both their logical role in strategic research, moving scientific resources downstream into development, and their declared policy of wider inter-centre collaboration and wider partnerships with regional and national organizations.

This is one factor that perpetuates the 'international' phenomenon, whereby institutions at the global level such as FAO (with agroecological zoning for policy planning) and the CGIAR centres (with FSR itself), have perceived their role as supplying answers on development process to less well-endowed organizations. The promotion of training and visit extension by the World Bank is perhaps an extreme case. Sasakawa Global 2000 is another pushing a single answer in modern high external input farming, a future victim will be FAO's Farmer Field Schools. It is a phenomenon that has caused international organizations to overlook the perspectives and assets of stakeholders closer to the ground, their potential partners and to take a top-down, dogmatic stance on concept and process. International centres with donors behind them easily bully national organizations to accept their paradigm and way of thinking. It remains a difficult equation to solve. New vision and fresh ideas are crucial to solving the development problem, yet even the best international organizations offer only one perspective on solutions, other stakeholders perspectives are needed to come to conclusion on their practicality and value. There are far too few places where national organizations can find facilitators of innovation and not flag-flying dogma.

12.13.1 Short-term promotion

The immediate need is clear. We must begin to improve the performance of existing FSR cadres, building capacity in the field and increasing credibility in the process. We must begin to raise awareness of the value of FSR among all the potential users by demonstrating its practical applications. Building new alliances with farmer organizations and advocacy groups will crucial here. Three activities seem particularly important:

- In-service training which includes regular refreshment in new methods and cross-site visits to innovative research and development projects.
- Bringing a wide range of potential users into decisions on the location for new projects, choosing areas where users perceive that understanding the farming system would benefit their own functions.
- By recording and disseminating successes, and promoting their replication through exchange visits and networking.

12.13.2 Long-term promotion

The 25 years experience of FSR as an innovation in research process and organization has much to offer on strategies for increasing institutional commitment and performance in developing countries. Chapter 7.1 on institutional change in Tanzania by Ann Stroud, and Chapter 7.2 by Stuart Kean and Chris Ndiyoi on the power components of the platform from which change is launched, are particularly informative. Some distilled experience is offered in a list of seven strategic principles for the promotion of FSR:

- Within a country a wide platform, across ministries, enabling institutions and interested NGOs, provides a diverse power base that is particularly valuable in managing changing policies and personalities in R & D institutions.
- Within a country support should be identified as high as possible in a number of interested institutions. It is important not to leave the power in the hands of the leadership of the sole implementing agency.
- Within R & D organizations leadership should be sensitive to the difficulties of smallholder development, have a broad view of where new opportunities might come from, have influence in wider circles both government and non-government, and

ideally have a long term ahead in post, and the loyalty and respect of their staff.
- Promote the process positively on the basis of improvement in performance and field results.
- Educate all R & D managers on the role of FSR in improving their own success rate.
- Sensitivity to existing lines of authority within the institution(s) implementing the FSR process.
- Protect the new FSR cadre from the 'institutional establishment' while building its capacity and seek credibility for them through their understanding of small farmers, and the relevance of this to the research and development priorities of the institution.

As Hall[28] points out, radical restructuring of government to take on board a systems approach to R & D is out of the question. More often than not it would just be another international flag waving for an external dogma. He advocates the idea of 'soft changes' which do not require either organizational upheaval or significant extra resources. Following this lead it is useful to see the FSR package as divisible into 'internal' and 'external' components. Major internal components might include: the introduction of social scientists into the establishment, the identification of groups of relatively homogeneous farmers as target clients, the movement of experimental work on to farmers' fields, the design of experiments to meet farmers needs rather than maximize yield, the evaluation of experimental results using farmer assessments, and the involvement of farmers in identifying local research and development priorities. Major external components might include establishing links between small farmers and research and development planning, and policy formulation, and credit and input supply institutions.

In institutional innovation, as with adoption by small farmers, those components most easily absorbed, and which bring benefits in their own right, can be identified and promoted first. The appropriate sequencing of component introductions is inevitably dependent on the circumstances of the institution and the country concerned. In an age of widening institutional collaboration, partnership with a NGO with a strong community constituency will often be a way forward at the local level. Inappropriate though it turned out, the T & V experience taught important lessons not least of which is that funds, focused management and in-service training can change institutions, even in the context of low motivation and weak management. The T & V experience underlines the importance of clarity in the organizational paradigm and single-minded pursuit of its implementation. At the same time it also demonstrates the importance of an evolutionary approach to change. The T & V process, certainly as originally promoted, lent itself to a hierarchical top-down culture, in this the new paradigm was congruent with what already existed in many countries.

Experience shows that creating a country platform for the implementation of FSR is one key to success and sustainability. If FSR presence is dependent on a single senior bureaucrat it can be very difficult to sustain institutional development as personalities change. In one African country a newly appointed director of research saw no role for FSR. Capacity in social science, built up over the previous 3 years was rapidly eroded. Three years after the change in directors, 10 of the 12 trainee economists recruited 6 years earlier had gained their MScs but none remained in the research department. The process of institutionalization was back to square one after 6 years of investment in building capacity. A wide platform of support from potential users across government and in the NGO and private sectors, where these are politically acceptable, is important for the management of changing priorities and personalities. Such platforms, however, rarely exist because there is little national or donor support for them. This is a tragedy, development needs countervailing powers to both donors and government. Nevertheless, this does not mean that multisector platforms for FSR cannot be constructed. Indeed, the various professional associations for FSR-E and other agricultural disciplines could provide the membership and convening power for such platforms to develop.

Promotion in-country can also be enhanced through regional efforts. One idea might be the establishment of 'role model' programmes in one or two countries in each major region of Africa, Asia and Latin America. Success in particular countries with the enthusiasm and commitment to press ahead with FSR as a development tool will, with promotion, proliferate the process. Regional consortia are perhaps the appropriate forum in which to detail promotional strategy. Such fora might choose 'role model' countries to provide them with support, from other interested countries through a regional network, from donors committed to FSR, from universities with appropriate expertise, and from the professional associations for each region. The regional professional associations have a special role to catalyse promotion in each region. In a messy world of development and development politics only the professional associations can provide the sense of purpose and an operational coherence to give leadership in consolidating the use of FSR. They need to focus their efforts, detail their promotional strategy, and meet donor expectations in governance and accountability in order to attract funding support.

Finally it remains to emphasize the key role for higher agricultural education in developing-country universities. This is perhaps the single most important vehicle for sustained progress in FSR. The degree curriculum is the most powerful tool for creating skills in FSR and a constituency for the use of FSR and other systems approaches in agricultural development in the long term. The curriculum needs to inculcate the concept of systems' hierarchies, the links between levels in hierarchies and links across hierarchies at particular levels. Techniques in systems thinking, analysis and modelling should be taught. Within a hierarchical framework experiential learning techniques need to expose undergraduates to the realities of resource-poor farmers, their decision processes and their importance to the economy of the country. Students should leave university with the right skills, attitudes and behaviour for conducting participatory research. Finally students should be made aware of the place of systems approaches, in our case FSR, in improving the productivity of resource-poor farming. Today's graduates joining national research organizations are expected to work in interdisciplinary teams. Universities should adequately prepare them for this mode of work. There is a huge role for universities in the developed world in bringing this about. It offers an opportunity to undo some of the irrelevance built in to their education system by metropolitan countries providing inadequate and inappropriate advice on curricula as developing countries gained independence and created their own universities.

12.14 CONCLUSION

FSR, computers and biotechnology are innovations that promised to influence the effectiveness of agricultural R & D more than they actually have. Biotechnology is having immense difficulty gaining both scientific and public credibility after more than two decades of experience. There are many examples of how institutions have adopted computers yet have failed to reorganize around them after decades of exposure. After 25 years we should not be too surprised that FSR has been slow to change institutions. It takes more to operationalize and capture the benefits of such potentially powerful innovations. Nevertheless, the large number of developing countries exposed to FSR, the large number of institutions experienced in it, and the large number of agricultural professionals better equipped to understand small resource-poor farmers and their systems, all demonstrate that FSR has had an impact on the big picture.

We have ascribed the lack of institutional impact to the general weakness of institutions, particularly public institutions, bilateralism across the international community and incoherence among professionals. We have tried to combat professional incoherence by redefining the scope of FSR and identifying contemporary best practice. We have tried to combat government and donor fatigue with short- and long-term promotion strategies. It is our judgement that the regional professional associations are the right vehicles to provide leadership and we have described strategies they might pursue in promoting FSR.

There is another side to the coin that, in our view, adds urgency to the issue. It seems clear that the planet *can* support its future population. It has become a question of how many people, how much of the environment and how far social cohesion will be damaged while learning how. Frustration with efforts to improve resource-poor rural communities is growing among developing country élites and among some of the international community. Two old strategies for the improvement of agricultural productivity are being resurrected:

- A selective strategy which identifies the best managers as a master farmer class and helps them to become rural employers.
- A transformation strategy which changes the structure of the sector to large-scale commercial farming, displacing much of the rural population to towns and cities.

Both of these imply that countries have no escape from the development path followed by the West with its attendant neglect of the environment, migration for work, the breakdown of the family unit and loss of social cohesion. Post Brundtland it has become clearer that for sustainable development the ecology and the economy need to be kept in balance. Yet the arithmetic of balancing economic gain and environmental conservation is still fuzzy and proving difficult to reconcile with the dominance of the market and the private sector. It is also becoming clearer that social capital is a third dimension necessary to the arithmetic, and that all three need to be kept in balance. Evidence for this is the degrading of family and community life in many Western countries and the burgeoning frequency of resource-based conflicts worldwide.

We believe we should be seeking a lifestyle consistent with the resource base and human ingenuity, with the private sector as a vehicle in the service of lifestyle rather than lifestyle in the service of the private sector. Two of the many attributes of small resource-poor farmers and their communities are their knowledge and respect for their local environment, ignored only under threat of survival, and their social cohesion. Experience tells us these are worth preserving and this underpins the need for an equitable rural development strategy in which, for cost-effectiveness, advice and information is tailored to groups of farmers for whom similar interventions will be appropriate.

In this chapter we have described what we think is an important niche for FSR. We have argued that FSR can make a contribution to the improvement of smallholder agriculture. Yet looking around today it is hard to see FSR realize its potential. Why? Much of the fieldwork that is called FSR fails to convince donors, policy makers and even research managers. They see research with very little of what can be called systems in it, they see farmer participation that looks like old-fashioned extension – demonstration trials that farmers can barely explain, and they see NGOs working with many farmers without technical know-how. When these key decision-makers look back on their experience in commodity or disciplinary research there is little to make them change their minds. Of course, they know that research should be doing more for the smallholder, but what they see of FSR fails to justify change. Relevant university education, professional rewards and personal promotions provide too little support to the sustained practice of FSR. The tragedy is with the smallholder. FSR offers researchers and developers a way to raise their eyes from the soil, the crop, the animal, and see the social decline, the environmental degradation and the economic stagnation that threatens to be the future facing the smallholder farm family.

REFERENCES

1. **Booz, Allen and Hamilton Inc., 1968**. A program for new product evolution. Quoted in Robert R. Rothberg (Ed.) Corporate Strategy and Product Innovation, 2nd edn. Free Press, 1981.
2. **Bunting, A.H. (Ed.), 1987**. Agricultural Environments. CABI, Wallingford.
   Oram, P., 1987. Combining socioeconomic data with biophysical and environmental data. In: Bunting (op. cit.).
3. **IBSRAM, 1998**. International Workshop on Resource Management Domains (RMDs). Proceedings of the Conference on Resource Management Domains. (Kuala Lumpur, Malaysia, August 1996). J.K. Syers & J. Bouma. Bangkok, (Eds). Thailand. IBSRAM Proceedings no. 16.
4. **Maxwell, S., 1986**. Farming systems research: hitting a moving target. World Development, 14, 65–77.
5. **Weber, G., 1996**. Heterogeneity and complexity in farming systems: towards an evolutionary perspective. Journal of Farming Systems Research-Extension, 6(2), 15–32.
6. **Bawden, R.J. & R.L. Ison, 1992**. The purpose of field-crop ecosystems: social and economic aspects. In: C.J. Pearson (Ed.) Field-Crop Ecosystems. Elsevier, Amsterdam, pp. 11–35.

7. **Chambers, R., A. Pacey & L.A. Thrupp, 1989**. Farmer First: Farmer Innovation and Agricultural Research. Intermediate Technology Publications, London.
8. **Biggs, S., 1989**. Resource-Poor Farmers Participation in Research: a Synthesis of Experiences in Nine National Agricultural Research Systems. OFCOR Comparative Study Paper no. 3. ISNAR, The Hague, The Netherlands.
9. **Scoones, I. & J. Thompson (Eds), 1994**. Beyond Farmer First. Intermediate Technology Publications, London, UK.
10. **Korten, D.C., 1990**. Getting to the 21st Century: Voluntary Action and the Global Agenda. Kumarian Press, West Hartford, CT.
11. **Biggs, S.D. & E.J. Clay, 1981**. Sources of innovations in agricultural technology. World Development, 9, 321–6.
12. **Berdegué, J., 1993**. Challenges of farming systems research and extension. Journal of Farming Systems Research-Extension, 4(1), 1–10.
13. **Jiggens, J., 1994**. Foreword. In: J.B. Dent and M.J. McGregor (Eds) Rural and Farming Systems Analysis – European Perspectives. CABI, Wallingford.
14. **Baker, D., 1991**. Farmer-participatory research: reorientation. Journal for Farming Systems Research-Extension, 2(1), 125–47.
15. **Rhoades, R., 1998**. Participatory Watershed Research and Management: Where the Shadow Falls. Gatekeeper Series no.81. IIED.
16. **van Eijk, T., 1998**. Farming Systems Research and Spirituality. An Analysis of the Foundations of Professionalism in Developing Sustainable Farming Systems. Doctoral thesis, Wageningen Agricultural University, The Netherlands.
17. **Hall, M., 1993**. Summary and conclusions, p. 323. In: Doppler, W. & Maurer, M. (Eds) Farming Systems Development in Academic Training. Institutionalization of a Farming Systems Approach to Development, p. 91. Farms Systems Management Series no. 5. FAO, Rome.
18. **Collinson, M.P., 1998**. Institutional and Professional Obstacles to a more Effective Research Process for Smallholder Agriculture. Invited paper to a joint AFSRE/IUFRO symposium, December 1998, Pretoria, SA.
19. **Richards, P., 1989**. Agriculture as a performance. In: Chambers, R., Pacey, A. & Thrupp, L.A. (Eds) Farmer First: Farmer Innovation and Agricultural Research. Intermediate Technology Publications, London.

    Cox, P.G., A.D. Shulman, P.E. Ridge, M.A. Foale & A.L. Garside, 1995. An interrogative approach to system diagnosis: an invitation to the dance. Journal for Farming Systems Research-Extension, 5(2), 67–84.
20. **Hollings, C.S., 1978**. Adaptive Environmental Assessment and Management. John Wiley, New York.

    McLain, R.J. & R.G. Lee, 1996. Adaptive management: promises and pitfalls. Environmental Management, 20(4), 437–48.
21. **Brown, K.J., 1963**. Rainfall, tie-ridging and crop yields. Empire Cotton Growing Review, Jan. 1963. Empire Cotton Growing Corporation, London.
22. **Haugerud, A. & M.P. Collinson, 1990**. Plants, genes and people:Improving the relevance of plant breeding in Africa. Experimental Agriculture, 26, 341–63.
23. **Lightfoot, C. & R. Noble, 1993**. A participatory experiment in sustainable agriculture. Journal for Farming Systems Research and Extension, 4(1), 11–34.
24. **Ewell, P.T., 1988**. Organization and management of field activities in on-farm research: a review of experience in nine countries. OFCOR Comparative Study Paper no. 2. ISNAR, The Hague, The Netherlands.
25. **Hall, M., 1993**. (Op. cit.).
26. **Tabor, S.R., W. Jansen & H. Bruneau (Eds), 1998**. Financing Agricultural Research – a Sourcebook. International Service for Agricultural Research, The Hague.
27. **Tims, W., 1995**. Response to Sections A and B, pp. 113–121. In: J. Bouma, A. Kuyvenhoven, B.A.M. Bouman, J.C. Luyten & H.G. Zandstra (Eds) Eco-Regional Approaches to Sustainable Land Use and Food Production. Proceedings of a Symposium on Eco-Regional Approaches in Agricultural Research 1994 (ISNAR). Kluwer Academic Publishers, The Netherlands, in association with CIP.
28. **Hall, M., 1993**. (Op. cit.).

Index

Note: Page numbers in *italics* refer to figures and tables